Understanding Changing Telecommunications

– Building a Successful Telecom Business

Understanding Changing Telecommunications

– Building a Successful Telecom Business

Anders Olsson
Teledrom AB, Sweden

Published by John Wiley & Sons Ltd, The Atrium, Southern Gate, Chichester,
West Sussex, PO19 8SQ, England

Telephone (+44) 1243 779777

Email (for orders and customer service enquiries): cs-books@wiley.co.uk
Visit our Home Page on www.wileyeurope.com or www.wiley.com

Other Wiley Editorial Offices

John Wiley & Sons Inc., 111 River Street, Hoboken, NJ 07030, USA

Jossey-Bass, 989 Market Street, San Francisco, CA 94103-1741, USA

Wiley-VCH Verlag GmbH, Boschstr. 12, D-69469 Weinheim, Germany

John Wiley & Sons Australia Ltd, 33 Park Road, Milton, Queensland 4064, Australia

John Wiley & Sons (Asia) Pte Ltd, 2 Clementi Loop #02-01, Jin Xing Distripark, Singapore 129809

John Wiley & Sons Canada Ltd, 22 Worcester Road, Etobicoke, Ontario, Canada M9W 1L1

Wiley also publishes its books in a variety of electronic formats. Some content that appears in print may not be available in electronic books.

British Library Cataloguing in Publication Data

A catalogue record for this book is available from the British Library

ISBN 0-470-86851-1

Typeset in 10/12pt Times by Laserwords Private Limited, Chennai, India
Printed and bound in Great Britain by TJ International, Padstow, Cornwall
This book is printed on acid-free paper responsibly manufactured from sustainable forestry in which at least two trees are planted for each one used for paper production.

Contents

Preface xi

About the Author xiii

References and Acknowledgements xv

Glossary xxi

1 Introduction 1
- 1.1 The Book in Brief 1
- 1.2 A Dynamic Situation 10
- 1.3 Success Factors for the Growth of Mobile Services 11
- 1.4 Comment on Terminology 12

2 End-User Needs and Demands 15
- 2.1 Objectives 15
- 2.2 The Role of the Unpredictable (?) End User 18
- 2.3 User Analysis and Segmentation 19
- 2.4 Basic Needs Model 33
- 2.5 Mapping of Needs and Services 35
- 2.6 The Human End User as a Traffic Generator and Receiver 41
- 2.7 The Future Most Common End User: A Machine 43
- 2.8 What are the Service Drivers? 45
- 2.9 User Perception 46
- 2.10 Summary 47

3 Networks and Technologies 49
- 3.1 Objectives 49
- 3.2 What is a Network? 51
- 3.3 What is a Vertical Network? 54
- 3.4 The Convergence (or Collision?) 57
- 3.5 What is a Horizontal Network? 63
- 3.6 Fundamental Plans 65
- 3.7 A Techno-Economic View of the Convergence 70

3.8 Adaptation of the Basic Triangle and FPs to the Converged Multi-Service Network 71
3.9 The Connectivity Layer 75
3.10 The Control Layer 78
3.11 The Service Layer 78
3.12 The Distributed Network Dimension 83
3.13 The Processing Dimension 87
3.14 Key Enablers 89
3.15 General Enabler Development 93
3.16 Enabler Overview 93

4 Telecom Business 99
4.1 Objectives 99
4.2 The Telemanagement Forum 101
4.3 Adopting a Telecom Business Perspective 105
4.4 Telecom Enterprise Strategy: Roles for Positioning 108
4.5 Tools for Profitability Calculations and Business Cases 122
4.6 Revenue 130
4.7 Cost Efficiency 135

5 Services 147
5.1 Introduction 147
5.2 The Service Plan 154
5.3 A Common Segmentation of Services for Mobile Internet 157
5.4 Service Segmentation for Planning 159
5.5 Value-added Services 165
5.6 Economy of Service by Means of Caching 166
5.7 Economy of Service by Means of Saving Bandwidth 166
5.8 Bandwidth Requirements 170
5.9 Security 172
5.10 Future Service Development 172
5.11 Pricing: Charging in the New Telecom World 174
5.12 The Service Plan versus the New Architecture 177
5.13 The Core Network and the Service Plan 177
5.14 The Access Network and the Service Plan 180
5.15 Telecom Management and the Service Plan 183

6 Security 185
6.1 Objectives 185
6.2 The Goals of the User and Actor. Terminology 186
6.3 The Problem 187
6.4 Non-Availability for Non-Security Reasons 194
6.5 Connecting Security Terms into Telecommunication 194
6.6 Main Ways to Implement Security 196
6.7 Integrity and Confidentiality by Access Control – Authentication 202
6.8 Integrity by Access Control – Authorization in Enterprises 205
6.9 Integrity by Access Control – Firewalls 205
6.10 Confidentiality: Encryption and Key Management 207
6.11 Confidentiality by Tunnelling 210

6.12 Confidentiality and Integrity by IPsec 212
6.13 Confidentiality and Integrity for Mail by S/MIME 214
6.14 Applications and Solutions 215
6.15 Summary with IPsec and FP Focus 219

7 Quality of Service 221
7.1 Objective 221
7.2 Introduction 221
7.3 Perception of QoS 224
7.4 Threats to QoS 229
7.5 QoS Enablers 237
7.6 QoS at the Application Level 243
7.7 Implementation of QoS in UMTS 244

8 Service Implementation 247
8.1 Objectives 247
8.2 Chapter Structure 249
8.3 Target Network 250
8.4 Development Tracks 254
8.5 Introduction to Packet Design 256
8.6 The Role of Fundamental Technical Plans in Packet Design 258
8.7 Top-Down Approach to Packet Design 259
8.8 Specific Fundamental Technical Plans 266
8.9 Convergence Between Fundamental Technical Plans 275
8.10 Traffic Cases 280

9 Service Network 285
9.1 Objectives 285
9.2 Connection to Preceding Chapters 285
9.3 What is a Service Network? 286
9.4 Service Network Domain and Principles 288
9.5 Terminology 290
9.6 The Architecture of Service Networks 290
9.7 The Needs of the User Domain 295
9.8 The Needs of the Service Network Owner 296
9.9 Service Network Implementation 299
9.10 The (IP) Service Network Support Entities 300
9.11 Examples of Service Implementation 301

10 Terminals 305
10.1 What is a Terminal? 305
10.2 Business Aspects 308
10.3 History 309
10.4 Terminals for Mobile Networks 309
10.5 PDA Development 311
10.6 Terminal Convergence 312
10.7 The Changing Role of Terminating Devices 312
10.8 What is a Customer Premises Network? 313
10.9 Some Enablers 315

10.10 Terminal Functionality – Example 317
10.11 The Future 318

11 Edge Nodes 319
11.1 Introduction 319
11.2 Access and Backbone Networks 321
11.3 MGW Interfaces 323
11.4 Media Gateway Tasks 324
11.5 Summary 329

12 Packet Backbone 331
12.1 Objectives 331
12.2 Service Plan versus Packet Backbone 332
12.3 Capacity Development 334
12.4 Control Functions in the Packet Backbone 336
12.5 The Distributed Dimension 339
12.6 Traffic 339
12.7 ATM Solutions 340
12.8 IP Routing 342
12.9 IP QoS 344
12.10 Multi Protocol Label Switching (MPLS) 347
12.11 Multi-Layer Control 348

13 Access Network 351
13.1 Objectives 351
13.2 Introduction 351
13.3 What is an Access Network? 352
13.4 Access System Fragmentation 357
13.5 Unification 358
13.6 The Distributed Dimension 359
13.7 The Layered Dimension 361
13.8 Fundamental Plans in Access Networks 363
13.9 Mobility 364
13.10 Access Technologies in Mobile Networks 364
13.11 System Evolution 366
13.12 Fixed Systems 374
13.13 Fibre-Based Systems 376
13.14 Ethernet 376
13.15 Combined ADSL over Copper and Ethernet Over Fibre Solution 377
13.16 Cable Modem 378
13.17 WLAN 379
13.18 Satellite Technologies 381
13.19 High Speed Fixed Radio 382

14 Control Network 385
14.1 Introduction 385
14.2 The Environment of the Control Network 387
14.3 Fundamental Plans in the Control Network 388
14.4 A Simple Target Control Network Signalling 390

14.5 Circuit Mode Domain 394
14.6 Packet Mode Domain 397
14.7 IMS Domain = IP Multimedia Subsystem 399
14.8 HLR/HSS for all Previous Domains 402
14.9 The Domain of (Voice and) Signalling Over IP 402
14.10 Common Support Functions 406

15 Interconnection 409
15.1 Objectives 409
15.2 Introduction 410
15.3 Interconnection in Tele-Centric Fixed Voice Networks 413
15.4 Definition of an Actor Interface Reference Point 414
15.5 Service Level Agreements 415
15.6 Service Interworking 416
15.7 QoS Interworking 417
15.8 PDP Context Activation for Connection to a Data Network 418
15.9 Security Interworking 419
15.10 Signalling Interworking 420
15.11 Routing 421
15.12 Mobility Management 423
15.13 Charging and Accounting 424
15.14 Possible Interworking UMTS–WLAN 426

16 Telecom Management – Operations 429
16.1 Introduction 429
16.2 The Management System 431
16.3 Basic Process Part 438
16.4 The TMN Functional Areas 441
16.5 Service Management 443
16.6 TM Operations from a Roce Perspective 445
16.7 Customer Care and Data Warehousing 448
16.8 Security Management 451
16.9 QoS Management 452
16.10 Terminal Management 453
16.11 Access Network Management 454
16.12 Management of Layered and Serial Interworking 454
16.13 Conclusions 457

Appendix 1 Web Services and a Service-Oriented Architecture 459

Appendix 2 Financial Calculations 463

Appendix 3 Development Tracks 473

Appendix 4 Dimensioning Media Gateways and Associated Telephony Servers 481

Index 499

Preface

Many readers can use this book as a 'standalone'. For other readers it might favorably be combined with a 'conventional' book/other learning material covering the particular technical area of the reader. This would create an excellent combination of depth and breadth.

The chosen broad context-oriented approach does not only require a number of books, articles, standards, etc. as input. In fact, finding a suitable book structure has been the most demanding issue. For the main author it has taken years to acquire a reasonable conviction. Often, the evolving conviction tends to successively upgrade business aspects on behalf of technology, in order to justify the title word 'Understanding'. Understanding is also firmly related to a pronounced pedagogical goal of the book.

Finding a good book structure for a broad subject is similar to building a house with a number of rooms for different purposes. Filling the chapters with content then corresponds to furnishing the rooms. Changing the rooms is difficult and expensive. Changing the furnishing is easier and mistakes can more easily be corrected. New modern furniture can replace older inventories.

To make the structure issue even trickier, the telecom complexity of today requires different types of 'rooms':

- **Functional** rooms devoted to a specific purpose, such as security or quality of service.
- More hardware and **implementation** oriented rooms, such as access and control equipment and its capabilities.
- **Process flow** oriented rooms, where we can identify the users, the various service providers (in a broad sense) and their planning, implementation and management, the technical development tracks and the related business models.

The *main* references mentioned in the reference section concern the important structuring of the 'house'.

The 'room furniture' may be continuously changed, following the new standards from bodies such as 3GPP, ITU, IETF, ETSI and many others. Evidently, a book with a fairly long production time like this one cannot include all the latest 'furniture'. However, the web often provides information on new standards.

About the Author

Anders Olsson has worked for many years as a network designer in Sweden (Televerket, nowadays TeliaSonera), Colombia (Telecom) and Tanzania (TP & TC), and in 1991 was appointed senior expert in Network oriented training at Ericsson.

References and Acknowledgements

FUNCTIONAL STRUCTURE AND RELATED ACKNOWLEDGEMENTS

The most important applied structure is called **Fundamental (Technical) Plans (FP)**. The plans can indeed also reach beyond pure technique, such as organization. The main author has worked within this area as a consultant and as a teacher for many years, starting in the 1970s. The students have represented operators from all over the world.

Historically, after a long stable period during the twentieth century, digitization brought the need for a synchronization plan in addition to a basic set of plans. The area of IN and mobility were the next challenges to the basic 'FP house design'. A main contribution was a mobility management plan. Then IP entered the overall telecom area, giving a new touch to the quality of service plan, urging for a separate security plan and for a controllable convergence.

The FP view supports no doubt an orderly convergence between technologies. After all, all technologies have a common target: to connect communicating parties in a satisfactory way, technically and economically.

Proceeding with the plans, the deregulation and the evolving global networks demanded more attention to the establishment of an interconnect plan.

In this book another two areas are included, without being necessarily labelled FPs: *service layer enablers* (other than those belonging to established plans; for example content-oriented functions, such as positioning) and *economy of service*. A main target for economy of service is to enable voice over IP with reasonable efficiency. Efficient voice and video coders, header compression techniques, many WAP features and concentrating and aggregating equipment can be included within the scope of economy of service.

A few more fundamental plans may be justified in practice, depending on the technical context.

Regarding FP I would like to refer to ITU and an extensive cooperation during the 1990s with Herbert Leijon. See for example:

http://www.itu.int/itudoc/itu-d/dept/psp/ssb/mpg/ch08.html
http://www.itu.int/itudoc/itu-d/dept/psp/ssb/planitu/plandoc/corplan.pdf
http://www.itu.int/itudoc/itu-d/dept/psp/ssb/mpg/ch00.pdf

The fundamental plans are parts of the 'ITU Master Plan Guide'. See for example:

http://www.itu.int/ITU-D/bdtint/baap/sec1_03.html

I have myself earlier treated the area in a couple of books, called Understanding Telecommunications 1 and 2, (ISBN 91-44-00212-2 and ISBN 91-44-00214-9), especially in section A.10.8.5.

If still available, see http://www.ericsson.com/support/telecom/ and proceed to A.10.8.5 Fundamental technical plans.

I would also like to mention my cooperation with Ludvig Widell, the father of the PTP (Particular Technical Plans) concept, mentioned in the second reference and also in this book. Ludvig covered both fixed and mobile networks during the 1990s.

FP and PTP can no doubt significantly support and integrate many operator processes.

The cooperation with Thomas Muth has also been very valuable. Thomas covers both functions and implementation. His first book is called *Modeling Telecom Networks and Systems Architecture: Conceptual Tools and Formal Methods*. See position 317 in:

http://www.isbn.nu/sisbn/telecommunication%20engineering::6

His next book is called *Functional Structures in Networks* (Springer Verlag).

Thomas has been a source of inspiration especially for parts of Chapter 3.

REFERENCES AND ACKNOWLEDGEMENTS FOR SEPARATE PLANS

Security

Acknowledgement: the cooperation with my daughter Helena Andersson on IT security has been very valuable. Many thanks! Helena is researching IT security from a legal point of view and will produce her thesis on that subject.

Säkerhet vid trådlös datakommunikation. SIG Security Studentlitteratur 2001.
Handbook i IT-säkerhet. Predrag Mitrovic.pagina.se.
Communications Security in an all-IP world. *Ericsson Review*, 2, 2000.
Johan Gustafson: Kroppen som lösenord. *Nätverk och Kommunikation*, September, 2000.
Intermec White Paper on wireless security.
http://epsfiles.intermec.com/eps_files/eps_wp/WirelessSecureWPWEB.pdf

Quality of Service

UMTS Quality of Service (QoS) – An End-to-End View. Brad Stinson, Narayan Parameshwar, Ramki Rajagopalan. See:

http://www.awardsolutions.com/downloads/Award_solutions_UMTS_QoS_wcnc1.3_(0801).pdf

The challenges of voice-over-IP-over-wireless. *Ericsson Review*, 1, 2000.

Interconnect

Interconnect acknowledgement: the cooperation with Håkan Åkerstedt, who participated in a group for interconnection within the European Commission, has been valuable. Håkan has in addition written the appendices on MGW dimensioning and development tracks.

Implementation Structure

The network structure applied in the book is called the all-access architecture, supposing a fairly complete convergence except for the access area. The convergence is partly based on layering and horizontalization.

This does not mean that such development is the obvious short-term one. With the present circuit-mode voice networks still being a cash-cow, there are strong conservative forces advocating a limited convergence, initially, for economical reasons.

Some references and acknowledgements are:

The stupid network. David Isenberg. See:
http://www.camworld.com/att.html.
Cooperation with Håkan Wolf who participated in ACTS (Advanced Communication Technologies and Services) Program of the European Commission CONVAIR Project. See:
http://www.etic.be/convair/Documents/teams-cv.pdf
Building the GII, *Telecom Journal of Australia*, 47(2), 1997.
Carrier Grade Voice Over IP. Daniel Collins. McGraw-Hill, 2001.
Network Evolution the Ericsson way. Erik Örnulf and Steinar Dahlin, *Ericsson Review*, 4, 1999.

Acknowledgement: Miltos Tricopoulos assisted significantly in writing the chapters on access network, packet backbone and control network.

REFERENCES AND ACKNOWLEDGEMENTS FOR SEPARATE NETWORK PARTS

Service Network

Here I would like to mention contacts with Christoffer Andersson, who wrote the book *GPRS and 3G Wireless Applications*. See:

http://www.wiley.com/WileyCDA/WileyTitle/productCd-0471414050,descCd-authorInfo.html

This book is one of the main references. Another contribution by Christoffer deals with Enablers – The link between Terminals, Networks and Applications. See:

http://www.ericsson.com/mobilityworld/sub/articles/other_articles/enablers_the_link_between?PU

Open Mobile Alliance is another input. The principles are described in

http://www.openmobilealliance.org/docs/OMA-principles.pdf

Many articles in *Ericsson Review* deal with the service network. See for example no. 1, 2003. A meeting with some of the involved service layer staff was indeed valuable.

MGW Chapter

Media gateway for mobile networks. *Ericsson Review*, 4, 2000.
http://www.ericsson.com/about/publications/review/2000_04/files/2000042.pdf. All *Ericsson Review* articles are accessible by means of the general part of the address above.

Control Network Chapter

Control servers in the core network. *Ericsson Review*, 4, 2000.

Packet Backbone Chapter

Reference article, example: The big question: carriers that want to build out their networks must first address another fundamental issue: What type of traffic will their networks carry – and why?, See *Telephony*, December 13, 1999.

Access Chapter

There is a huge number of books and web articles available on both mobile and fixed access. Let us mention two web examples:

Universal Mobile Telecommunications System (UMTS):
Real-Time Multimedia in UMTS (UMTS 22.72 version 0.0.0)
http://www.3gpp.org/ftp/tsg_sa/WG1_Serv/TSGS1_02-Edinburgh/Docs_All/Tdocs_All/S1-99126.pdf

UMTS 3G IP Multimedia System
System Description. Summary of 3GPP System Standards, Network Architecture, and Applications and Services (based on 3GPP Standards)
http://www.seas.upenn.edu/~hphuang/Networking/PCS/UMTS%203G%20IP%20Multimedia%20System.pdf

PROCESS FLOW ORIENTED REFERENCES AND ACKNOWLEDGEMENTS

Process Flow Structure and Techno-economics

A general source is TeleManagement Forum: info@tmforum.org
FlowThru: Co-operative Secure Management of Multi Administrative and Technology Domain Network and Service Management Systems: AC335 TeleManagement Forum Business Process to TINA-C Business Role Mapping. See:

http://www.cs.ucl.ac.uk/research/flowthru/content/bmp-role-map/bmp-role-map.pdf

Acknowledgement: the cooperation with my daughter Christina Teden regarding a wide area of techno-economics has been very stimulating and valuable, indeed, and a lot of thanks go to her.

Playground

New Telerica, a fictitious country used for the process of network design, has been extensively used in network-oriented courses. Although not explicitly mentioned it has also been a source of inspiration for this book It is mentioned or utilized in:

http://fc.it.kth.se/~asa.skog/studie/KS-projekt.pdf

Group work at Royal Institute of Technology in Stockholm (access network), and in

http://www.fek.su.se/ProfUpdate/botco.htm

Chapter 2

Acknowledgement: my cooperation with Jonas Selen at Ericsson Consumer lab has been very valuable. Thanks a lot! Regarding Consumer lab see for example:

http://www.3gnewsroom.com/3g_news/apr_02/news_2062.shtml

Håkan Wolf has written much material related to this chapter, and also brief parts in a couple of other chapters.

Chapter 4 and Other Chapters

Analysts such as Kearney, Arthur D. Little, Ovum, Forrester etc. provide much information. Contacts with Bengt Alm, teacher in techno-economics, have been very valuable.

Chapter 16 Telecom Management

A very valuable cooperation and co-writing with Adrian Faduaga, Buenos Aires, on telecom management and other subjects should be mentioned. Adrian conducted a seminar on Voice and Data Convergence and Multi-Service Networks at the University of Buenos Aires in 2001.

http://www.aefconsultores.com/PDF_Files/Espanol/AEF-Corporativa-Marzo_2003.pdf

Appendix 1

My partner within this area has been Teledrom AB.

Appendix 2

A partner within this area has been Chris Fletcher.

Appendix 3

This appendix is written by Håkan Åkerstedt.

Appendix 4

This appendix is written by Håkan Åkerstedt, assisted by Göran Ekstedt. Göran has also contributed to chapters 11, 12, 14 and 15. Many thanks!

GENERAL REFERENCES AND ACKNOWLEDGEMENTS

Platform Knowledge

This area is for example represented by *Understanding Telecommunications 1 and 2*

The books have been available for years at http://www.ericsson.com/support/telecom/authors.shtml (Telia-Ericsson) but are now probably removed. ISBN 91-44-00212-2 and ISBN 91-44-00214-9

IP Internetworking Learning Product

See www.ericsson.com/services/globalservices/training.shtml

Acknowledgement: the stimulating cooperation with Hans Nihlen, who led the development of this product should be mentioned.

Chapter 1 provides some ***basic terminology***. Among supporting material is for example:

LoL@: a UMTS location-based service. Günther Pospischil, Harald Kunczier, Alexander Kuchar.

http://lola.ftw.at/homepage/content/a40material/LoLa_a_location_based_service.pdf

This book has been much inspired from the course Understanding the New Telecom, with Miltos Tricopoulos as one of the main teachers.

http://www.ericsson.com/ie/training/courses/prd138.shtml

Miltos has worked a lot with the figures in the book.

IT Support to the Author

Many thanks for a professional computer support go to my son, Staffan Olsson. Thanks a lot!

Other Acknowledgements

Other acknowledgements go to my patient and supporting wife Marianne, Erik Oldmark, who initially ordered the book project, Tessa Hanford, who did an excellent job as my copyeditor in the U.K, and finally Lars Jansson, Leif Karlsson, Georg Lewin, Mikael Lundgren and Christer Mildh who have been my most important contact persons at Ericsson.

Glossary

3G	Third generation (mobile systems)
3GPP	Third generation partnership project
3GPP2	Third generation partnership project 2
AAA	authentication, authorization, accounting
AAL	ATM adaptation layer (AAL1, AAL etc)
ACL	access control list
ADM	add-drop multiplexer
ADSL	asymmetric digital subscriber line
AES	advanced encryption standard
AF	assured forwarding
AH	authentication header
AIN	advanced IN
AMR	adaptive multi rate
AN	access network
AP	access point
API	application programming interface
APN	access point name
“APZ”	symbolizes processor part of a telephone exchange
ARPU	average revenue per user
ASP	application service provider
ASuS	application support services/server
ATM	asynchronous transfer mode
AV	adding value
AUC	authentication centre
B2B	business to business
B2C	business to customer
BD	bridging distance
BER	binary encoding rules
BER	bit error rate
BGCF	border gateway control function
BGP	border gateway protocol

BGW	billing gateway
BICC	bearer independent call control
B-ICI	broadband inter-carrier interface
BSP	business service provider
BSS	base station system
CA	certificate authority
CAGR	compound annual growth rate
CAPEX	capital expenditure
CAMEL	customized applications for mobile network enhanced logic
CAGR	compound annual growth rate
CAP	CAMEL application part
CATV	cable TV
CAS	customer administration system
CAU	customer access unit
CC/PP	composite capabilities preference profiles
CDPD	cellular digital packet data
CDMA	core code division multiple access
CDR	call detail record
CE	circuit emulation
CHAP	challenge-handshake authentication protocol
CLI	calling line identification
CMIP	common management information protocol
CMN	call mediation node
CMTS	cable modem termination system
CORBA	common object request broker architecture
CPN	customer premises network
CR-LDP	constraint based routing – label distribution protocol
CRM	customer relations/relationship management
CSCF	call state/session control function
CSF	call serving function
CWDM	coarse wavelength division multiplex
DCN	data communication network
DEN	directory enabled networking
Des	data encryption standard
DESP	development environment service provider
DHCP	dynamic host configuration protocol
DMZ	demilitarized zone
DNS	domain name system
DOS	denial of service
DOCSIS	data over cable service interface specification
DRM	digital rights management
DSCP	diffserv code point
DSL	digital subscriber line
DSLAM	DSL access multiplexer

DSS1	digital signalling system no. 1
DTMF	dual tone multi-frequency
DWDM	dense wavelength division multiplex
DXC	digital cross-connect
E1	2.048 Mbit/s transmission capability/rate/facility/path
e2e	end-to-end
EAP	extensible authentication protocol
EDI	electronic data interchange
EDGE	enhanced data rates for global (or GSM and TDMA) evolution
EF	expedited forwarding
EIR	equipment identity register
EMS	enhanced messaging service
EMSP	e-mail and messaging service provider
ERP	enterprise resource planning
ESP	enterprise service provider
ESP	e-business service provider
ESP	encapsulating security payload
eTOM	enhanced telecom management map
ETSI	European telecommunications standards institute
E(V)LL	emulated (virtual) leased line
FAB	fulfillment, assurance and billing
FAST	fast active queue management scalable TCP/IP
FDD	frequency division duplex
FDMA	frequency division multiple access
FEC	forwarding equivalence class
FTTH	fibre to the home
FTTA	fibre to the antenna
FP	fundamental technical plans
FNR	flexible numbering register
FSC	fibre switching capability
FR	frame relay
FTAM	file transfer access and management
FTP	file transfer protocol
GCP	gateway control protocol
GERAN	GSM/EDGE radio access network???
GGSN	gateway GPRS support node
GIX	global interconnect point
GIF	graphics interexchange format
GMPLS	generalized MPLS
GPRS	general packet radio service/system
GPS	global positioning system
GRE	generic route encapsulation
GRX	GPRS roaming exchange

GSM	global system for mobile communication
GSN	gateway serving node
GSTN	general switched telephone network
GTP	GPRS tunnelling protocol
HE	home environment
HFC	hybrid fibre-coaxial
HLR	home location register
HSCSD	high speed circuit switched data
HSP	hosting service provider
HSS	home subscriber server
HTML	hypertext mark-up language
HTTP	hypertext transport protocol
IAM	initial address message
IBSP	Internet business service provider
ICMP	Internet control message protocol
ICT	information and communication technologies
IETF	Internet Engineering Task Force
IGMP	Internet group management protocol
IMAP	Internet message access protocol
IMS	IP multimedia subsystem
IMSI	international mobile subscriber identity
IN	intelligent network
IP	Internet protocol
IPBCP (BICC) IP	bearer control protocol
IPsec	Internet protocol security
IRR	internal rate of return
ISDN	integrated services digital network
IS	information systems
ISUP	ISDN user part
ISP	Internet service provider
ISN	interface serving node
ITU	International Telecommunication Union
JAIN	Java API for integrated networks
JDBC	Java data base connectivity
JPEG	Joint Photographic Experts Group
L2TP	layer 2 tunnelling protocol
LDAP	lightweight directory access protocol
LEO	low earth orbit
LER	label edge router
LMDS	local multipoint distribution system/services
LSC	lambda light-wave switching capability
LSP	label switched protocol/path
LSR	label switch router

M3UA	message transfer protocol 3 – user adaptation layer
M2M	machine to machine
MG	media gateway ?
MAC	media access control
MExE	mobile execution environment
MGC	media gateway control
MGCF	media gateway control function
MGW	media gateway
MIB	management information data base
MIME	multipurpose Internet mail extensions
MM/IP	multimedia over IP
MMS	multimedia messaging service
MO	managed object
MP-3	MPEG-1 layer 3
MPC	mobile positioning centre
MPEG	Motion Picture Experts Group
MPLS	multi-protocol label switching
MRF	media resource function
MRFC	media resource function control
MSC	mobile switching centre
MSISDN	mobile station international ISDN number
MSN	multi-service network
MSP	management service provider
MSS	mobile satellite service fig 13.28/13/29
MSSP	managed security services provider
MT	mobile terminal
MVNE	mobile virtual network enabler
MVNO	mobile virtual network operator
NAS	network access server
NAT	network address translator
NE	network element
NGN	next generation network
NGOSS	new generation operation support systems
NIST	National Institute of Standards and Technology
NMC	network maintenance/management centre
NPV	net present value
OFDM	orthogonal frequency division multiplexing
OMA	Open Market Alliance
OMC	operation and maintenance centre
OOP	object oriented programming
OPEX	operating expenditure
OSI	open systems interconnection
OSA	open systems architecture
OSPF	open shortest path first

OSPF-TE	OSPF traffic engineering
OSS	operation support system
OTA	over the air
OXC	optical cross-connect
P2C	person-to-content
P2P	peer-to-peer
P2P	person-to-person
PAP	push access protocol
PABX, PBX	private automatic branch exchange
PCM	pulse code modulation
PDA	personal digital assistant
PDP (context)	packet data protocol (context)
PDC	personal digital communication
PDCP	packet data convergence protocol (fig 5.14)
PDP	packet data protocol
PDSN	packet-data service node
PG	peer group
PGI	peer group identifier
PGL	peer group leader
PGP	pretty good privacy
PHB	per hop behaviour
PKI	public key infrastructure
PLMN	public land mobile network
PNNI	private network-network interface
POI	point of interconnect
POP	post office protocol
POTS	plain ordinary telephony service
PPP	point-to-point protocol
PPG	push proxy gateway
PPTP	point-to-point tunnelling protocol
PSC	packet switching capability
PSEM	personal service environment management
PSTN	public-switched telephone network
PTD	personal trusted device
PTP	particular (fundamental) technical plans
PXC	photonic cross-connect
QoS	Quality of Service
QAMS	Quantitative assured media playback service
RADIUS	remote authentication dial-in user service
RAB	ratio access bearer
RANOS	radio access network operation support
RBS	radio base station
RFC	request for comments

RIP	routing information protocol
RLC	radio link control
RNC	radio network controller
ROI	return on investment
ROCE	return on capital employed
ROHC	robust header compression
RLC	ratio link control
RSVP	resource reservation protocol
RSVP-TE	RSVP – traffic engineering
R-SGW	roaming signalling gateway
RTCP	real time control protocol
RTD	round trip delay
RTP	real time protocol
RTSP	real time streaming protocol
RTT	round trip time
SAT	SIM application toolkit
SDP	session description protocol
SCM	supply chain management
SCP	service control point
SCS	service capability server
SCTP	stream control transmission protocol
SDH	synchronous digital hierarchy
SDP	service description protocol
SGSN	serving GPRS support node
SGW	signalling gateway
SHDSL	single pair high bit rate digital subscriber line
SIGTRAN	signalling transport (working group)
SIM	subscriber identity module
SIP	session initiation protocol
SIP	strategy, infrastructure and product
SLA	service level agreement
SLS	service level specification
SME	small and medium enterprise
S/MIME	secure multipurpose Internet mail extensions
SMPP	short message peer-to-peer protocol
SMS	short message service
SMTP	simple mail transfer protocol
SNMP	simple network management protocol
SOAP	simple object access protocol
SOHO	small office/home office
SOP	service-oriented programming
SONET	synchronous optical network
SPC	stored program control
SPI	security parameter index
SQL	structured query language

SS7	signaling system No 7
SSH	secure shell
SSID	service set identifier
SSL	secure socket(s) layer
SSP	storage service provider
SSP	service switching point
T1	1.544 Mbit/s transmission capability/rate/facility/path
TCA	traffic condition agreement
TCAP	transaction capabilities application protocol
TDD	time division duplex (fig 13.28/13.29)
TDM	time division multiplex
TDMA	time division multiple access
TD-SCDMA	time division – synchronous code division multiple access
TE	traffic engineering
TeS	telephony server
TLS	transport layer security
TMN	telecommunications network, telecommunication management network
TOM	telecom operations map
TOS	type of service
TRAM	tools for radio access management
TSC	transit switching centre
T-SGW	transport signalling gateway
TSN	transit serving node
TTC	time to customer
TTM	time to market
TTS	time to service
UDDI	universal description, discovery and integration
UDP	user datagram protocol
UNI	user-network interface
UMTS	universal mobile telecommunication system
URI	uniform resource identifier
USIM	universal subscriber identity module
UTRAN	universal terrestrial radio access network
VASP	value-added service provider
VC-4	virtual container at level 4 (140Mbit/s in PDH, plesiochronous digital hierarchy)
VDSL	very high bit rate digital subscriber line
VHE	virtual home environment
VLR	visited location register
VML	vector mark-up language
VoATM	voice over ATM
VoFR	voice over frame relay

VoIP	voice over IP
VPN	virtual private network
VSP	vertical service provider
WAN	wide area network (as opposed to local area network)
WAP	wireless application protocol
WASP	wireless application service provider
WCDMA	wideband code division multiple access
WEP	wired equivalent privacy
WI	wireless Internet
WIM	WAP identity module
WISP	wireless Internet service provider
WLAN	wireless local area network
WSDL	web services description language
WSFL	web services flow language
WSP	wireless session protocol
WTLS	wireless transport layer security
WWW	world wide web
xDSL	(any) digital subscriber line
XML	eXtensible Markup Language
XHTML	HTML reformulated as an XML application

1

Introduction

1.1 THE BOOK IN BRIEF

This book is follow-up of *Understanding Telecommunications 1* and *2*, published in 1996–98 by Telia and Ericsson. The main author of all the books is Anders Olsson. The basics of various networks (PLMN, PSTN/ISDN, ATM, Internet etc.) are not repeated here.

This book instead provides an integrated holistic view of users, telecom business and networks at a time when telecom, datacom and media industries are converging. See Figure 1.1.

It offers the reader a cross-scientific, generic and well-structured overview of the telecom area. A main objective is to understand what competencies are required to become successful within the converging telecom world of today. The technical content is fairly generic but with the main focus on an all-access network and also some focus on UMTS.

1.1.1 THE PROBLEM

The multidimensional paradigm shift and an increased overall complexity create a significant need for new competencies, which go beyond only technique. Some of the main

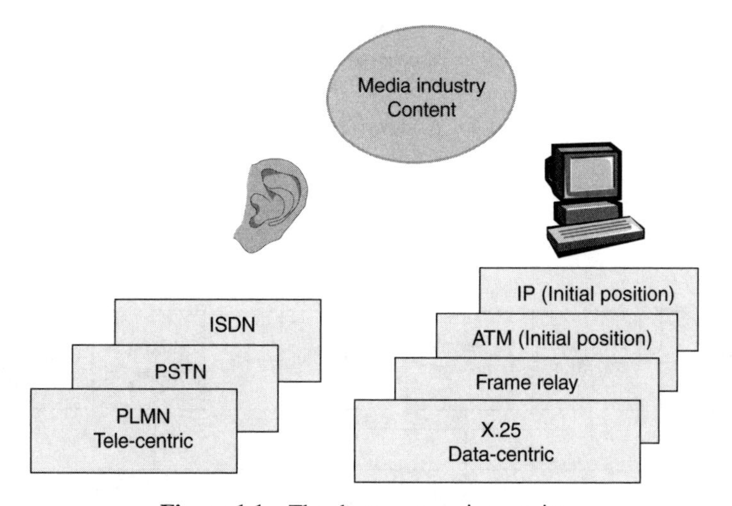

Figure 1.1 The three converging parties

Understanding Changing Telecommunications – Building a Successful Telecom Business. Edited by A. Olsson
 ISBN: 0-470-86851-1

changes are shown in Figure 1.2. The overall change is even bigger when the integration of the media industry is included. The datacom industry is concerned when involving real-time voice and multimedia transfer, priority mechanisms and a more sophisticated charging in the data transfer.

1.1.2 SUPPORTING COMPETENCE CREATION

The telecom market is indeed large with a number of players. The main ones are indicated in Figure 1.3. The focus in this book includes stakeholders 1 to 3 and the network. The supplier (4) is represented by new enabling technologies. The approach is top-down, which means that the network is seen as a tool.

The stakeholders are connected by means of a logical series of questions:

- What services are needed or demanded? The unpredictable user is the queen/king.
- How can the services be implemented to the satisfaction of the main stakeholders? This includes business aspects.

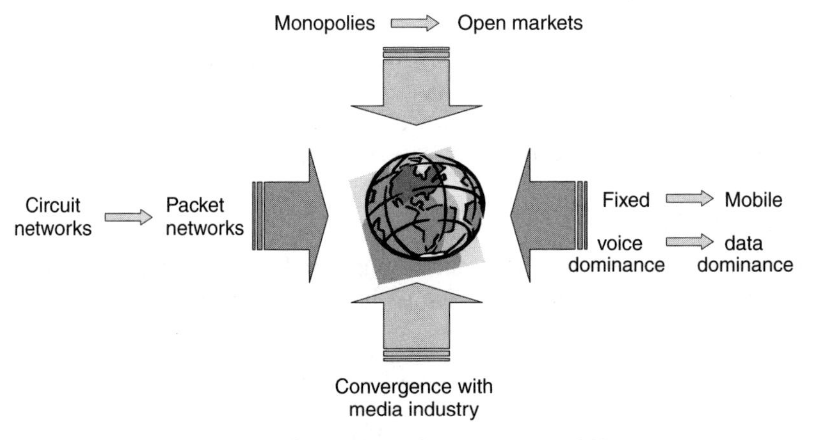

Figure 1.2 A view of the paradigm shift

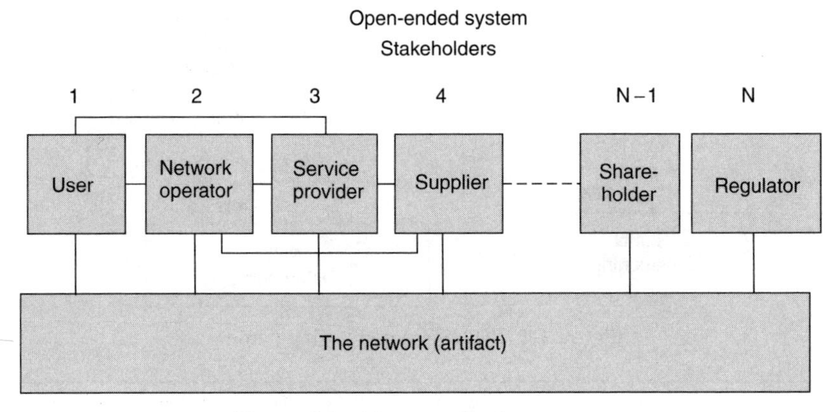

Figure 1.3 The overall telecom area

- How do/can the various available technologies cooperate in supporting this target? The chosen approach is the all-access network with a universal service-handling approach.

The book thus applies a holistic horizontal integrating view of the changing telecom world, as seen from the telecom business, user and network angles (see Figures 1.4 and 1.5). The prime focus is on principles and context rather than on profound technical descriptions: 'The trend in the future should be a more holistic view of things'.

In order to connect subjects and contexts in an understandable way, a number of generic techno-economic and technical reference models are used or highlighted. Examples are a network segmentation model (see part of Figure 1.10 below), a triangle model (Figure 1.5), a fundamental plan model, a profitability model, a process model (Figure 1.6) and the return on capital employed (ROCE) definition. Consequently, reference models act as maps that identify the chapters in the book.

The book is written at a basic level enhancing the context orientation. A number of figures are used more than once as a kind of glue between different chapters. As we know, some repetition has learning advantages.

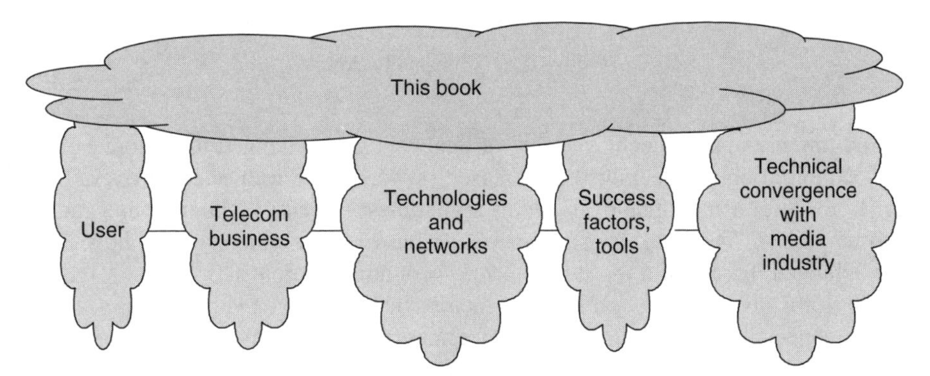

Figure 1.4 The horizontal approach is similar to the use of the IP protocol

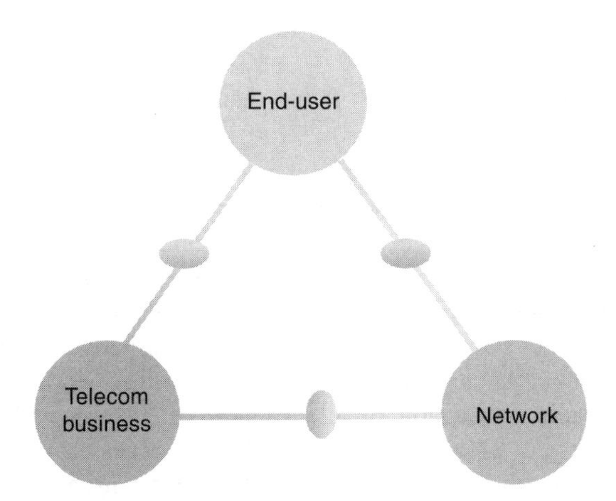

Figure 1.5 The multi-angle view of this book

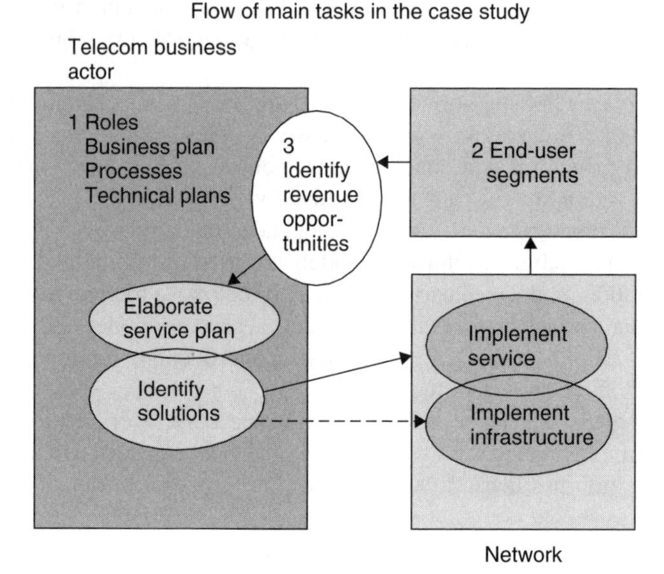

Figure 1.6 Case study flow

The commonly applied technical and vertical focus has changed to a more horizontal techno-economic one, reflecting the increased power of the user and the overall uncertainty. In the prevailing situation it seems sometimes wiser to be humble and indicative than to be precise. The book then becomes a framework for understanding how progress can be achieved, based on a top-down techno-economic approach.

In line with this it focuses on success factors in the user network and user–business interface, especially services including implementation, charging, quality and security. Services and the service layer are described in detail because of their increasing importance from a business point of view.

Fundamental technical plans are the most important tool used to explain the technical context, that is, when vertical networks turn into a horizontal network. Such plans offer a layered end-to-end approach.

The book focuses on the core network rather than on the access network. The reasons are the following:

- The paradigm shift is faster in the core network parts than in the access parts.
- At the same time the core converges whereas the number of access alternatives increases as new technologies are developed for existing media (new mobile systems, xDSL (a family of digital subscriber lines), power-line systems, WLAN (wireless local area networks) etc.). It is not possible to cover all access alternatives with any depth within the frame of this book. An access system overview is provided in Chapter 13.
- The goal is to enable connection of all different accesses including WLAN to the evolving new core, the 'all-access network'. The all-access network becomes a distributed gateway between the different accesses. People with a telecentric background may possibly regard the new core networks as huge group switches.

On the other hand the access chapter is closer to standardized systems than other chapters in the book. It could therefore be useful to have an early look at this chapter when reading the book.

The overall management of the paradigm shift implementation is a challenging task, and different approaches are possible. Leading telecom operator managers recommend much more business-oriented thinking at the expense of focusing on the very latest hyped services. When appropriate, the book applies a business-oriented view in line with this statement.

Can the shift be measured? Of course not, but a way to make the change more concrete is to write a book and compare its content with the previously issued books, written five to seven years ago.

Another approach is to use a traditional thoroughly tested network design course from the 1990s and apply its sequence of tasks to the environment we have today. After working with both the vertical and the new multimedia network the conclusion is clear: *It **is** a paradigm shift.*

The network course mentioned above is called 'The Middleton case study'. It covers everything from a business plan to a fairly complete high-level network design that supports the business plan. The case study extended over five days and was attended by operator managers and planning staff from all over the world. Within Ericsson the main target group was marketing staff, managers and 'system engineers'. The case study focus, as well as the focus of this book, is on functions, not products.

The choice of technology in the case study was a pragmatic one, based on what best served the business plan of a specific actor in his specific environment.

There are no dedicated chapters in this book on specific systems, although UMTS has been used a number of times to exemplify for example traffic cases, numbering and quality of service. A main reason for this approach is that plenty of system-oriented material already exists in books, on the Internet etc.

The book is by no means a handbook. A handbook requires a deeper level. However, by applying a sequence of subjects similar to a case study we also get a checklist and a natural sequence of the chapters. The rough case study sequence is illustrated in Figure 1.6.

This could be compared with a 'real' operator network planning cycle, which may look like Figure 1.7. When managing the process sequence in a professional way you are *'Building a Successful Telecommunications Business in the 21st Century'* or *'Becoming a Successful Telecommunications Operator in the Multimedia Society'*. These are two interesting book title proposals from reviewers.

The three main stations in the flow form a kind of triangle. Each corner has a basic chapter of its own in the book. See Figure 1.8.

Some of the most critical components of the telecom business are the interfaces to the users, especially the user perception. Figure 1.9 shows (with thick rings) subject areas that have been given particular attention. Three of them have been given separate chapters:

- Services
- Security
- Quality of service

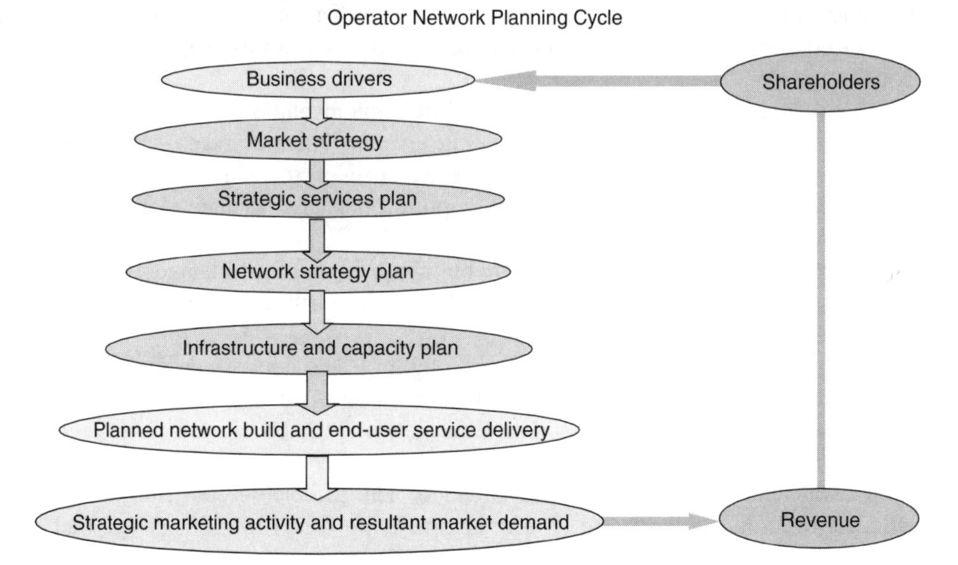

Figure 1.7 Planning flow

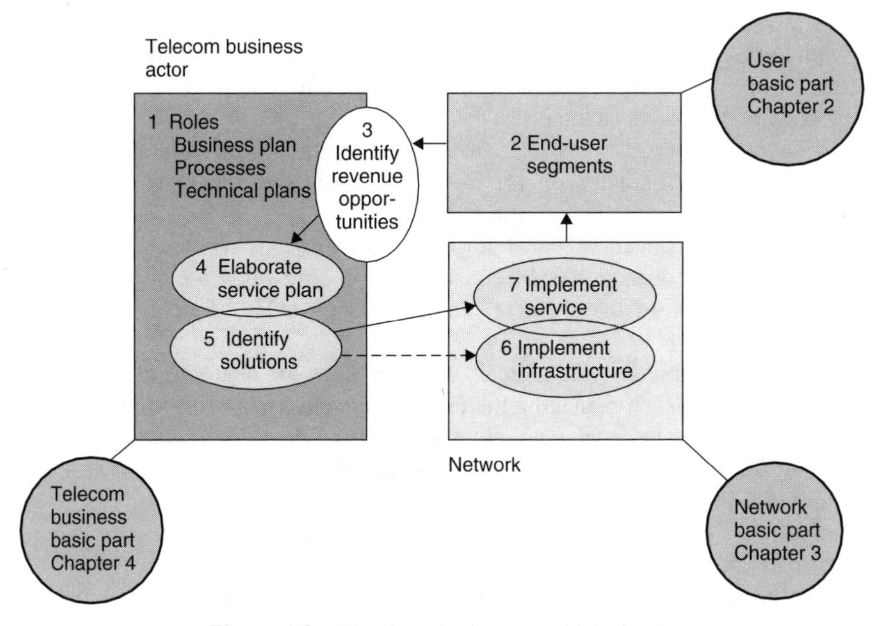

Figure 1.8 The three basic parts (thick rings)

Regarding the remaining chapters, three are partly process oriented:

- Service implementation
- Interconnect
- Telecom management – operations

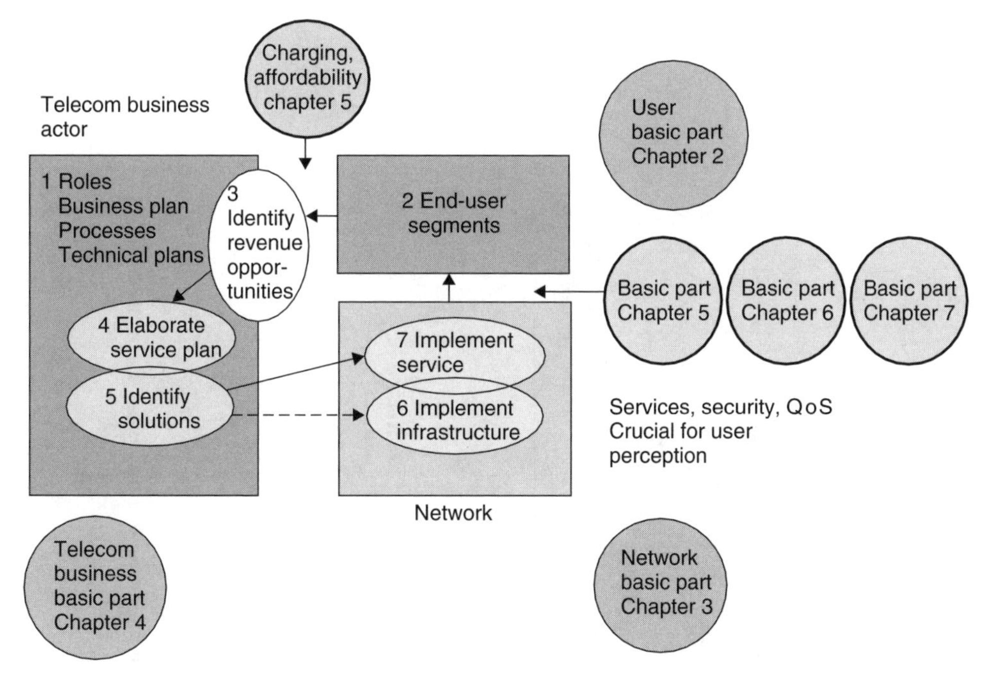

Figure 1.9 Chapters that relate to the user interface

whereas the others are segments of the new network. The network is split into three dimensions: the layered dimension, the distributed dimension and the processing dimension. The chapters embrace the following network segments:

- Service network
- Terminal devices
- Edge nodes
- Packet backbone
- Access network
- Control network
- Processing dimension

The related book reference model is shown at the bottom part of Figure 1.10, with the three dimensions on the right-hand side and the triangle at the upper left.

A main intention of this book is to use it in courses. We know that many courses are product oriented. It is easy to complement the content of this book with products, for example using Figure 1.11 as a reference. Please observe WLAN as one of the access alternatives.

The overall chapter structure becomes:

- Chapter 1: Introduction
- Chapter 2: End user needs and demands
- Chapter 3: Technologies and networks (Appendix on web services)

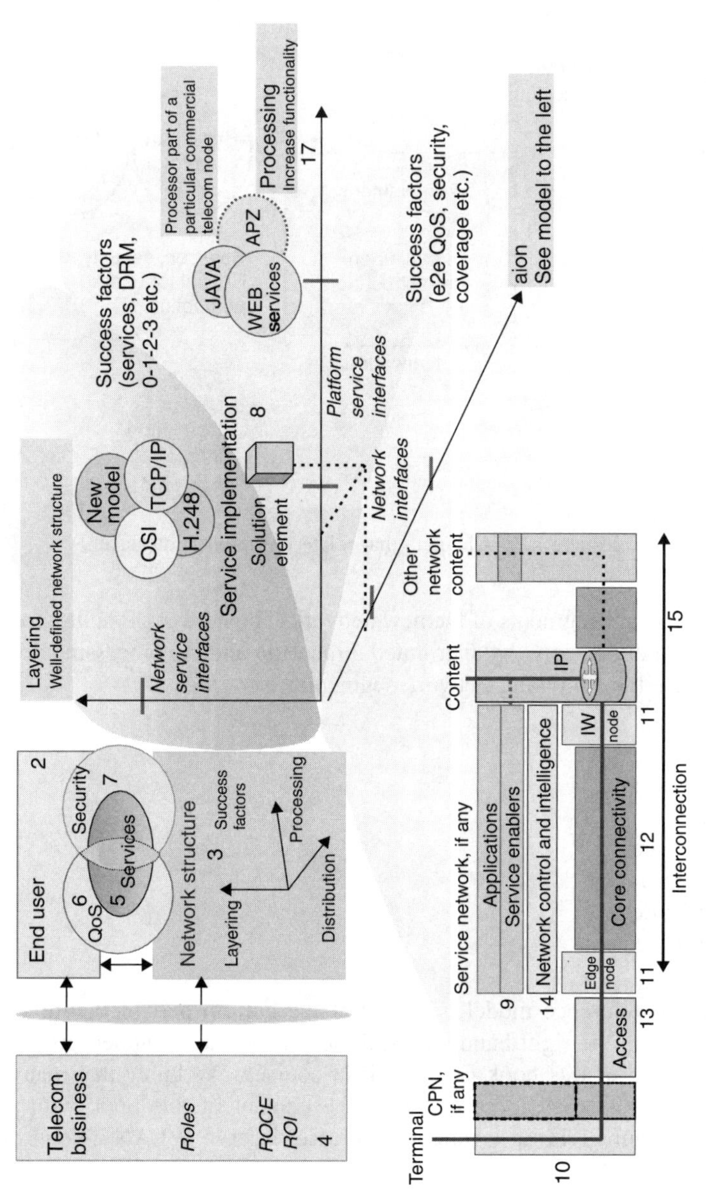

Figure 1.10 A comprehensive view of the book contents with the distributed network view at the bottom

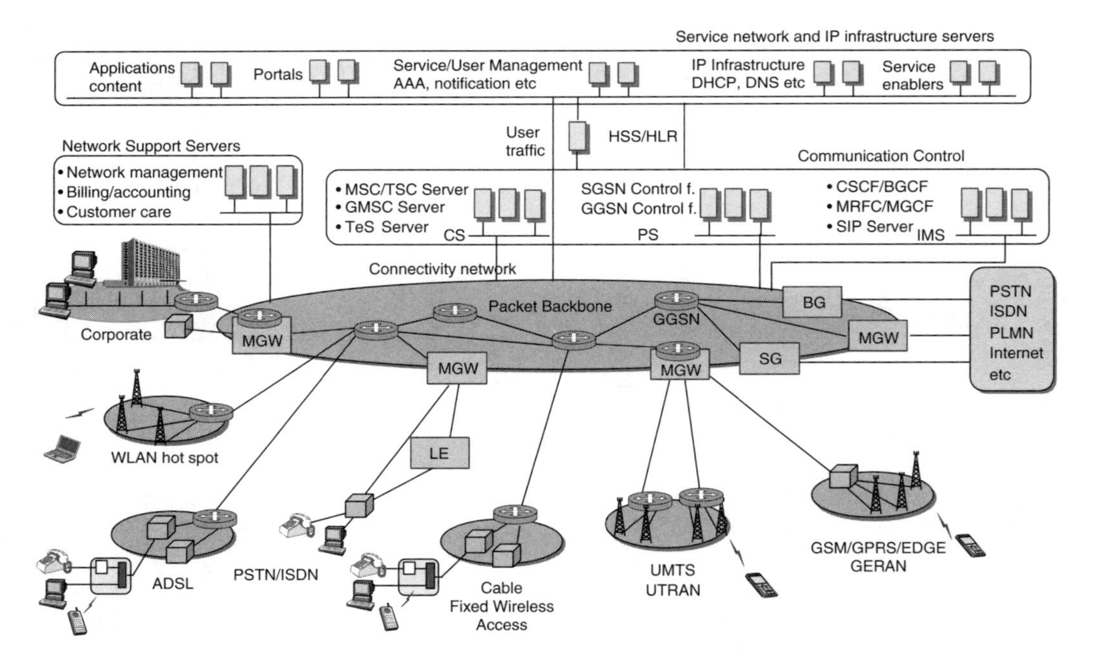

Figure 1.11 Example of a somewhat more product-oriented view of the new network

- Chapter 4: Telecom business (Appendix on Financial calculations)
- Chapter 5: Services
- Chapter 6: Security
- Chapter 7: Quality of service
- Chapter 8: Service implementation (Appendix on Incumbent operator migration)
- Chapter 9: Service network
- Chapter 10: Terminals
- Chapter 11: Edge nodes (Appendix on MGW dimensioning)
- Chapter 12: Packet backbone
- Chapter 13: Access
- Chapter 14: Control network
- Chapter 15: Interconnect
- Chapter 16: Telecom management – operations

1.2 A DYNAMIC SITUATION

1.2.1 OVERVIEW

Figure 1.12 describes the dynamics in the evolution. Here we can see three loops:

- User and services, which is a market-driven loop
- Enabling technologies
- Regulations and society loop

All these inputs will change with time, creating a constantly changing market environment.

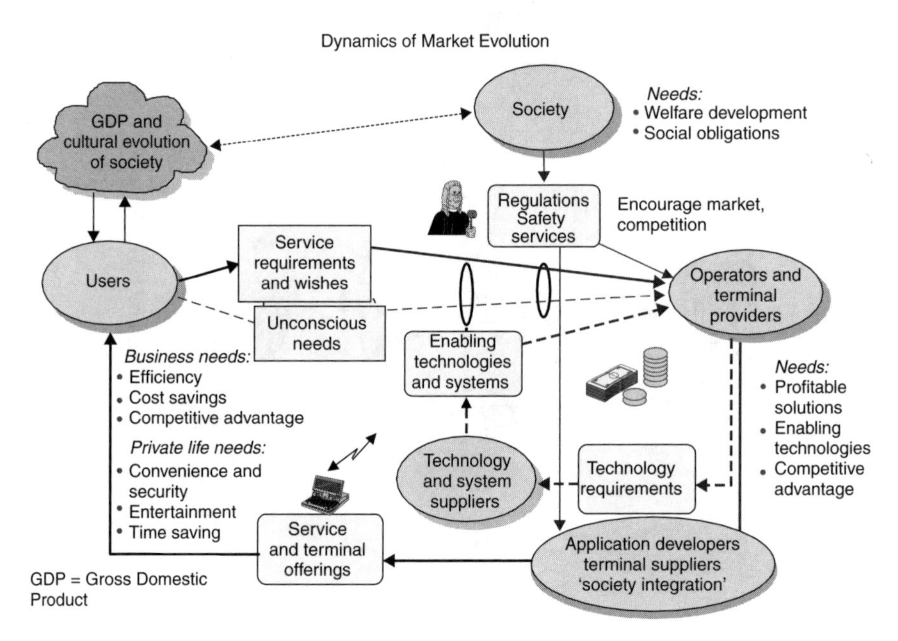

Figure 1.12 The overall dynamics make the situation even more complex

This dynamics is a particular problem when writing a broad-based book. The vast underlying range of standards and technological developments cannot be updated, unless there are new editions. Therefore, the approach is aimed at avoiding details, identifying trends and being as context oriented as possible, with business aspects as (often invisible) drivers.

1.2.2 ENABLERS FOR CHANGE

The paradigm shift needs a lot of new standards, sometimes called enablers in this book. Once you know the enabler name a lot of information can be found elsewhere, for example on the Internet. Chapter 3 contains a (non-complete) list of important enablers in Sections 3.14 and 3.15. Readers who want to go deeper than what is provided in this book can of course use the huge Internet treasure of information.

Other Pedagogical Products

A web-based training flow 'IP Internetworking' and publications such as the two previous books (*Understanding Telecommunications*, parts 1 and 2) contain significant basic material. The 'IP Internetworking' flow can be considered quite data/IP-centric, complementing the more telecentric approach of this book. The flow is accessible at www.ericsson.com/services/globalservices/training.shtml.

1.3 SUCCESS FACTORS FOR THE GROWTH OF MOBILE SERVICES

At the end of the 1990s there was no need for a heading like this because of the optimistic telecom atmosphere. The first years of this millennium have been gloomy, and many have been looking for ways to get back on track again. There are no officially recognized success factors, but many respected persons have expressed an opinion. And such opinions have admittedly had a certain impact on this book. Below a couple of Swedish voices follow.

Allan Malm and Bertil Thorngren, Sweden, both professors in business economics (from the Swedish newspaper *Svenska Dagbladet*, 6 February 2003):

- Create the market first, using the existing systems.
- Market the new services in a user-friendly language, based on customer benefit, not with expressions like GPRS.
- Create a standard for secure identification, based on the mobile device.
- Launch public safety services for elderly people and for the medical service.
- Minimize investment cost (per user) for the new radio access equipment.

Östen Mäkitalo, Senior Vice President, Mobile Business, Telia Sonera, Sweden- Finland (from Networks Telecom in Stockholm, September 2002):

- Sufficiently attractive, affordable services, making life easier.
- Sufficient coverage (supports GPRS, UMTS).

- Sufficiently low congestion (roughly quality of service).
- Sufficient security.
- Sufficient availability of devices.
- Sufficiently easy to use.

Let us also mention the Open Market Alliance (OMA). The OMA delivers open standards for the mobile industry, helping to create interoperable services that work across countries, operators and mobile terminals and are driven by users' needs. To expand the mobile market, companies supporting the OMA will work to stimulate the fast and wide adoption of a variety of new, enhanced mobile information, communication and entertainment services. The OMA includes all key elements of the wireless value chain and contributes to the timely and efficient introduction of services and applications. For more information, please visit www.openmobilealliance.org.

A more long-term success factor was presented in *Computer Sweden*, 22 August 2003 by Jens Zander, professor at the Royal Institute of Technology in Stockholm: 'The issue of radiation from radio equipment could be finally solved when small (down to matchbox size) and cheap radio base stations are available and abundantly installed. This will heavily reduce the required radio power also in mobile terminals'.

1.4 COMMENT ON TERMINOLOGY

The terminology within telecoms can be confusing. In a time of convergence the converging parties want to promote their own terminology. One example of possible confusion is the relationship between the words 'telecom' and 'datacom'. We have chosen to use the word 'telecom' for the converged communication, since telecom really means distant communication rather than telephony communication, which makes it generic.

In a time when telecom and datacom are converging with the media industry it is also natural to examine the words 'media' or 'media stream'. Media is used a lot, for example in multimedia and media gateway. The specific term multimedia is explained and defined in Chapter 5 and media gateway is allocated a chapter of its own. Media stream is the user information (such as speech, tones, announcements, MPEG video, JPEG pictures, XML data etc.) carried on a bearer. In multimedia calls, a number of media streams are carried on a (possibly different) number of bearers.

Another example is the interpretation of the word 'service provider'. Since a *service* is a very general word, you would expect also 'service provider' to be quite general. However, it often has a narrow meaning, for example compare Internet service provider, or 'service provider' as opposed to 'network provider' or 'network operator'. In this book we will use 'network operator' quite often. Therefore 'service provider' will normally be used in its narrow sense, and does not include 'network operator'.

There are other problems with the word 'service' as well. Supplier units offer professional services to network operators/service providers. Often the word 'professional' is deleted and just 'services' remains. 'Service' is also used to indicate technical resources.

In the telephony world the word has got a flavour of IN (intelligent network) services. This is normally a kind of network intelligence supporting fairly dumb terminals, such as ordinary telephones. This network intelligence mainly offers advanced routing, charging and number conversion tasks. Unfortunately the term 'application' is also sometimes used for such network-based services. In another context 'service' could refer to frame relay, ATM (asynchronous transfer mode) or IP (Internet protocol) communication facilities. A better expression is 'network services', as opposed to 'applications' located in servers outside the network. The expression 'multi-service gateways' used in this book is network service oriented.

Applications normally consist of a terminal component and one or several cooperating application components in servers. Applications are not involved in the media transfer. See Figure 1.13. Some common resources for various applications ('generic application components') are:

- Content converters
- E-commerce servers
- Business logic

The conclusion is that the word 'service' should preferably not be used on its own.

The latest meaning of the word 'service' is: 'A service is a contractually defined behaviour that can be implemented and provided by any component, based solely on the contract'. This definition applies to web services and service-oriented programming, a future trend.

Having talked about services the next step is to discuss *networks*. This term will be defined (for the book) at the beginning of the network-oriented chapter (Chapter 3). *3G* and *next generation networks* (NGN) are specific network solutions. The book objective

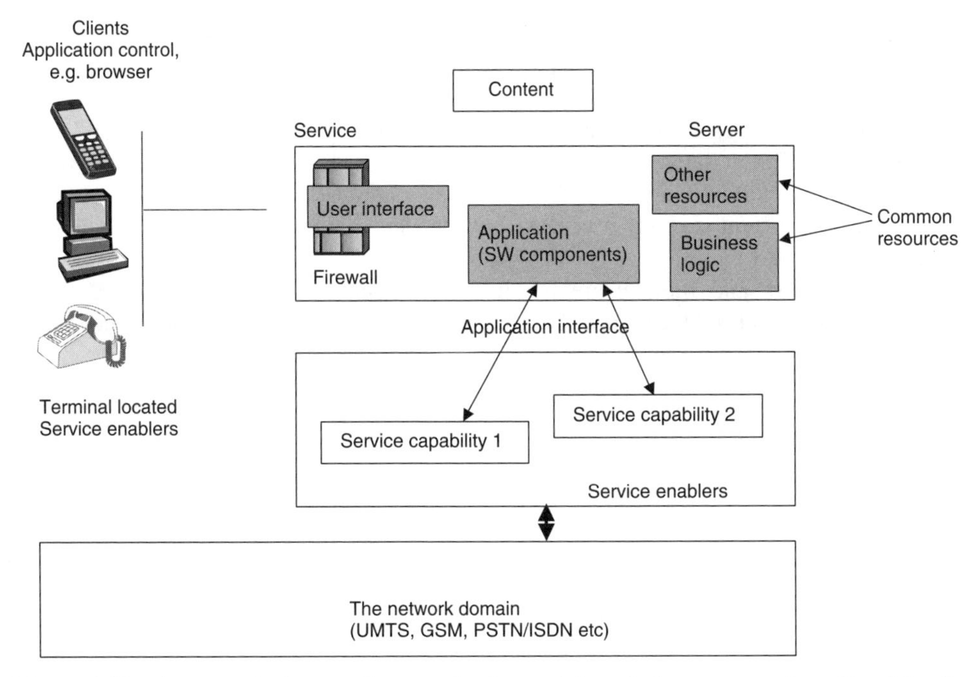

Figure 1.13 An overall picture of an application and its relation to a service offered by the application and its support functions

to be general and holistic favours other expressions such as 'all-access networks' and 'horizontal networks'. Therefore 3G and NGN are not used very much.

Another common term is '*broadband*'. How broad is broadband? Originally (ITU-T recommendation I.113) it was faster than primary rate ISDN (in Europe about 2 Mbit/s, in the USA about 1.5 Mbit/s). Nowadays network operators advertise broadband at 256 kbit/s and higher. FCC in the USA has chosen 200 kbit/s since this bit rate is considered sufficient for full-motion video and changing web pages as fast as one can flip through the pages of a book.

The term '*enabler'* may also cause interpretation problems. In this book it is used with a meaning close to 'new standards'. In the literature sometimes other interpretations can be observed, where enablers reside only in the service layer. To avoid misunderstanding the term *service enabler* is used in such a context. By 'service enabler' in this book we mean service supporting elements in the service network and the terminals. It embraces another tricky term, '*service capability servers*' (SCS). An ETSI definition of SCS is 'a functional entity providing open systems architecture (OSA) interfaces for an application'. An SCS adds value to the application. Services offering added value are sometimes referred to as value-added services. See Chapter 5, section 5.

Finally, the term '*middleware*' is important to understand. When two people with different native languages use a third language, such as English, for communication, the English language becomes a kind of middleware. When going from vertical network architectures to a horizontal one, integrated with the media industry, a middleware layer becomes convenient and indeed necessary in order to mediate between applications and heterogeneous underlying networks.

The following definition comes from the book *GPRS and 3G Wireless Applications*: 'Middleware is software (often denoted as a platform) that mediates between the network and the application and that enables seamless communication over heterogeneous networks'.

Middleware provides a semi-manufactured functionality for adaptation to and optimization over different networks and even presentation to the user. Middleware decreases flexibility but shortens development time. It seems natural to also allocate service economy features (efficient coding, user data and header compression etc.) to middleware rather than to added value.

Service enablers and middleware are just the beginning of a service layer terminology situation that initially may be somewhat tricky to sort out for 'connectivity oriented' readers.

2

End-User Needs and Demands

'People often ask how I know what people want. I tell them that the difficulty is to know what people don't know they want.'

2.1 OBJECTIVES

The user is a key component in the flow shown in Figure 2.1. The main objective of this chapter is to efficiently integrate this component into the flow by means of an improved user understanding. This helps us improve business planning, product development, marketing and sales, and customer service.

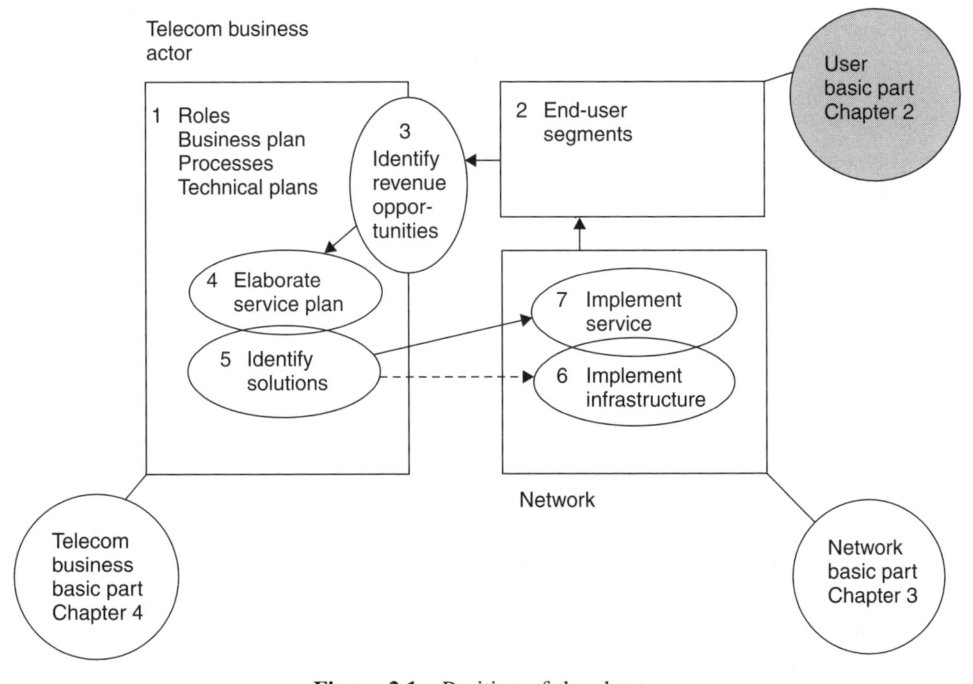

Figure 2.1 Position of the chapter

Understanding Changing Telecommunications – Building a Successful Telecom Business. Edited by A. Olsson
 ISBN: 0-470-86851-1

Users have sometimes (normally?) been regarded as merely consumers of technical innovations, such as ISDN, WAP (wireless application protocol), GPRS (general packet radio service/system), 3G (third generation) wireless etc., with little understanding of how they can benefit from such innovations. It came as a surprise to the innovators, both suppliers and operators, when the expected positive response from users did not occur.

Let us put it the other way around. Isn't it the telecom industry that should adapt to the user and to the society we live in? And all the new possibilities that appear, especially as a result of the convergence of media, telecom and datacom, shouldn't they be seen as elements in a continuous iterative process in finding an optimal everyday user life? And further, how to handle the fact that there is not just a few types of users, but indeed a spectrum of private users and organizations with different needs, values, life styles and life situations? Finally, how to segment this spectrum? What parameters?

A number of conclusions can be drawn:

- There is a need for a better overall user understanding within the telecom sector. This understanding must include attitudes, behaviour and the human inclination of resistance to change.
- There is a need for a better insight regarding the new power balance, with the user in a central position as a queen/king. The expansion of the IT/telecom area and the deregulation/re-regulation of the telecom business during the latest decade or more have significantly strengthened the role of the end user.
- Together with the first point this should lead to a more service-minded approach with improved flexibility and agility as important characteristics.
- More and more, the telecom business needs a 'societal integration', which includes alliances with other businesses, like content suppliers (to provide the new services and applications), applications developers (to enable the provision of contents), banks and finance companies (for e-commerce developments), and so on.
- The marketing and other parts of the user interface must be more 'human'.

Deregulation has enabled the user to choose between service offerings from competing actors (see Figure 2.2).

As mentioned in Chapter 1 many respected persons have analysed the present (2003) gloomy situation within the telecom industry. Let us take a sample of their proposals on needs to satisfy in order to get back on track again:

- The need for services that are attractive and/or make life easier. The need for safety services for elderly people, the need for more medical services.
- The need for easy-to-use devices.
- The need for a user-oriented language, avoiding expressions like WAP, GPRS, ISDN in marketing and other user-related activities.
- The need to create, market and test new services in the established systems.
- The need for security and related standards, especially for transactional services.
- The need for accessibility, with sufficient coverage, and roaming between different systems.

The user orientation is clear. The obvious need for a bottom-line quality of service adds to the list.

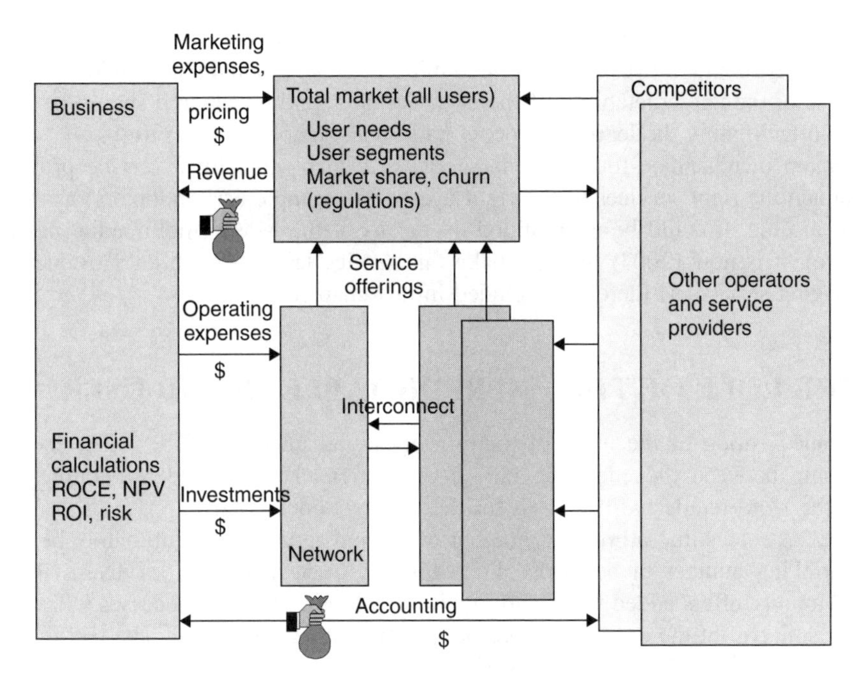

Figure 2.2 Service offerings from competing actors

The statement that users live in a dynamic environment is recognized. Yet, let us take some examples from the last 10 years:

- More *interactivity* is offered, such as games and of course surfing the web.
- *Multimedia* is offered. Telecom is heading towards a media channel like radio, TV, books.
- *Mobility for services other than voice* is offered.
- *Business support* has increased significantly, helping enterprises to focus on core business.
- *P2P (peer to peer)*, such as Napster, was unknown before.
- *M2M (machine to machine)* is coming
- Security is increasingly important.
- Quality of Service (QoS) was perceived unconsciously before, and it was more network internal. Now it is an all layer, end-to-end embracing issue.

The above list also gives quite a good picture of the changing interface between the user and the network, and describes the current main requirements of the network. This interface will be treated in more depth in Chapters 5, 6 and 7. The business–user interface along with pricing, charging and billing is treated in Chapter 5, Section 5.11. This chapter will instead concentrate on user segmentation and the needs and demands of the segments as an approach to improved user understanding. User services are in principle treated in Chapter 5, but there are some reasons to include this area in this chapter:

- In order not to lose continuity some rough matching between segments and services is included. It should, however, be borne in mind that forecast and reality seldom coincide because of the unpredictability of the user, and this kind of material should be seen as tools in facilitating the learning process more than (a non-existent) truth.
- The main merchandise for a user is services. A main goal for a service provider is to launch the right services to the right users (for example early adopters) at the right point in time. It could be exemplified by means of the MMS (multimedia messaging service). It is now (2003) time to make this service take off. A brief introduction on messaging services is therefore included in this chapter.

2.2 THE ROLE OF THE UNPREDICTABLE (?) END USER

The strong position of the users of today is illustrated in Figure 2.3, which shows the relationship between the end user and (one) retailer/service provider/operator. In this section the word retailer will be used for the end-user counterpart.

Figure 2.3 shows that a growing number of offered services will ultimately be carried by a shrinking number of networks. For example, more and more services will be IP based. Mobility offers added value and mobile (or at least wireless) accesses will conquer an increasing part of the overall number of calls (including data calls). The fact that users in Europe during 2002 sent more SMS (short message service) messages than e-mails supports this statement.

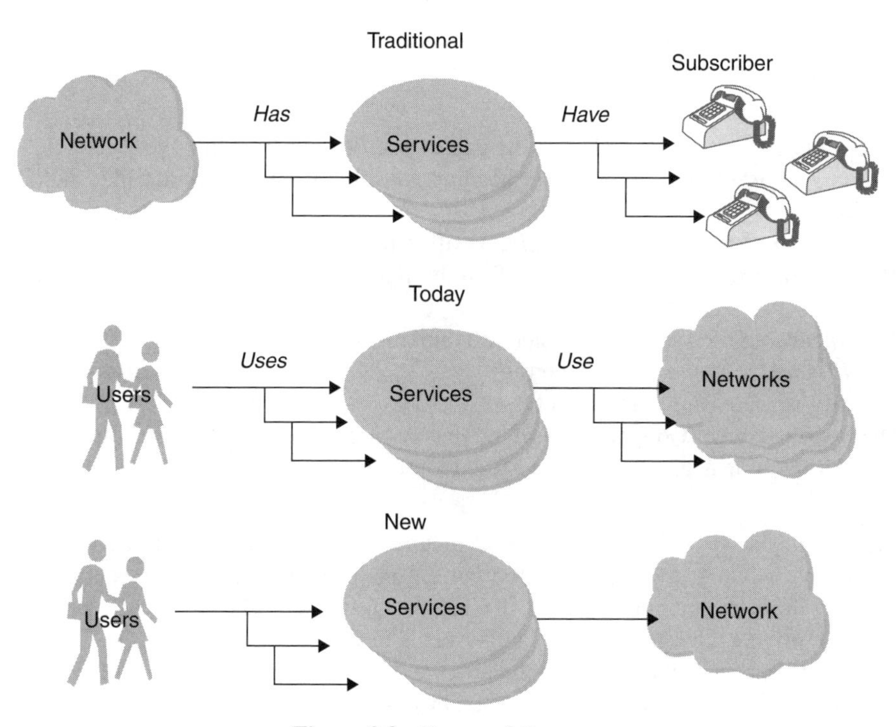

Figure 2.3 Power of the user

In addition to the illustrated change, the end user can easily change retailer, making his position even stronger. These changes lead to a faster, less controllable and less predictable development than what was previously the case. 'Churn' is a key word, indicating the degree of subscriber movements from one operator to another.

The fact that the end user is the main (final) financing source for telecom development gives a further dimension of user power. It also explains the increasing interest in the development of average revenue per user (ARPU) for various user segments. In general the expense patterns for needs such as food, entertainment and communication have been fairly stable. ARPU is further dealt with in Chapter 4.

Figure 2.3 shows another way to describe the new power of the user. With many operators there is also churn between parallel networks. The development towards a technically unified single network is maybe a vision.

Today we are in the middle of an unparalleled transitional period. Although the increasing abundance of possible services contributes to end-user power, it also creates uncertainty. A main reason for this is the (virtual?) unpredictability of the user. A good example is SMS. The technical solution and the user interface are far from perfect, but SMS fulfils an obvious need. In Sweden in 2001 more than one billion SMS messages were sent, despite a cost of 1:25–1:50 SEK per mail. Videophone, on the other hand, has been a potential service since the 1970s with little success so far. It can be said that a successful service is an affordable service that fulfils a need.

Yet, it is no guarantee of revenue or retention of customers. For the actors with an interface to the end user, a true understanding of the end-user needs, professional customer care and appropriate service level agreements are keys to retaining existing end users and to acquiring new ones. Thus, customer focus is more important than ever before. But the questions are: what is really needed from a human point of view, considering that the human being is roughly the same as it was thousands of years ago, and how fast can the perception of the end-user change? Other business sectors, such as the car industry, have no doubt been more successful than telecom here, although selling telecom services should not be very much different from selling a car, theoretically.

Östen Mäkitalo at Telia Sonera in Sweden says that there is a need for considerably more research within this area. 'The behavioural research is lagging behind. The mobile terminal should become everyone's personal assistant. A simple use, with voice control and voice recognition, is an important interface feature. The assistant should primarily help us in common everyday tasks'.

'Life enhancement' is another way to describe this goal. Others express it like this: 'Take control over the electronic environment by using your mobile as a personal universal remote control'.

Hopefully this book can shed some more light on the vast and crucial end-user area.

2.3 USER ANALYSIS AND SEGMENTATION

2.3.1 THE PRIVATE AREA ENVIRONMENT AND ITS END-USER CHARACTERISTICS AND SEGMENTS

To be understandable, the basic needs should preferably be combined with a basic environment, which is separate for private people and enterprises.

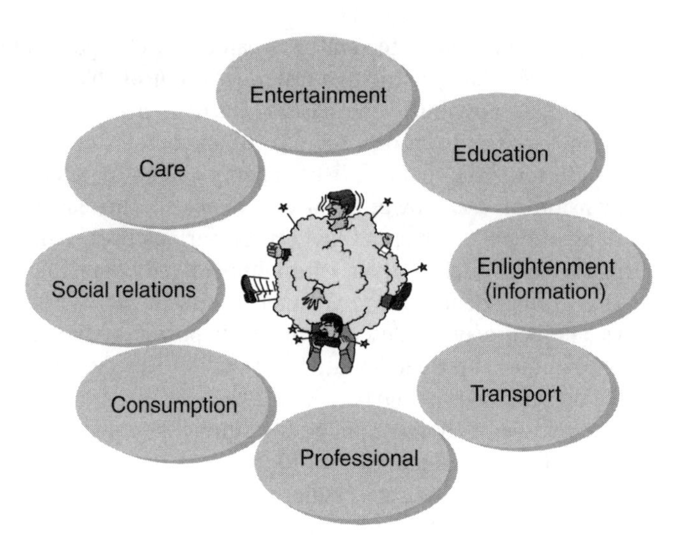

Figure 2.4 Potential areas for time and money savings

The private and professional environment can be illustrated by means of Figure 2.4. The private sector is not homogeneous. It can be segmented in various ways, two of which will be described in more detail as examples of possible segmentations.

The number of possible segmentations is virtually unlimited. Let us start with a very simple approach, which is exclusively based on the 'common life stage' segmentation. This segmentation was applied by Ericsson Australia as part of a comprehensive computer program for modern network design.

The segments are:

1. **Singles.** Singles are people of all ages, although dominantly young people not yet settled in a fixed relation and where the friends are important. This group is also called 'Free Ones 1' (Copenhagen Institute for Future Studies, CIFS).
2. **Couples.** Couples focus more on the internal relation and home although still maintaining relations with friends.
3. **Couples with children.** This group focuses even more on internal relationships than the time without children. Time has now, if not before, become a critical factor.
4. **Single parents.** The children have left home. Freedom and economy improve. They now want to do a lot of things they did not have time for before. This group is also called 'Free Ones 2' (CIFS).
5. **65+.** Retired from regular work, but may take on temporary or project tasks 'for fun'. New generations of '65+' are expected to have good health and to be very active. They will increasingly take advantage of the emerging range of new services.

Users within each segment have a lot in common, but due to personal life styles, values and ambitions the variations within each segment may be considerable. One user may prefer a quiet home life, another within the same segment may prefer a hectic life with adventures, parties and friends. One user may regard mobile phones and PC purely as necessary tools, another user may love to buy, use and show all his friends the hottest communicators with the latest multimedia applications. Therefore there are also other criteria for segmentation, such as values, attitudes, behaviour, demography and cultural differences. The conclusion is that there is a strong need for a refined model that goes beyond the pure life stage. Such a model has been developed by Ericsson Consumer Lab. It is called the 'Take five model'. Both Western and Asian trends are taken care of in this segmentation. The business and society sector can also benefit if looking at an organization as a number of individual users/employees.

The 'Take five model' covers the life stages by means of **age** (youth, middle-age, older), but also by means of **education level**. As an example, young/middle-age educated is an important life-stage group.

2.3.2 THE TAKE FIVE MODEL AND ITS 'MAP'

Consumer Lab has developed a mapping concept called the MarketReality™ Monitor, in order to cater for values beyond life stages, in particular socio-cultural values and attitudes to telecom including more general aspects of human communication. The Monitor describes the telecom market from a consumer perspective.

The monitor space can be used for plotting each individual. It is also possible to plot different segments, behaviours, attitudes etc. The Take Five segments will be presented below. Even more interesting in the context of this book are perhaps the attitudes to telecom in many respects:

- Positioning of services, such as mobile text messaging (SMS/e-mail), ring tones, sync services etc.
- Spending level – average revenue per user (ARPU).
- Prepaid versus postpaid.
- Mobile versus fixed.
- Usage: heavy–low, business–private.
- Triggers for diffusion of services to the sectors of the model.

The Monitor consists of two dimensions: 'Exploration vs. Stability' and 'Lasting Benefits ('we/efficiency oriented') vs. Instant Gratification (me/entertainment oriented)'. See Figure 2.5.

The first dimension, 'Exploration vs. Stability', is the most important differentiator between consumers. It also measures at what pace people need and require changes in their lives.

People near the Exploration end are characterized by:

- Continuous personal development
- Positive to changes and innovation, flexible
- Explorative and mobile, get the most out of life
- Look ahead rather than looking back
- New technology is interesting, fun and has to be conquered
- High tech and new mobile services are interesting
- Brand loyal if brand lives up to their need for constant development

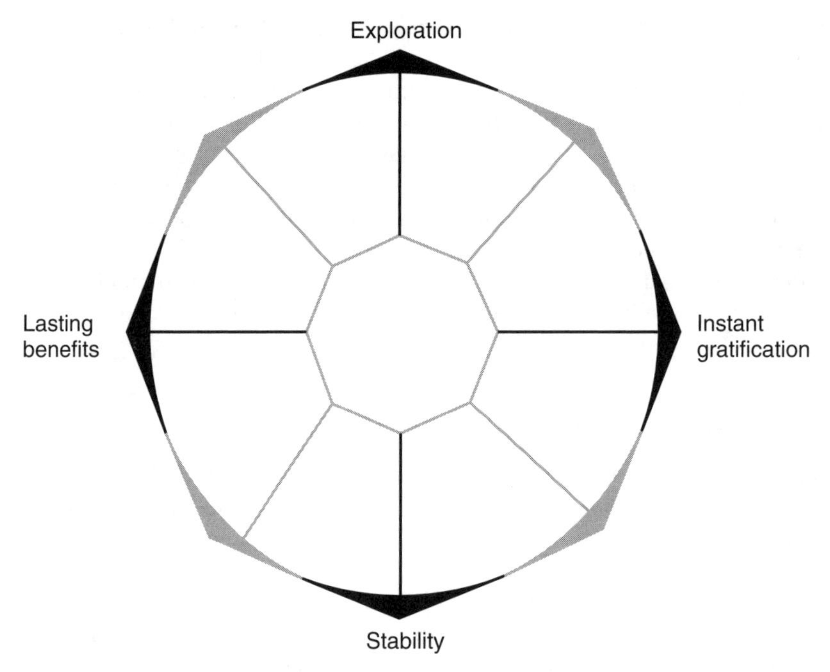

Figure 2.5 The Monitor 'map' model

People near the Stability end are characterized by:

- Present situation more acceptable than what may come tomorrow
- Clear opinion of how things should work, not very flexible to change
- Seek comfort and security
- Follow routines
- More sceptical towards new technology, which they find scary
- User friendliness, transparency and late adoption of new technology
- Loyal to brands perceived as reliable, stable and traditional

Lasting Benefits vs. Instant Gratification is the second dimension, and has to do with the relationship between one self and others.

People near the Lasting Benefits end are characterized by:

- Strong concerns for others/the world
- Strong sense of morals
- Equality for everybody
- Not impulsive buyers
- Have high expectations: of people, brands/products, want quality
- New technology is only accepted if it gives them a truly rational benefit
- Often more work oriented in their adoption of new technology
- Brand loyal if satisfied with the products and if the brand stands for the right thing

People near the Instant Gratification end are characterized by:

- Me-oriented, strong individual desires
- Strong appetite for pleasure
- Live for the moment
- Spontaneous and impulsive consumption
- Mobility, technology and communication means to enjoy life
- Want new technology to give them immediate satisfaction and enjoyment
- Brand loyal if it serves their purpose
- Want personalized products and services

The value/attitude segments described in the Take Five model are also necessarily general. An individual certainly has features from more than one segment with a certain profile, but by taking the most prominent features it is possible to 'label' individuals into just a few classes. Thus five characteristic groups have been identified. They are:

1. Pioneers
2. Achievers
3. Materialists
4. Sociables
5. Traditionalists

Pioneers

The world of the Pioneers is a world of action and motion: to imagine is to do. Therefore they are a very interesting target group for the marketing of new services.

The pioneers lead a multitasked existence – life is too short to do only one thing at a time. They are also perfectly in line with the latest trends in society. They enjoy cutting-edge technologies and use them in the most creative ways imaginable.

'It's new, it's advanced, it's fun and it's the future, I want it.'

Pioneers are challenge-seekers, devoted to pleasure and they often feel at ease in creative and dynamic settings. Due to their strong sense of individualism, they want flexibility in life, in relations as well as at work.

Pioneers are very active and interested in new experiences, both intellectually and physically. Since they often are too restless to tie themselves up with just one thing, they have probably tried a wide variety of leisure activities and hobbies. They like everything that is fun and enhances the quality of life, like surfing the Internet, spending time with friends, practising sports, going out to restaurants, theatres, art exhibitions, dancing etc.

Consumption habits:

- Enjoy looking at advertising
- Like innovative design and products
- Aware of a brand's history, values and innovation capabilities
- Look for unique objects and/or unique usage context
- Willing to purchase a new untested brand if its product is of high quality and utility

They like very innovative, avant-garde products and brands. Therefore, their attitude towards a brand depends on its current activities and image.

The preference for Pioneers' brands is very often taken over by Materialists. A brand must renew its image constantly in order to be appreciated by Pioneers.

How to seduce Pioneers:

- High tech, original and daring design
- Stress innovation and potential for creativity
- Allow for personalization
- Communicate by using brand values rather than functionality

Pioneers were the first Internet users and they remain connected today. The Internet fulfils their need for a constant renewal of information, for contacts with different people, and also allows them to use their talents as explorers. Innovation makes Pioneers feel in contact with the future. For them change is good. Unforeseen events or complex situations do not confuse them.

Achievers

Achievers are rational people, who like to see their world as a well-functioning machine. They are also status seeking and status to an Achiever means having the 'right' job and affording to buy all the products and brands they regard as status markers. They want to live a life in luxury and make sure that other people notice it.

Achievers are not the kind of people to bother with petty details and often go straight to what is important. They trust themselves and are often confident in their judgements. They privilege productivity over creativity. Showing-off, 'getting there', status seeking, having power, having money are typical features.

Achievers don't read much, but they sometimes have a look at car or men's fashion magazines.

They are searching for products and services that stress convenience. They will pay for quality product categories that convey the idea of success and act as status markers.

Materialists

The world of the pleasure and opportunity-seeking Materialist is a world of amusement and leisure combined with moments of hard work to get wherever he or she wants. The looks and acts are very important to the Materialist because they are symbols of their rank and status. The clothing, manifested by choosing the right brands and labels, becomes the uniform that announces their position as members of the club.

Having fun, spending time with friends and taking care of the body are essential hobbies within the segment. But don't let the leisure and pleasure seeking fool you, the Materialist is also very ambitious when it comes to getting ahead in working life and to reach the position they want. Nevertheless, they might not always be the ones who are aiming at the highest posts in society, but might well be satisfied with something less 'glorifying' but that still is important enough for them. And of course their work has to pay well so that they can maintain their high level of consumption, leisure and travelling.

Materialists are anything but serious. Their leisure time is dedicated to pure entertainment: watching TV is their main activity and they like parties. If reading, let it be a cartoon strip, if going to the cinema make it funny, full of action, in a word, entertaining. Materialists prefer to watch films, including movies with adult themes and all kinds of series – detective, family, comedy, action. They also show interest for musicals, variety programmes and shows for young people.

Another way to keep an eye on a screen is to play video games. Of course, if there is a possibility to make easy money while having fun, why not? Materialists like gambling.

The favourite activity is shopping. For Materialists, shopping is a real leisure activity: they think that it is fun and they like to see the new products on the market. They are compulsive, and at the same time smart buyers, who often find what they want at the lowest price.

The image is a travel agency ad that promotes last-minute bookings by offering two advantages: good value for money and the possibility to improvise a trip at the last moment.

People sometimes ask what the differences are between Materialists and Achievers. Chiefly, Materialists are more pleasure seekers, who are attracted to fun and amusement. They like to get the most out of their immediate environment and be associated with trendy products that promote appearance and convey fashion. Materialists tend to have a limited selection of in-depth relationships. In contrast, Achievers strive for efficiency and are more interested in success factors. As a result, they are attracted by symbols and relationships that might increase their chances for success. For example, Achievers tend to have a broader network of acquaintances, but perhaps less close relationships.

A key difference between Materialists and Achievers are the resources they use to make decisions. Materialists tend to be intuitive individuals who trust their feelings. Achievers are more rational individuals who justify their actions using logical reasoning.

Materialists prefer to feel close to the products they use and the activities they choose to undertake, and therefore may take a longer term, more independent outlook on the nature of their involvement in a given activity. In contrast, Achievers are more interested in benefiting from a series of short-term gains, and they are more apt to take risks based on their situation at a given moment.

Sociables

The world of the Sociable is a sophisticated world. Sociables have strong inner values and beliefs and can often set their own desires aside in order to do something that is good for society. Many Sociables feel a strong responsibility towards the environment. Sociables are interested in learning new things and to discover new cultures. They like travelling, not only abroad but also within their own country, in order to find the small and refined treasures within their own culture. They have many cultural interests and appreciate cultivated things like going to the theatre, opera or to art exhibitions. You can also find Sociables active in local associations like a local theatre group for example.

Sociables do not strive for 'having fun' as a Pioneer or Materialist would do. The Sociables often associate 'having fun' with shallow leisure. The Sociable can find leisure in reading a good book, going to the public spa or spending some quality time with the family, which for the Sociable is something more precious than just 'having fun'.

'I think it's important to disconnect every now and then, I just love pottering around in the garden or reading a good book, for example'.

Sociables are resistant to impulsive or status-oriented consumption. They tend to prefer product brands with ethical values.

Sociables are highly rational consumers who expect ease of use and attractive design, even in new products. They approach novelty by looking at the potential needs and benefits they feel will be gained from such products. As a result, their purchases reflect a drive towards value for money, but once they have made their decision, they can be very loyal to the brand they select. Technology is regarded a means for comfort. Sociables demand a high level of quality in a product.

Traditionalists

Traditionalists are what we call 'late followers'. When it comes to technology it means that they buy new products or concepts only when they have become a natural part of people's everyday life. When they decide to buy a product they are often aiming for quality, because once they have bought something they often want to keep it for quite a long while. The exchange rate for technical products among Traditionalists is quite slow. Traditionalists often feel most comfortable in their close surroundings and their country often marks the limits of the world they use as a reference. Traditionalists are very attached to 'their' neighbourhood, where they feel at ease with the familiarity that it offers them.

For the Traditionalist their own culture and national habits are often seen as the best. By acting in their immediate surroundings, they do not have to interact much with the unfamiliar or unknown. Their habits are respected and they have more chances to avoid surprises. An established routine helps maintain the illusion of safety. Familiarity is a main asset for this segment.

Keywords for Traditionalists are

- Known surroundings
- Quiet life
- Security
- Roots and tradition
- Conformity

Increasing the Granularity by Means of Life-Stage Parameters

The matrix shown in Figure 2.6 outlines the effect of combining the life-stage dimension and the value/attitude dimension in the Take Five Model. The output is the Take Five life stage providing us with nine distinct segments:

P1	Pioneers – Youth	P2	Pioneers – Middle-age
M1	Materialists – Youth	M2	Materialists – Middle-age
S1	Sociables – Young/middle-age educated	S2	Sociables – Basic educated/older
A1	Achievers – Young/middle-age educated	A2	Achievers – Basic educated/older
T	Traditionalists		

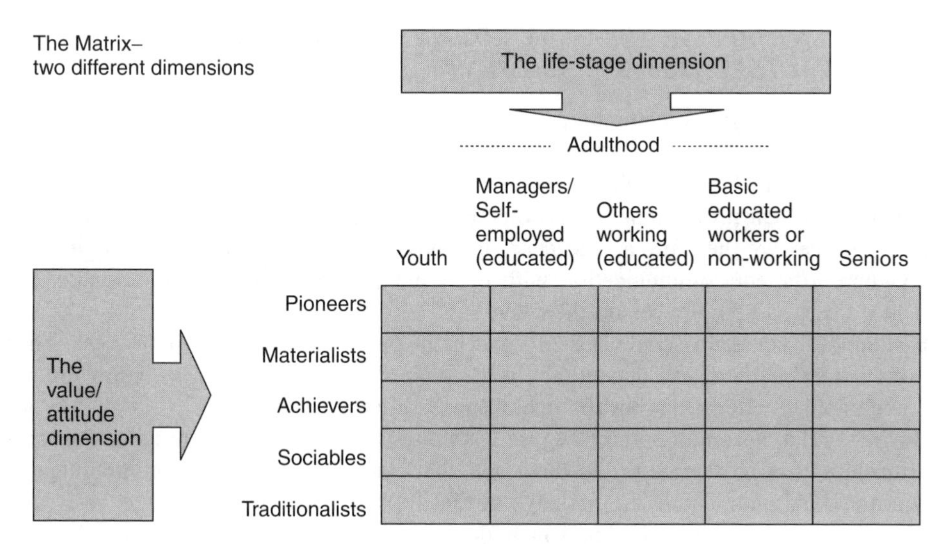

Figure 2.6 The life stage–value/attitude matrix

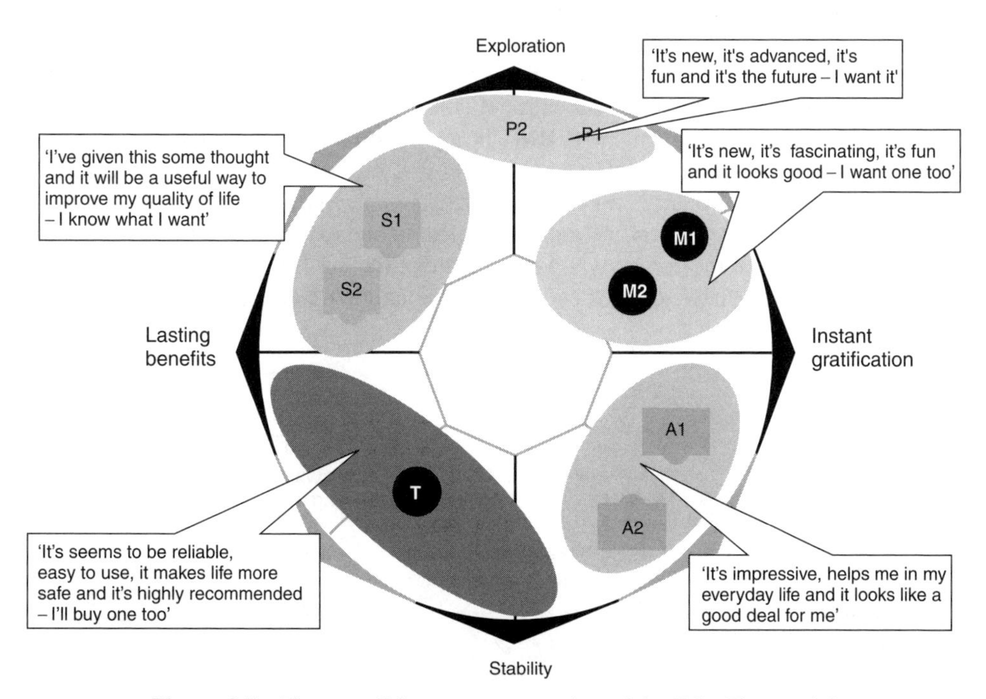

Figure 2.7 The overall human segmentation of the Take Five model

Positioning the Nine Segments into the Map

See Figure 2.7. Together with the description of the monitor axis, the value/attitude dimension and the life-stage dimension, and the location of segments into the figure, we have a useful human map.

2.3.3 THE ENTERPRISE AREA ENVIRONMENT AND ITS END-USER CHARACTERISTICS AND SEGMENTS

- What are the enterprise communication needs?

Business users, including both companies and individuals, need to be effective and competitive in their professional operations. Effective communication and essential information services are therefore a vital need. Business users must maintain an anytime and everywhere efficient communication with customers, partners, information sources and branch offices, and employees on the move.

Fast anytime access to people and information is a must for a successful business. There are of course specific needs depending on the type of business operations, company size and geographical allocation, but the overall needs are the same.

Business relations are important. Figure 2.8 shows some important external and internal relationships that most enterprises have and that demand for effective communication services and communicators. Let us have a brief look at these relations.

'The customer is king'. 'The person with the money' or the customers are the most important asset to the enterprise as they are the very origin of revenues for the enterprise. For interaction with customers, enterprises need call-centres, *e-business* and *m-business* (electronic business and mobile e-business) enabled company web sites or portals for business-to-business (B2B) communication, and effective means of bit transport with sufficient bandwidth for external and extranet communications.

Customer relations management (CRM) is an important part of IT-based processes within many enterprises. The customer must feel that the enterprise is always at hand to support with services, products and consultancy. Customer support by manned helpdesks, which may be outsourced to call-centre companies, is one example of an important need. The helpdesk portal at the company's web site where the customer can put questions, get information and order new services and products is another example.

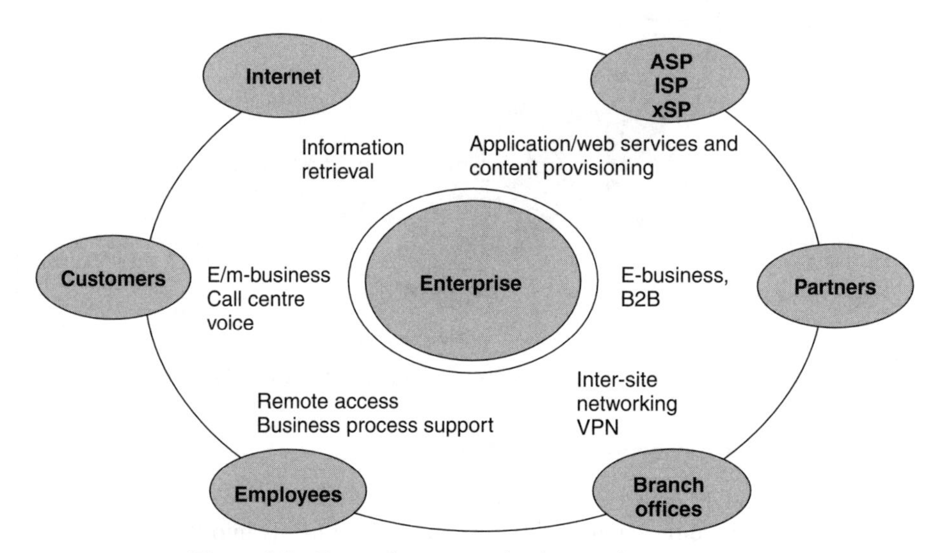

Figure 2.8 Enterprise communication needs – customers

- Employees

Employees are frankly a heavy investment for enterprises. It is therefore of utmost importance that they can work effectively and don't feel frustrated by lack of access to people or information. The employees of today are normally mobile in their operations. They may move around in the same city where the enterprise is located or travel nationally or internationally. In all cases they must be in anytime, anywhere contact with their company. *Remote access* is crucial. They need to talk with secretaries, experts and so on within the company. The mobile phone or communicator is their tool. They need to get access to information in company databases inside the firewalls that sort out authorized persons from hackers and other non-authorized persons to get access to sensible information.

We should also note that enterprises increasingly strive for a 'slim' organization, hiring experts for time-limited projects. The core competence is, however, kept within the company by permanently employed people. Typically if the permanent employees are counted in hundreds, experts in the enterprise's contact network may be counted in thousands. The experts may work for more than one company in parallel. This poses the question 'Which persons should get access to what information and when?' Security mechanisms and administration of these are vital in order to avoid information leaks.

- Branch offices

Branch offices operate as a local business partner to customers and are also a sensor for what is happening in that market. Effective *inter-site networking* for communication and access to information is critical. The ideal situation is when a branch office does not see the distance to the head office and other branch offices as a problem.

- Who owns the office buildings?

We should note that enterprises increasingly hire office spaces in large office buildings, maybe sharing a common communication network with other enterprises. Virtual private network (VPN) functions, security mechanisms and administration of these are also vital in order to avoid information leaks.

- Partners

Effective communication with partners is essential. Thus enterprises need call-centres, e-business enabled company web sites or portals for B2B communication, and effective means of bit transport with sufficient bandwidth for external and extranet communications. The partners and their employees normally have access to large parts of an enterprise's intranet, its facilities and information. Some information is of course confidential and may only be accessed by the enterprise employees.

- Application services providers (ASP) and Internet service providers (ISP)

The general trend today is that enterprises want to focus strictly on the core business, that is, what is expressed as a business idea in the business plan. The reason is that they want to increase effectiveness and competitiveness. Non-core activities and assets, for example housing and management of housing, communication networks, cleaning and commodity manufacturing with no strategic content is outsourced to other companies. In doing this the enterprises gain two things. One is that they don't need to waste management attention and other company resources on non-core business activities. The second gain is that other companies that are specialized in the respective areas probably will do it much cheaper and better due to large-scale operations and specialized experts.

This is a reason for outsourcing, in this case outsourcing communication and Internet services. *Services and content for specific applications* are sourced from specialized providers. Many enterprises want to keep in-house control over sensitive or mission-critical communication resources. Still, the largest part of communication resources and services can be outsourced, especially as security functions such as firewalls, encryption and anti-virus programs are continuously evolving. See further discussions in Section 2.3.4 below.

- The Internet

The Internet represents a globally anywhere and anytime information and communication resource. A main use is for *information retrieval* in databases and Internet portals all over the world. Initially only text was transported, later still pictures, and now multimedia combining text, sound and still or moving pictures are used. Another main use is for e-mail. In the evolving IP-telephony or voice over IP (VoIP) service, voice or rather sound information in telephone calls is transported as IP packets within Internet.

- Major forces are changing the scene for enterprises

There are some major market trends that are now having a great impact on business operations and consequently the type of applications and communication needed. These trends are themselves spurred by the worldwide Internet and mobile communication, also called 'mobile Internet'.

In principle, any person or any database-stored information can be reached independent of time and geographical position.

Globalization

Most enterprises are today working in a global environment, with suppliers and/or customers in other countries; in some cases the company itself becomes a global entity, directly or with partners in an international cluster.

Restructuring, Outsourcing and 'Clustering'

Increased competition forces enterprises to focus on their 'core business'. On top of this we see the drive to keep capital costs low with 'just-in-time' logistics strategies. This creates many more intercompany connection points, with an increased need for quick, efficient, reliable and information-heavy communication, between people and between systems.

Towards Knowledge-Based Organizations

As the educational level of employees increases and the decision making become more distributed and delegated, internal communications within enterprises increases in strategic importance. Companies are putting more efforts in keeping their people up to date with on-line conferences, web-casting, internal TV-channels, newsletters, multi-location workshops and other means.

To address the three forces mentioned above (globalization, restructuring and new types of organizations), enterprises need effective communication solutions that support individual productivity and their interaction with customers and business partners. Increased and improved person-to-person communication capabilities are a key success factor for this development. For further increased individual productivity, employees require local and wide area mobility with secure anytime, anywhere access to information and applications. Thus employees need:

- Communication and computing devices, for example a mobile phone, a laptop and maybe a personal digital assistant (PDA) equipped to communicate and synchronize activities, and possibly also a fixed phone at the office desk.
- Service and application packages personalized for the duties and job of the individual.
- Remote access to enterprise network and resources when on the move or working from home.

2.3.4 OUTSOURCING AND NEW SERVICE PROVIDERS

As we mentioned above, enterprises want to focus strictly on the core business and outsource non-core activities and services. In principle, everything except for core business can be outsourced.

A new market for a great variety of service providers is here and evolving. *It can be regarded as evolving new enterprise segments for the network operators, service providers and suppliers.*

In non-communication areas we have for a long time seen outsourcing of cleaning, housing, catering, and production of commodity components, for example printed boards.

In communication we have seen telecommunications being outsourced, first for long distance then for internal communication. A main example is Centrex (centralized exchange), where all internal communication is handled by the network operator's public switches, instead of an enterprise-owned PABX (private automatic branch exchange) for intra-site communication. In the same way VPNs (virtual private networks) may replace private networks or leased lines when different sites are to be interconnected.

Some examples of new roles for service providers are presented in Table 2.1. In some cases a service provider takes on only one role, but in many cases two or more roles are taken on. One example of the latter is the former telecom administrations; that is, the national, government-owned telecom operators that have a long and broad experience from running networks and services although in a monopoly situation. The example list is very dynamic; there is no static truth! Some roles may disappear, many new roles will arise and we will also see a lot of mixtures of roles as long as it promotes business.

Table 2.1 New roles for service providers

Role	Description
ASP, HSP	Application service provider and hosting service provider. They provide and host services, often physical, but also applications. See also the Role section in Chapter 4.
WASP	Wireless application service provider is similar to ASP, but for wireless communication (mobile communication).
SSP	Storage service provider providing storage services for programs and data including back-up services with extremely high safety. Data warehousing may also be offered, i.e. storing and management of data that can be used in the enterprise's business operations or for internal process management.
MSSP	Managed security services provider. Provision of firewalls, intrusion detection and monitoring of access attempts to the enterprise communications network and databases in order to detect and prevent hackers and crackers from access.
IBSP	Internet business service provider. Offers access and hosting of Internet-related services to small and medium enterprises (SMEs).
MSP	Management service provider. Management of information systems (IS) including for example monitoring and operations.
ESP (1)	Enterprise service provider. Offers Internet services to enterprises.
ESP (2)	E-business service provider. Provision of services for e-businesses including e-commerce both B2B and B2C.
EMSP	E-mail and messaging service provider. Provision of services and support for e-mail, messaging and scheduling.
BSP	Business service provider. Offers services such as reporting of activities, results of enterprises, accounting and human resource management.
DESP	Development environment service provider. Offers software development environment and tools.
VSP	Vertical service provider. Offers a complete service including in principle everything to run a certain service. Examples are services, networks, terminal operations and maintenance of these. One example is a global video conference system at an enterprise site. Another example is 'Call Centre' where calls from customers are directed to a specific call-centre answering and acting on behalf of the enterprise. In the ideal case the calling customer should feel that it is the addressed enterprise that is answering the call.

2.3.5 ENTERPRISE SEGMENTS

In this book, three 'Enterprise segments' are defined below, primarily based on size, which is one of the key parameters that determine the type of service needs. The needs for external and internal communications and also administration and business processes are to a large extent size dependent. The large enterprise, counted in number of employees, normally has a larger need for applications in order to maintain a complex network of customers, employees, local offices worldwide, subcontractors and so on. However, within each segment there are considerable variations in the type of services used, communication intensity, complexity, and so on depending on the type of business operation.

A hightech company can be anticipated to have other needs than a commodity manufacturing company. It is also important to remember that a small company working for example with front-line technology and products might have a much bigger need of advanced multimedia communication and services than a big company depending on the type of business operation. As said before, all segmentations and characterization of the segments must by necessity be quite generalized.

The three enterprise segments defined in this book are:

1. Small office/Home office (SOHO)
2. Small and medium enterprise (SME)
3. Corporate

2.4 BASIC NEEDS MODEL

Needs by individual users, organizations and enterprises is a main point when discussing what type of telecom services are important to design and implement. There are basic needs that are common to parts of the society; some are even common to the whole society. Consumer Lab at Ericsson suggests a model for basic needs. The model was originally intended for consumers ('end users') but has in this book been slightly modified to suit any type of user, including organizations as business enterprises, societal institutions and all other types of organizations we can think of.

Let us now have a look at each basic need slice of the triangle in Figure 2.9.

Safety is a need for all individuals and organizations. For an individual it means personal security, to possess the means to avoid being hurt or becoming trapped in any dangerous

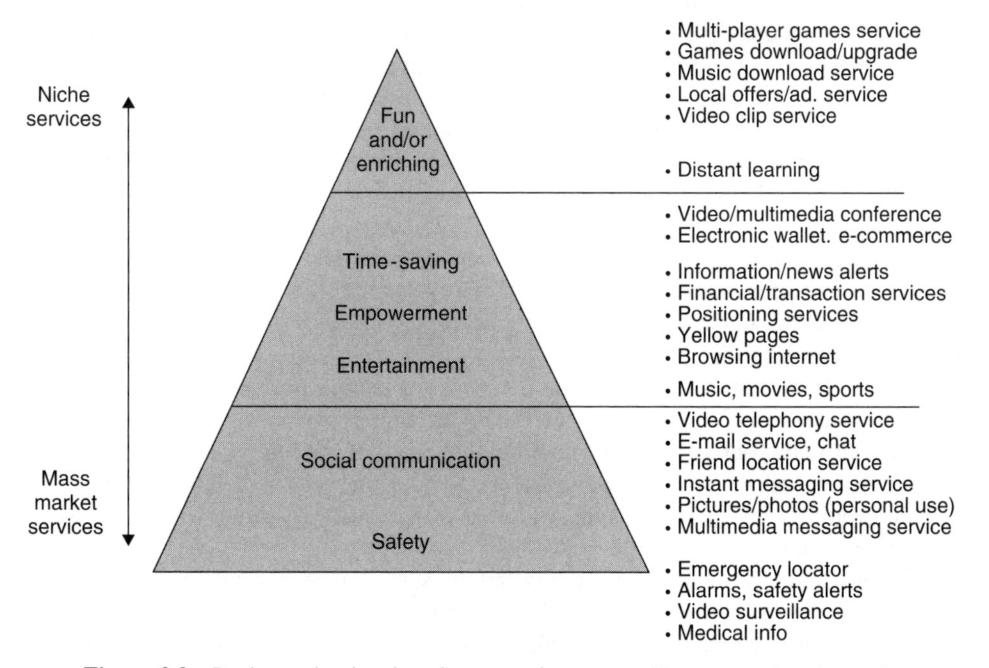

Figure 2.9 Basic needs triangle – Some services may address several main needs

situation and to have a rescue tool if some of these dangers should occur. For organizations safety means to be safe against sabotage, theft of secret information, fire or anything else that can violate the organization's existence. Personal safety includes safety for family and relatives and also personal properties.

Social communication is vital for almost all individuals and organizations. For individuals it means to be accepted and to be 'inside' social groups in society, for example at school or work and in the community. Similar reasons can be used for organizations, which want to be respected and appreciated for their activities, be it business enterprises in a competitive market or governmental institutions or (almost) any other type of organization.

Messaging meets several human needs, one of them a need to be 'existing and visible' in social relations. With ever more powerful messaging, such as combined cameras and mobile terminals that offer possibilities to send video clips, messaging will be a significant traffic and income generator. Personal electronic Christmas cards are a predecessor.

Empowerment is important to most individuals and organizations. It means in practice to provide tools for adequate actions, often under time pressure. A main tool is appropriate information to enable individuals and organizations to make the 'best' decisions and take the appropriate action. For an individual it may mean getting the appropriate information to feel sure on the 'right' choice at political elections or having the necessary information when buying a house or a car. For organizations it means to perform effective operations, for business enterprises it could be a matter of survival and flourish on a competitive market or to disappear.

Time saving is important for many people and organizations, that is, 'time is money'. This is especially true for routine activities. For individuals it means maximizing the disposable time for desired (pleasant) activities. For organizations it means effective operations that save money (salary costs). For enterprises in competition it is often crucial for survival and prosperity to be first, for example with a new product. Another example is journalists and newspapers that always strive to be first with hot news.

Special interests are specific for smaller groups of people or communities. For companies there may be specific branch or company interests due to unique conditions and contexts. Numerous such different communities exist.

A telecom service that fulfils several or all of the basic needs above has the potential to become a success. Some examples of existing services are voice communication, e-mail, general information services and directory services. Emerging 'success services' are positioning-based services and imaging-based services that are foreseen to satisfy all the basic needs described above.

Example 1: a positioning service or a service containing positioning capabilities can be used for constantly informing your family or friends where you are (= safety), to keep contact with your friends (= social communication), to help you to find out where you are in a foreign city (= empowerment), to find the nearest hotel that fulfils your specific requirements (= time saving) and to play a hunting game with your friends (= fun/special interest).

Example 2: photos/pictures can address safety (surveillance cameras), social communication ('a picture says more than words'), entertainment ('Check this out'), empowerment (especially in professional usage like taking pictures of the construction site and sending back to the office etc.), time saving (especially for professional usage, but also personal usage, no need to go to the photo shop and develop film, you just click and the photo is there!), fun and enriching (as a way of communication: 'Look at us at the football game 'we are having a blast!').

2.5 MAPPING OF NEEDS AND SERVICES

Experience from market studies gives a list of around 30 end-user services that can be defined as default services, i.e. they are always relevant for household and/or business segments. These services are a mixture of application-oriented and network-oriented services. To each service a note is made regarding what basic need(s) the service fulfils. The more basic needs a service satisfies, the higher probability that it is or will be a 'killer'.

Personal safety including safety for family and relatives and also personal properties (e.g. home) is important for everyone. Social communication and information in all types of situations for everyday life and leisure time is important to most people. For many groups time saving is important, perhaps especially for working parents with small children, in order to accomplish everything that has to be done within the day.

As mentioned, the basic needs are represented as:

1. Safety
2. Social communication
3. Empowerment
4. Time saving
5. Special interests (including fun)

The mapped services are:

- Voice communication (mobile, fixed and IP) (1–5)
- Unified messages/Text messaging (with attachments) (2–5)
- Video telephony/conferencing (3–5)
- E-learning (later evolving into mobile e-learning, i.e. M-learning) (3–4)
- Directory services (3–5)
- E-trading and portfolio management (later evolving into M-trading) (3–4)
- Online banking (3–4)
- Pay TV (3, 5)
- Interactive games (2, 5)
- Audio on demand (3–5)
- Video on demand (3–5)
- Downloading audio/video (3–5)
- Interactive TV (3–5)
- Remote (device) control (1, 3, 4)
- Home security online (1, 3, 4)
- Virtual shopping (online shopping) (3, 4)
- Tracking services (1–5)
- Extranet (1–5)
- Intranet (1–5)
- VPN (virtual private network) (1–5)

- Information gathering services (1, 3–5)
- Virtual classroom (3, 4)
- Dispatch services (3–5)
- Field sales automation (3–5)
- Management services (1, 3–5)
- Remote diagnostics (1, 3–5)
- B2B e-commerce (3, 4)
- B2C e-commerce/m-commerce (3, 4)
- Security and surveillance services (1, 3, 4)

2.5.1 WHICH USERS LIKE WHAT? COMBINING USER SEGMENTS AND SERVICES

How Do We Know What the Users Like?

User needs and demands depend on the user age, occupation, society culture and evolution stage, context and a great number of other parameters. There are for instance considerable differences if a user is on holiday in a foreign country or working in an office, or at a supermarket needing price information on a product in other (competing) shops, just to mention a few practical examples. Age is a factor that we know affects behaviour and preferences.

Figure 2.10 shows the result of a study performed in Sweden on users' willingness to pay for mobile Internet services. This points to young people as the first adopters for mobile Internet, and an average spend of $15 per month across all groups. The willingness to spend changes over life time, and you can see that the teenage market is the highest potential spender, reinforcing evidence from i-mode and SMS in Europe that it is the youth market that currently driving the mobile Internet.

The increase in the 55 years age group is perhaps due to parents whose children have now left home leaving them with more time and money to spend on themselves!

Figure 2.10 Willingness to spend on mobile Internet by age group (example from Sweden)

Private Life Segmentation, First approach

The first approach makes use of the segments presented in section 2.3.1

- Singles
- Couples
- Couples with children
- Single parents
- 65+

2.5.2 USERS AND THE MOST PRIORITIZED SERVICES

The services listed in Figures 2.11 and 2.12 have been evaluated using two important parameters. One is 'attractiveness', e.g. perceived value and importance by the users and likelihood of take-off regarding future services. The other important parameter is 'timing', e.g. when the mass market is ready for a certain service with the necessary technology to get comfortable and affordable services are available.

The most prioritized services are listed under each segment in an approximate order starting from 'Very attractive' and 'Want it now' to 'Medium attractive' and 'Can wait' meaning a time of about three years. We should note that the existing services 'voice calls' (= telephony and evolving variants) and 'Internet browsing' are considered highly attractive by all eight segments. These services are therefore not listed here.

- **Singles.** The attractiveness level for these services is high, between 4 and 6 on a six-grade scale.
- **Couples.** The attractiveness level for these services is rather high, between 4 and 5 on a six-grade scale.

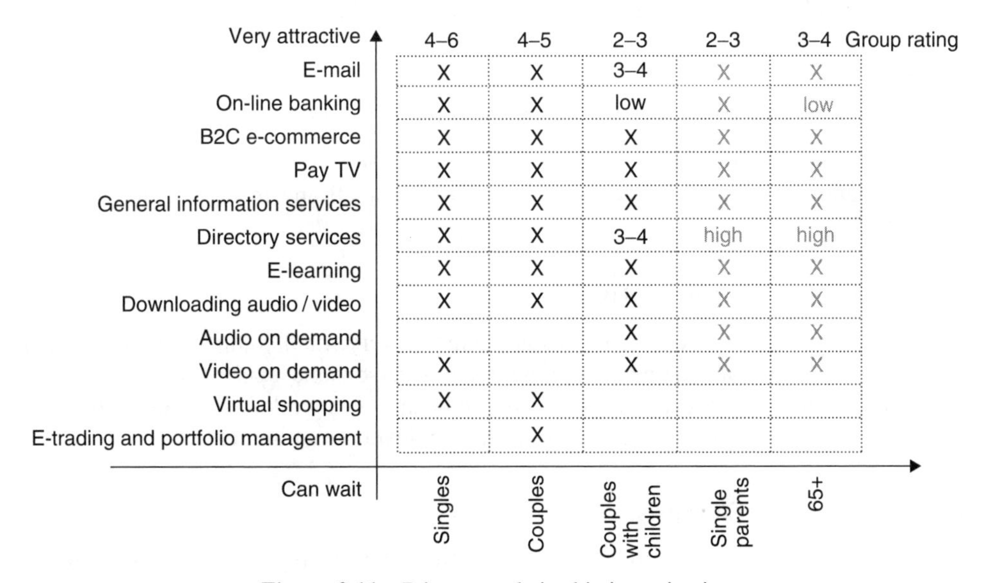

Figure 2.11 Private needs in this investigation

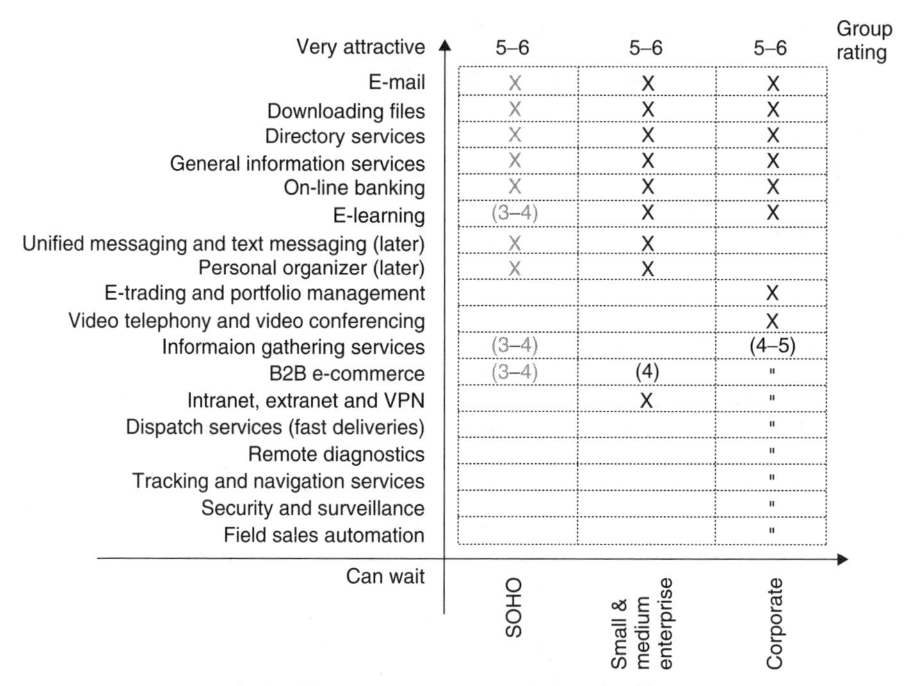

Figure 2.12 Business needs in this investigation

- **Couples with children.** The attractiveness level for these services is low, around 2 and 3 on a six-grade scale apart from directory services and e-mail that are rated 3–4.
- **Single parents:** The attractiveness level for these services is low, between 2 and 3 on a six-grade scale.
- **65+:** The attractiveness level for these services is rather low, between 3 and 4 on a six-grade scale.

2.5.3 THE ENTERPRISE SERVICE AND COMMUNICATION NEEDS

Let us now look at the enterprise segments and their desired services. The three enterprise segments are small office/home office (SOHO), small and medium enterprise, and corporate.

Small Office/Home Office (SOHO)

SOHO means roughly one or a few persons running a very small company with a limited, maybe only one, specialized operation. One or a few persons must perform all operational, business and administrative activities. It can also be an office workplace at home, financed by an employer and equipped according to the employer's requirements.

Small and Medium Enterprise

This is a medium-sized company, typically located at one or a few geographical localities. The company has typically a limited product or service portfolio. The business operations may vary from very local business, e.g. a chimney-sweeping firm, to worldwide operations such as specialized fashion clothing. The telecom needs may also vary from a basic portfolio of mobile telephony, e-mail and call-centre services to a broad variety of services, e.g. Intranet/VPN, video-conference facilities and other services that are also normally used by large companies.

Corporate

This is a large company with diverse business operations and a comprehensive product or service portfolio. The company has typically many employees, customers, local offices worldwide and a comprehensive administration.

- **Small office/Home office (SOHO).** The attractiveness level for these services is very high, between 5 and 6 on a six-grade scale. B2B e-commerce, information-gathering services and e-learning are, however, rated 3–4 on the scale.
- **Small and medium enterprise.** The attractiveness level for these services is very high, between 5 and 6 on a six-grade scale. B2B e-commerce is, however, rated 4 on the scale.
- **Corporate.** This segment has a great interest in a wide variety of services. This depends partly on the great variety of business operations. Therefore 16 services are listed instead of 10. The attractiveness level for the first eight services on the list is very high, between 5 and 6 on a six-grade scale and for the last eight services is rather high, rated 4–5.

The Mobile Enterprise (Source: Wireless World Research Forum)

Employment of wireless communication technologies such as UMTS, wireless LAN and Bluetooth can support business processes, such as B2B, by means of increased flexibility and dynamics and an improved information flow. This applies both internally and in relation to customers and business partners.

Among the specific enterprise processes that may benefit from wireless communication are supply chain management (SCM), enterprise resource planning (ERP), CRM and production processes.

In the area of SCM, it is important to have a holistic view of the whole transport chain in order to quickly overcome changes due to unknown situations, for example accidents or traffic jams. With the knowledge of the whole chain, available transport and handling resources including infrastructure networks could be better used and managed.

ERP and CRM processes might benefit from mobile access to databases, as mobile decision support and mobile workplaces will become important success factors for companies.

A final area of interest for the enterprise is production systems. Existing production structures can be improved by the application of combined information and communications technologies (ICT), ad hoc networks, etc. to obtain increased usability and improved information access for the user.

Enterprises Important to Service Providers

As an average, services to enterprises represent today about 50 % of telecom operator's revenues. In particular they represent more than 90 % of the ordinary operator's datacom revenues. They have always been important drivers of new applications and services for the mass market, i.e. the private end users. Some major reasons are that enterprises have seen the potential of large cost reductions and an increase in operations effectiveness, which motivate costs for new applications and services.

2.5.4 COMBINING USER SEGMENTS AND SERVICES AND FINALLY TIME. TAKE FIVE MODEL

The Take Five segmentation offers interesting target groups for service offerings and directed marketing such as 'single pioneers'. The operator must decide accordingly which segments are his main priority and how he intends to package and launch the applications to each segment, and when.

The final relation refers to *time factors* when developing a market. See Figure 2.13. It is important to address the technology enthusiasts first. They can also more easily overlook initial technical problems with a service of the type that hit the WAP service. Marketing for WAP was, however, broad, something that eventually brought about a bad reputation.

Identifying the *early adopters* is therefore critical.

Each market is different, but Ericsson Consumer Lab has identified in most markets two main key groups of people. Using the Take Five terminology these are:

- Young Pioneers and Young Materialists
- Educated Pioneers and Educated Sociables

The early adopters tend to be interested in all media such as video, and all service types such as entertainment and professional services.

The early majority tends to be most interested in work-related services and transactional services, but the media possibilities are also interesting.

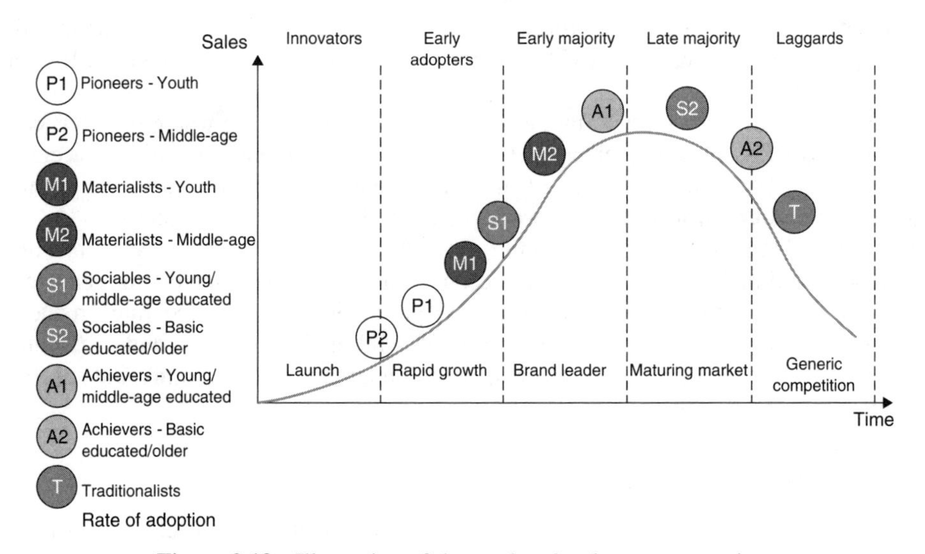

Figure 2.13 Illustration of the market development over time

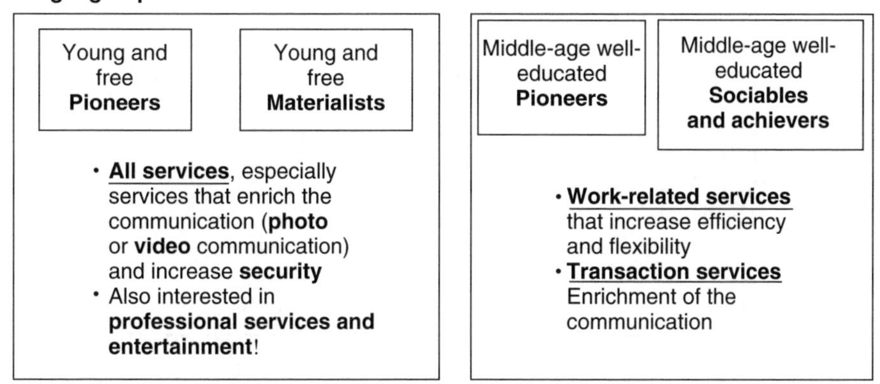

Figure 2.14 Combining age and the Take Five model to analyse services

Combination of Take Five Segments and Telecom

The following relationships between segments and telecom illustrate the usability of the model for telecom purposes. Figure 2.5 is used as a watch dial to indicate the most typical users. It is reproduced as Figure 2.15.

Mobile services – For mobile services see Figure 2.14.

Positioning of prepaid versus postpaid users – Young people and low usage people are generally pre-paid users, while professionals are usually a combination of pre- and post-paid users. To maximize initial take-up, it is therefore imperative that new services (such as MMS) are available to both pre- and post-paid users from launch.

Positioning of mobile text messaging (SMS/e-mail), ring tones and sync services – Messaging (M1, P1) half past one, ring tones half past two, sync services nine o'clock on the dial.

Spending level (ARPU) – Overall 12 o'clock, mobile phone half past one, mobile for work the area of sociables (about 10 o' clock).

Mobile versus fixed services – Mobile to the right, fixed to the left.

Usage: heavy–low – About 12 o'clock.

Business: private use – Business users to the left; private users to the right.

Triggers for diffusion of services to the sectors of the model – Example of a trigger for traditionalists: social communication and safety, basics.

See more about users in general and their charging and marketing in Chapter 5.

2.6 THE HUMAN END USER AS A TRAFFIC GENERATOR AND RECEIVER

2.6.1 THE RECEIVING SIDE

Several studies have shown that our senses in conjunction with the brain have fundamentally different capacities regarding information reception translated into 'information

Table 2.2 The information reception 'bandwidth' for different human senses

Senses	Received by brain	Consciously experienced
Eyes	10 Mb/s	40 bit/s
Ears	100 kb/s	30 bit/s
Skin	1 Mb/s	5 bit/s
Mouth	1 kb/s	1 bit/s
Nose	100 kb/s	1 bit/s
Total	11 Mb/s	75 bit/s

bandwidth'. Moreover, what the brain consciously experiences is far lower than the amount of information received by the senses (see Table 2.2). One example is that the brain can receive sight information via the eyes corresponding to 10 Mb/s but only has the capacity to experience 40 bit/s! Combination of information received by two or more senses ('multimedia') improves the total experience capacity. Finally, the eye is the most effective way to receive information. Some calculate that a multimedia impact is up to three times more than the impact of words alone. Compare the saying 'One picture tells more than 1000 words'.

The limitation in human reception means of course also a limitation in traffic that the end user is willing to pay for.

2.6.2 THE GENERATING SIDE

All traffic is in some way generated by the human end user. Let us take a brief look at some of the services where the end user plays an active role in traffic generation. The generation of voice traffic is well known. For a population the volume can be calculated in terms of Erlang per area unit.

Video

For the business corporate sector video conference and video telephony are listed. Videoconferencing can obviously often replace travelling. Except for fixed devices mobile 3G terminals will be able to participate. Video clips are supposed to become a very successful application, indeed.

Messaging

Messaging has developed to a really important service area with the human being as the main traffic generator. Since it traditionally has been a slow service it fits quite well to the human reception ability.

Photo generation is already possible by means of integrating a camera into the mobile terminal or connecting a camera to the terminal. A considerable generation of traffic, for example during vacations, is expected. Christmas cards are another example. Instead of integrating camera and mobile terminal, the mobile terminal can serve as a transmitter connected to ordinary digital cameras. In this case bandwidth and/or compression are vital. Pictures and clips can be sent as e-mail attachments as well.

The global collection of private photo albums represents one of the largest information stores at present. When digitized, the growing treasure of photos can of course be easily interchanged electronically, constituting a significant potential traffic generator. The need to send photos might be compared with 'remote viewing of photo albums'.

According to an investigation done by Ericsson Consumer Lab two out of three pioneers want a camera in the mobile terminal/device.

2.7 THE FUTURE MOST COMMON END USER: A MACHINE

In 1980, the old mainframe world had 100 users per mainframe. Around 2000, there is roughly one PC and mobile phone for every user. There are predictions that by 2020 there might be 100 computing devices per person. Most people will have a wireless device as a last link. With a world population of 10 billion and an assumed penetration of 40 per cent, this will imply that future generation mobile networks will have to accommodate, in total, 400 billion wireless devices of all shapes and sizes. Consequently, somewhere after 2010, there will be no shortage of computers and radio links to create 'ad-hoc' networks.

The opposite of human-generated traffic would be M2M (data) traffic. This is considered to be an end user that changes lives. See Figure 2.16. However, the business impact for operators will not be significant during the next few years.

Even M2M is normally in one way or another human related and might embrace product to product, product to human, and human to product. Robust industrial applications come first, such as remote monitoring of elevators and remote supervision of train boogie

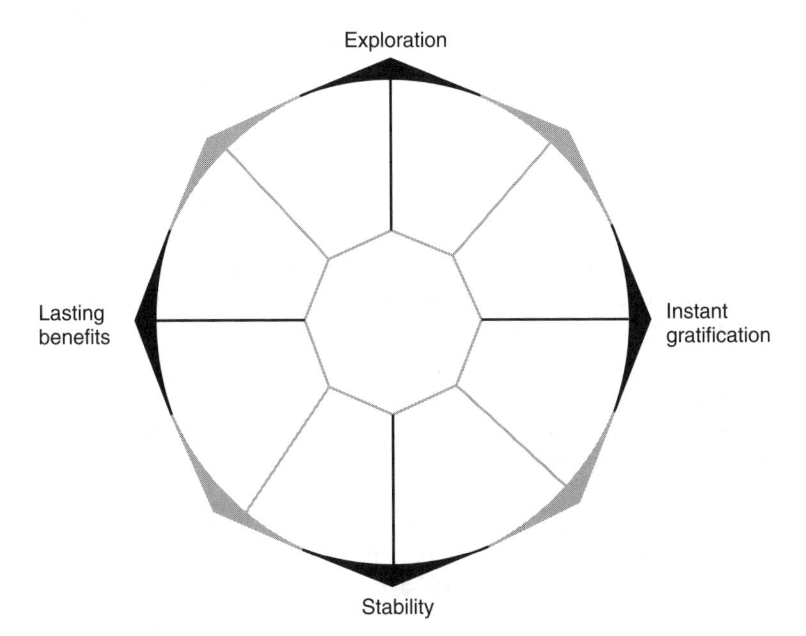

Figure 2.15 The 'dial' used in this section

- What will the world look like when each person has >100 computers?
  - Online, position-dependent Internet connectivity in your car
  - A coffee pot that knows when you are home
  - Mail on your watch

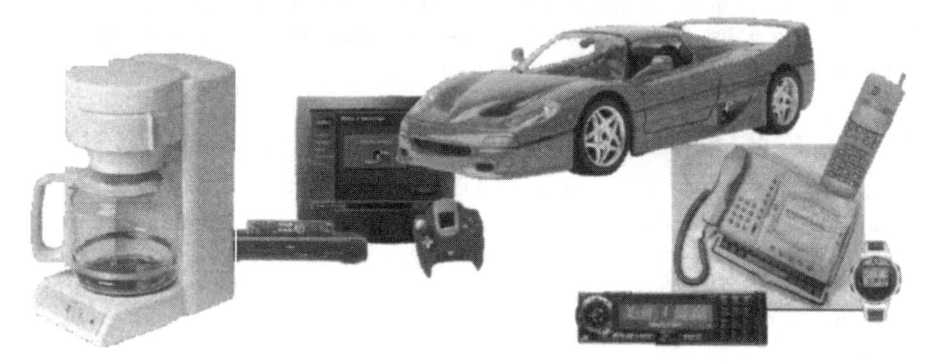

Figure 2.16 The new types of users

systems. Later on home services are foreseen, such as remote electricity meter readings, alarm systems and remotely controlled heating.

Cheap, small, discrete processors are key enablers. They already exist in cars, stereos, videos, washing machines and clock radios, so far not connected. In the car there is a lot of sensors connected to processors. The total possible market of remote performance management and fault management is waiting. Let us therefore look a little closer into M2M. In many cases systems communicate without manual involvement unless errors occur that require manual intervention. In other cases M2M communication is needed before a user can use a service. Some examples are shown below:

- File transfer, e.g. machine–machine information when taking back-up copies of a system content or when system upgrades are done. Another example is when a user orders a download of music files, like Napster or download of 'video on demand' content.
- 'Intelligent' M2M communication and interaction for 'autonomous' systems. One example is telecom networks with switches, routers and communication links handling fluctuating traffic streams, overload situations and correction of malfunctions.
- Remote control data, e.g. when an operating system controls traffic systems in the example above. The 'manual' variant has already become hype, i.e. 'control your toaster, house heating/cooling and light from wherever you are'.
- Remote notifications, for example transmitting alarms, responses and other status information for autonomous or manual actions.

The media types are transported as 'data' except for 'classical telephony', which is treated explicitly as voice communication. This is slowly replaced by packet-oriented transportation, i.e. 'IP telephony' or 'voice over IP'.

The typical development goes from large communicating machines towards smaller ones. The first wave was a network of IP-addressed computers that swelled to encompass hundreds of millions of systems. The Internet connected machines and people tapped in when they could.

The second wave is the one we are riding now. It could be described as a network of things that embed computers. It is mainly made up of wireless phones and other handsets, game-players, teller machines and cars. The number of possible connections is more than a billion. An example of application is remote control of ventilation plants, which saves energy.

The third wave is on the way. More or less all things (and animals) might be involved: light bulbs, sensors, notes, beer glasses, bees. Miniature tags are added to all kinds of products, and can tell the story of the thing, such as temperature fluctuations. The pace of introduction depends partly on the business case. Today the case is unclear. Examples of the third wave:

- A chip in the beer glass calls for a refill when the amount falls below a certain level.
- Chips in bank notes to prevent 'money washing'.
- Chips on bees checking land mines.

Some potential M2M services, for example those including positioning, imply obvious integrity risks. 'You are watched'.

2.8 WHAT ARE THE SERVICE DRIVERS?

For more and more activities IT/telecom is a cheaper or at least comparable tool in relation to traditional alternatives. Consider for example e-commerce and commerce, e-banking and visits to the bank. Often the IT/telecom alternative is time saving. By the way, Swedes spend on average six hours a day with different media such as TV, books, radio, Internet etc. More than 60 % of Swedes use a computer in their work.

The rational time and money-saving reasons for the increased use of IT/telecom can be proven by means of end-user business cases. These aspects are supported by Table 2.3. And many people are really short of time. Parents of small children are a good example.

Money-saving IT alternatives often include an elimination of steps in a delivery or production chain, based on communication mobility and accessibility. An example is saving money by ordering tickets from the source.

By investing in an Internet adaptation of business systems (such as SAP, Scala, TEADS and Primavera) the workload on the company will decrease. Customers and suppliers can perform administrative routine tasks. Internally employed staff can register themselves

Table 2.3 Examples of savings by using the Internet

	Traditional system ($)	Internet ($)	% savings
Airline tickets	8	1	87
Banking	1.08	0.13	89
Bill payment	2.22–3.32	0.65–1.10	67–71
Life insurance policy	400–700	200–350	50
Software	15	0.20–0.50	97–99

and collect data without involving intermediary administrative functions. Increased efficiency can for example be achieved by means of remote access of data in the business system, such as tenders and sales data, complemented with possibilities to enter orders and transactions when travelling.

The outcome of the business case depends on the size of the company. In general, large companies with heavy administrative routines can more easily justify the investment of an Internet adaptation. Interesting administrative areas are B2B, B2C and Intranet. Examples are an Internet-based product catalogue, with possibilities to order via the Web, customer order handling via the Web, to follow the order status, employments, applications, and just-in-time stock control.

The crucial question is this: is the human being rational? Thus, do the rational reasons inevitably lead us towards an expanding IT/telecom sector, bringing all of us into the information society with an accelerating number of offered services? This would affect enterprises, private people and society.

Is the human being just a short time planner? What about long-term needs?

According to experts a shortage of oil may be expected before 2020. Since IT enables distant communication without travelling, the oil shortage could possibly be postponed by means of an increased use of the economical IT/telecom resources in relation to physical travelling. Such good long-term planning needs of course political initiatives.

Most experiences point in the direction that the human being is not very rational. He might buy a car based on its design rather than space for goods from the store and the baby carriage. Entertainment is one of the interesting non-rational areas, where business cases do not fit very well. Regarding games, the interactivity offered by Web based services, as opposed to film and newspapers, could be a killer feature. A combination of interactivity, multimedia and mobility could be used in many contexts, for example in distant learning. The combination offers a possibility to an active dialogue instead of just a passive reception of information.

The video capabilities of 3G can be used in combination with success factors according to Chapter 1 and mass market safety needs including medical info according to Figure 2.9.

Let us take an example:

Many sick children can be at home instead of at hospital, with the same quality of care. The solution is cheaper for the hospital and cheaper for the parents. No traveling is necessary for the parents. There is no need for babysitters for brothers or sisters to the sick child. GPS or a simpler positioning service is used to easily find the patient address when visiting the sick children.

Video enhances significantly remote check of the condition of the sick child. Case-book access supports the remote diagnosis/contact.

Stockholm and Åland are the first application locations.

Similar remote applications could be used for grown-up people and especially old people.

See further Chapter 5, sections 5.2–5.5.

2.9 USER PERCEPTION

Finding the optimum range of services for a target group is of course not enough. There are at least three areas where user perception comes in. These are pricing, quality of service and security. There is a kind of relationship between them. The user may for example be able to choose between lower priced, lower quality voice over IP and 'ordinary' telephony. She may also be able to choose between a number of operators/service providers offering a particular service with different quality of service and security at different price levels. This is of course one of the reasons for churn.

Quality of service has been important since the beginning of telecommunication, when loudness was a parameter in networks with a heavy loss. The ear was the critical receiver. Today one can add the eye and the computer. See Chapter 7, which is devoted to the subject of quality of service.

Security concerns basic human needs, primarily integrity and confidentiality, explained in detail in Chapter 6. The Internet has meant a revolution in communication, but also significant increased security risks.

Pricing is dealt with in Chapter 5, section 5.11.

2.10 SUMMARY

A number of user segmentations have been presented. Some matching between segments and services has been presented as an illustration of the methodology. A general feeling from the main author is that the user is the most complex element in the overall telecom context, and it is not easy to make this key element fit into the larger picture. Requests for more research have been voiced by telecom actors.

See also the introduction to this chapter, referring to security, quality of service and mobility and treated in Chapters 6, 7 and 3, respectively.

User unpredictability and impatience require a flexible network, which is easy to manage (Chapter 3). The service area needs further attention (Chapter 5).

3

Networks and Technologies

We've had a lot of new technology to digest in recent years

- no wonder we're struggling a bit!
- once we've got the hang of it, we'll start demanding more.

3.1 OBJECTIVES

The main objective of this chapter is to create an improved understanding of networks as an input to business planning, network design including network segmentation and telecom management. The chapter clarifies the ongoing convergence, network layering,

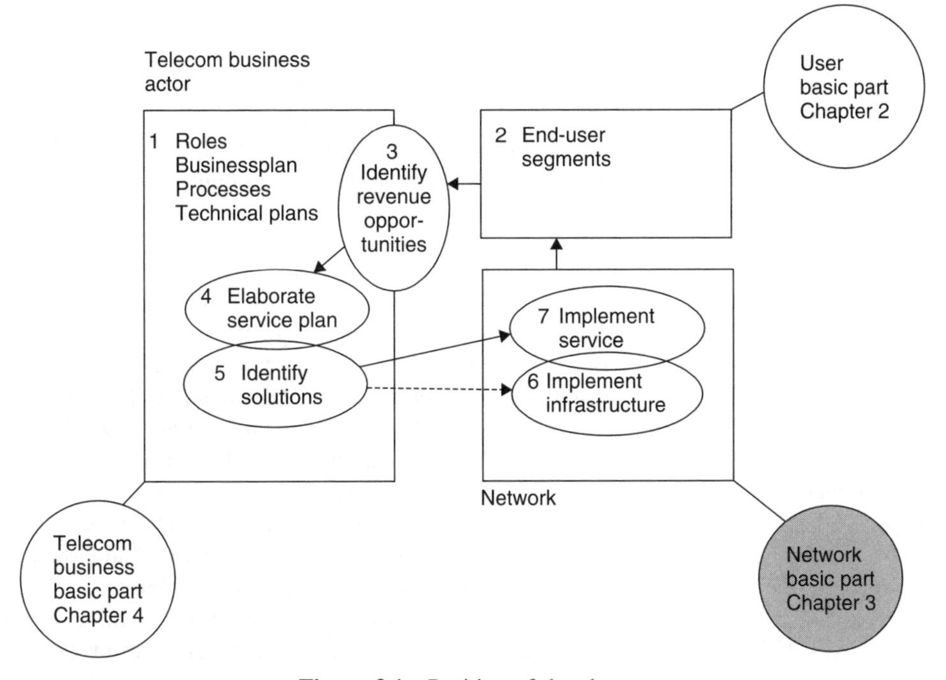

Figure 3.1 Position of the chapter

Understanding Changing Telecommunications – Building a Successful Telecom Business. Edited by A. Olsson
 ISBN: 0-470-86851-1

a target network structure, the related fundamental technical plans and the main enablers for the network transformation. Some key objectives are:

- To understand the basics for the present development such as what is normally meant by a *network*, including fundamental technical plans, and the difference between a vertical node-oriented view and a horizontal layered view.
- To understand the convergence between the networks of today and the media industry into a layered and stratified target network with its segments.
- To understand the need for control and management of the network, to make it respond optimally regarding service type, QoS, security, economy and other needs of the subscribers and/or operators.
- To understand the difference in control between connectionless and connection-oriented networks.
- To understand the simple reference model(s) used in this book to segment the network into 'chapters'.
- To understand the range of enablers that are required to transform the vertical networks into the converged horizontal target network, especially API, H.248 and IPv6.
- To explain the functionality distribution in the network. This will be treated in more depth in Chapters 9–13.
- To explain the relations between roles in the business part and layers in the layered model. This will be treated in more depth in Chapter 4.

References are:

- Thomas Muth, *Functional Structures in Networks*, to be published in 2004.
- Christoffer Andersson, *GPRS and 3G Wireless Applications.*
- *Guidelines for the Elaboration of a Business-Oriented Development Plan from ITU*, in particular the contacts with one of the main authors Herbert Leijon.
- *Third Generation Partnership Project* (3GPP).

Networks are defined here as the service production tools, or machines if you like, of many telecom actors. This book focuses on public services and the corresponding basic functions, whether it is a mobile voice-oriented network or the Internet that is used. All networks, whether vertical or horizontal, can be described with a number of basic functions, called fundamental technical plans (abbreviated FP to avoid conflict with file transfer protocol, FTP). In the ITU document *Guidelines for the Elaboration of a Business-Oriented Development Plan*, the subject of fundamental technical plans constitutes one of the main chapters.

Figure 3.2 shows a classical map of fundamental technical plans. The service production should be as cheap as possible without jeopardizing the quality and security demanded by the end users. A low production cost is based on investments as well as operation cost. Thus, the network must also be a user-friendly tool, easy and cheap to operate. Dependability parameters describe occurrence rates of different error or fault events. Typical events are set-up failures, bit error rate and unwanted or unsuccessful connection clearings.

It must also be possible for the actors to cooperate in the production of services, by specializing into various niche activities. Obviously this demands clear interfaces within

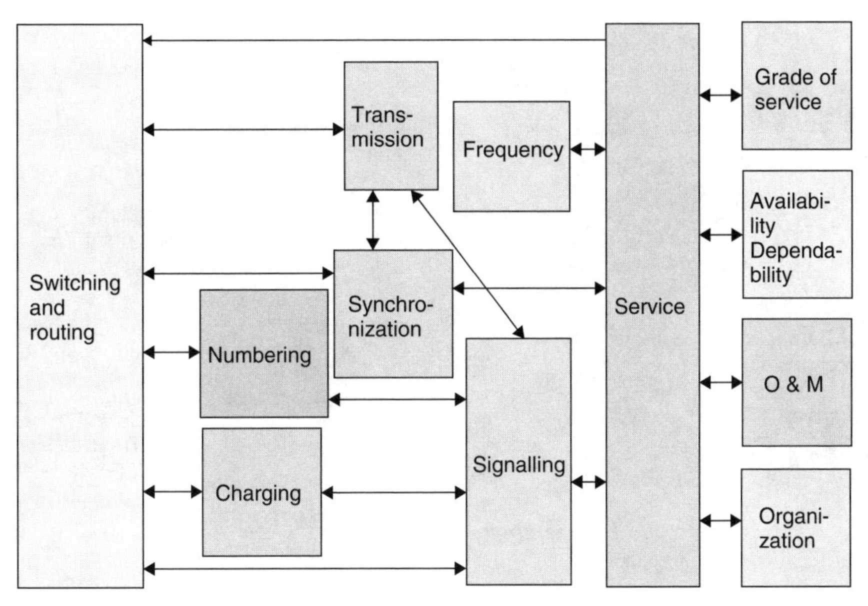

Figure 3.2 A classical map of fundamental technical plans

the network. So far we are used to interconnection via vertical interfaces, where two actors meet, for example in a network interconnect point. Now a horizontal interconnect or interworking is becoming more common.

Considering the service production aspect it would be natural to start this chapter with a comprehensive section on service requirements. However, the ongoing paradigm shift implies quite a lot of technical changes of the networks, calling for a basic technical chapter in this book, before going into service optimization of the network. In this first network-oriented section we look therefore more at the architectural options of the network rather than on service implementation aspects, which are covered in Chapter 8.

3.2 WHAT IS A NETWORK?

3.2.1 THE FUNCTIONAL CONTENT OF A SERVICE PRODUCTION UNIT

To make it simple, let us start with the traditional telecom networks and split them into two logical systems: the *traffic system* and the *management system.*

The traffic system serves the end-users, and the management system supports the traffic system with the activities that can be found in Figure 3.3. This is a general system model, which is valid for all types of network architectures.

The management system is at least partly a function of the traffic system, that is first we have to understand the traffic system before defining how to manage it. Adding the users we get the triangle which is found throughout this book (see Figure 3.4).

Any traffic system is characterized by two main functions: resources and resource control. This is a programmed control, not to be confused with management. In ISDN, the concepts of *user plane* with the user traffic, and *control plane* were introduced and

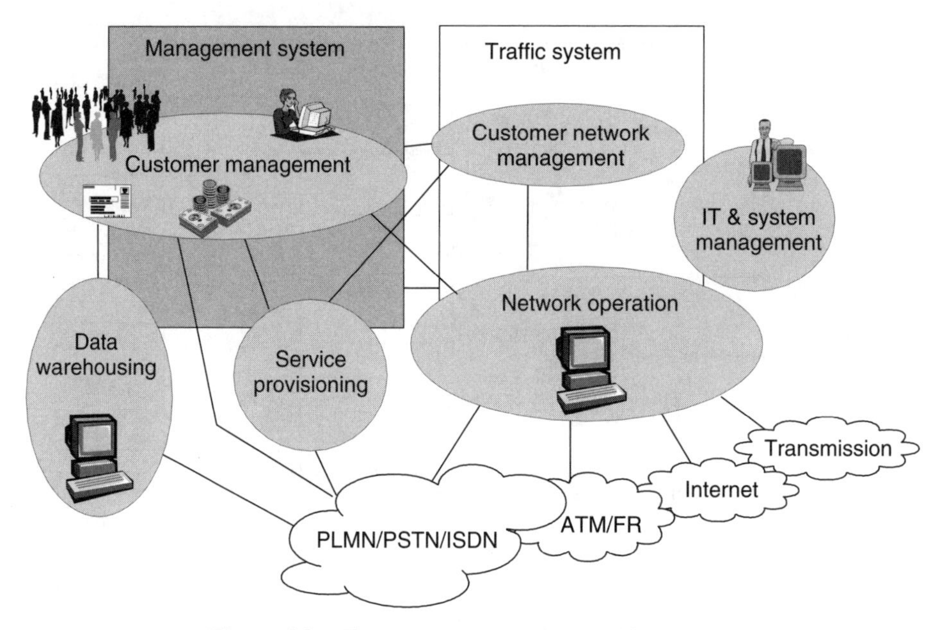

Figure 3.3 The management and the traffic systems

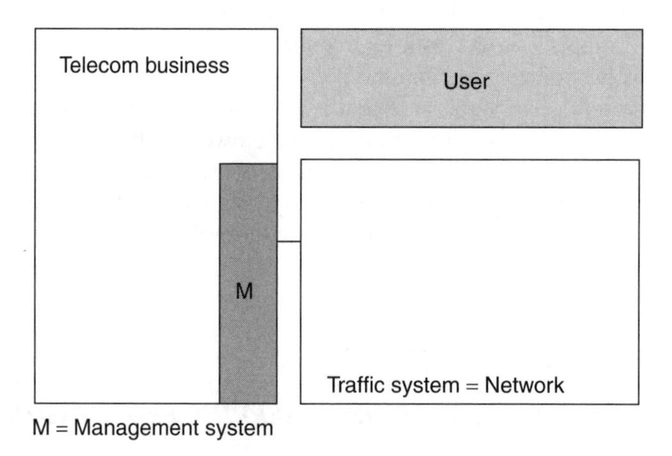

Figure 3.4 The triangle

defined. In PSTN the planes are not described separately, and consequently there are no standardized interfaces between them. In the evolving horizontal network, the distinction between these planes is pronounced. An example is the split into a control layer and a connectivity layer.

Packet-switched networks (or packet-mode networks) are sometimes regarded as having no control plane. Especially in connectionless networks, the control layer is difficult to identify, since for example routing is done for every single packet. (Connectionless = 'Any packet can theoretically take its own path to the receiving end'.) However, there is always a kind of control plane. In packet-switched networks it just partly deals with other issues than the tasks of the control plane of circuit-switched networks. For IP networks this is a heritage from the original US defence-based requirements on robustness, with quite a lot of intelligence located at the terminal devices, and a less intelligent but more autonomous network part in comparison with the circuit-mode networks. Among IP control functions are traffic supervision (ICMP, Internet control message protocol), routing (BGP, border gateway protocol, for routing to/from external domains; OSPF, open shortest path first, for internal routing; and IGMP, Internet group management protocol for multicast routing), QoS (different alternatives such as RSVP, resource reservation protocol), DiffServ and MPLS (multi-protocol label switching) and security (IP security). IP control functions are further dealt with in Sections 3.14 and 3.16 and in Chapters 6, 7 and 12.

IP network nodes are able to learn the network topology by themselves and perform routing autonomously. This minimized network administration is in line with network robustness and reliability policies so that the network might automatically recover from any intentional damage.

The importance of control is related to business. Most markets are deregulated today. In the resulting competitive environment business models and charging possibilities become central questions. What we can charge for depends on the services and network control we have. Control means awareness. And awareness means charging possibilities. The awareness might concern duration, calling and called IP address, bandwidth, security and QoS requirements.

The telecentric operators are used to a firm control of the traffic through their networks. The networks are fairly intelligent, often with sophisticated charging possibilities. David Isenberg in the USA wrote a famous article in 1997, called 'The rise of the stupid network', see http://www.rageboy.com/stupidnet.html, contrasting the intelligent telecentric networks with a more advantageous future network where most intelligence was located at the terminals. The contemporary Internet was considered a rudimentary form of this future network.

Service and *QoS assistance* from the network were however regarded as necessary.

Another possible area for network assistance is *security*.

So, also in a 'stupid' network there is a need for assistance to the end user from the network operator. This assistance means at the same time that the operator gains control of the subscriber calls.

A key component for awareness in the new horizontal networks is called the call session control server (CSCF). The CSCF will gather charging information and send it to a billing server. See Chapter 14 and Figure 3.5.

Control signalling is transported in very different ways in circuit and packet-mode networks. In circuit-switched networks control messages are normally handled and transported separately from user traffic by the well-standardized separate sub-network Signalling System 7 (SS7). For further study of this important standard see *Understanding Telecommunications* books 1 and 2. In packet-switched networks data of the control plane and the user plane are transported in the same messages. This fact may have added to the impression that packet-switched networks lack control planes.

In the horizontal networks the separation is not just logical (as in ISDN) but there is also a physical implementation defined called the gateway control protocol, GCP/H.248.

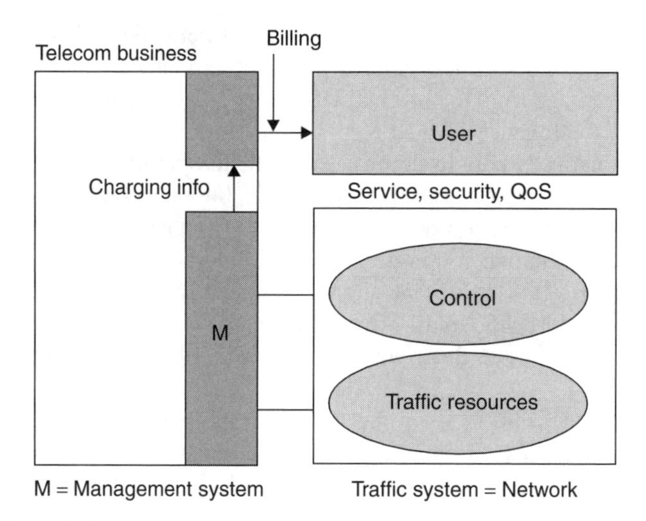

Figure 3.5 Control, traffic resources and charging based on control

3.3 WHAT IS A VERTICAL NETWORK?

Today, almost all networks are 'vertically integrated' single-service networks where the operator offers everything from subscriber access to service creation and service delivery across a wholly owned network infrastructure, optimized for a particular service category. Each vertically integrated network has its own protocols, nodes, end-user equipment/terminals and builds on different principles and practices to ensure reliability of a single service.

These vertical 'monolithic' networks have basically their own fundamental plans with some exceptions: PSTN and ISDN are strongly related, and the transmission facilities are often shared between different networks. To a more limited extent an IN (intelligent network) network service layer is also common for a number of networks. An obvious difference between them is the user service-oriented top layers, which are much more developed in the IP networks. See Figure 3.6.

The new network has to interwork with the networks already present. Therefore a good knowledge of current 'vertical' networks, in particular their network properties, benefits and drawbacks, is an excellent starting point (Figure 3.7).

The various network types were originally dedicated to various media type services according to Table 3.1.

GPRS is considered a step between GSM and UMTS but is no doubt supporting mobile access to the information society media sources. The telephony and data service domains are still more or less kept separate.

Historically, most functionality was located on telephone exchanges with vertical interfaces offering a number of signalling interfaces for adaptation to the existing network environment. Taking Sweden as an example, hundreds of signalling systems and variants have been used. Apart from signalling the interfaces were fairly homogeneous. The carried voice traffic mainly had only two shapes: either analogue with a bandwidth of 300–3400 Hz or digital with 64 kbit/s bandwidth. It should be borne in mind that both

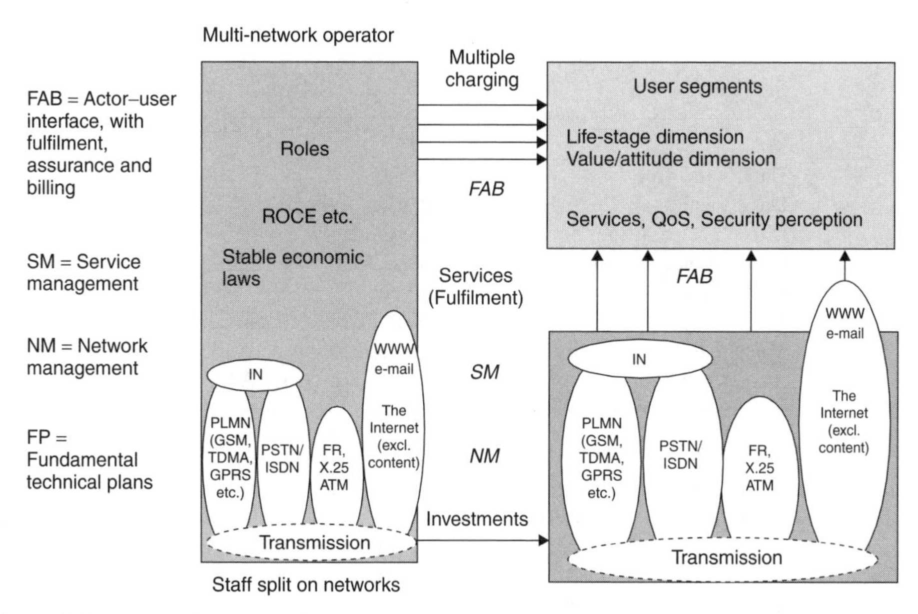

Figure 3.6 The vertical 'monolithic' networks demand a staff split on networks, multiple subscriptions and multiple charging

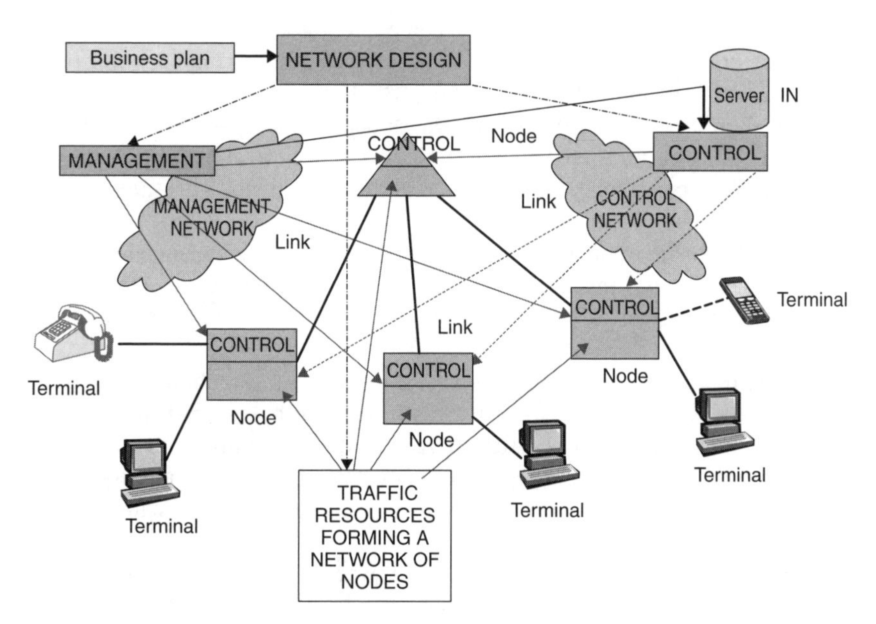

Figure 3.7 A simple distributed and node-oriented tele-centric view of a traditional network

forms represent the circuit transfer mode. Today, the situation is different, voice can be carried by many transfer modes (VoIP, VoATM, VoFR) with many different bandwidths.

In order to introduce new functions into the 'historical' PSTN it was necessary to upgrade the exchanges or build a more service-rich overlay network, such as an ISDN

Table 3.1 The vertical networks and the targeted services

Service	Network
Voice	PSTN
Data, wide area	X.25
Data, local area	Ethernet, Token Ring
Data, 'Faster X.25'	Frame relay
Data + voice 'dream'	ISDN, ATM
Data internetworking	Internet
Voice with mobility	GSM, TDMA, CDMA, etc.
Voice + data + mobility + information society	Next generation network, 3G (e.g. UMTS, WLAN)

network. It is easy to realize that individual upgrades of exchanges is a very time-consuming way of improving network functionality. However, it was in the monopoly era, and the risk of losing subscribers was very low (and the waiting lists were often very long!).

The IN architecture brought the first 'server' into the fixed network at the end of the 1980s. This was before the convergence between voice and data networks, and the term 'server' was not used by telecom staff. The IN concept with a service control point (SCP) node on top of a group of exchanges also became integrated into the mobile systems. The SCPs offer a centralized way to introduce new services all over the network, proving the advantage of a server-supported 'layered' network. The time to customer for the operator could now be shortened. However, the interfaces between telephone exchanges were not simplified. On the contrary, since the IN concept adds new signalling in the network, the overall interface complexity grew. Managing an IN network has also offered more problems than expected.

It should be borne in mind that the IN services to a large extent are 'network oriented', often meaning advanced routing, advanced charging or advanced numbering. Such services have little or no content to offer, as opposed to for example WWW services.

A more system-oriented approach was necessary when the mobile telephony systems were standardized. Here the functionality was more distributed, with many 'servers' in the network (such as HLR, VLR, EIR, AUC in GSM) starting with the second generation of systems from the beginning of the 1990s. IN was also introduced in this generation of mobile systems. A mixed fixed–mobile system came with 'fixed cellular systems'.

The data-oriented networks, on the other hand, were layered from start, which enables a smooth transition to the new horizontal network. However, except for ATM, they are optimized for data only.

The introduction of TCP/IP networks meant something new, especially a possibility for global communication, since IP could 'surf' on other networks. The service offerings on top of TCP/IP, headed by WWW services and e-mail, were another strong feature. So it is not curious that the future network might lend many features from TCP/IP networks. See Figure 3.8.

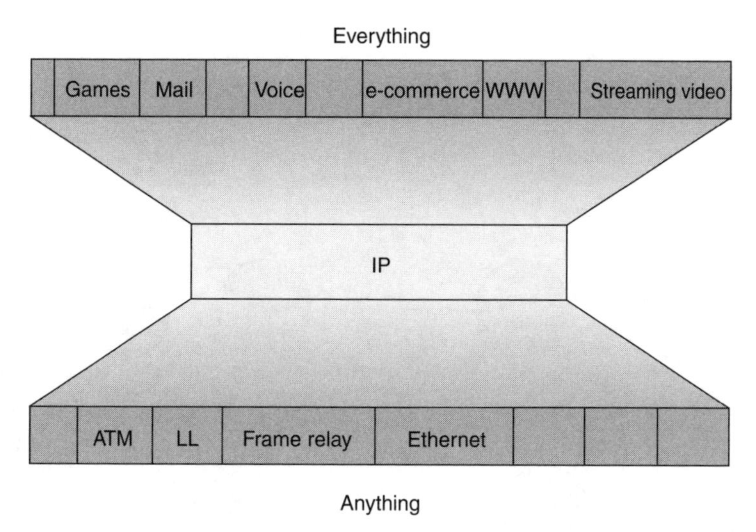

Figure 3.8 The 'secret' behind Internet success and its world-wide connectivity

ATM was designed to become a multimedia-capable type of network. ATM has mainly found its niche in the network backbones as an important layer 2 technology. Closer to the subscribers Ethernet is a very strong candidate at layer 2.

3.4 THE CONVERGENCE (OR COLLISION?)

3.4.1 TRANSITION FEATURE 1: CONVERGENCE

The previous books in this series ended by describing a number of vertical networks, almost all of them developing to manage both data and voice transfer.

It is not difficult to realize that such a technological competition with abundant almost parallel technical solutions is far from optimal. The marriage between telecom and the media industry, enabled by the Internet in particular, is now guiding us towards a desired convergence. It should be noted that the word convergence can be used in a broader sense than just technical convergence. It means that there are many interpretations, four of which are quoted below.

> 'Convergence is a global trend, not a specific technology. Besides technology, convergence forces operators, service providers and suppliers closer'.
>
> 'Convergence is a way of making more money from the networks. The networks become multi-service to adapt to the convergence and generate more money.'
>
> '. . . the case for convergence of network technologies is clear. The result is increasing openness and sharing of technologies, systems and knowledge across different networks. The end result is improved economy of scale with lower costs, and a larger room for other criteria than technology to drive operators' and service providers' network strategy decisions.'
>
> 'Convergence is the key word for communications in the 21st century. Operators are converging through mergers and acquisitions, often across national boundaries.

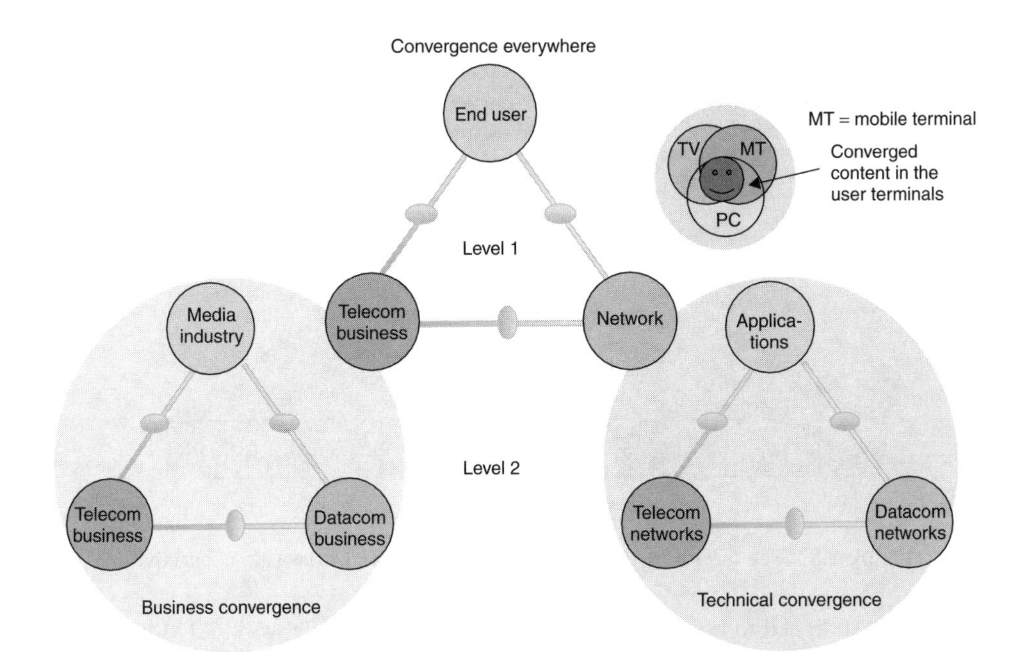

Figure 3.9 The convergence affects business, network and content

> Infrastructure is converging, with the established telecommunications technologies transitioning to solutions based on IP. Services are converging in a variety of ways. Operators are offering bundled services so that customers can obtain all their communication needs – data and voice, mobile and fixed line – from one provider. To make these bundled services work, operators are converging services onto a single multi-service network. No operator is immune from the changes that are taking place – all are facing increased competition, customer churn, and financial challenges.'

When using this broad interpretation and the triangular model, convergence can be illustrated according to Figure 3.9. Observe the convergence that is also taking place in the end-user terminals, with alternative channels for the same content.

The convergence between a PC and a mobile terminal (MT) is sometimes called wireless Internet. The mobile device also becomes an entertainment channel, exemplified by a convergence TV-MT. This convergence is sometimes called mobile media. The terminal convergence is applicable at the service and application layer. The underlying transport part becomes increasingly fragmented into a number of existing and new technologies.

Bearing this broader view in mind, this chapter focuses on the *technical convergence*.

3.4.2 TRANSITION FEATURE 2

Digitization: A Necessary Convergence Enabler

Let's come back for a while to the world as it was in the 1960s and early 1970s. All the activities related to telecommunications, represented by telephony, radio and television, were based on analogue transmission. Images, newspapers, books, music records and films were just media and nobody would have ever thought to associate them with telecommunications, except when TV and radio broadcast them.

Computers were the only equipment that was clearly digital. Big mainframes mostly took care of the repetitive and rather boring office jobs that many of our fathers and grandfathers had done manually. The associated telecommunications was in its early childhood. The environment was centralized, and the bit rates required to interconnect a remote dumb terminal had a speed from 50 to a maximum 300 bits/s. Each industry branch had its own techniques, patterns, codes and major players. The world was analogue, as it had always been, and the media industries were completely independent.

Today the telecom services are mainly digital, that is, they are represented by ones and zeroes. What is not yet digital will be converted to digital sooner or later. 'The IT world is made of bits'. Thus the ultimate substance of a service is bits.

The bits represent voice, TV pictures, books, newspapers, radio, music and data merely by combinations of ones and zeroes. Of course, in order to understand the pattern you must know what standards are used: MP3, GIF, Real Player, voice coding. The common shape of digitized services is obviously an important explanation of the convergence between telecom, datacom and media. The road to multimedia is open, indeed!

For economical reasons the traditional networks will not disappear over night. The new network must be 'backward compatible' with the traditional networks, to use a term from the design world. As mentioned before, the traditional network services and their new enablers and the network convergence, give us in a way enhanced traditional networks intertwined, or integrated if you like, within the framework of the horizontal networks.

This makes it easier to understand the new network and the migration into it. A large advantage is that interworking becomes easier. However, later versions of the new network for multimedia differ significantly from the existing networks.

The traditional data-centric networks are primarily used for computer networking in the business area. See Figure 3.10.

The traditional voice-centric networks are primarily used for public and business-oriented voice calls and lately dial-up to the Internet. However, PSTN has been used

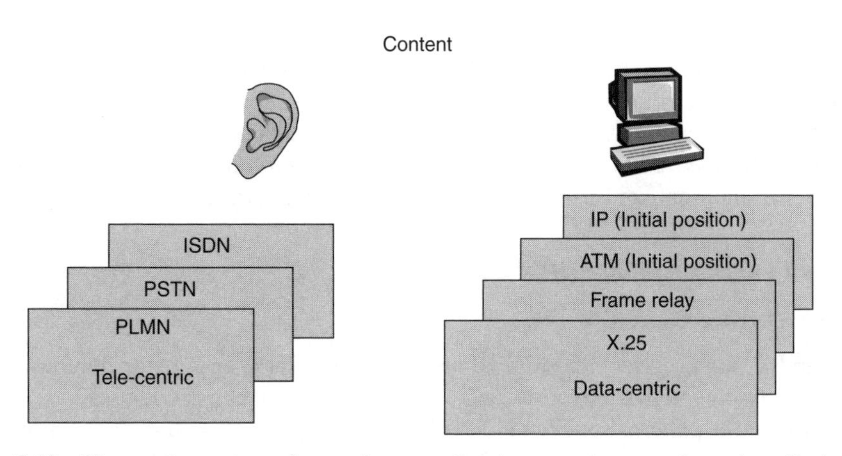

Figure 3.10 The starting point: tele-centric networks, data-centric networks and media industry content – no convergence

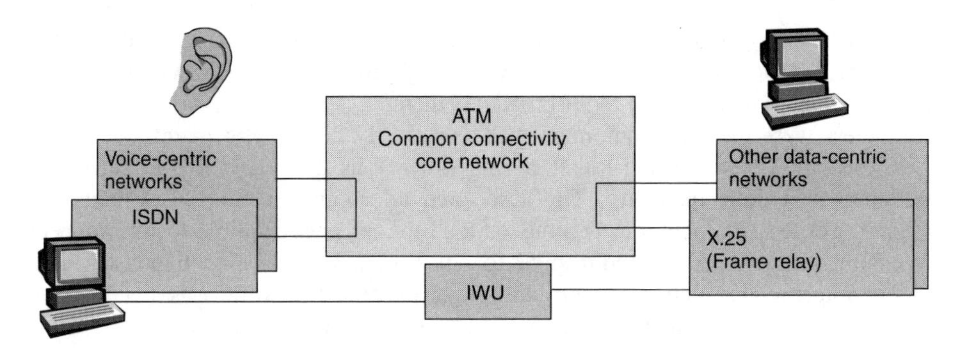

Figure 3.11 The first signs of integration and convergence: multi-service application of ISDN and ATM

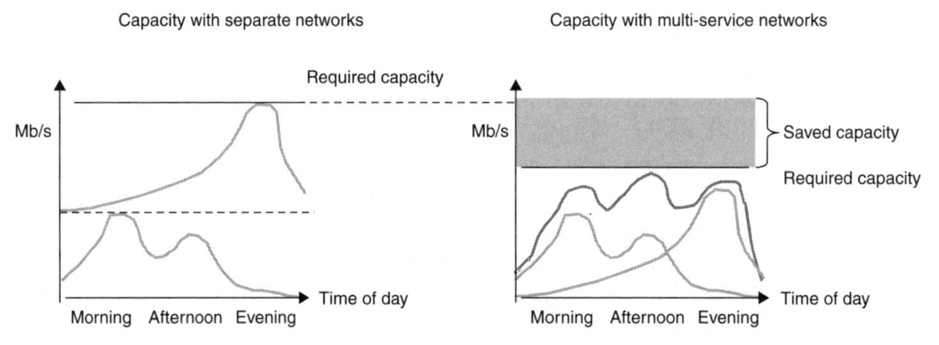

Figure 3.12 When the bearers can manage both tele- and data-centric services, capacity can be saved

for other data traffic for more than 30 years, supported by an impressing development of modems.

ATM has offered efficient multi-service use of the transport capacity. See Figures 3.11 and 3.12. In Figure 3.12, the upper curve on the left-hand side represents data traffic and the lower curve represents voice traffic. Both curves and the sum of the two curves appear on the right-hand side. In this situation we want to use the investments as much as possible. Further we need an adaptation between the purpose-built networks and a service/application layer, which is independent of available access/bearer networks. Roughly we get the traditional voice-centric and data-centric networks converging with a service/application-oriented part, offering information retrieval and streaming services.

3.4.3 TRANSITION FEATURE 3

The Convergence Glue: IP

From a traffic point of view the original interactive and symmetric voice-dominated communication is complemented with, and in fact overtaken by, a client-server IP-based asymmetric traffic pattern. See Figure 3.13. This bursty pattern usually implies more traffic from the server than from the client. Behind the server-generated traffic one can find the media industry and other parties in the IT society as content providers. The various ways

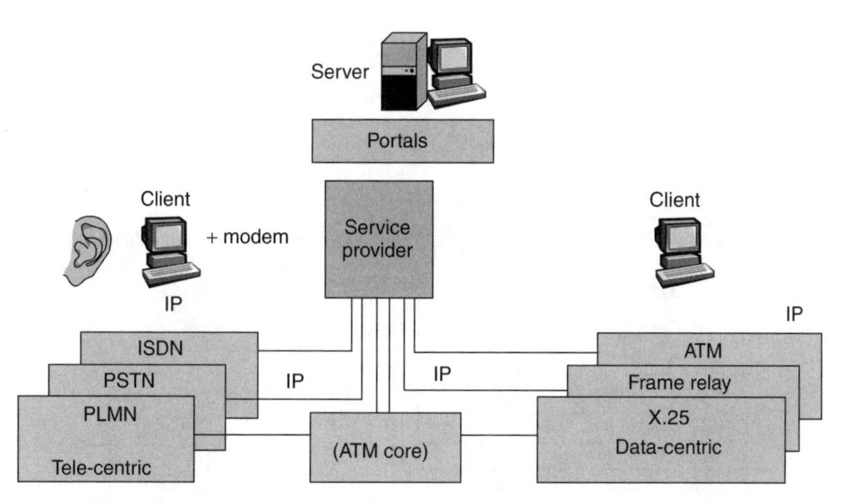

Figure 3.13 The converging glue: IP, service providers and (often) portals. The ATM core is not compulsory, but offers high quality to the IP traffic

to deliver content 'infotainment' are usually called applications with WWW/HTTP as the first 'killer application'. Accessibility to applications from any end user independently of the existing type of access media connection is a main driver for the network evolution.

Except for client-server relationships there are also peer-to-peer relationships, such as Napster downloads.

This traffic pushes the need for 'broadband'. The reason is the fast Internet WWW and video services or other bandwidth-consuming services that can be downloaded from the servers, or from peers, such as the above-mentioned Napster.

We are now experiencing a dynamic time: the perception of the end users changes over time, new standards are produced continuously and implemented into the networks, new network solutions are designed by vendors.

New network solutions are introduced in existing (or sometimes new) networks by the operators/service providers. New services are produced, maybe by third-party developers. New roles in the value chain are created or disappear. New services are marketed towards the end users.

The end users show their power by choosing services and operators.

There are other dynamics as well, if regulators and society are taken into consideration. Some of the new standards can be seen in Figure 3.14. There are two drivers: broadband for IP-traffic (or indeed only wideband (up to 2 Mbit/s) in some cases) and voice convergence (VoIP, VoATM, VoFR). This is where we are today. The next step is likely to be MM/IP (multimedia over IP), combined with a full-fledged mobile Internet, in addition to the fixed Internet. See Figure 3.15.

Summing up the convergence so far IP has a key role in connecting everything together, by means of its capability to surf over other networks, and some killer applications like WWW and e-mail.

Figure 3.16 shows an important feature of IP networks, which is to utilize and achieve internetworking across other networks. By the way IP is no longer just global; it is used for interplanetary communication between the earth and Mars.

Servers
Portals
Service provider
Adaptation to information society:
Extended use of the Cu-pair
PLMN 2G→2.5G→3G
Packetized voice for
multi-service and multimedia
xDSL
GPRS packet data channel
UMTS WCDMA path
IP
(VoIP
(MM)
VoIP
VoATM
VoFR
Cable TV → Hybrid fibre coax

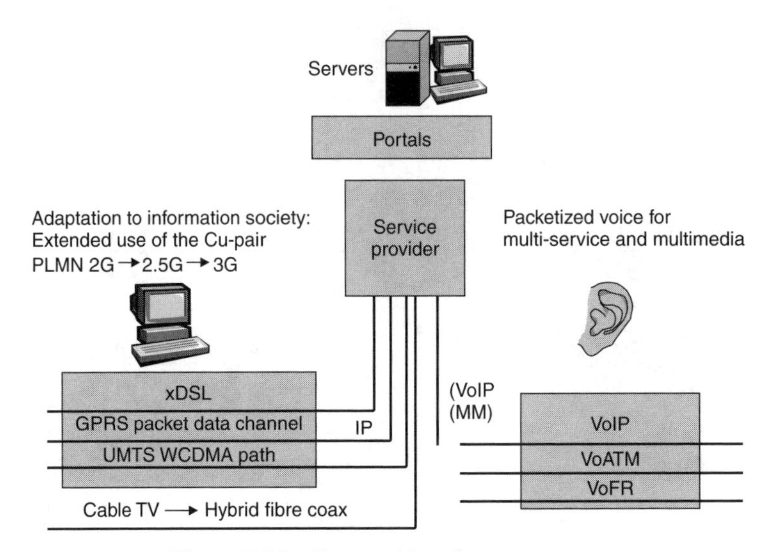

Figure 3.14 New enablers for convergence

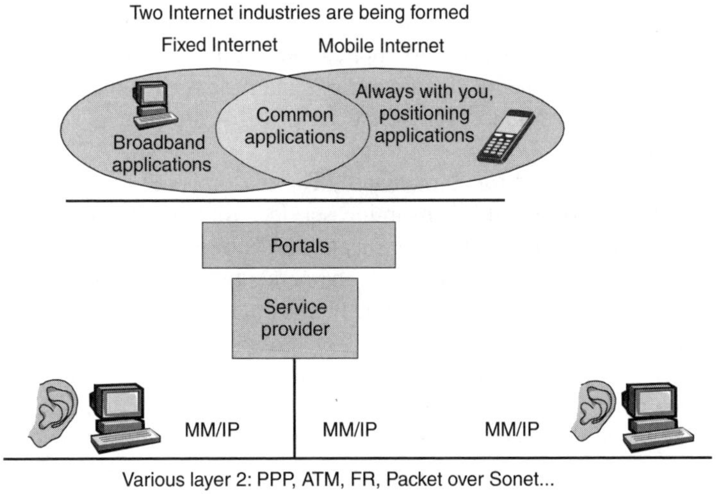

Two Internet industries are being formed
Fixed Internet
Mobile Internet
Broadband applications
Common applications
Always with you, positioning applications
Portals
Service provider
MM/IP
MM/IP
MM/IP
Various layer 2: PPP, ATM, FR, Packet over Sonet...

Figure 3.15 Multimedia over IP combined with mobile and fixed Internet

Content provider
and other roles
Multiple 'virtual' IP networks
User
Access
Transport
Service network, Internet, corporate network
IP host/client
IP routing
FR
ATM
IP host/server

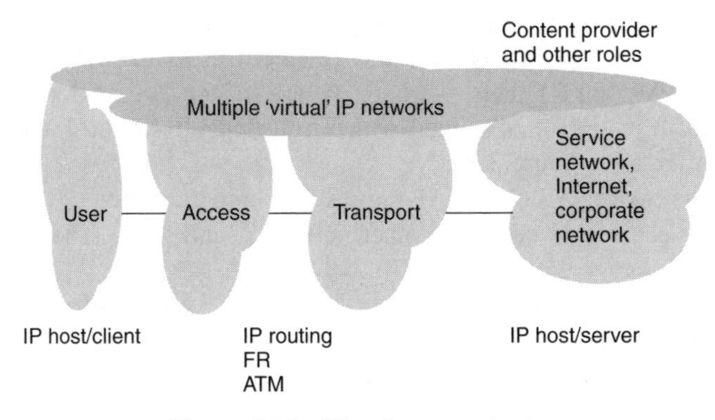

Figure 3.16 IP – the network glue

3.5 WHAT IS A HORIZONTAL NETWORK?

This concept of layered network architecture is already in use in datacom networks, so it is not brand new.

The development within the telecom industry during the last decade has underlined that the future range of successful services is impossible to predict. Bearing this lesson in mind, network architects are highly aware of the need to build the networks as flexible service platforms. The number of services increases quickly, especially because of the paradigm shift when the telecom, datacom and media industries converge. Together with fast technical development, this provides a background to the changing architecture of networks in the telecom world, from what is frequently called the vertical networks of the past and present, to the horizontal networks of the future. The development is supported by operators, service providers, vendors and standardization bodies. A truly horizontal network is built up by autonomous layers. Each layer has its own management, control (if any) and resources. Each layer is built on standards-based components and the interfaces between the layers are also standardized.

A layered architecture allows each layer to evolve independently as technology evolves. Multiple service networks can share the same transport network.

The horizontal network architecture can provide all types of communication and application services, only restricted by the capabilities of terminals and the used access. Applications will/should thus be independent of any core/radio technology/fixed access technology. In such a service-converged situation users from GSM, UMTS, PSTN/ISDN, CATV etc. will be able to use the same service, but with different capabilities.

The *implemented* new networks will probably lend many features from the described customer optimal architecture, but many actors want to retain a view of being special, and will continue with partly vertical features.

A differentiation is sometimes necessary, especially in the complex control layer, in order to cope with different existing signalling systems that are connected to the new backbone. See Figure 3.17. Differentiation in the service layer arises when one operator deliberately tries to differentiate itself from the others. There are, indeed, also some problems with open networks regarding confidential or company-sensitive information about the user (location, personal etc.), which may impact the transition.

The first breakthrough for a layered view within telecentric networks was the introduction of SS7, the connectionless data network for signalling within PSTN, ISDN and PLMN. The separation of signalling and user traffic was also a step to split the network in a more functionally oriented way.

The real 'vertical' networks of today are also layered to a considerable extent, in particular at the transmission level. This makes the migration process cheaper and simpler. Let us take IP networks as an example, looking from the application down the stack.

The power of TCP/IP applications is that the application generally can assume a simplistic view of the network (a basic packet-carrying network). This is because the complexity of the network is hidden from the application: the error-free end-to-end transmission channel offered by TCP and the fast, 'best-effort' packet-sending interface offered by UDP are there for the application to use. The complexity of addressing, routing, re-transmissions, etc. is implemented in the TCP/IP protocol stack.

The migration process implies a convergence process, since the target network is quite similar for the various vertical networks. When shifting from vertical to horizontal, we

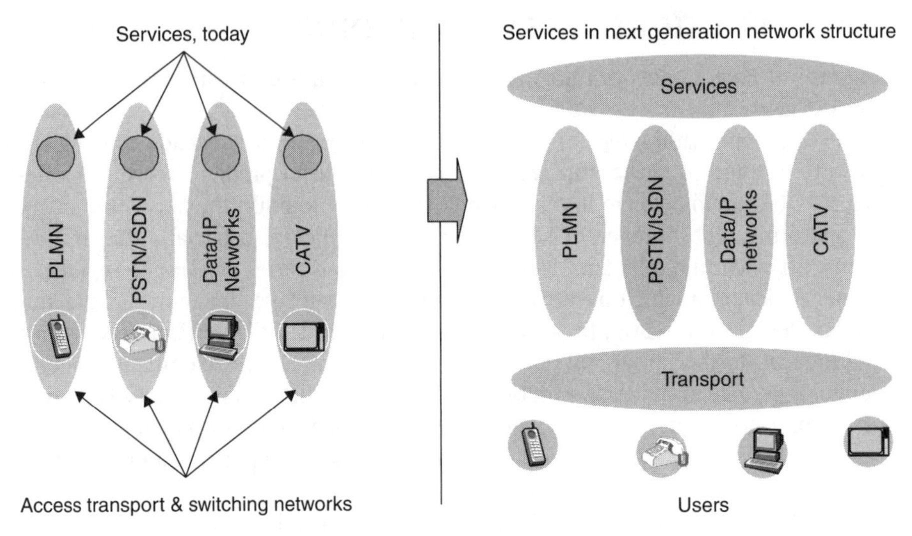

Figure 3.17 The layering will take time in the control layer because of the existing signalling systems

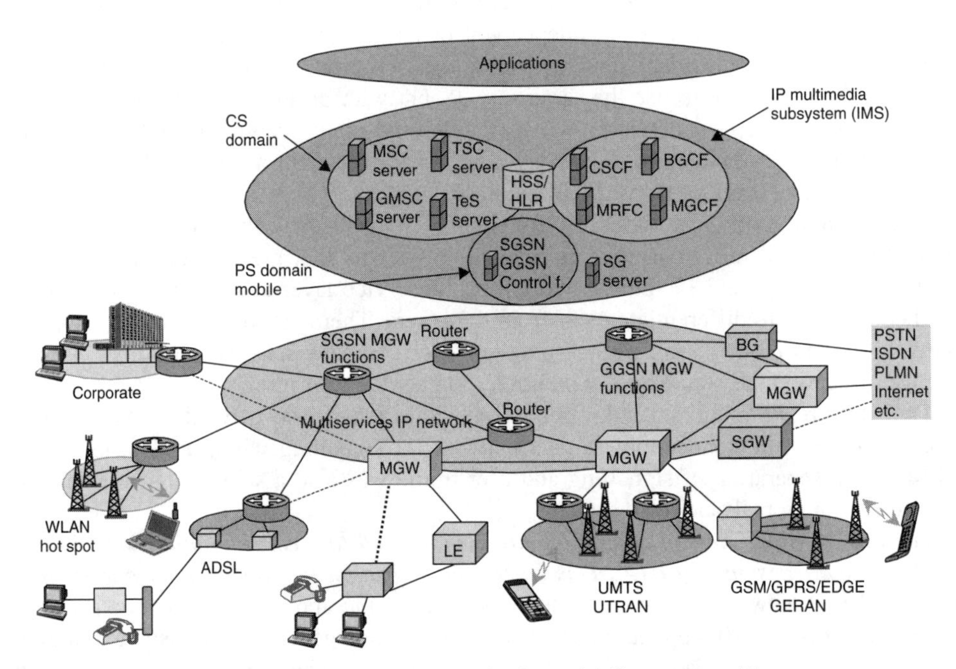

Figure 3.18 Example of the new horizontal 'all-access' architecture

tend to speak about the main layered functions such as communication control, services and applications rather than nodes in PSTN, GSM or the Internet. A node can be seen as an implementation option of functionality location, usually comprising more than one layer.

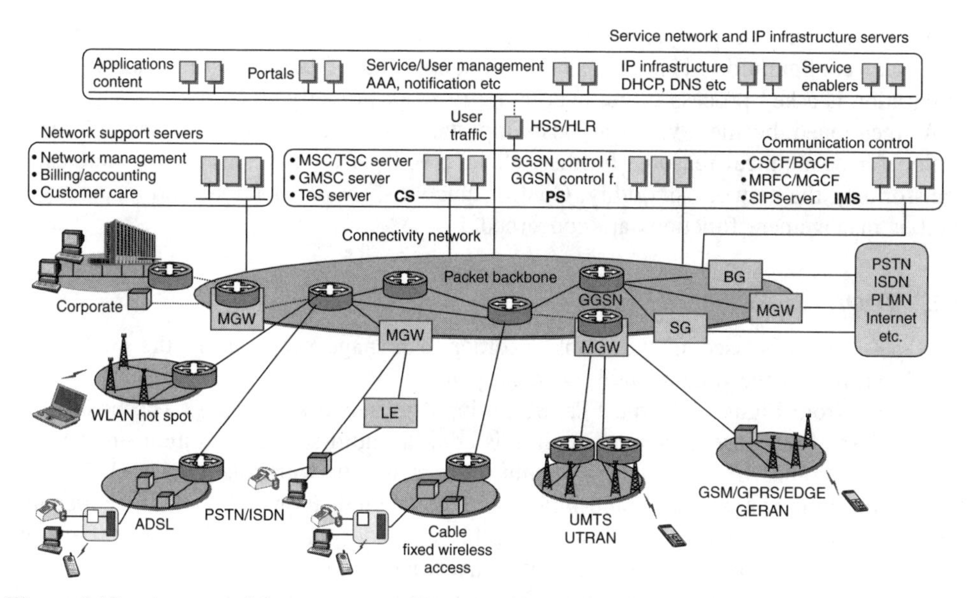

Figure 3.19 A server-rich target network where all new media industry-based applications can be distributed to many access networks/different users

Layering and functional split can be achieved in many ways, usually based on some kind of layered architecture model (OSI, TCP/IP). The development is also a matter of convergence as described earlier in this chapter.

Relationships between business, network design and management are further elaborated in Chapter 4. The architecture will be further explained in this chapter, with the concepts of network, convergence, layering, stratification and fundamental technical plans as pedagogical cornerstones.

A network architecture for all types of accesses could look like Figure 3.18, assuming a significant legacy of circuit-mode voice services. Here media gateway controllers are needed in the control layer for the signalling for all the service access types supported. The IP multimedia system could eventually control all services from wherever they are accessed (home, work, visited location) or over whatever access (fixed broadband, digital TV, WCDMA, WLAN etc.).

The new architecture is very server rich. See Figure 3.19. The control layer and the service layer are typically built up by servers. Server overviews are presented in Chapters 8, 9 and 14.

3.6 FUNDAMENTAL PLANS

3.6.1 TARGET – PROBLEM

The general goal is to build a network that produces the desired services with sufficient end-to-end QoS and security at lowest cost. This may not sound overwhelmingly difficult. However, in a time of transition there will be many interconnection points between technologies and between actors, both vertically and horizontally. The weakest link will decide the overall quality and security.

All nodes must be programmed according to required functionality, and obtain the necessary configuration. This work requires some kind of fairly detailed rules. System integration is a key process to get everything to work properly.

As mentioned, business views are also important, and over-provisioning may not be an option for economical reasons. Further, all involved parties have a common interest in transforming the network smoothly. Control functions, traffic functions and resources, as well as management functions are concerned.

3.6.2 TOOL

The tool, which is used in this book in order to manage successfully the challenging transition, is thus the fundamental technical plan or FP.

As seen from Figure 3.20 the FPs are an input to network planning. The boxes with broken/dashed lines are an adaptation in this book to multi-service and multi-media. The basic figure comes from ITU-BDT/Planitu. Planitu is a network planning tool provided by ITU for optimization and dimensioning of telecommunications networks. The meaning of the term telecom management in the figure is broader than the abbreviated TM used in this book (which is service + network management).

The use of this overview (Figure 3.20) is intended to contribute to a better overall understanding.

The FP is thus used to make an overall end-to-end requirement specification for the network and then, as a second step, to distribute the requirements and functional capabilities to the various network parts and platforms. The chosen IETF, ITU and 3GPP (and other) standards should be indicated in the FP specifications.

The language in an FP is normally supplier independent. Let us call it 'operator language'. FPs must be updated and adapted to changing conditions. When entering a new

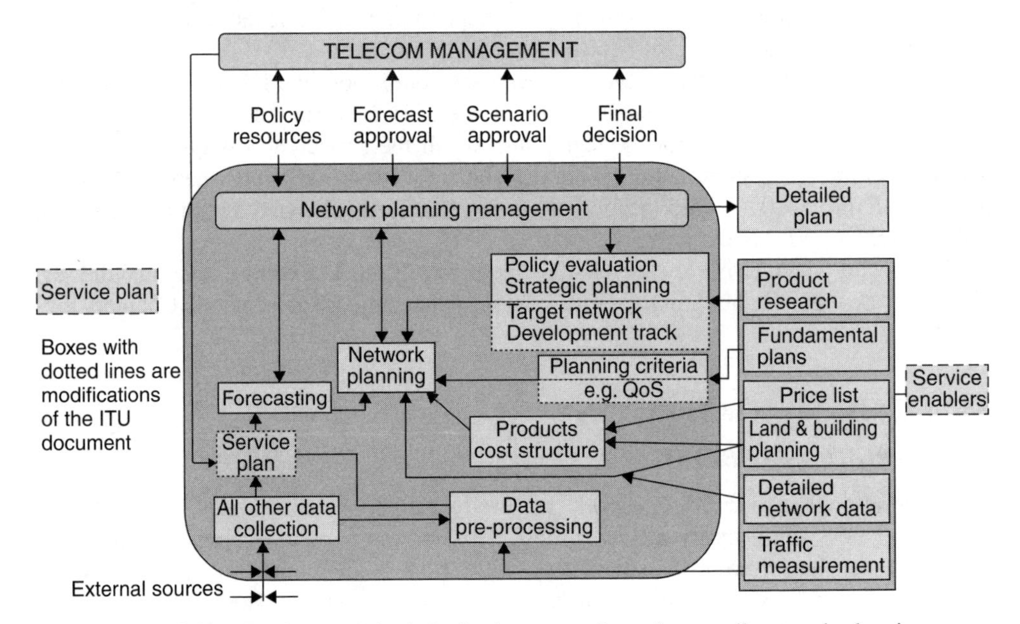

Figure 3.20 Fundamental (technical) plans as an input in overall network planning

architecture with a common bearer for voice and data the network design principles must consequently be reviewed.

The plans constitute a cornerstone for the network design plans, for procurement of equipment, system integration and other processes.

3.6.3 SPECIFIC PLANS

When looking in more depth at control and traffic resources behaviour we used to find the plans shown in Figure 3.21. Dependability includes measures to maintain availability when faults have occurred.

There are often dependencies/relationships between the plans, which was illustrated by Figure 3.2.

In a converging phase like the one that is prevailing right now, there is a technological war regarding optimal plans for the converged network. For example, a selection of numbering plans are in use, which is unfortunate. The numbering issue is discussed in Chapter 8, Section 8.8.

Example of content of the plans is:

- Switching and routing plan: network hierarchy, routing rules etc.
- Numbering plan: trunk codes, subscriber numbers etc.
- Charging plan: principles for charging of traffic and services etc.
- Transmission plan: quality requirements and technique for transmission.
- Synchronization plan: rules for synchronization of nodes in the network.
- Signalling plan: signalling systems etc.
- Frequency plan: use of available frequency bands etc.
- Service plan: service range etc.
- Mobility management plan: anchoring ('call responsible MSC'), hand-over, roaming etc.

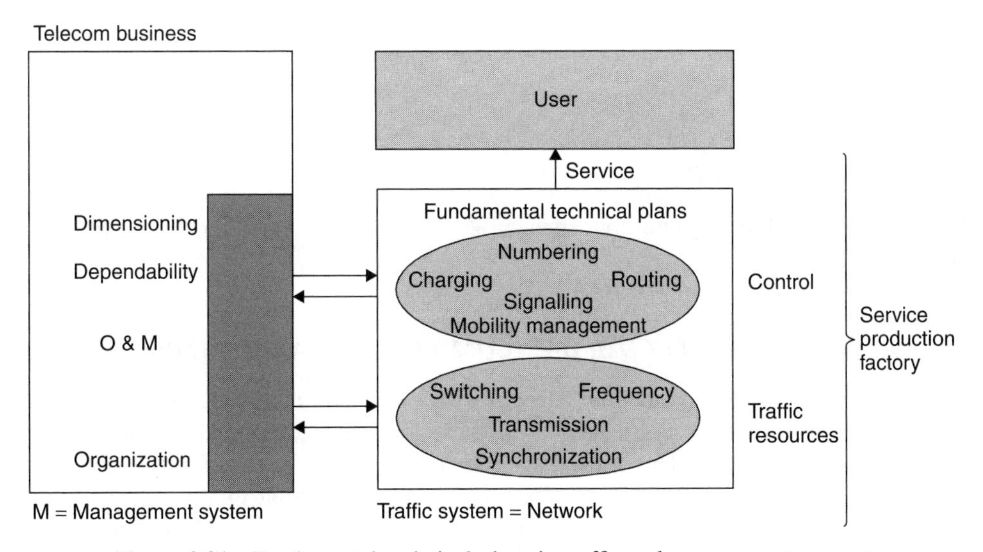

Figure 3.21 Fundamental technical plans in traffic and management systems

In the new horizontal networks there are a lot of possible traffic cases to consider, with different transfer modes in/out at the media gateways and fairly complex layered radio interfaces. This makes these networks more complex to design than the 'monolithic' vertical networks of today.

A smooth interconnection between a large and increasing number of operators can only be achieved if the operators agree on common rules at their common border, on signalling gateway protocols, routing, charging, security, management information transfer, transfer of transit calls between other operators etc.

Historically, the area of FPs has developed during the two last decades from a focus on loudness (transmission quality) in analogue PSTN networks to a focus on security and QoS for real-time services in IP networks, with call control servers, policy servers and application servers. That is why this book contains dedicated chapters on these subjects. Charging is another important area. This subject is covered in the Chapter 5, Section 5.11.

The new systems for horizontal networks will be rolled out in many states of revision. In order to be somewhat independent of the various implementation steps, a layered model is preferable. In such a model the particular node implementations are less important. Since FPs are normally seen end to end they fit excellently into layered models.

3.6.4 FP VERSUS PTP

PTP stands for particular (fundamental) technical plans. The particular characteristics are based on the capabilities of a specified supplier solution and related network design. Thus they describe supplier-specific network design capabilities using operator language. Such plans could be used widely, starting with a statement of compliance to operator specifications. Network integration is another process supported by PTP. The PTPs can accompany a procurement from tender to ready for service, enabling a consistent process for a particular contract. 'What is ordered should also be delivered'.

Historically, the local exchange has been the main network element for the use of PTP at Ericsson. Later on the concept was adopted for other network solutions. With PTP it is also easier and faster to respond to an operator specification, particularly regarding the statement of compliance.

3.6.5 CONVERGING FP

To a large extent the ongoing convergence is in fact a convergence between FPs. See Figure 3.22.

3.6.6 CONVERGING TELECOM, DATACOM AND MEDIA INDUSTRIES

In principle three worlds collide: the telecom world, IP world and media world. To make the converged multimedia world work well there is a need for a kind of glue or middleware. See Figure 3.23. This requires an additional part (sometimes implemented as a network: the service network) to cater for a proper utilization of the new possibilities offered by the media industry. The new network component implies a need for a coordinated development in the terminal devices and service networks.

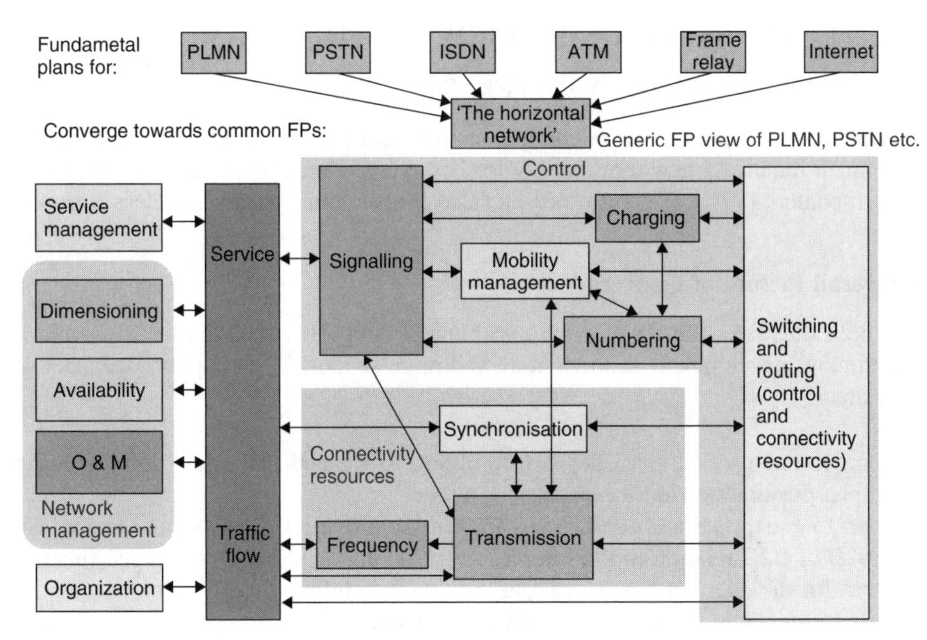

Figure 3.22 A main point in network design is to cope with converging technologies. The generic FP view uses a mirrored Figure 3.2 to adapt to the shape of Figure 3–22

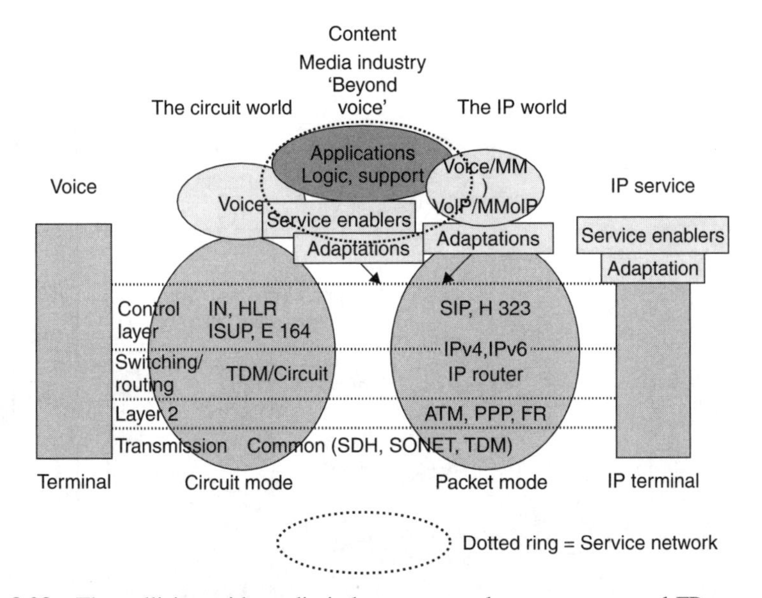

Figure 3.23 The collision with media industry, network convergence and FP convergence

Further, the service network develops into a key enabling part for new or enhanced applications, with *service enablers* as a main component. (Service enablers are service-supporting elements in terminals and/or service network such as positioning functions.) See also next section in this chapter and Chapters 5, 8 and 9.

3.7 A TECHNO-ECONOMIC VIEW OF THE CONVERGENCE

3.7.1 NEW COMMUNICATION PATTERNS

A service network appears in the communication path between person and content. The development requires a new terminology for communication patterns: peer-to-peer (P2P), person-to-content (P2C), person-to-person (also P2P), mobile media, wireless Internet.

The Overall Picture of Convergence

Figure 3.24 illustrates the development within P2P and P2C. P2C is most dynamic and the continuous development requires more and more support. The support is concentrated in four areas:

- Support to end-user services beyond traditional voice and plain data communication. Example: positioning (service enabler).
- Support to enterprises and content and application providers or any other external actor in a B2B or B2C relationship. Example: suitable interfaces and e-commerce functions.
- Support for the operator business processes, such as billing.
- Access to functions in the core and access networks

The P2C support materializes into a service layer with a service framework as a core. See Figure 3.24.

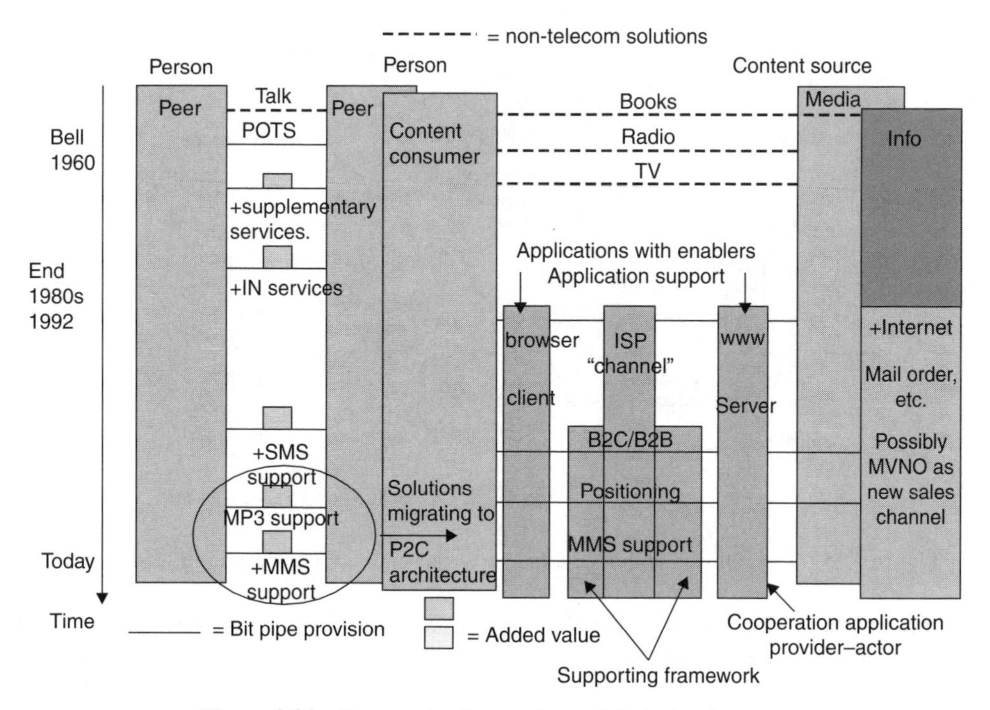

Figure 3.24 Communication modes and their development

P2P is often used to denote exchange of music and messages among end users.

Peer-to-peer communication also exists between computers. Peer-to-peer computing is the sharing of computer resources and services by direct exchange between systems. These resources and services include the exchange of information, processing cycles, cache storage and disk storage of files. Peer-to-peer computing takes advantage of existing desktop computing power and networking connectivity. In a peer-to-peer architecture, computers that have traditionally been used solely as clients communicate directly among themselves and can act as both clients and servers, assuming whichever role is most efficient for the network. This reduces the load on servers and allows them to perform specialized services (such as mail-list generation, billing, etc.) more effectively.

Person-to-content communication was given a real push by Internet WWW services at the beginning of the 1990s. Person to person is used to discriminate between P2C and the previous traditional situation with exclusively person to person. Content can be split into media and information. Translated to mobile communication it becomes 'mobile media' and 'wireless Internet'.

3.8 ADAPTATION OF THE BASIC TRIANGLE AND FPs TO THE CONVERGED MULTI-SERVICE NETWORK

The new functions on top of bit pipe provision will also require additions to FPs. The functions are more or less installed in the user–service interface. This means that the end-user perception must be examined.

The users have a broad scope of needs and demands as shown in Chapter 2. The success factors used in this book justify choosing the user–network interface as a focal point. The book focuses on three issues in this interface: services, security and QoS. The service area is introduced in this chapter, security in Chapter 6 and QoS in Chapter 7. Chapter 5 is the main chapter on services.

In line with Figure 3.25 the basic model must be complemented with a server part for content provision and similar server tasks. It could also be another peer or a mailbox according to Figure 3.24.

The convergence of telecom and datacom leads (in the optimal solution) to the creation of one single network replacing all vertical networks after a long transitional period. The convergence with the media industry is shown by means of content provision and the additional functional area on top of the network. The new model represents a *person-to-content* relationship, as opposed to the previous person-to-person model.

A service layer adds to the fundamental plans. See Figure 3.26.

In order to meet the user demands many devices and servers have been developed. Application developers design applications based on user needs and the possibilities offered by the network.

The FPs shown in Figures 3.2 and 3.21 are now insufficient and a number of service-oriented plans are added. The service, security and QoS plans embrace functions in all layers (e.g. security in numbering and routing, QoS over the air). QoS and security can be seen as attributes to the service.

In fact, certain services, security and QoS requirements have existed before. One could for example argue that ordinary voice and associated supplementary and/or intelligent network services have been around for a long time. Security may have been represented

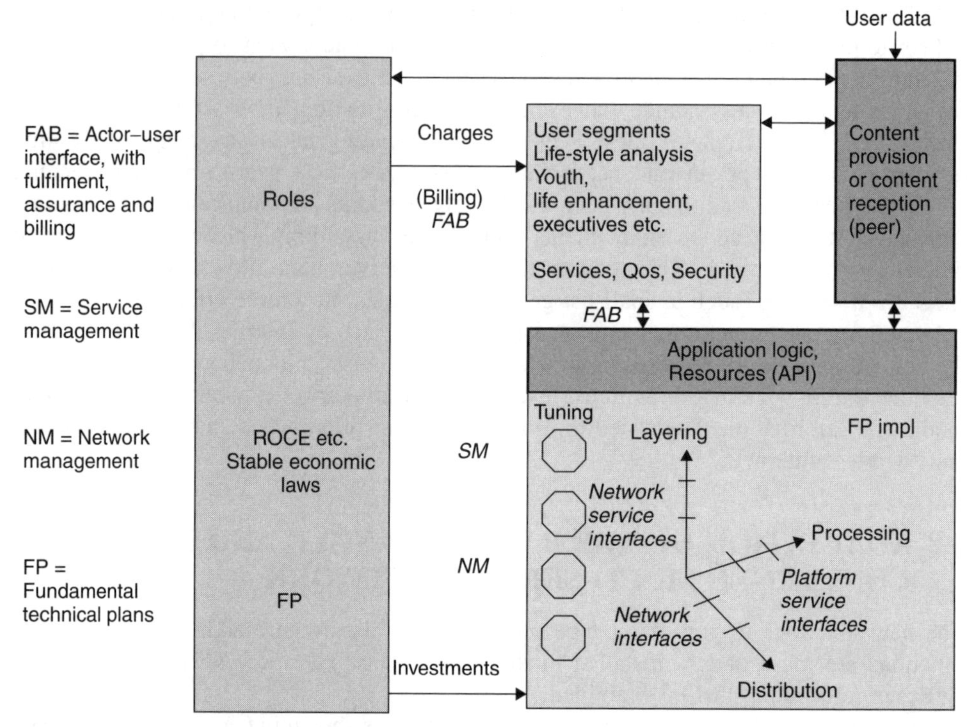

Figure 3.25 A service part is added for client-server communication (the horizontal network will further be presented using a three-dimensional view)

by for example intelligible crosstalk requirements. In the old analogue days an acceptable 'historical' quality of service required amplifiers to make it possible to reach a sufficient voice level without shouting. In the non-layered days of telecom this QoS type of requirement resided in the transmission plan, since the transmission part was the deteriorating part worth mentioning.

Considering the historical background, we only need to make the incorporation of the three plans more formal. A fourth plan may occur here: the charging plan. Initially both the control layer and service layer provide charging. On the management side a service management plan is now vital. It is therefore also time to redraw the classical FP diagram in Figure 3.2. When complemented with 'the new layer' and adapted to the structure of Figure 3.26 it gets a look according to Figure 3.27.

When looking at Figure 3.27 the following view should be applied:

- The figure shows a service factory with a service production line. The service production (= service plan) extends to all layers. All other FPs have a direct or indirect contribution to and impact on the service product.
- Many FPs have QoS features, such as transmission, switching and routing, which impact on the service. See also the Chapter 7. QoS adaptation protocols are provided at the top of the service production line.

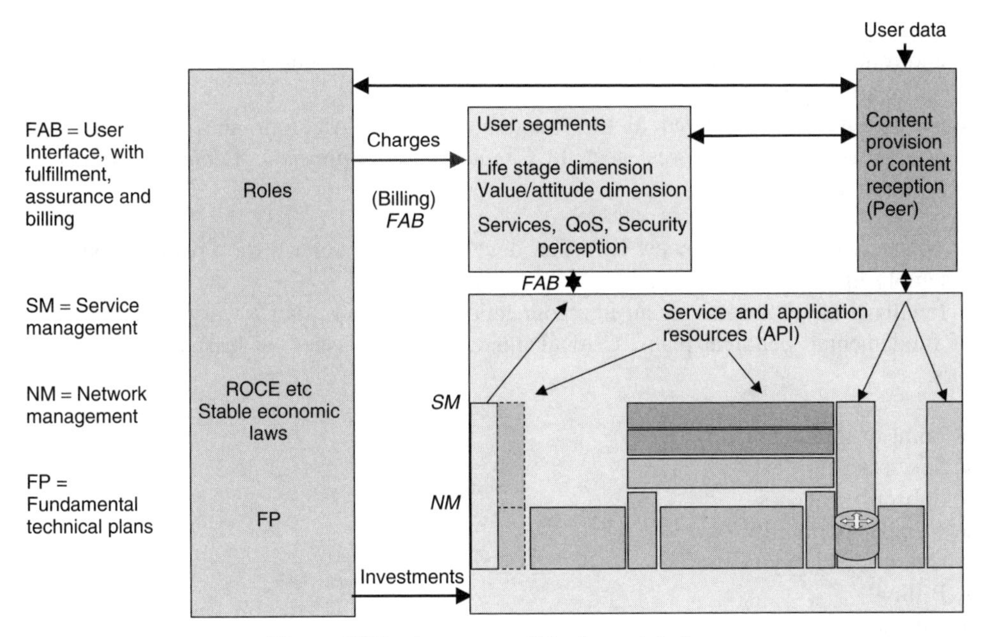

Figure 3.26 A new set of fundamental plans

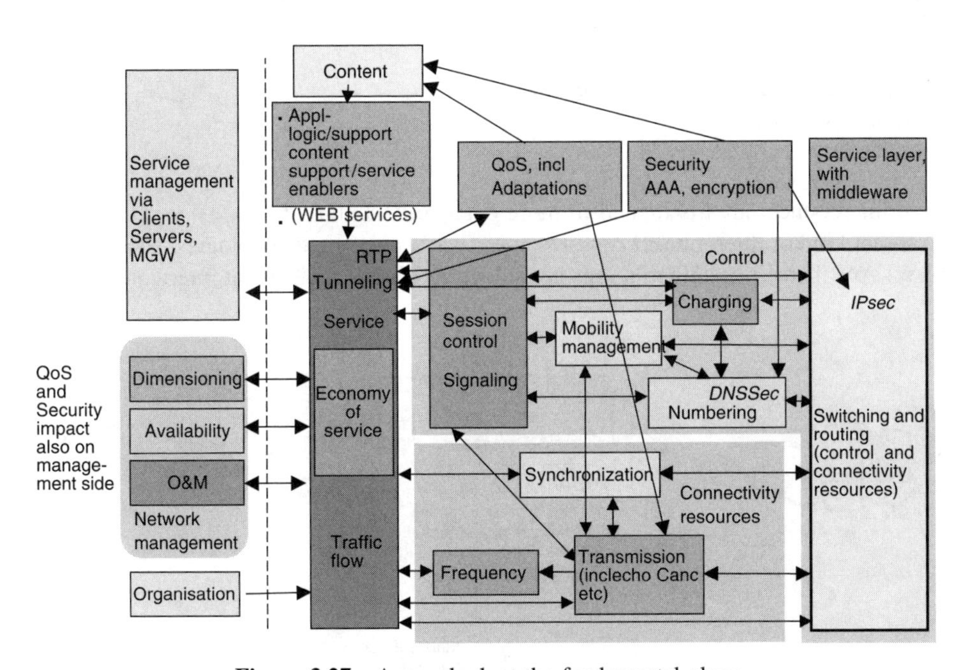

Figure 3.27 A new look at the fundamental plans

- Many plans have security features, because of threats such as attacks on numbering and routing as described in the Chapter 6. Also the management side must be protected (compare Figure 6.10).
- Cost elements are related to the functional content. An optimal solution offers an acceptable service at lowest cost. In Chapter 5 the expression 'Economy of service' is introduced.

This service production view is further developed in Chapter 8 for a packet production assembly line.

Finally there is a need for an interconnection plan. Such plan is very similar to a set of fundamental technical plans. Critical interconnection issues, at least for fixed operators, are:

- Point of interconnect (POI)
- Services
- Subscriber access
- Number portability
- Charges and payments
- Billing
- Numbering
- Signalling and interface standards
- Routing principles (e.g. emergency)
- Synchronization
- Installation, operation and maintenance
- QoS and security
- Service level agreements

The interconnect area is treated in more detail in Chapter 15.

Another kind of interconnect is represented by cooperating technologies, such as UMTS and WLAN. Three possible relations are shown in Figure 3.29. Tight interworking means

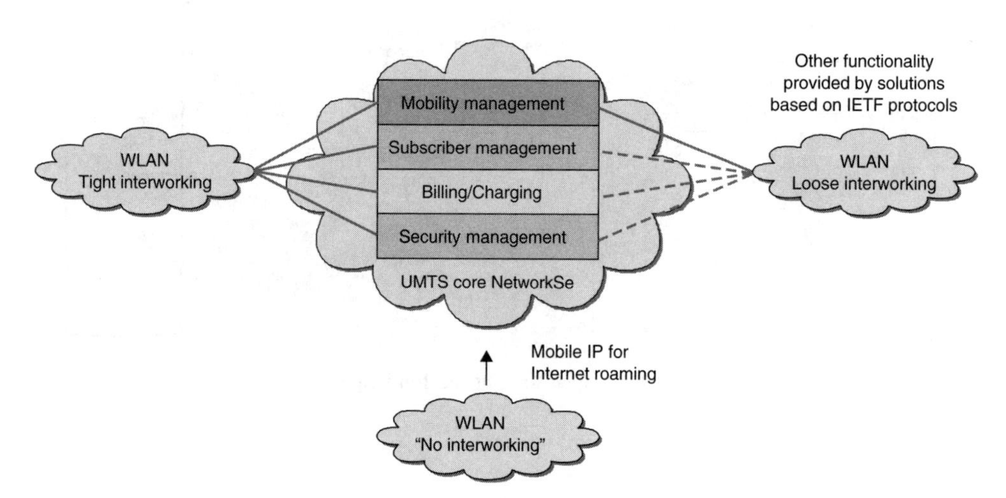

Figure 3.28 Technical interworking

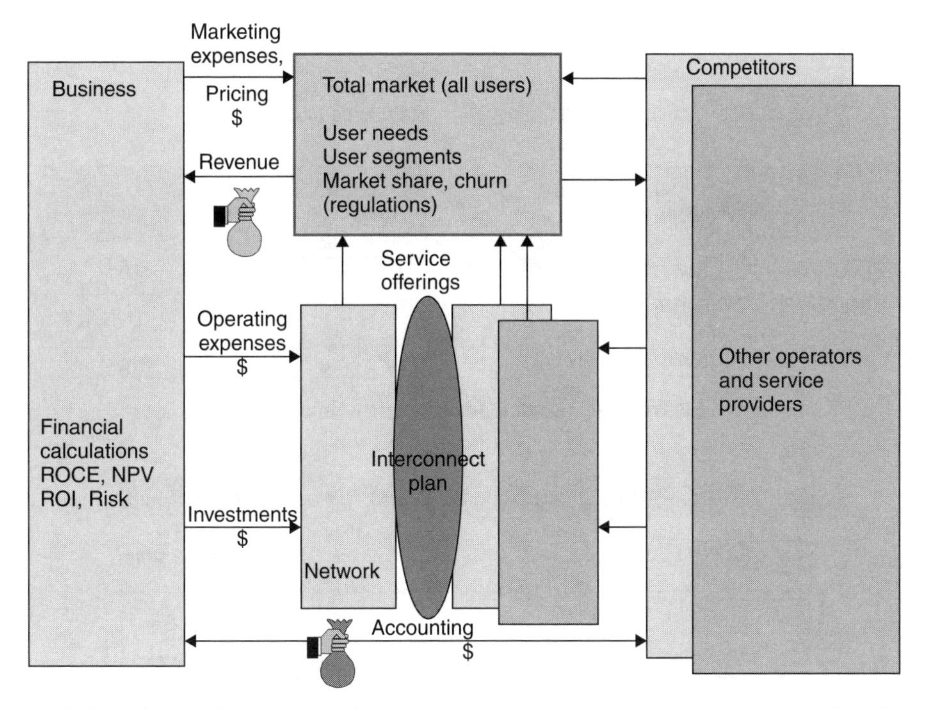

Figure 3.29 The multi-operator environment also requires an interconnect plan which embraces many of the other plans, such as routing, QoS and numbering

that UMTS provides the functions for both systems. Loose interworking means that the technologies work more autonomously.

For the new generation network the following list gives examples of further plans under review:

- Technical framework/Network modelling
- Traffic handling
- Call processing
- Service management
- Call data management

3.9 THE CONNECTIVITY LAYER

The three basic layering models or stacks are shown in Figure 3.30. From a management point of view a minimum number of networks is optimal. The possibility of handling all services in one network should justify the choice of just one network, shouldn't it? The only possible common transfer mode is IP, which is installed in practically all computers.

The demands from voice services and data services are, however, very different and IP was designed for data. IP voice packets become quite different from IP data packets with much overhead (at least before header compression). See Figure 3.31. The answer to the question could be: 'One network eventually, but the convergence from many tailored

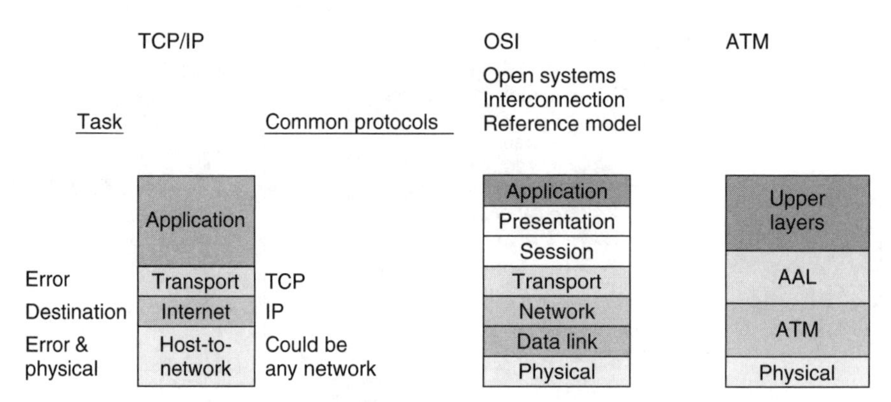

Figure 3.30 Models for layered communication

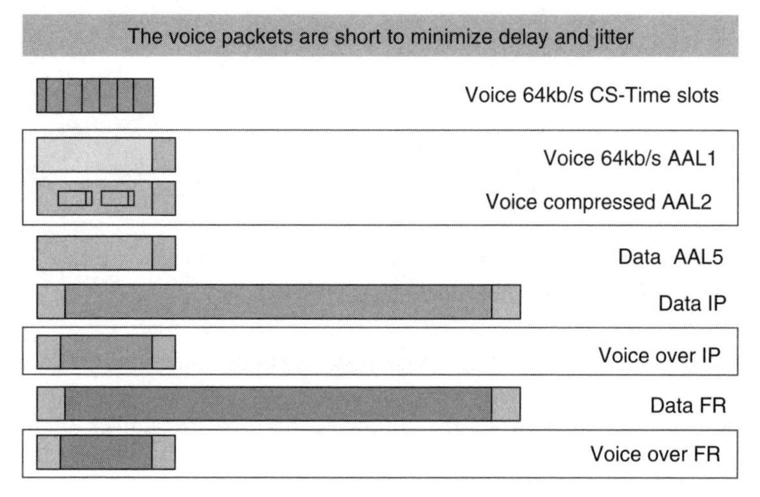

Figure 3.31 Various transfer modes

networks into one network will take time'. See the CS (circuit switched or circuit mode) part in Figure 3.18. In any case protection of investments already made is likely to call for a kind of migration path. Examples of paths are included in the book. See Chapters 8, 13 and Appendix 3.

The uncompressed IP header (especially IPv6) for a voice packet is in reality larger than the payload (IPv6/UDP/RTP 60 bytes, voice payload 32 bytes). For ATM there is no need for a tail since the length is always the same (53 bytes).

With an increasing number of services and fewer networks the adaptation and enabling area becomes a key part from a business point of view.

3.9.1 STRATIFICATION

When using a functional horizontal split of the network, each area represents a unique role or a logical network in the overall network. This role extends over the whole network.

Conceptually, when applying a layered functional approach, the network is dissolved into horizontal slices, one for each functional area. Note that these slices also comprise functionality in customer premises equipment/networks!

Layering opens up for stratification. Stratification means that different technologies with almost the same function in the network can be layered on top of each other. Stratification is commonly used in networks employing ATM in the packet backbone, for transport of TDM, X.25, IP and frame relay network services. In fact one network makes up the infrastructure for one or more other networks. See Figure 3.32 and Figure 3.33.

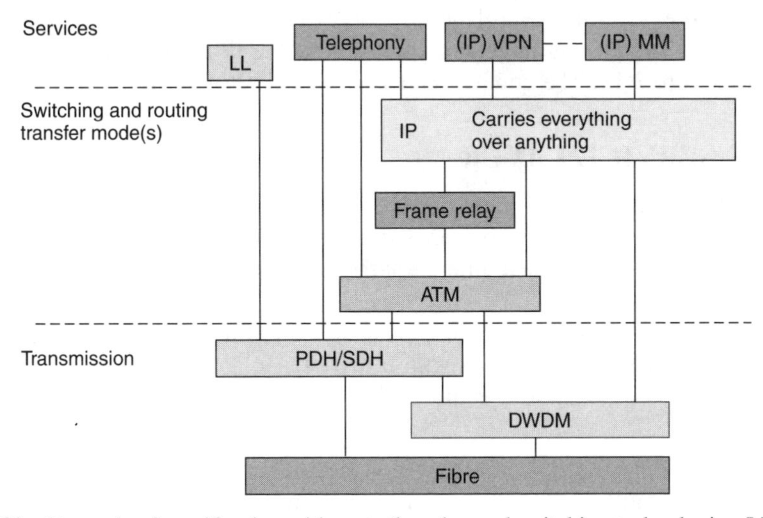

Figure 3.32 Example of stratification with up to three layered switching technologies. LL = leased lines

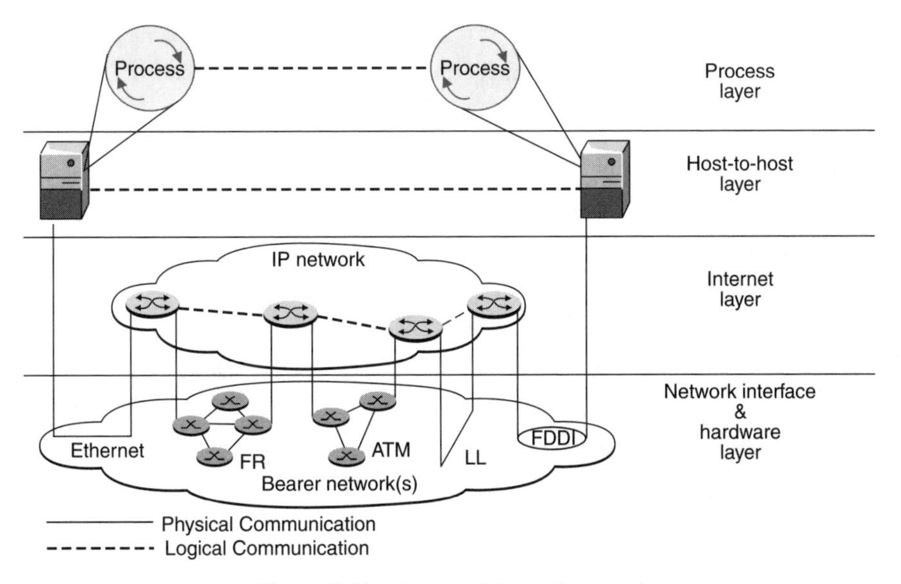

Figure 3.33 A typical layered network

As shown in Figure 3.33 the expression 'Network interface and hardware layer' can be used for underlying connectivity layers in IP stratification.

A similar idea is used for *tunnelling*, a layering and stratification technique for protection of data integrity and confidentiality in IP networks. Tunnels require 'extra' layers to encapsulate the often sensitive end-to-end user traffic. Tunnels are described in the Chapter 6.

3.10 THE CONTROL LAYER

As seen in Figure 3.17 the control network will initially be more heterogeneous than other core layers. The present target control design is the IP multimedia subsystem, IMS, according to Figure 3.18. The control layer is further dealt with in Chapter 14.

3.11 THE SERVICE LAYER

From the convergence point of view the service network has the most central position. See Figure 3.34. An obvious task becomes integrating contents with users, their terminal capabilities and the network capabilities. Another task is to successively support new sets of applications.

The service layer role for service implementation is further dealt with in Chapter 8, and the service network itself is dealt with in Chapter 9.

The service layer responds to a number of needs:

Contents

- Needless to say, multimedia is an important requirement for content.
- The final receivers are the eye, the ear and the computer, which could be a mobile device. For mobile devices it should be possible to update the content according to position. Other changes of content may be synchronization of content with other terminals employed by the user.
- It might/should be possible to download applications to the terminal, in order to offer new content. (Java, .NET).
- It should be possible to receive real-time, streamed or stored content.

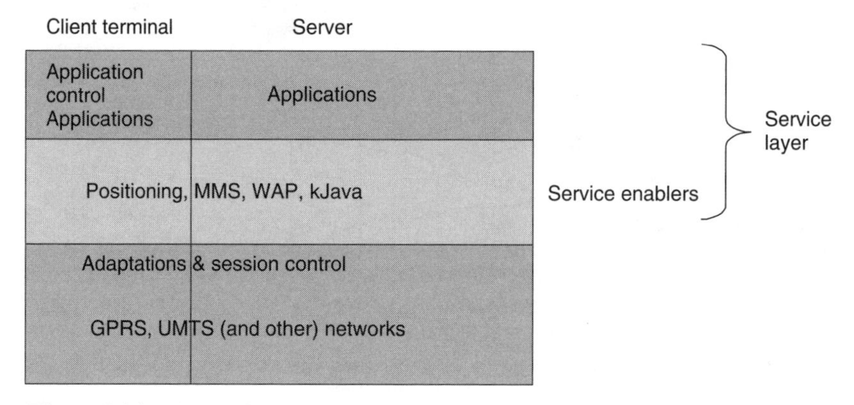

Figure 3.34 The position of service enablers – they also extend to the terminals

Terminal capabilities and requirements

- The content can be coded in different ways, such as AMR, MPEG-4, XML and VML. The receiver must be able to decode the received standard.
- The small presentation format of mobile terminals must be considered.
- Security and QoS protocols might be necessary (tunnels, RTP, RTSP, RTCP).
- Header compression protocols are an advantage, especially for IP calls.
- New signalling protocols may be required in the terminal (SIP, SDP, possibly H.323 type of protocols.)
- Configuration over the air is important, as well as device management.

Network capabilities

- It should be possible to select capabilities in the network that correspond to the desired application demands regarding quality and security. Advanced (IN type) FP control functions are necessary in voice/multimedia type networks.

Solution in brief

- In order to handle client and server-specific features, a functional block is allocated in the client–server interface to the network layer. This block must be able to adapt between fairly unpredictable user needs, interpreted by the application developers, and the network(s). It has another important task, namely as location for various application logics and their associated developer-facing service enablers and a service delivery platform architecture. A corresponding service management function is found on the management side. The service delivery platform or framework contains middleware functionality that can support multiple applications in the mobile devices. Examples of platform functions are network access and device mapping, and device management including configuration over the air.
- A good relationship with a systems integrator is valuable.

Some key enablers are the application programming interface (API), which is explained below, and service enablers. New sets of applications are designed by means of new service enablers. See Chapters 8 and 9.

In order to be flexible in responding to user needs without changing existing devices, flexibility by means of Java (or similar) enabled devices is critical. This represents a processing dimension in the description of the network. As seen from Figure 3.25 the other dimensions are a layered and a distributed dimension.

What is called service management in the book corresponds roughly with the management of the solution above.

3.11.1 APPLICATION PROGRAMMING INTERACES AND SERVICE CAPABILITY SERVERS

The application programming interface (API) is one of the key enablers, which enables the convergence and the creation of the new architecture.

The network resources should be possible to define and access by the application developers. This has led to the introduction of APIs in the network. The first application is service enablers and other service network components with API interfaces.

The API is defined as a set of technology independent interfaces in terms of procedures, events, parameters and their semantics and it is based on distributed computing concepts such as CORBA, Java and other technologies.

A good API makes it easier to develop a program by providing all the building blocks. The programmer puts the blocks together.

While open interfaces at the application level are the prerequisite for an open service architecture, it is important that applications' requests can be mapped appropriately onto the behaviour of the wireless network. Therefore open APIs are also needed at lower layers to enable customizations. Programmable networking aims at opening low-level access to network elements (routers, switches, base stations) by defining appropriate interfaces. These interfaces can then be accessed by various entities (protocols, agents) to offer advanced, customizable and collaborative network services. This concept can then be extended with mechanisms for distributing and executing code, which programs the interfaces on behalf of individual applications (including those running on end systems). The new paradigm has been called active networking. See Figure 3.36. A more telecom-oriented interpretation of API is application interface

Many of the APIs correspond to service capability servers (SCSs). SCSs are service enablers that provide the applications with service capability features, and are abstractions from underlying network functionality. From the viewpoint of applications, an SCS can be seen as a resource gateway to the core network.

Examples of service capability features offered by the SCSs are call control and user location. A common standard that provides a model of the network capabilities is called Parlay. See Figure 3.35.

The 3GPP is committed to open interfaces for SIM cards with defined APIs making it possible for application developers and network operators to develop services.

In an optimal situation the programmer has API access to all independent parts. See for example Figure 3.36.

The applications should thus be separated from the core network, leading to 'service convergence' at application level. SCS/service enablers with a resource management

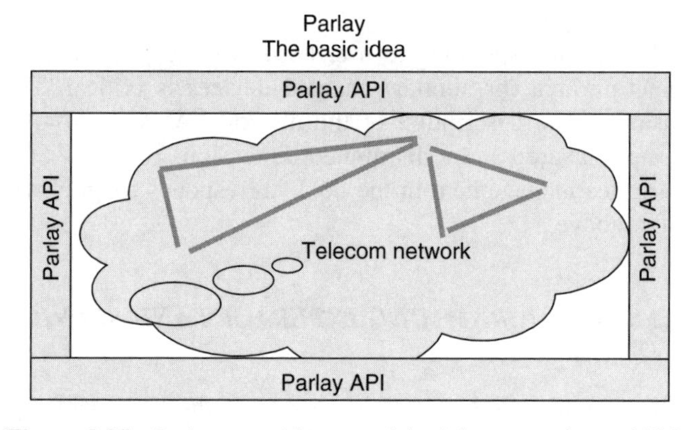

Figure 3.35 Parlay: provides a model of the network capabilities

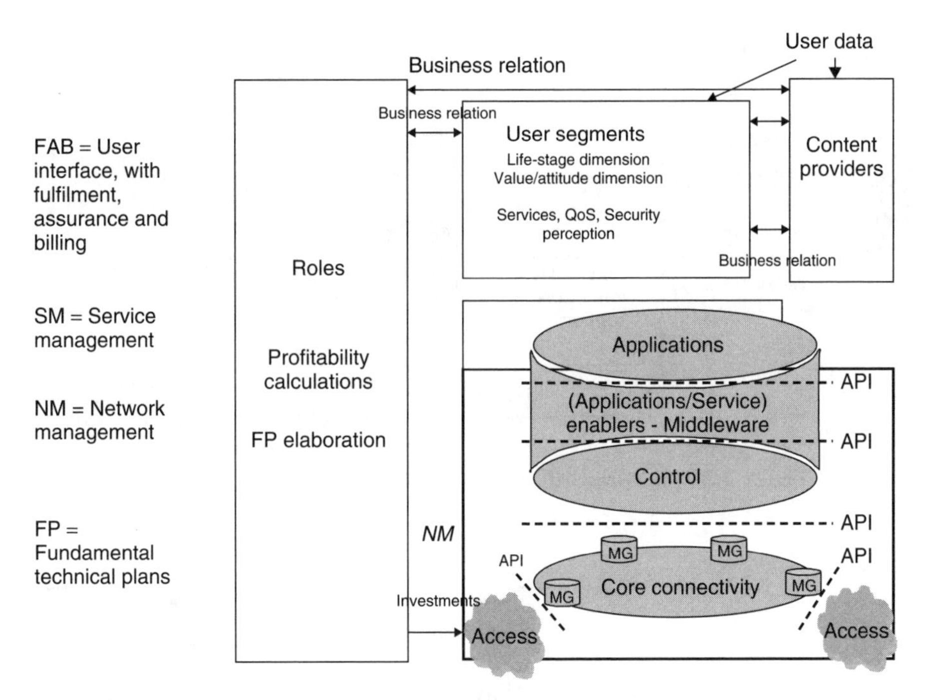

Figure 3.36 Possible API interfaces in the future network

function cater for the access to various network resources. See also the Chapter 9. The application level then becomes the 'melting pot' for all types of IP services/service combinations, also providing an 'end-user personalized service environment' independent of access type.

3.11.2 MIDDLEWARE

Another important part of the solution is the re-use of *middleware* for different applications. Like service enablers it is positioned roughly between applications and the network connectivity. See Figure 3.37. This area is one of the key success areas. The term can be used in many contexts. For example, for B2B relationships such as e-commerce, the XML (eXtensible Markup Language) middleware may become the standard format. See Figure 3.38.

IP is not middleware, rather an overlay technique at the top of the connectivity network. But the impact is similar to middleware: global networking across different underlying networks and a platform for applications.

There are several core components in the service layer that can be reused in many solutions. The middleware ties terminals, networks and applications together.

The middleware can vary in functionality from 'total middleware' making life simple but inflexible for application designers, to development studios with more work and more flexibility for the designers. For services delivery, the strong direct dependencies between the terminals and a service network are clear, and needs of combined application development for these are obvious.

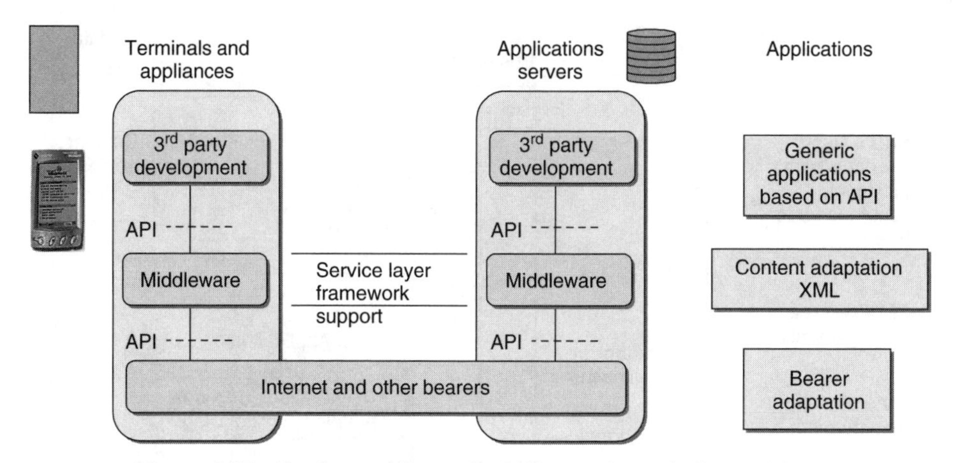

Figure 3.37 Service enablers and middleware have similar positions

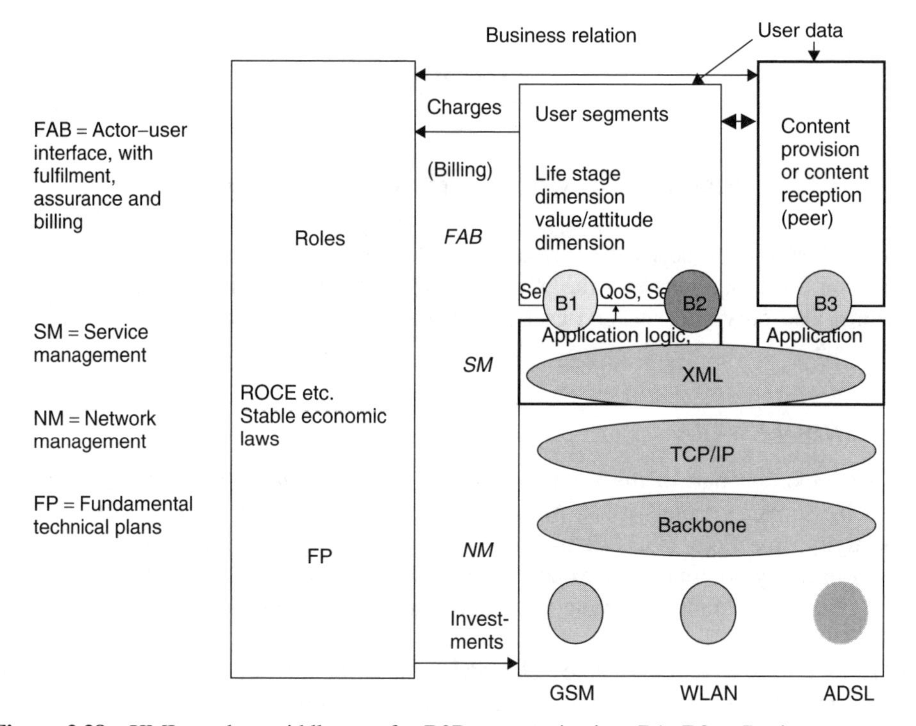

Figure 3.38 XML used as middleware for B2B communication. B1–B3 = Business processes

3.11.3 LAYERING DRAWBACKS

When trying to satisfy everyone's needs the result can be that nobody is happy. In optimized vertical networks with a minimum of overhead and a firm control over the whole stack, the quality tends to be better. Another problem is that with open horizontal layers there will be no differentiation among the actors. Therefore there is also a resistance, especially in the service layer. Considerable layering also initiates management challenges. See for example Chapter 12, which presents a layer interconnect protocol called GMPLS.

3.12 THE DISTRIBUTED NETWORK DIMENSION

The distributed dimension of the network, for example according to Figure 3.41, must be designed for a multi-service offering with end-to-end security and QoS. Further the deregulated multi-operator environment requires agility and flexibility and a number of interconnection points. Interworking between nodes is provided by means of control signalling or management/built-in control.

3.12.1 ENABLER EXAMPLES: SIP, SDP

The *session initiation protocol* (SIP) is an application layer protocol. It is also a signalling protocol for setting up sessions between clients over a network, i.e. the Internet. These sessions do not necessarily have to be Internet telephony sessions. SIP could just as well be used for setting up gaming sessions or for distance learning where a lecture is streamed out to the participants.

SIP can be transported over ATM, frame relay and X.25 as alternatives to TCP/IP.

The service description protocol (SDP) communicates several session descriptors:

- Session name and purpose
- Media type (e.g. audio, video etc.)
- Media transport protocol (RTP/UDP/IP, H.320 etc.)
- Media format (e.g. for video, H.261 or MPEG, etc.)
- Information on how to receive media (e.g. address, port, format etc.)
- Period of time the session is active

SIP and SDP are described in more detail in Chapter 14.

From Figure 3.39 and Figure 3.41, the terminal devices and the service network are at the top. The connectivity is split into access and core connectivity, since there are so many access types. On top of the backbone there is a control network. In the border access backbone an edge node is inserted. On the border to another operator another edge node is inserted. This distributed view is very useful to show particular implementations, in Figure 3.39 a future IN implementation. Although IN basically belongs to the control layer, it can be implemented like an application with application logic = IN logic, a network resource gateway, service capability servers from the service layer, and suitable interfaces according to the figure. The control servers are the controlled elements in this case.

The server in Figure 3.40 belongs to another network. An 'own' server should be included, but it is not very easy to find a good position. As an example it could be inserted close to the edge node to other operators. See Figures 3.41 and 3.42.

The present situation with quite a number of networks in operation is a kind of nightmare for the operators. And more networks are appearing. WLAN is already fairly widespread and 3G appears on the mobile side. The number and possible types of interconnection between different networks become quite large. This situation generates a need for network elements with one network interface on one side and another network interface on

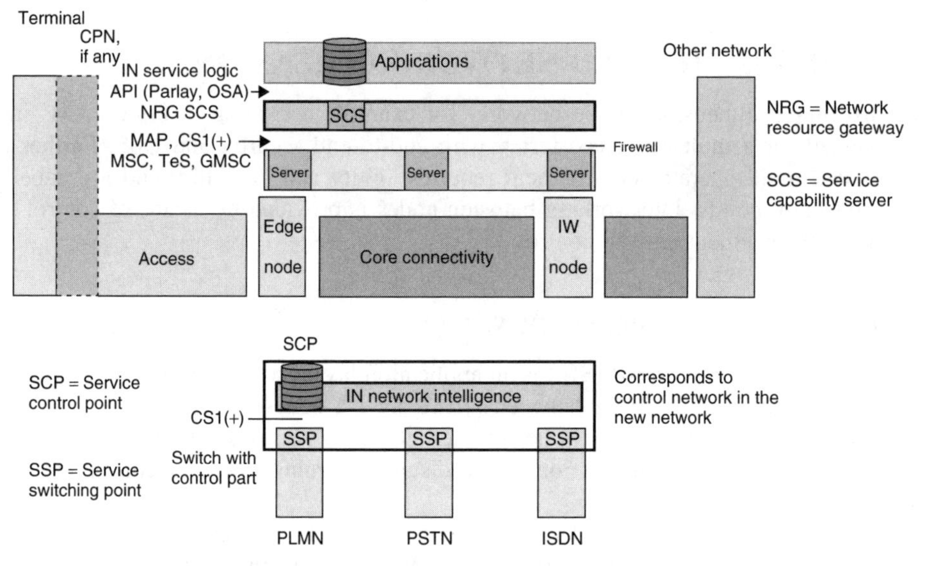

Figure 3.39 A distributed view

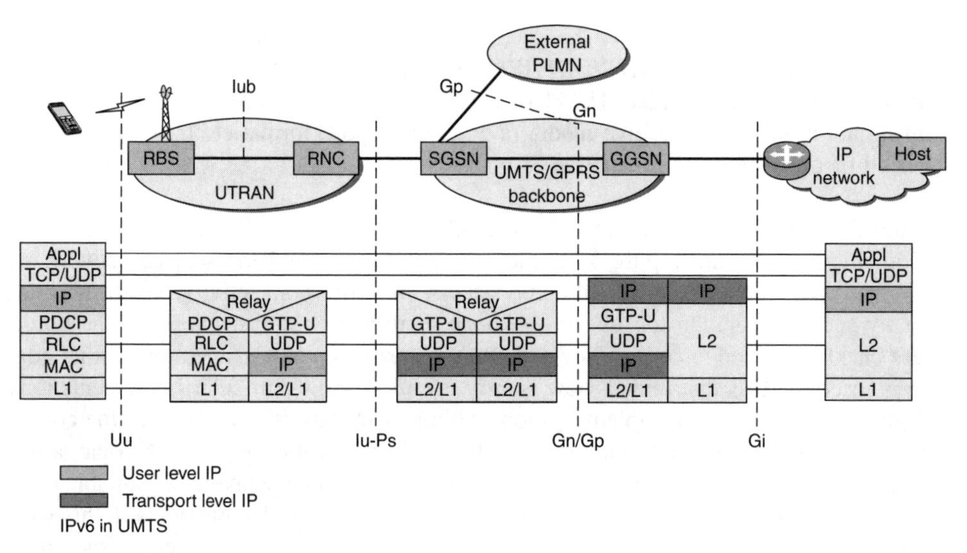

Figure 3.40 Example of server connection. The IP service network is normally on top of the core network SGSN-GGSN

the other, that is to say *gateways*. The gateways tend to attain a key 'spider' role in many types of networks. The gateways are also denoted edge nodes and interworking nodes.

The nodes enable a successive migration from the networks of today to the horizontal network. A typical situation is the combination traditional 'vertical' access and horizontal

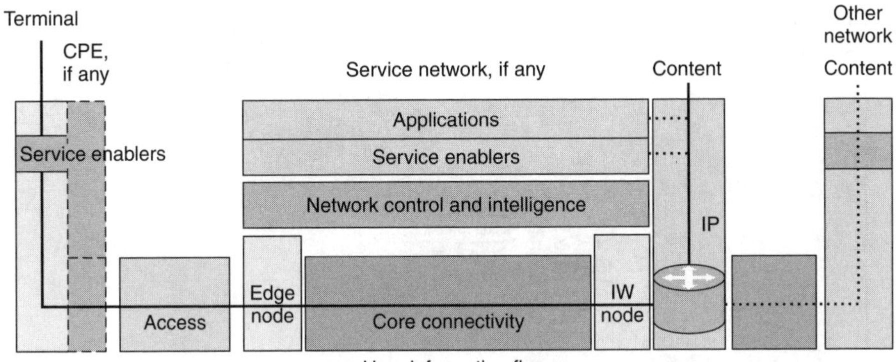

Figure 3.41 The server inserted at the 'outer' edge node

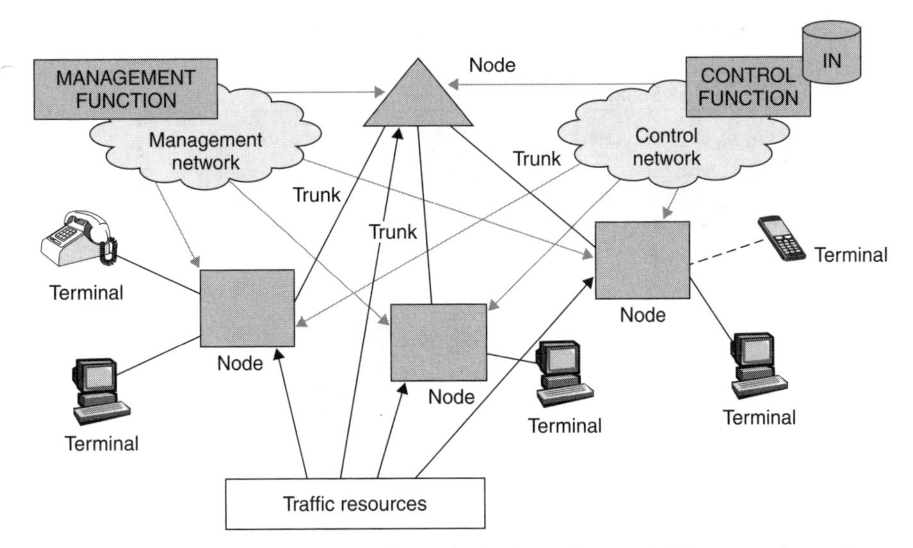

Figure 3.42 A network element based view of a horizontal network. The control parts have now been split from the nodes and located in a control network

core. In UMTS the opposite is foreseen during an initial period, until the core upgrade can catch up with the new layered access in UMTS.

Of course the new network can also be described by means of a network element oriented view. See Figure 3.42, which gives an example. We leave to the reader to conclude what view fits the horizontal network best.

The new network is sometimes called the next generation network (NGN), when talking about network architectures in this book.

3.12.2 THE BANDWIDTH BOTTLENECK

The distributed dimension contains the expensive access part. In addition access is the bandwidth bottleneck. However, the role as a bottleneck can be reduced with new access

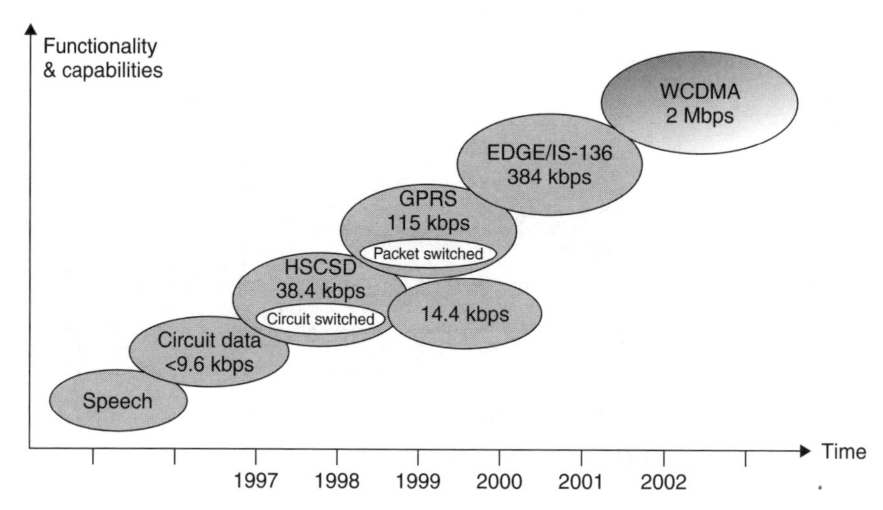

Figure 3.43 Bandwidth evolution for GSM-type systems

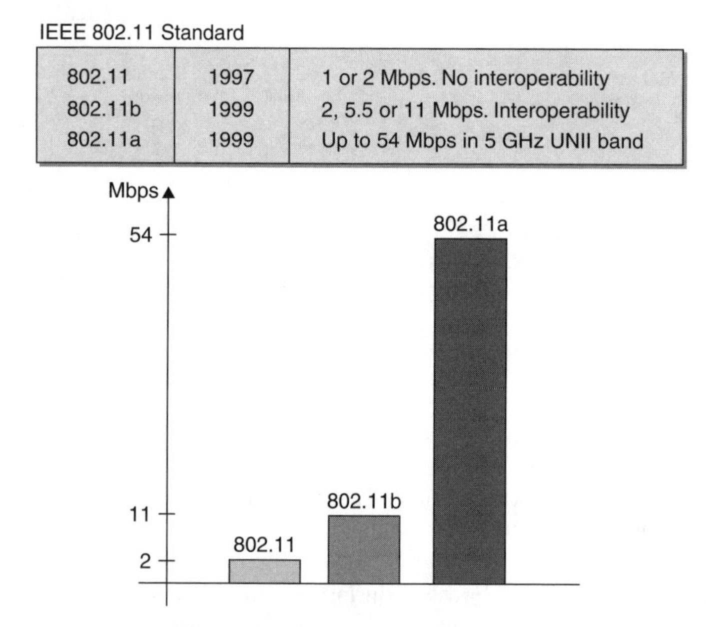

IEEE 802.11 Standard

802.11	1997	1 or 2 Mbps. No interoperability
802.11b	1999	2, 5.5 or 11 Mbps. Interoperability
802.11a	1999	Up to 54 Mbps in 5 GHz UNII band

Figure 3.44 WLAN access bit rates

systems, such as 3G, WLAN, VDSL and later 4G. Cost of provisioning wide-area bandwidth is (almost) directly proportional to the bandwidth. On the other hand the willingness to pay for bandwidth is hardly proportional to the bandwidth. This is the dilemma of the access network provider.

For mobile GSM-type systems the bandwidth development is given in Figure 3.43.

For WLAN systems the various bit rates for some main standards are shown in Figure 3.44. A further development of 802.11b in the 2.4 GHz band (802.11g) provides up to 54 Mb/s. The WLAN technology is shown schematically in Figure 3.45.

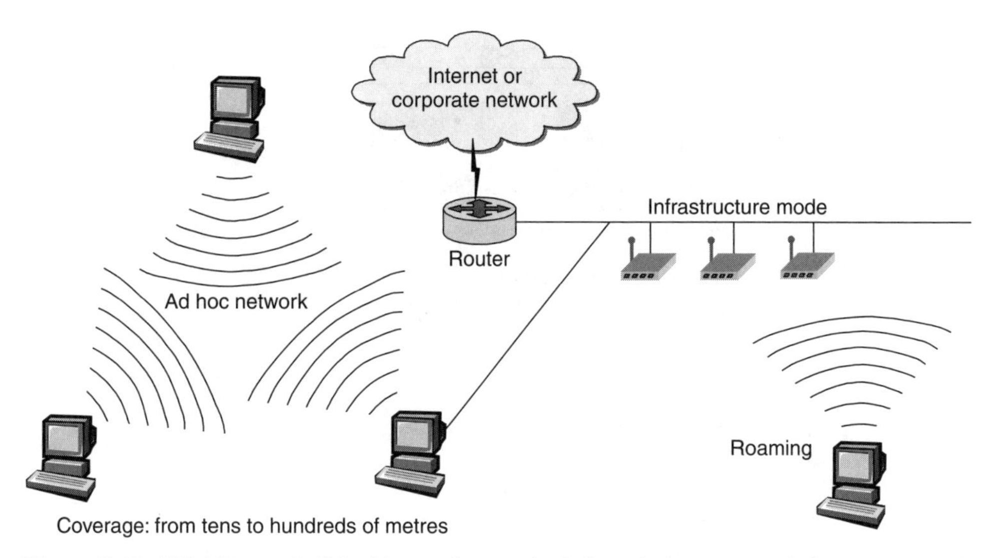

Figure 3.45 WLAN as a LAN without wires to the left, and as an access infrastructure to the Internet to the right

Table 3.2 VDSL speeds

Upstream	Downstream	Distance
1.6–2.3 Mbit/s	13 Mbit/s	1500 m
	26 Mbit/s	1000 m
	52 Mbit/s	300 m

3.12.3 VDSL

The fastest established technique over copper at present is the VDSL, 'Very-high-bit rate DSL'. VDSL is just as ADSL based on the use of advanced modems.

For symmetric use the full bit rate is 26 Mbit/s in each direction. For asymmetric use there are different speeds downstream and upstream for example according to Table 3.2.

3.13 THE PROCESSING DIMENSION

Ever since the introduction of stored program control (SPC), processor dimensioning is a common, important task within telecommunications. An example is given in Appendix 1.

The introduction of web services in for example a Java or .NET environment opens new possibilities. In future, rather than software components being developed and bound together to form a single rigid solution, systems will be developed as a 'federation' of services at the point of execution. When it becomes possible to utilize services to create new ones the power of networking will be fully exploited. This will enable alternative software components to be substituted between each use of a system.

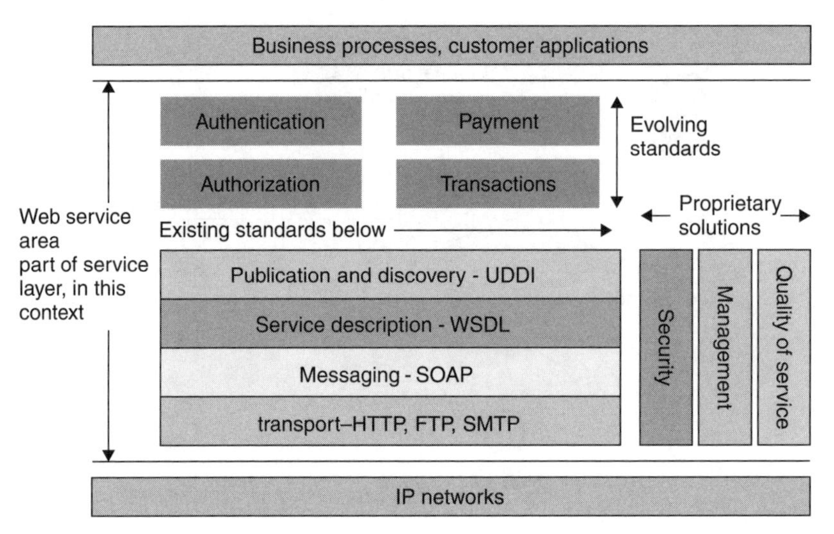

Figure 3.46 Web services standardization

Web services perform functions, which can be anything from simple requests to complicated business processes. Once a web service is deployed and published, other applications (and other web services) can discover and invoke the deployed service. The main web service area in the context of this book is shown in Figure 3.46. Clearly, it is positioned close to service enablers and middleware. This makes the area above IP a dynamic one, indeed.

Examples of component services that are reusable building blocks are currency conversion, language translation, shipping and claims processing.

The new era can be called 'software as a service', using service-oriented programming (SOP). SOP is a paradigm for distributed computing that supplements object oriented programming (OOP). Whereas OOP focuses on what things are and how they are constructed, SOP focuses on what things can do.

The way to work is roughly like this (for web services):

- **The developer –** Create application logic, generate a service description and register it in the publication and discovery library.
- **The user –** Call the service from the discovery library. The software is then transported as a simple XML-based protocol to let applications exchange information over HTTP. Insert the application logic into a process flow.

To boost the use of web services there is a need for standards (or possibly proprietary solutions) for AAA, payments, QoS, security and management.

3.13.1 SIM APPLICATION TOOLKIT (SAT)

The SAT function (SCS) is used for downloading applications on the SIM card in the mobile terminal.

3.13.2 TRANSPORT

The hypertext transport protocol (HTTP), which is used for browsing applications on the Internet, provides the semantics of requesting and transferring information between servers and clients in a distributed and collaborative way. Most often, the information requested from the server consists of HTML objects. However, HTTP is not restricted to HTML – essentially any information can be requested and transferred between the client and the server, including media objects. HTTP is used on top of TCP.

3.13.3 AGENT TECHNOLOGY

Intelligent software agent technology is one of the most important emerging IT technologies of the last years. The technology plays the role of an important enabler for highly distributed and complex solutions, self-organizing and collaborative systems. In particular agent technology is predestined for the integration of web services being supported by suppliers.

3.13.4 RELIABILITY ISSUES

Future networked systems will run applications whose behavior cannot always be anticipated, and this is a problem that did not exist in traditional telecom. These applications are composed by dynamically composed services which may have bugs that in addition run in an unpredictable environment.

3.14 KEY ENABLERS

3.14.1 H.248

(H 248. ITU 1999. Megaco1999. Final approval of a common protocol 2000. IETF RFC2885, RFC2886.)

H.248 is a new protocol that enables a layering split between connectivity and control. See Figure 3.47. We are used to 'horizontal' protocols for signalling, management and other purposes, but this is a master–slave 'vertical' type of protocol. The horizontal protocols might then be called peer-to-peer protocols. A master/slave model permits short time to market. Gateways can be dumb.

By means of H.248 a suitable number of control nodes can control switching nodes in the connectivity network. The control nodes could be arranged in pairs for increased reliability. See Figure 3.48. Alternatively, one switching point can be controlled by more than one server, provided that the networks utilize the same technology in the connectivity layer.

The control nodes do not serve all nodes in the connectivity network. The served nodes are edge nodes (towards access) and interworking nodes (towards other networks).

Especially in tele-centric networks the nodes are called media gateways. For a call set up the communication control part handles the 'call' phase and the connectivity part handles the 'connect' phase. By such a separation the connectivity becomes more independent of the network type.

H.248 defines a connection model, which is a central concept for describing the logical entities within the media gateway that can be controlled by the server. Thanks to this model, different transmission media can coexist, and media streams can be processed in the connection. H.248 allows an authenticated server to establish, move, modify, remove and obtain events that have been reported on a connection or group of connections. A server can audit the media gateway to determine the extent of its capabilities.

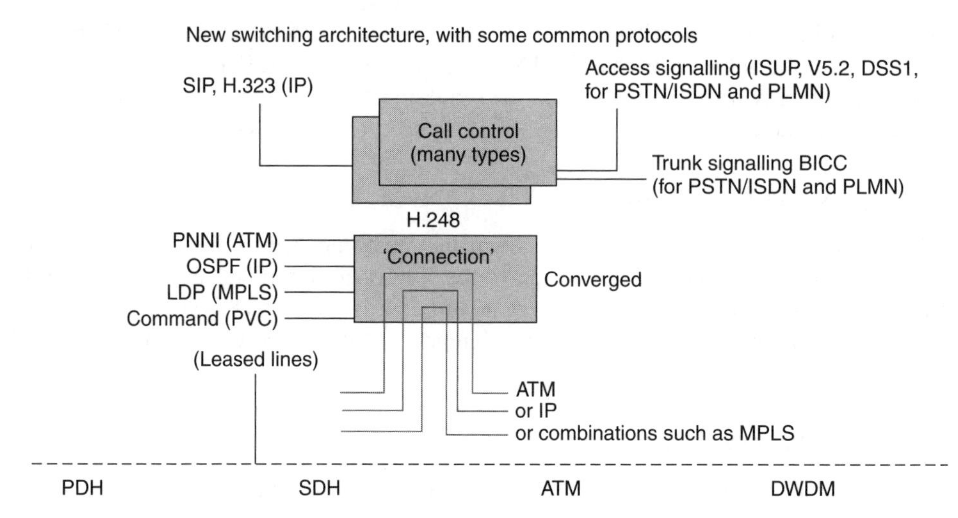

Figure 3.47 By means of H.248 call and connect functions can be split. A horizontal interface appears

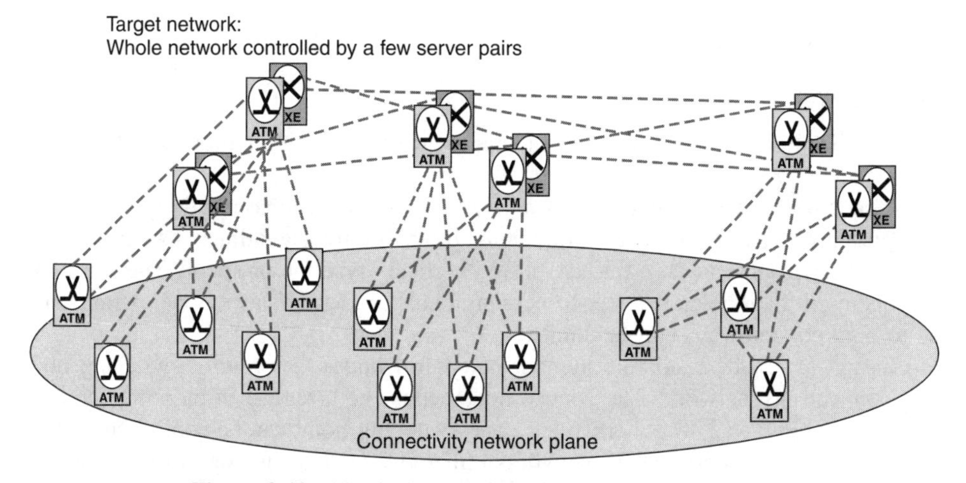

Figure 3.48 Physical separation of call and connect nodes

Communication control can comprise both peer-to-peer and master/slave elements, with each type of approach used where it is optimal. (Peer-to-peer models are SIP, H.323 and master/slave models are H.248/Megaco, MGCP. Compare the IN model, which is also kind of master/slave.)

Call control is contained in the functional control servers (master), which also have peer-level protocols to interact with other functional elements in the system, and drive a large number of slave devices.

H.248 is developed jointly by IETF and ITU, based on simplicity, flexibility and cost effectiveness. See also Chapter 11 on edge nodes.

H.248 uses a simple, powerful connection/resource model to describe the logical entities or objects within the media gateway that can be controlled by the MGC. It is fundamentally based on two key concepts: termination and context. Terminations identify media flows or resources, implement signals, generate events, have properties and maintain statistics. All signals, events, properties and statistics are defined in packages, which are associated with the individual terminations. Controlled terminations are for example PSTN trunking gateways, ATM interface, analogue line, Internet telephones and announcement servers.

3.14.2 IPv6

The initial IPv4 numbering design for IP networks did not anticipate the following:

- The recent exponential growth of the Internet and the impending exhaustion of the IPv4 address space.
- The growth of the Internet and the ability of Internet backbone routers to maintain large routing tables.
- The need for simpler configuration.
- The requirement for security at the IP level.

To address these concerns, the Internet Engineering Task Force (IETF) has developed a suite of protocols and standards known as IP version 6 (IPv6). New services, such as multimedia, require globally unique addressing. IPv6, with its very large address space will guarantee a unique IP address for each device. The characteristics of IPv6 can be expressed and mapped by means of the FP concept.

Numbering Plan

- Hierarchical addresses.
- New services, such as multimedia, require globally unique addressing. IPv6, with its very large address space (over 3.4×10^{38} possible combinations) will guarantee a unique IP address for each device. IP addresses will also be allocated to interplanetary traffic.
- Network address translation is not necessary (IPv4 addresses have become relatively scarce, forcing some organizations to use a network address translator (NAT) to map multiple private addresses to a single public IP address. While NATs promote reuse of the private address space, they do not support standards-based network layer security or the correct mapping of all higher layer protocols and can create problems when connecting two organizations that use the private address space.)
- IPv4 headers and IPv6 headers are not interoperable. A host or router must use an implementation of both IPv4 and IPv6 in order to recognize and process both header formats. The new IPv6 header is only twice as large as the IPv4 header, even though IPv6 addresses are four times as large as IPv4 addresses.

IPv6 has a 128-bit address space	3×10^{38} addresses
IPv4 has a 32-bit address space	4 billion addresses

Security Plan

- Private communication over a public medium like the Internet requires encryption services that protect the data being sent from being viewed or modified in transit.

Although a standard now exists for providing security for IPv4 packets (known as Internet Protocol security or IPSec), this standard is optional and proprietary solutions are prevalent.
- Support for IPSec is an IPv6 protocol suite requirement. This requirement provides a standards-based solution for network security needs and promotes interoperability between different IPv6 implementations.

Mobility Management

- Built-in mobility support (MIPv6).

Routing Plan

- Because of the way that IPv4 network ID numbers have been and are currently allocated, there are routinely over 70,000 routes in the routing table of the Internet backbone routers. The current IPv4 Internet routing infrastructure is a combination of both flat and hierarchical routing.
- In IPv6 more hierarchical routing enables shorter routing tables.

QoS Plan

- There is a need for better support for real-time delivery of data – also called quality of service (QoS).
- While standards for QoS exist for IPv4, real-time traffic support relies on the IPv4 type of service (TOS) field and the identification of the payload, typically using a UDP or TCP port. Unfortunately, the IPv4 TOS field has limited functionality and over time there were various local interpretations. In addition, payload identification using a TCP and UDP port is not possible when the IPv4 packet payload is encrypted.
- New fields in the IPv6 header define how traffic is handled and identified. Traffic identification using a flow label field in the IPv6 header allows routers to identify and provide special handling for packets belonging to a flow, a series of packets between a source and destination. Because the traffic is identified in the IPv6 header, support for QoS can be achieved even when the packet payload is encrypted through IPSec.

Configuration

- Most current IPv4 implementations must be either manually configured or use an address configuration protocol such as dynamic host configuration protocol (DHCP). With more computers and devices using IP, there is a need for a simpler and more automatic configuration of addresses and other configuration settings that do not rely on the administration of a DHCP infrastructure.
- Hosts on a link can automatically configure themselves with IPv6 addresses for the link (called link-local addresses) and with addresses derived from prefixes advertised by local routers.

Management Plan

- Features for simplified network management.

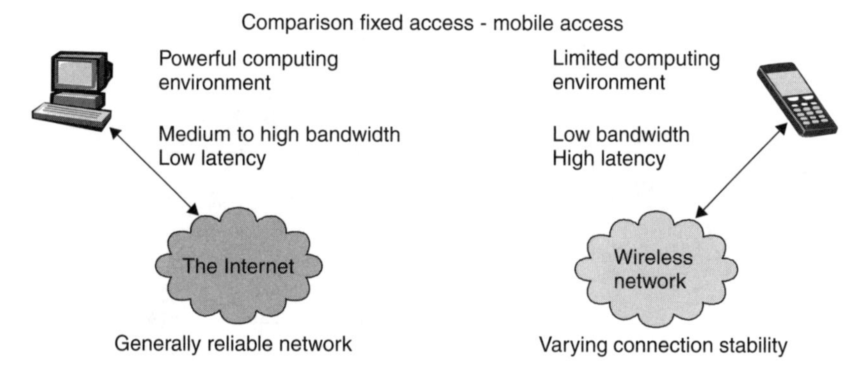

Figure 3.49 The initial view of computing in mobile devices

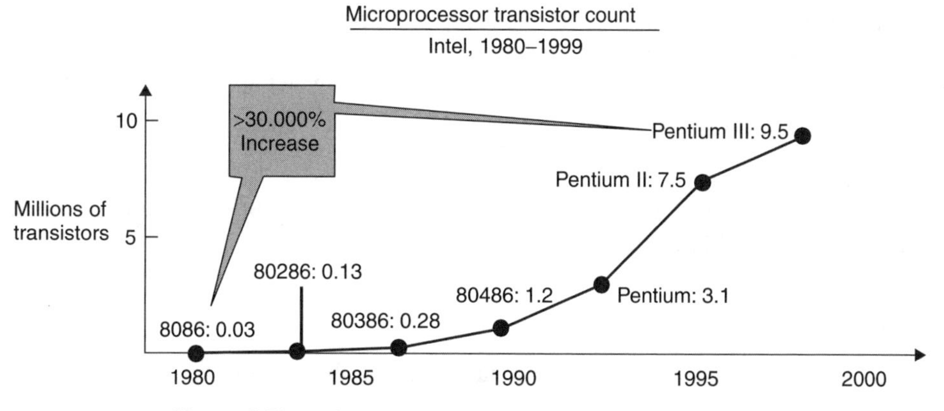

Figure 3.50 Microprocessor transistor density development

3.15 GENERAL ENABLER DEVELOPMENT

Technology development enables intelligent terminal devices. Moore's law says that the capability of electronic circuits doubles every 18 months. The processor power doubles at present faster than Moore's law, enhancing extended functionality in devices. As seen in Figure 3.49 this is crucial for mobile communication.

The basic technological advances can be illustrated with the development of memory cost and microprocessor transistor count. See Figures 3.50 and 3.51. The development offers increased processing power and possibilities for downloading applications and increased storing in the devices.

It also offers drawbacks: To combat computer-related crime the police sometimes copy the computer disk of a suspect. With local disks with capacities exceeding 100 Gb this job takes many hours.

3.16 ENABLER OVERVIEW

Major support for the development is provided by a number of new standards and other technological development.

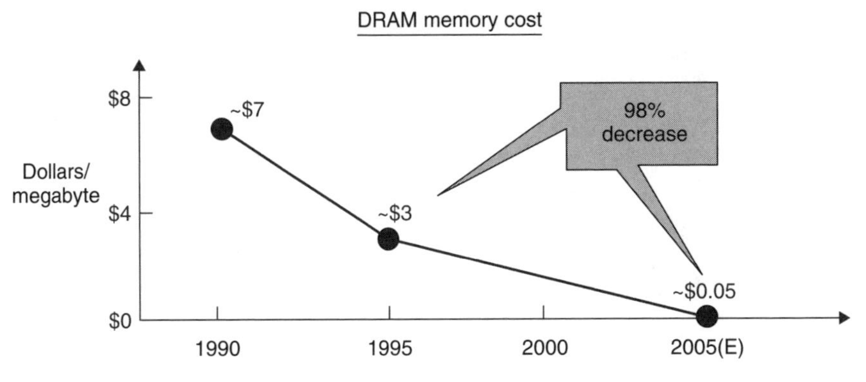

Figure 3.51 Memory cost development

Many enabler explanations can be found on the Internet.

The list only includes major new or fairly new enablers. The overview is included as a service to the reader. Such list is very hard to keep complete and updated. Two dimensions are used:

1. What happens within the IP area.
2. What happens within the various main network building blocks, according to the segmentation used in this book.

Chapter 3

The main developments respond to FP requirements at the network level:

- IPv6 for an extended/new numbering plan and some other improvements, such as security.
- Mobile IP for numbering in IP networks.
- New standards for VoIP framing to reduce overhead.
- MPLS – A standardized technology that provides connection-oriented switching based on IP routing protocols and labelling of data packets.
- Diffserv, RSVP for IP QoS.
- Sigtran – A protocol architecture for SS7 signalling transfer using IP as a transport technology (BICC/ISUP, MAP, etc. over IP).

Level 3: IP routers and combinations with other technologies – Below is a series of protocol stacks from real routers, most of them not dealt with in the book:

- Routing protocol: OSPF, RIP/RIPv2, EIGRP, ISIS, BGP(4), DVMRP, PIM-D, PIM-S, MBGP, MPLS, IGMP, IPX, PPTP, IP policy routing, NAT, Appletalk, SW and HW routing.
- Management protocols: SNMP, RMON, HTTP, Telnet, CLI, TACACS+, Radius, TFTP, RCP, LDAP client for policy-based management, built-in DHCP server, dynamic DNS, embedded browser interface.

- Connection: OSI 1–2: (roughly transfer mode): 10/100/1000 Ethernet, ATM, frame relay, X.25, ISDN, POSIP, DPT, FDDI, channelized E1, SMDS, STM-1, STM-4, PPP, HDLC, token ring, RS232, X.21, V35.

Chapters 5 and 9

- **AMR** for adaptable voice coding.
- **MPEG** for coding of moving pictures. Each MPEG encoder must compress, and each decoder must decompress both audio and video.
- **JPEG** 2000 for coding of still pictures.
- **WAP** for adaptation of mobile terminal units to the Internet.
- **API** for fast creation of new applications An API is a set of routines, protocols and tools for building software applications. A good API makes it easier to develop a program by providing all the building blocks. A programmer puts the blocks together.
- Positioning, GPS global positioning system.
- Messaging over IP protocols.
- Parlay versus JAIN. See Table 3.3.
- **JAIN** – Java API for integrated networks. It is an interface that provides a uniform interface to wireless systems, traditional Internet access, PSTN/ISDN and ATM systems. JAIN creates Java technology based APIs which enable the rapid development of products and services on the Java platform. The JAIN initiative integrates fixed, wireless and packet-based networks.
  The JAIN network topology provides carriers with the ability to deploy services on devices inside or at the edge of the converged network, including any Java technology enabled end-user device. Furthermore, support for all the necessary telephony protocols that are used between the different network elements in IN, and IP-based (telephony) networks is mandatory (MAP, TCAP, INAP, ISUP, SIP). A key aspect of the JAIN component architecture is to move the signalling layer away from proprietary switches to MGCs.

The API defines a programming interface to the converged networks in terms of an abstract, object-oriented specification. As such it is designed to hide the details of the specifics of the underlying network architecture and protocols from the application programmer to the extent possible.

Table 3.3 Parlay vs. JAIN

Parlay	JAIN
A generic model of a telecom network	A software component factory
Provides a model of the network capabilities (and means to manipulate the model)	Provides a set of conventions for building software components for the telecom industry
Independent of the network internals	Can be used by both network and service developers
Meant to be used by service developers	Both network software (inside the network) and applications
Services using Parlay are meant to reside outside the network trust boundaries	Software (outside the network) can be developed using JAIN components

- **MP-3 –** MPEG-1, layer 3. A standard for compression and storage of large sound files.
- **OSA –** Open systems architecture. A concept defined by 3GPP and Parlay. It specifies services available to applications on application servers over standardized programmable interfaces.
- **XML –** Used for exchange of data between applications. It is a markup language much like HTML. XML is designed to describe data in a universal format on the web. XML is used for documents containing structured information. It enables operations to be performed on content. New languages can thus be defined more easily. Moreover SOAP, simply speaking remote procedure call over XML, is predestined to overcome the different component-based architectures and supporting open infrastructures.
- **XHTML –** HTML reformulated as an XML application.
- E-commerce standards.
- **CORBA –** Common object request broker architecture. A specification defining how programmers describe interfaces for cooperating objects. The objects may have different origins regarding programming language and operating system.

Chapter 6

- IPsec.
- LDAP.
- MIME, MPLS VPNs, VPN/IPsec with encryption, firewall services.

Chapter 7

- MPLS, Diffserv, RSVP, TOS based on user, application, time etc., Intserv, MPLS with RSVP traffic engineering.

Chapter 8

VoIP

Some short facts about H.323:

- Collection of standards.
- H.225 for session administration.
- H.245 for multimedia control, T.120 for sharing of data (whiteboard).
- H.235 security (new).
- H.450 additional services (new).
- Complex standard.

Some short facts about SIP:

- Simple, flexible and generic, signalling protocol.
- HTTP style, text-based.
- SIP handles call signalling, SDP and others handle data transfer and more.
- Handles all types of conversations.

- Scales better than H.323.
- **MMS**

Chapter 9

- OSA, SCS, SAT, service enablers.

Chapter 10

- Plug and play technologies enable ad hoc networking of different devices. Based on existing technologies like TCP/IP, HTML, Java VM or XML.

Chapter 11

- H.248

Chapter 12

- IPv6, MPLS, GMPLS, Diffserv, RSVP (see IP standards above).
- DWDM, CWDM

Chapter 13

- Bluetooth
- WLAN
- GPRS, EDGE etc
- UMTS
- LMDS
- ADSL
- Ethernet
- SHDSL

Chapter 14

See Figure 3.52 regarding interfacing standards.

- H 248 – A gateway control protocol standard, defining the interface between servers and edge node.
- BICC – A standard for signalling between servers in the horizontal network.
- SIP, H.323 – See IP standards above.
- Sigtran, M3UA, SCTP.

Chapter 15

- VHE – A concept for personalized service portability across network boundaries and between terminals.

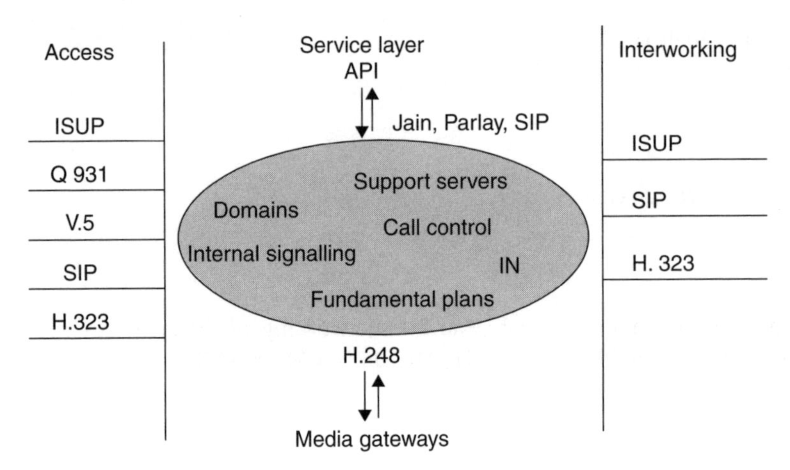

Figure 3.52 Interface standards to the control layer

- CAMEL – Customized applications for mobile network enhanced logic. An IN service that enables utilization of home services in visited environment.

Chapter 16

- SNMP, CMIP, FTAM, CORBA, EDI, XML.

4

Telecom Business

'Technology changes. Economic laws do not.'

4.1 OBJECTIVES

A main purpose of this chapter is to introduce basic market and business-oriented thinking for the changing telecom area. No previous business knowledge is presupposed. Among more detailed objectives are to:

- Know what a telecom operations map and strategy, infrastructure and product processes stand for.

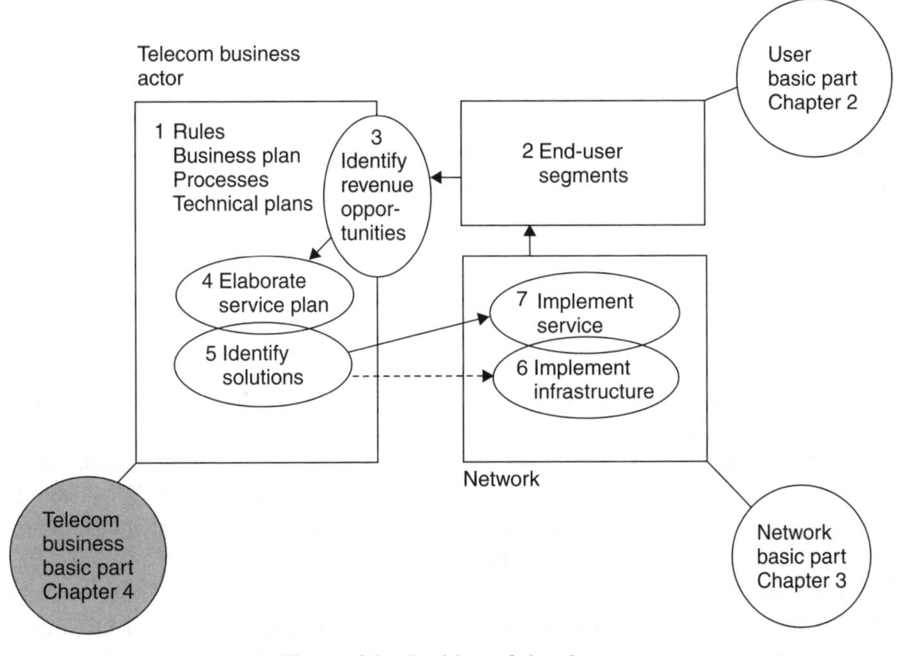

Figure 4.1 Position of the chapter

Understanding Changing Telecommunications – Building a Successful Telecom Business. Edited by A. Olsson
 ISBN: 0-470-86851-1

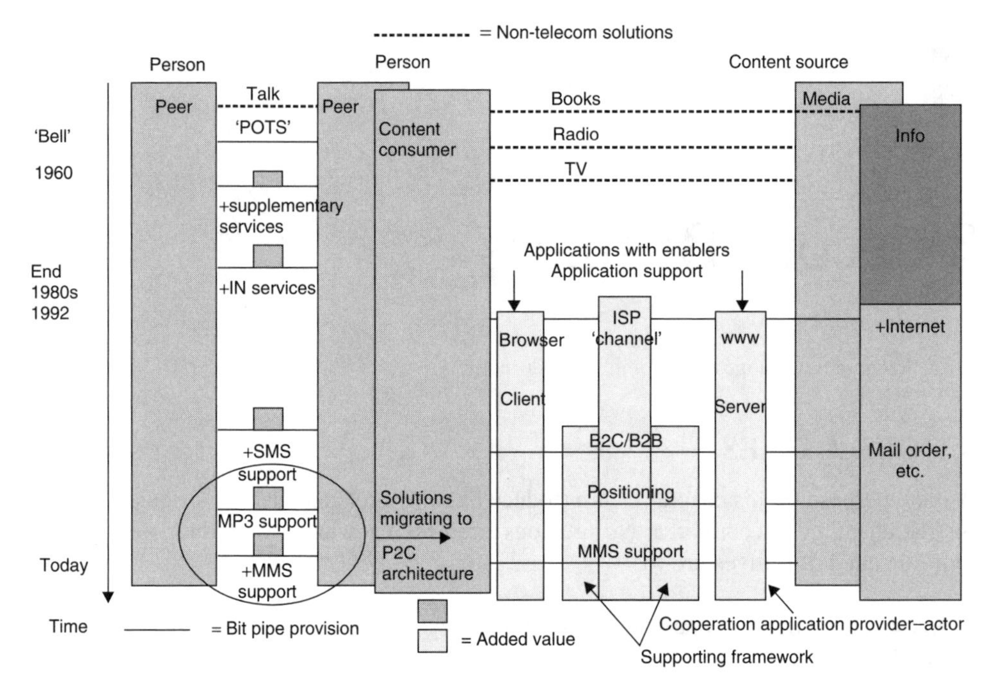

Figure 4.2 Business convergence between the three worlds

- Explain ROCE.
- Give examples of roles/positions in the value chain.
- Explain how business cases can be used as a tool to evaluate investments.
- Explain access unbundling and peering.
- Explain OPEX and CAPEX and their relationship to the main operator process.
- Explain ROI, payback period, NPV and IRR (see Appendix 2)

Much of the changes in telecom business are related to new techniques, especially the convergence, the IP paradigm and a new layered structure.

The development leads to new roles, such as portals, content providers and application service providers. Are the roles profitable or not? Or will they be? What about the ARPU (average revenue per user) development? How can we measure if investments are profitable, using means such as ROCE or ROI?

With the telecom market turbulence in mind let us immediately state that the area is affected by much uncertainty. This makes it impossible to answer many of the questions raised above, but this does not reduce their importance to a telecom area actor.

The chapter is fairly generic. The main topics covered are roles and business cases for investments. Figures 4.2 repeats Figures 3.24 and Figure 4.3 is composed by Figures 3.25 and 3.41, forming a bridge to the previous chapter.

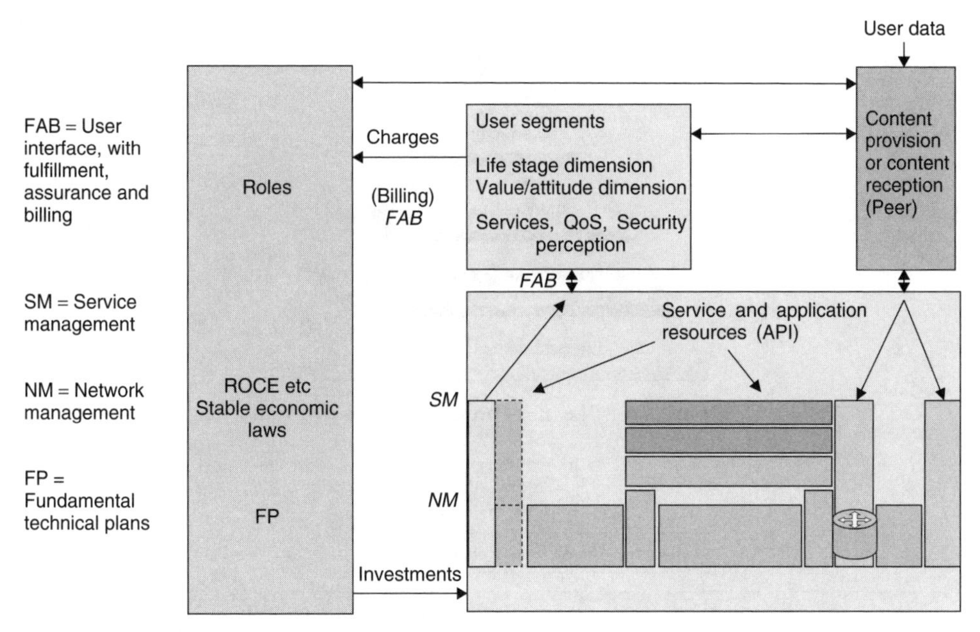

Figure 4.3 Telecom business, users, content provision, service and application provision and network

4.2 THE TELEMANAGEMENT FORUM

The telecom business can be looked at from different angles. Let us first refer to the considerable work of the TeleManagement Forum (TM Forum) www.tmforum.org, a non-profit global organization that provides leadership, strategic guidance and practical solutions to improve the management and operation of information and communications services. Among about 400 member companies are incumbent and new-entrant service providers, computing and network equipment suppliers, software solution suppliers and customers of communications services.

The TM Forum provides management tools for smoothly running the operational work towards profitability in a company. The telecom operations are structured into maps with interfaces towards end user and network.

One of the goals is a management methodology – a system of principles and procedures applied to the operation of telecom networks.

Roughly the TM overall view development started with ITU-T Recommendation M.3000 in 1985 with its four-layer structure according to Figure 4.4. The concept presented was called the Telecommunications Network, TMN (for further studies see for example *Understanding Telecommunications*, Book 1, chapter 8 on network management).

The TM Forum developed the four-layer TMN structure into a more detailed map during 1995–98, which was called the telecom operations map (TOM). The TOM horizontals correspond with the TMN layers. TOM is a reference map of an idealized organization. In TOM the end-user operational interface is denoted 'FAB':

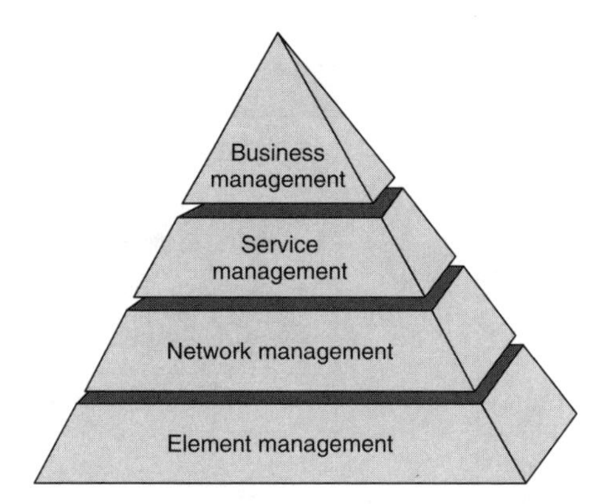

Figure 4.4 The TM Forum four-layer structure

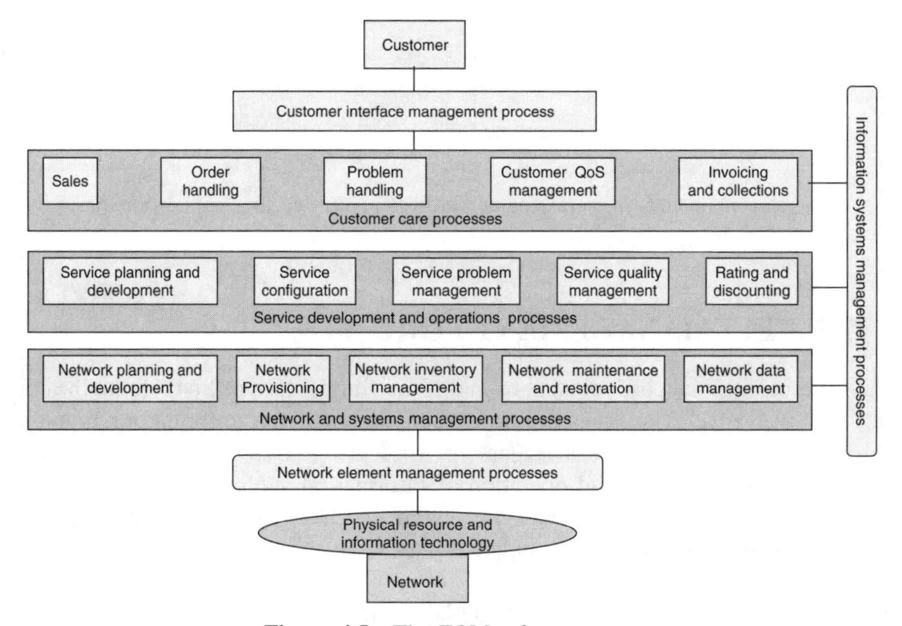

Figure 4.5 The TOM reference map

- F = Fulfilment towards the customers, with sales and service provisioning.
- A = Assurance, which includes e.g. QoS and problem handling.
- B = Billing.

See Figure 4.5.

The next step started in 2000 with the development of the eTOM, the enhanced Telecom Operations Map. Then TOM became 'TOM classic'. eTOM is wider than TOM with three main areas:

- the operations area;
- the strategy, infrastructure and product (SIP) area; and the more internal
- enterprise management area.

Roughly, the TOM processes are captured in the 'FAB' area of the eTOM map. An alternative name for eTOM is 'new generation operation support systems business map', or NGOSS business map.

The focused areas are:

- New generation operation support systems (NGOSS)
- Business process modelling and automation
- Managing next generation network technologies
- Service management
- Web-based customer care (e-care) and customer relationship management (CRM)
- Systems integration

The operational challenges, as seen by the TM Forum, are:
Service development at Internet speed

- Real-time flow-through service delivery
- QoS 'guarantees' across multi-service/multi-technology, multi-provider/infrastructures
- Proactive, real-time, content-based, location-based billing

The three eTOM areas are shown in Figures 4.6–4.8.

The relationships between eTOM and this book are roughly the following: to the operations area belongs especially Chapter 16 on Telecom management – Operations.

To the SIP area belong Chapters 3, 8 and the second half of this chapter on business cases and investments.

To the enterprise area belongs the first part of this chapter, which deals with roles.

Concluding this section on the TM Forum it should be mentioned that the processes can be mapped to roles such as broker, retailer, service provider, connectivity provider and other roles mentioned later in this chapter. However, there is no clear one-to-one relationship and the subject is not further treated in this book.

The success factors for the telecom business are pretty similar to what was listed for the users. A success in the user interface means more money to the telecom sector. However,

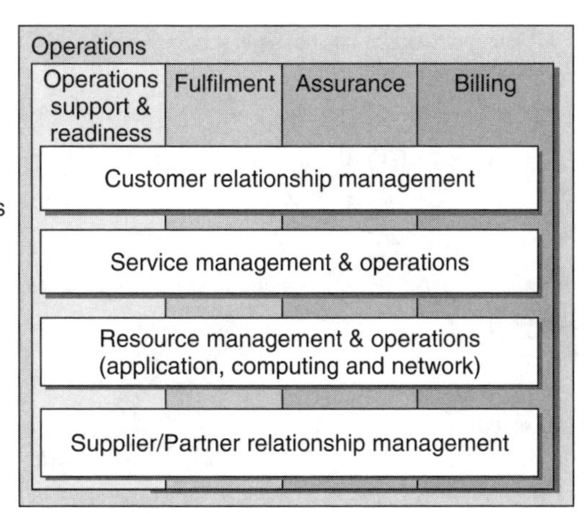

Figure 4.6 The operations area

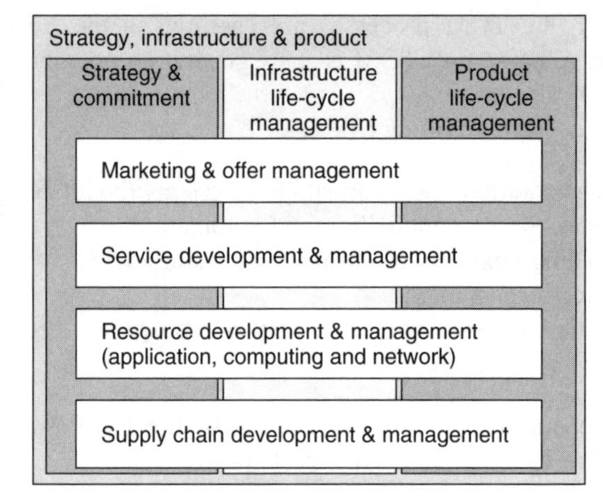

Figure 4.7 The strategy, infrastructure and product area of eTOM

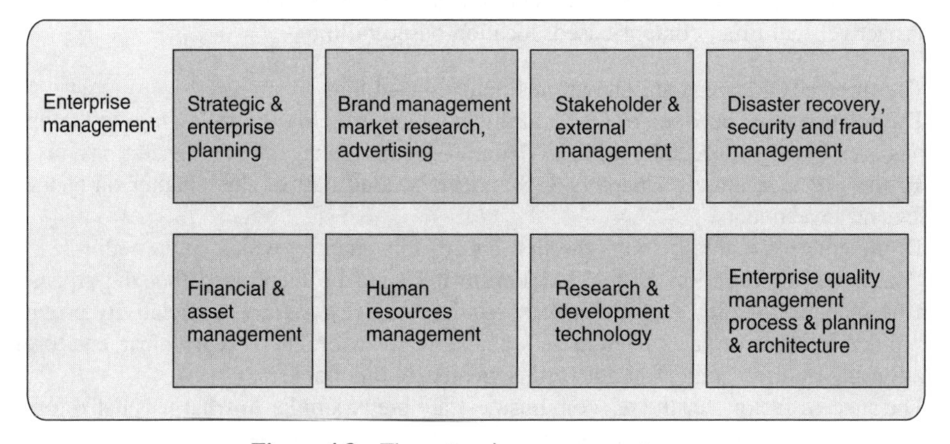

Figure 4.8 The enterprise management area

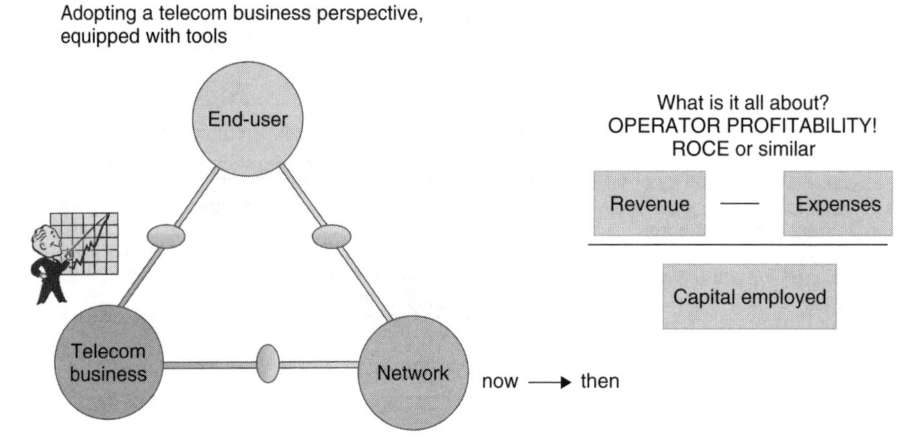

Figure 4.9 Profitability is a main goal

there is also an important cost side, although it is not very common to speak about success factors here. Notwithstanding this fact both components are equally important when calculating profitability, for example by means of the ROCE method according to Figure 4.9.

4.3 ADOPTING A TELECOM BUSINESS PERSPECTIVE

4.3.1 THE END USER CORNER

As said before, the end user is king/queen, and therefore must be treated accordingly. This explains the importance of customer care, flexibility and agility, with short time to customer and short time to market for new services. Useful and attractive services with affordable price, QoS including bandwidth, security and service level agreements (SLAs) are other important end-user values.

A common business related question is: 'Where is the money?' This gives us indirectly two important parameters: the need for end-user segmentation and the need to assess ARPU. Figure 4.10 illustrates the view.

4.3.2 THE NETWORK CORNER

Network Evolution

The network is evolving towards support of the information society, for example by offering more bandwidth and enabling profitability for a new range of services.

Convergence has led to more complex networks, covering more than one single service and using mixed and sometimes superimposed technologies. Layered stacks have

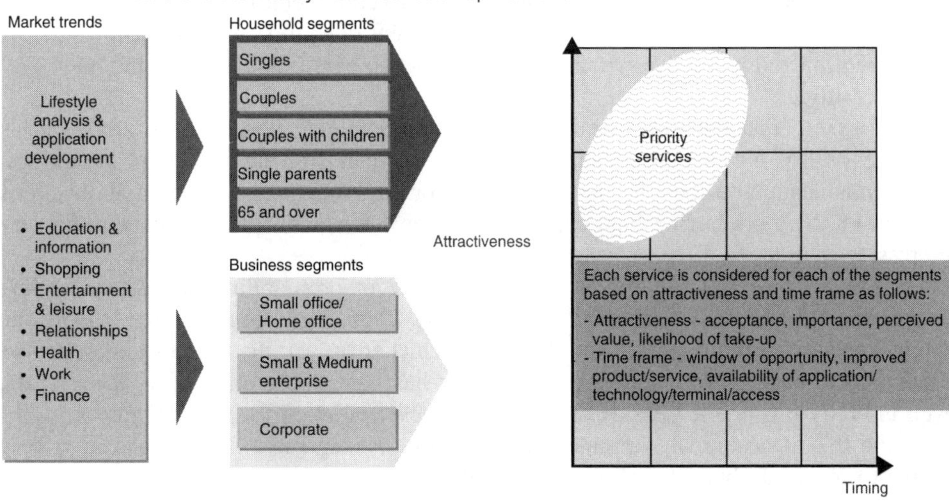

Figure 4.10 A business-oriented view towards the end user

become taller due to horizontal stratification and tunnelling. Networks in one technology behave like bearers or tunnels (physical and data links layers) for higher layer networks in a different technology. This is very usual with IP (i.e. IP over ATM, IP over frame relay, etc.).

In combination with the ongoing technical paradigm shift from a number of vertical networks to a horizontal network, increased attention is justified for a business-oriented view towards the network corner. The view should include a comparison between optional architectures regarding:

- The revenue potential, for example expressed in a service implementation overview. Chapters 5 and 8 treat this subject.
- The investment cost aspects. A common expression is CAPEX, capital expenditure, described in this chapter.
- The operation cost aspects. A common expression is OPEX, operational expenditure, described in this chapter.
- The performance versus cost aspects. This is regulated in rules for network design and dimensioning. Chapter 7 on QoS brings up the issue.
- The security aspects when the IP technology is widely employed. See Chapter 6.

The time frame is also important for this corner, because of the fast development.

4.3.3 THE TELECOM BUSINESS CORNER

Since the early 1930s to the late 1980s telecommunication markets have been regulated monopolistic national markets, ruled by the national PTT (Post Telegraph and Telecommunication), and usually state owned.

Deregulation has now reached most parts of the globe and monopolistic state-owned operators are becoming rare. Most countries have several operators competing in the big market areas and many calls have to traverse different operators' networks.

Competition and privately owned operators create an environment where the main goal is profitability.

Most issues related to the social impact of telecommunication such as service costs and accessibility, e.g. for the handicapped, lie in the hands of the regulatory authorities. Telecommunication business has nowadays little or no differences with any other business activity. Furthermore having high capital investment demands, operators have to be very aggressive in order to drive investor support.

The present time is harsh for the operators. Lack of profitability may cause funding difficulties and therefore marketing weakness and technology obsolescence, leading the operator towards a future possible collapse. As a matter of fact, the mortality of operators either due to bankruptcy, mergers or take-over has been substantial along these years, even in pretty favourable periods of overall business evolution. See also Figure 4.11.

Within this environment, the operator's corporate management has a goal that is very simple to formulate but difficult to implement:

> 'To keep both end users and investors simultaneously happy'.

The requirements of the end user have been covered above.

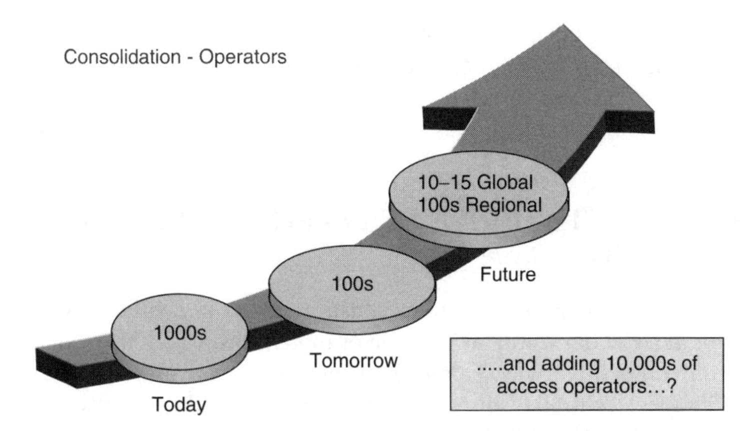

Figure 4.11 The restructuring and consolidation of operators

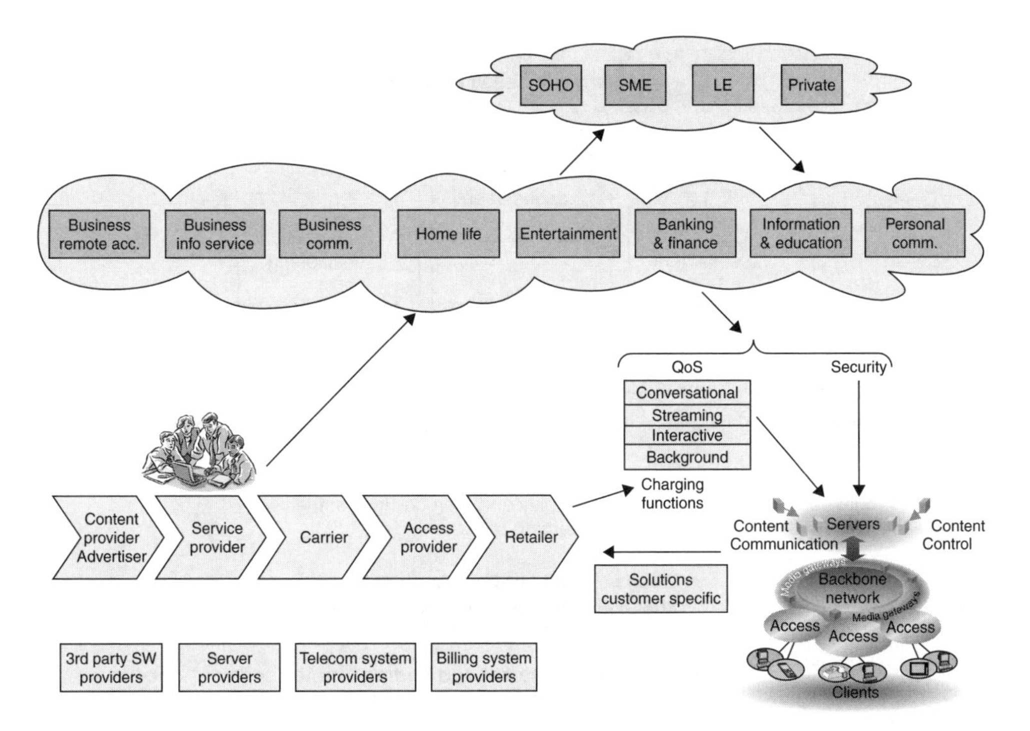

Figure 4.12 Example of an overall view

Investor focus concentrates on a few leading (ROCE, market share) and lagging indicators (branding, innovation).

Another basic fact for telecom business is that the corner is getting segmented into a large number of roles. A particular actor can take on many roles.

The overall view is illustrated in Figure 4.12, using the triangle model. A number of actor segments offer a number of service types to a number of user segments. The network must as a consequence fulfil many diversified requirements, and the complex user–network relationship must be managed by the many roles in cooperation.

Different service roles are directed towards other roles in the value chain, and not towards end users. The subject of roles will be treated in the next part of this chapter.

4.4 TELECOM ENTERPRISE STRATEGY: ROLES FOR POSITIONING

The choice of role is related to an operator situation analysis. See Figures 4.13 and 4.14.

The actors in the telecom world take on one or more roles, which they consider possible to perform in a competitive way. Competitive means for example that they can charge properly for the services to the subscribing end users or to other roles and also that they can achieve a profitable overall business. Roles, which are impossible to make profitable, will disappear or change. Speed is important.

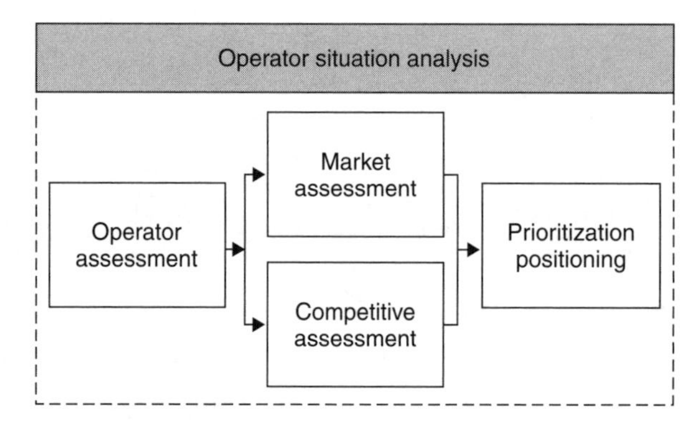

Figure 4.13 Operator situation analysis – specialization into particular roles

Actors should define the role(s) they want to play and then quickly set their stake in the ground in the battle of the converging space

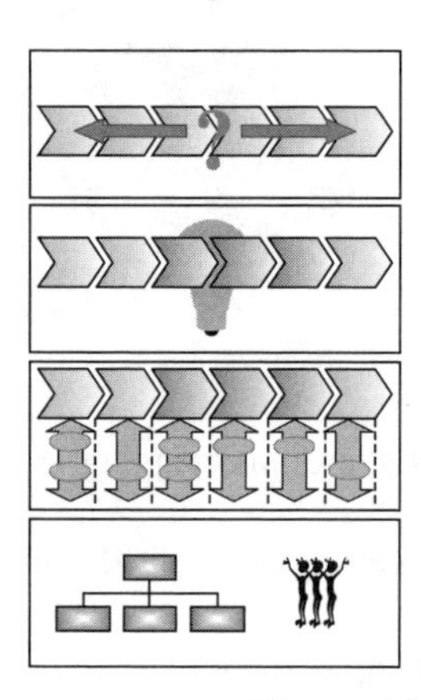

- **Define overall strategy for the convergence space and elaborate the role to play** based on one's market and resource position
- **Segment the market and understand the segments' needs** considering both industry and functional areas
- **Develop value proposition and services for the segments** consistent with the chosen role(s)
- **Engage in partnerships or conduct acquisitions** in order to acquire and/or supplement the required skills (especially in the content and IT/system integration area)
- **Align the organization and the staffing to the new requirements** esp. marketing/sales, service provisioning, partner management and network/technical competence

Figure 4.14 Speed is important in a competitive environment

The roles are partly formed by the market, but perhaps more by the changing technologies and the architecture of the new network. In Figure 4.15 a basic role split into five sub-roles is shown. The roles to the right are sometimes called content and application (C&A) provider roles In fact the interfaces between roles need corresponding technical interfaces, in order to enable charging and/or accounting between the roles, and other parts of a service level agreement. Figure 4.15 also shows three business relationships. Between user and C&A could be everything from no relationship to being the main user relationship. Suitable interfaces are defined in the reference model shown in Figure 4.16.

Since the technical evolution is dynamic the roles must be flexible. Compare the present shift from 'vertical' to 'horizontal' networks. For large actors with many roles this might lead to a number of re-organizations in order to successively tailor the staffing to the new situation. There is also a volume relationship in order to be competitive and profitable. If the roles are over-established, the market share for new entrants will not be sufficient for profitability.

Terminology

The terminology within the area is not very clear. Common expressions are *value chains, value webs, service provider, content provider, network provider, actors* and *roles.* A value chain is needed to create the end-user services. When there is a lack of a clear consecutive chain (which is the case in the Internet service area), 'value chain' might be replaced by 'value web'. 'Service provider' is not a very precise term, but it is used a lot, also in

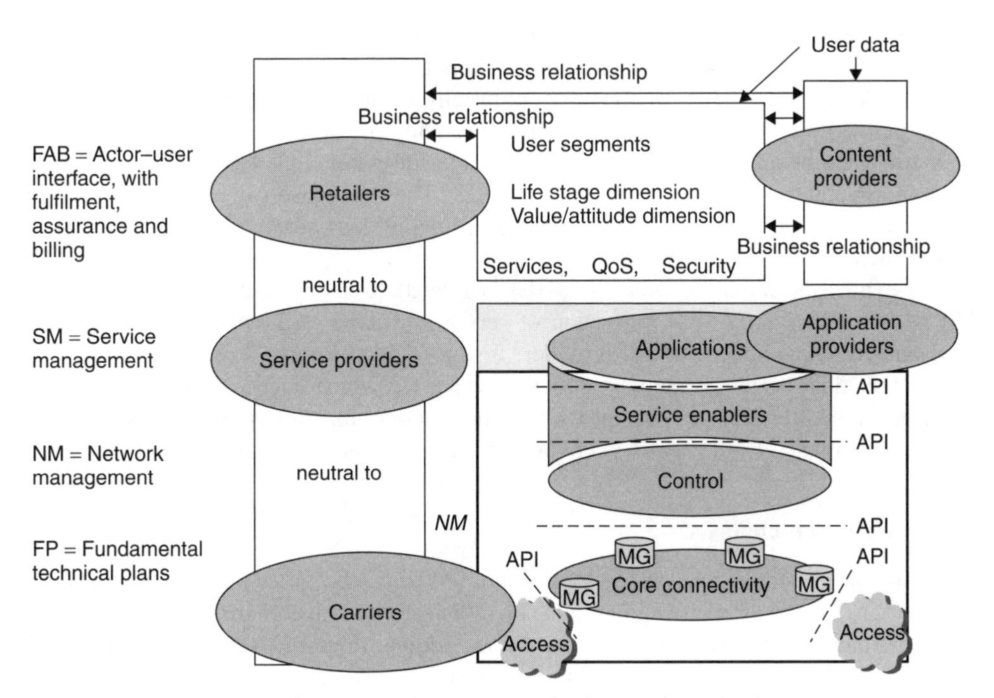

Figure 4.15 A horizontal network gives horizontal roles

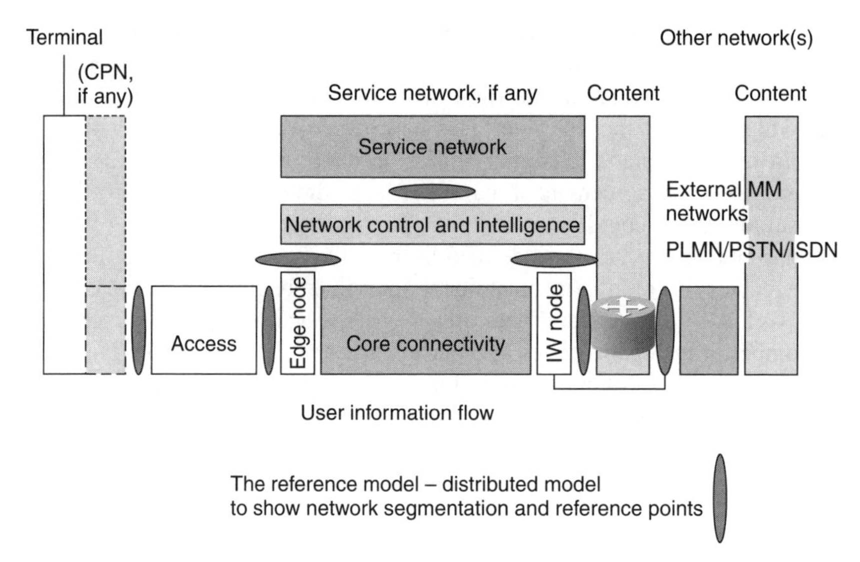

Figure 4.16 Reference points enhancing actor separation and cooperation – the target network interfaces are preferable

this publication. Adding a further word, for example *Internet* service provider, creates clarification.

One actor, e.g. Telia Sonera in Sweden-Finland, can take on many roles.

Network providers or *network operators* own and operate access and core networks. They sell network capacity and quality of service to service providers *and mobile virtual network operators* (MVNOs). This relationship is explained in Section 4.4.1.

Network providers have often global ambitions in order to leverage on economy of scale. Service providers own user subscriptions. They package services and sell them to their users. To be able to offer, handle and charge for compelling services they normally own and operate some kind of service networks. They purchase capacity and quality of service from network providers if the network provider does not provide this role as well within the company.

The expression *home environment* (HE) is related to actors with user subscriptions. HE is responsible for overall provision of services to users. HE-VASP stands for home environment value-added service provider. A value-added service provider provides services other than basic telecommunications service for which additional charges may be incurred. A VASP has an agreement with the home environment to provide services. VAS services are related to the service network and its enablers. See also Section 5.5.

Three Main Sub-chains

One sub-chain of actors handles the transport of bits. It can be called the 'bridging distance (BD) chain' or 'bit pipe provision chain' or 'connectivity value chain'.

Other roles can be described as part of the 'adding value (AV) chain/web' or application/content value chain/web'. Content provision is necessary for TV, videos, games, shopping, advertising, education, banking, music and information.

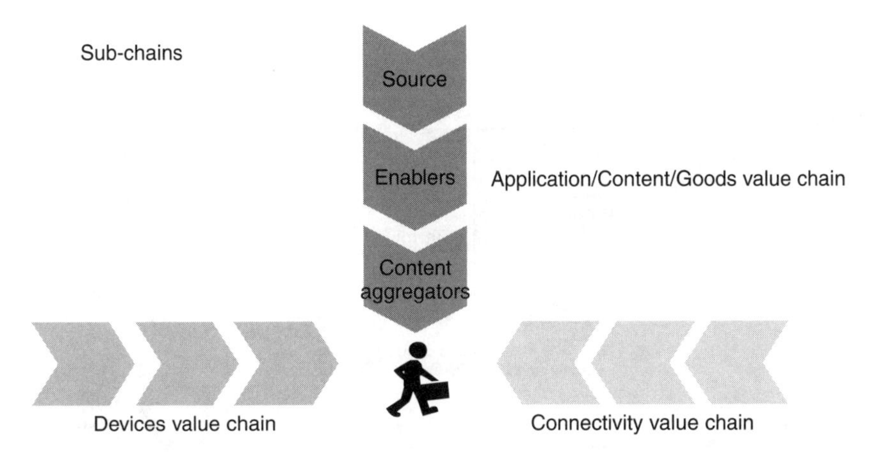

Figure 4.17 The chain of roles can be divided into three sub-chains

A third chain embraces terminating devices and other devices that we can refer to as forming a 'device value chain'. See Figure 4.17.

The BD actors are in principle service providers to the AV roles. The BD role might seem straight forward, but it is in fact fairly tricky. One reason is that the service explosion requires multimedia and multi-service traffic with different requirements regarding QoS and security. Another reason is a considerable possible layering in the BD area.

4.4.1 EXAMPLES OF INDIVIDUAL ROLES

A number of possible value chains have been presented in the literature. The area of roles including terminology and business relationships is not yet stable.

Some common features are: the convergence between media industry and telecom demands, as a minimum, a content provider role, a network operator role, and a role in between, to connect the end user to the content; the Internet service provider (ISP) or the wireless Internet service provider (WISP) using the WLAN technique (explained in Chapter 13). See Figures 4.18 and 4.19. Charging and security are important ISP tasks.

The connectivity or communication part of the chain can be split for example according to the reference model in Figure 4.16. The number of competing actors in the access role increases. In other parts mergers are expected.

Apart from the access part the reference network shows three networks with associated roles:

- a backbone connectivity network, operated by carriers, in its turn carrying;
- core networks (such as a GPRS or UMTS core network, a national PSTN/ISDN core network, or a network of a regional or global operator) and IP service networks;
- the service network provider role connects a large number of servers to an IP network to link the content residing in servers to the clients in a client-server or peer-to-peer relationship (application servers, application support servers, service capability servers, IP infrastructure servers).

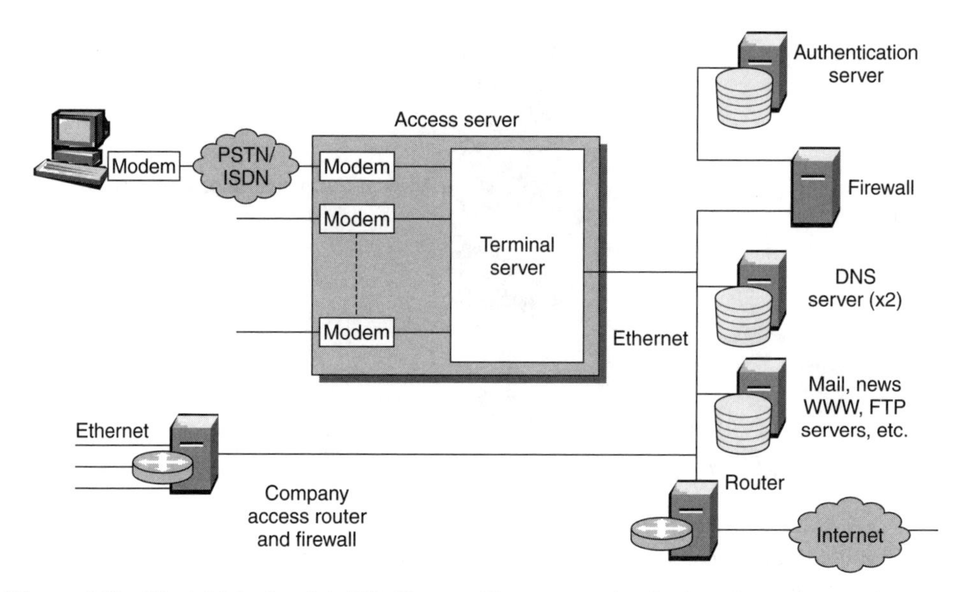

Figure 4.18 The initial role of the ISP illustrated by an example of point-of-presence configuration (the access server is sometimes denoted NAS, network access server)

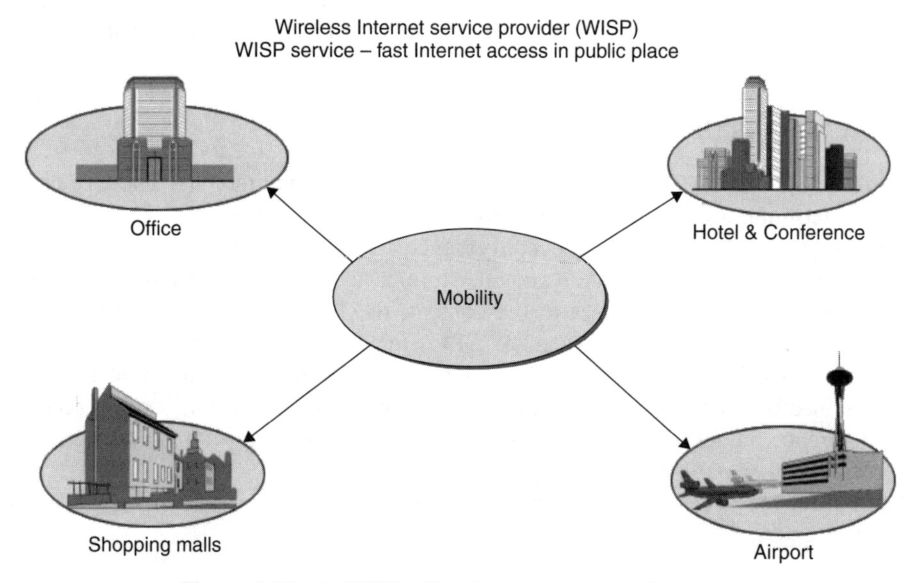

Figure 4.19 A WISP offers Internet access at hot spots

In networks with network-centric intelligence there are normally supplementary services, which include IN services, and information-type value-added services ('Telephone Bank'). These information-type services are a predecessor to the content provision of today. The supplementary services (including IN) are special, often providing *increased accessibility* by means of rerouting, renumbering, number portability and similar services.

This increased accessibility ('IN') role has been dynamic during the 1990s, but its development is now slower with some exceptions, such as prepaid and number portability. The role is normally integrated with a core network operator role for voice traffic.

The *content-oriented roles* are not stable. The links between the roles are complex. Many roles are quite new and still developing. The relations do not form a clear value chain but rather a 'web'. The question of who is paying whom, including the overall distribution of revenue between roles, will be touched upon in Section 4.6. The ISP role, as an example, can be quite simple with basic Internet access, e-mail and AAA functions. It can, however, expand to embrace web hosting, I-VPN, VoIP and possibly also commercial services. A white paper (common language) from Ericsson indicated some 10 typical roles around the task of content provision in the mobile Internet. See Figure 4.20. VAS/IN service provision has been added to the original list.

These roles can be split into general roles (left) and commercial roles (right).

General AV roles:

- ISP (on the border to connectivity)
- Content provider
- Information broker
- Portal
- Application provider (providing not running a service)
- Application service provider
- VAS/IN service provider
- xSP (see Section 2.3.4)

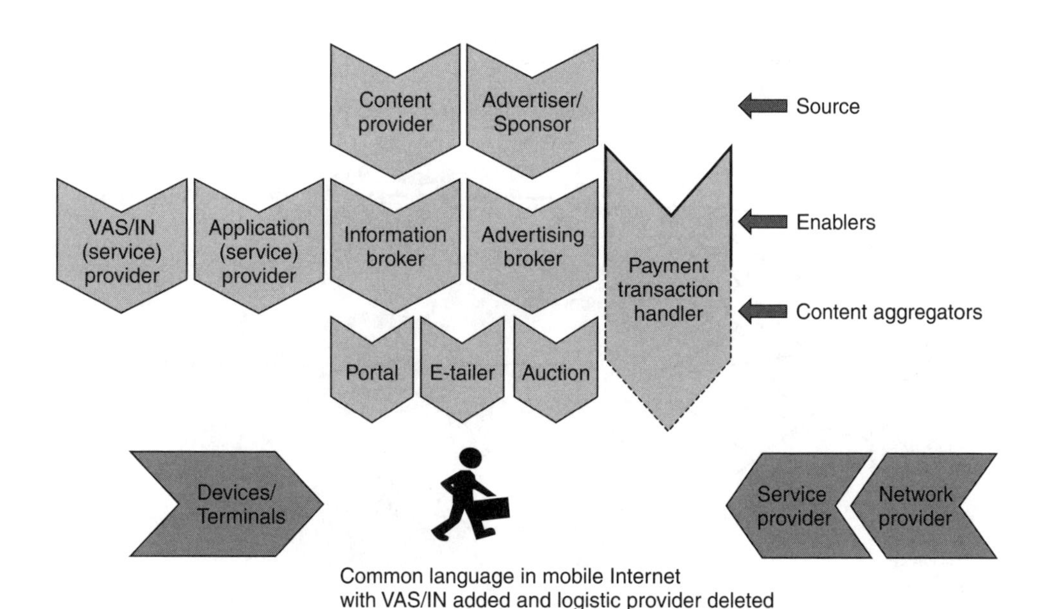

Figure 4.20 Roles in mobile Internet – example of architecture

The Portal Role

'A portal is an access point for services to the end-user, provided over Internet/Intranets'.

There are many different types of portals from more specific horizontal to broader vertical ones, serving the consumer as well as the business market. The access point offers a collection of hosted or linked content items, applications, customer groups and services. The portal role is a very diffuse role in its present form. Most of the players that have an end-user interface/site might also offer and play a portal role. A concrete example would be Amazon.com. Amazon.com is an e-tailer but in order to provide services it has its own site/portal.

One can, however, argue that some players have a more pronounced portal role. Such an example is Yahoo! With that in mind, we can say that there is a kind of 'pure portal role', that would be the 'main entrance' to communications and services in the Internet.

The portal role is situated between the end user and the content provider. The customers are:

- End users, such as enterprises and consumers
- Content providers, suppliers
- Advertising and information brokers

See Figure 4.21.

Portals can be regarded as competing 'information switches' and content aggregators between the end users/networks and the content. Many portals have been suffering from the gloomy development within IT at the beginning of the twenty-first century.

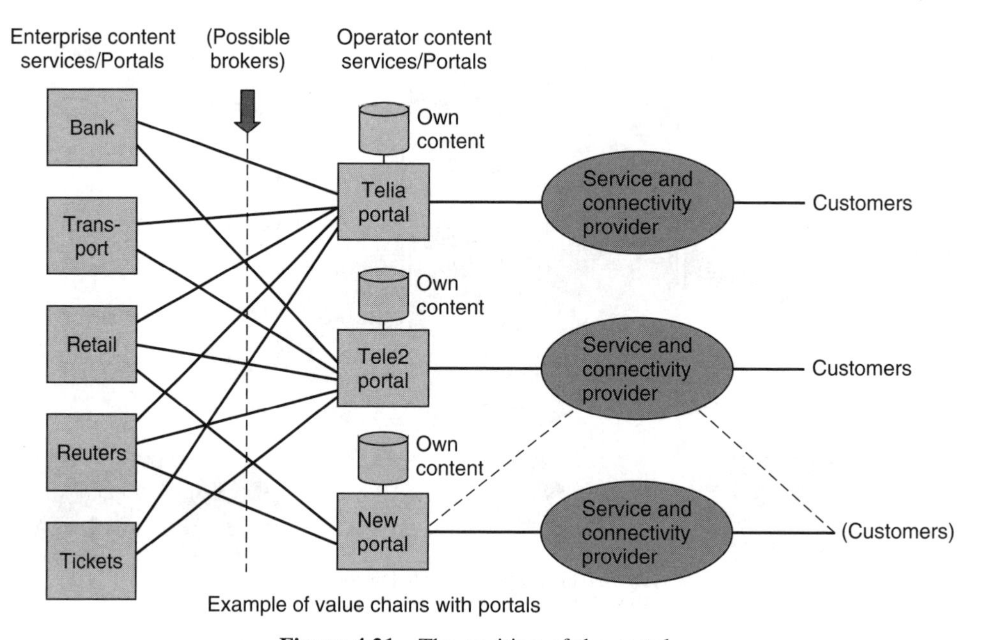

Figure 4.21 The position of the portals

Application Service Provider

The terminology is not quite clear regarding application service providers. For example, sometimes the content provider role is combined with a very general application (service) provider role. Sometimes the interpretation is narrower and business oriented, according to the following description.

An application service provider offers access to applications (such as software) and related services on a rental basis via the Internet. This role has evolved and also fits in the context of web services, as 'software-as-a-service' provider.

ASP services are an important alternative, not only for smaller companies and individuals with small technology budgets, but also for larger companies as a form of outsourcing. There will be a transition from locally installed applications to network-enabled applications and software. See Figure 4.22. Another interpretation sees ASP as a role between a content provider and a service provider in a value chain serving for example the mobile market.

The original role was defined: to deploy, host, manage and lease packaged application software to customers from centrally managed data facilities. The role is not yet firmly established.

The application services are installed in centralized server farms in safe data centre buildings. In a way it is a step back to the mainframe network type, with thin clients, but based upon Internet technology. Any application can be accessed from any terminal, if admitted by the contract/service level agreement between ASP and the customer (normally an enterprise). The ASP is responsible for the end-to-end application performance, including the BD role, which is solved as a leased line, a frame relay PVC or an I-VPN.

Only the user interface is sent over the network. Security for the various customer traffic flows is a main issue. Often different customers use the same physical server.

Other names used for an ASP are managed service provider and value-added service provider.

Example of an ASP: a grocery may outsource much of its IT activities to an operator. The operator will then provide centrally located applications and a link to the grocery operating centre. This link might be an I-VPN with back-up facilities (often dialled-up).

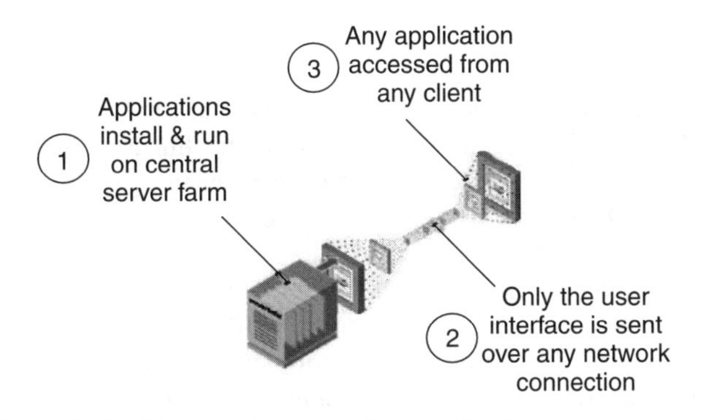

Figure 4.22 The principal role of an application service provider

Commercial AV Roles

- Advertiser/sponsor. Services for free against accepting individualized advertising
- Advertising broker
- E-tailer
- Auction
- Payment transaction handler

Advertising broker – The advertising broker is the link between the advertiser and the user that makes sure that the right advert reaches the right user. The key to success for advertising over the mobile phone is to make it as personalized and targeted as possible.

A mobile operator is well positioned to take on the role as advertising broker, by using some key assets:

- The operator has an extremely valuable customer database, which many advertisers would see as a gold mine. In addition, the customers trust the operator – they already have a relationship with the operator.
- The operator has access to profile information – information that allows targeting of advertising to specific end-user segments or even individuals.
- The operator owns the network, a media channel that reaches individuals wherever they are, at any time.
- The operator can identify the location of the mobile phone and can thus target advertising based on a user's geographic position.

Some people are reluctant about receiving mass-market advertising on the Internet. Advertising over the mobile phone is different – it is personalized and targeted. Research by the Boston Consulting Group shows that 70 % of consumers worldwide are positive towards receiving targeted advertising. This is true if they get some kind of incentive in return, for example free airtime or cash incentives.

And, there is money in advertising. The market for online advertising is growing faster than advertising in any other media. Operators can get a piece of the cake by acting as advertising brokers. Advertising also encourages follow-on voice calls, with more revenues from additional traffic as a result.

Payment Transaction Handler

The payment transaction handler provides secure payment services for e-commerce transactions on the Internet. The handler manages also the financial aspects of transaction handling in a standardized way, also taking the financial risk.

In the grocery case the operator could charge the end user/buyer for her/his purchases together with the telecom bill.

VISA is another player in this role. The transaction handling is similar to traditional signalling schemes. However, all communication is IP based. See Figure 4.23. Note that payments can also be arranged through the credit card company instead of e-Pay.

Apart from all these communication and content-oriented roles there are in fact other types of role.

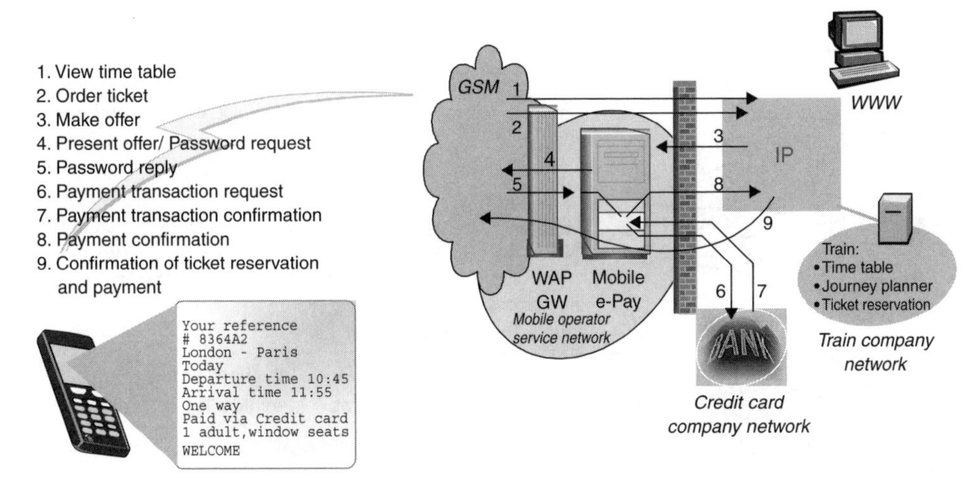

Figure 4.23 Payment transactions, example

Example 1: The retailer role. The retailers focus on the relationships with the end users: subscriptions, marketing, sales, after-sales, bundling of offerings, customer care and invoicing. They sell handsets and other customer premises equipment in shops.

Today this role is often part of the activities of the network operators, but it is supposed to become more and more independent. One-stop shopping is often part of the business idea.

The TINA-C international body includes this role. The TINA business model from 1999 defines the following business roles:

- The consumer role.
- The broker role, which enables other business roles to locate service providers.
- The retailer role, which is concerned with providing services to the consumer role.
- The third-party provider role, which is concerned with providing services to retailers or other third-party providers, but not directly to consumers.
- The connectivity provider, which owns a network and provides connectivity services over it to other business roles.

Example 2: The Mobile Virtual Network Operator (MVNO). An MVNO offers mobile subscription and transport service to end users, but does not have a licensed allocation of spectrum. The host network operator owning a spectrum license is supposed to be willing or forced to open his access to other players. See also section 4.4.2.

An MVNO can have any background: fixed operator; mobile operator going for new markets; media group content provider; portal provider; content aggregator; bank etc.

An MVNO competes with service differentiation and a branded name as the main weapons.

Example 3: Horizontal layer players in the core connectivity network. Common names are carriers and transport provider. See Figure 4.24. An example is the carrier MCI, offering as an example ATM PVC ABR, ATM PVC VBRnrt (variable bit rate, non real time) and ATM CBR (constant bit rate). This is an example of network internal services.

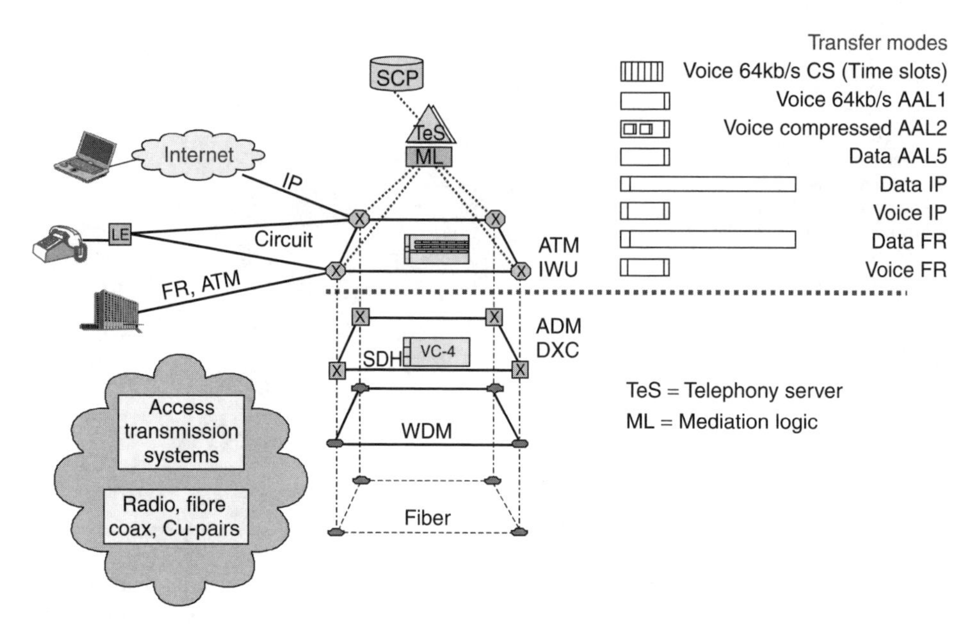

Figure 4.24 A layered network permits layering of roles

Another example is an IP-based network. The network can be used for a service network, VoIP networks, I-VPN and for the public Internet. We can use the expression retail for services towards end users and wholesale for services between actors.

Example 4: Competing operators are often denoted after their historical background:

- Incumbent operators, the old monopoly operator.
- Intruders, which often are incumbent operators in another country (Telia Denmark as an example).
- Colonists, which have consumer knowledge and take retailer and maybe access operator roles Hem–el in Sweden, Energis in the UK, other utility operators (water, gas). Such companies already have a billing relationship with the end user. Adding another item to the bill costs very little.
- Expansionists (expanding from one role/present roles to other roles).
- Combines (financially strong enterprises, business oriented. Example: Kinnevik in Sweden, Mannesmann in Germany).

In the access network new operators appear with a retailer background

- Utility operators
- Electricity companies
- Heating companies
- Water companies
- Gas companies
- Condominium associations (Example from Sweden: HSB, Svenska Bostäder)

- Municipalities, each would like to become the 'IT community', to attract business and as a consequence offer new jobs

Example 5: Wireless Internet Service Providers. See Figure 4.19.

Example 6: Manufacturers, suppliers. The area can be split into terminal manufacturers, software suppliers and network manufacturers. The supplier roles are also changing towards a partnership with operating actors.

4.4.2 CHOICE OF ROLES

In the good old 'vertical' days the positioning was mainly a matter of deciding what networks to operate: PLMN, PSTN/ISDN, or data networks. Each network represented a complete value chain for a particular service. Convergence between mobile and fixed network services or between voice-centric and data-centric networks was not common.

For an actor today it is important to define the overall strategy for the possible space in the converging telecom world, and elaborate the role(s) to play based on market and resource position. Further to understand the needs of the roles and develop consistent value propositions. Multimedia is often among the needs. See Figure 4.25.

In the horizontal era some key questions are:

- Where is the money? And particularly in the unstable situation: Where is the development taking us?
- How is the income apportioned within the value chain that produces the services? What about the future situation?
- How can I get a significant market share in order to enjoy economy of scale?

There is no final answer to such questions in advance because of the uncertainty in the market development, but a general trend is to move upwards in the value chain.

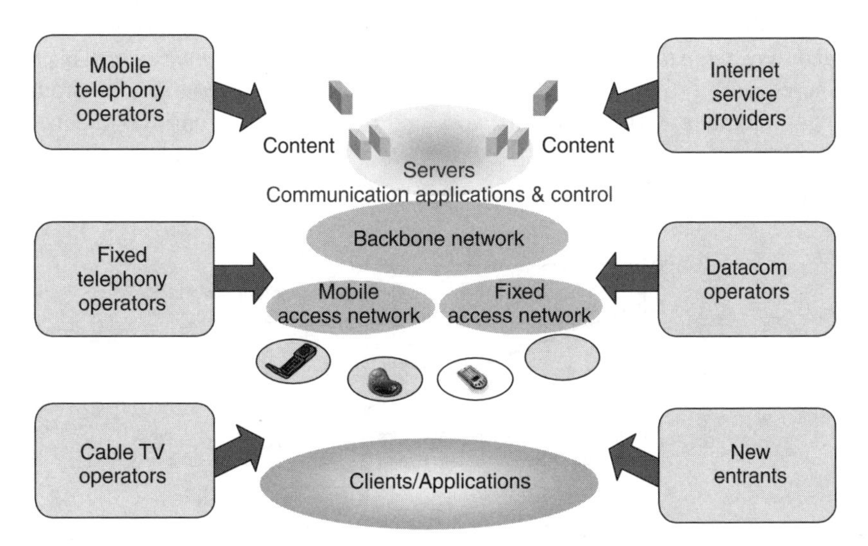

Figure 4.25 All roles go for multi-service and multimedia

A market study on the European market (Kearney, Ericsson from 2000–01) indicated, if extrapolated another year, roughly a doubling of the overall market in five years. It indicated also that the whole growth would go to the 'adding value' (AV) part of the value chain, and consequently that the connectivity value chain with the carriers and access service providers would not exhibit any growth at all in spite of a heavily increasing traffic.

Such figures are already history, but the trend is still valid. See Figure 4.26.

The operators have a number of options for role expansion.

A prosperous Internet requires acceptable business cases for all roles in the Internet value chain. Example: if advertisements and direct Internet payments are not sufficient for content providers, a fairly empty Internet could appear. Since such a situation is a threat also for the network operator, some kind of payment system for revenue sharing is a key. Win-win is better than lose-lose.

As mentioned, there are quite a lot of roles within the AV area to choose between. The roles are still unstable and so are the money streams between them. Even the direction of the money flow between the roles is sometimes unclear. A strong position, such as a strong brand that attracts end users, gives a good negotiation position when deciding the charging between roles.

Regarding economy of scale the telecom market is expected to follow the horizontalization of the computer market, creating economy of scale in hardware (HW) and software (SW) layers, like Compaq (HW) and Microsoft (SW) have done within the computer market.

Mobile Actors, Chains

In response to the increased importance of services, mobile network operators are transforming their organizations into separate entities for network and service provisioning allowing the different parts to focus on their core competencies. The focus for the network provider is *cost control, efficiency* and *economy of scale* for the network operation. The focus for the service provider is to compete for users by *service offerings* and *differentiation* and to *manage the user interface* including *billing* and *customer care*. This division allows the network provider to offer its network capacity to several service providers in order to get maximum traffic in the network and optimize return on investment.

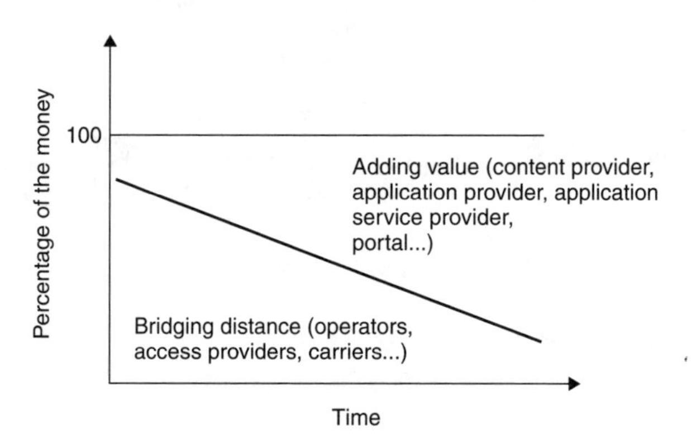

Figure 4.26 Trend in development of revenue distribution

The service layer business environment (Figure 4.27) is getting increasingly more competitive as new roles are emerging. Telecom operators wanting to expand their business to other markets or technologies, and non-telecom companies wanting to extend their current business through mobile services are all partnering with network providers to launch themselves as MVNOs. See Section 4.4.1. The business models for the different roles are centred on their core businesses. Each type of role may have local or global ambitions.

Many MVNO candidates lack wireless service expertise and are wary of the operational costs and complexities involved in running a wireless business. MVNOs benefit when a mobile virtual network enabler (MVNE) provides a turnkey solution platform, such as a complete technology and services platform that offers everything from customer service and billing to custom content and value-added services.

A typical mobile operator takes on many roles.

It is natural that the operator should continue in the roles as *network provider/access provider* and *service provider*. It is of strategic importance for the operator to offer services to its customer base via an own-branded *portal*.

Acting as a *payment transaction handler* and taking some percentages of each transaction is one way to increase revenues.

Another way is to act as *advertising broker*, to help advertisers target adverts to the operator's customer base, and get a fee for doing that.

For distribution of mobile media a value chain could look like Figure 4.28. The operator can be the sole actor regarding access, ISP and distribution. The operator or its partners can fulfil all the other roles in Figure 4.28.

The content provider owns the content. The wireless caster provides the actual service. The service is developed by the application developer. The owner of the application owns and probably also brands the service. Portal providers set up portals from which the end user can find and access on-demand mobile media services.

On a service-by-service basis, the operator may decide which of these roles it wants to keep for itself. There is a trade-off between providing all the elements in a complete service (thus reaping all the revenues and carrying the entire risk) and offering a product that is a component in a service (reduced risk). I-mode has shown that only a fraction of services is successful.

The sequential value chain will often be combined with horizontal roles, especially transport providers. See for example Figure 4.24. This gives service providers on top of transport providers.

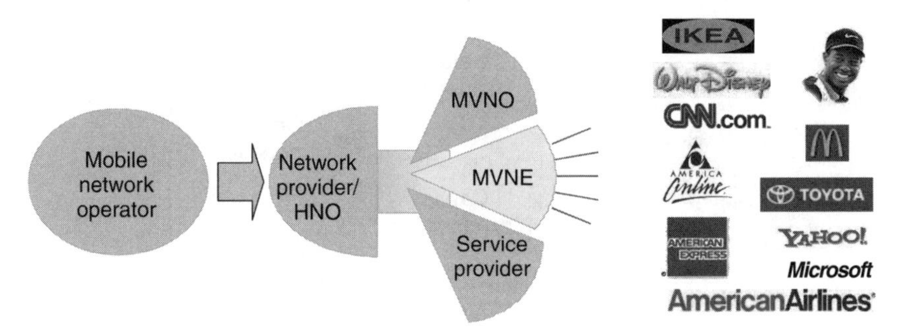

Figure 4.27 The service layer business environment

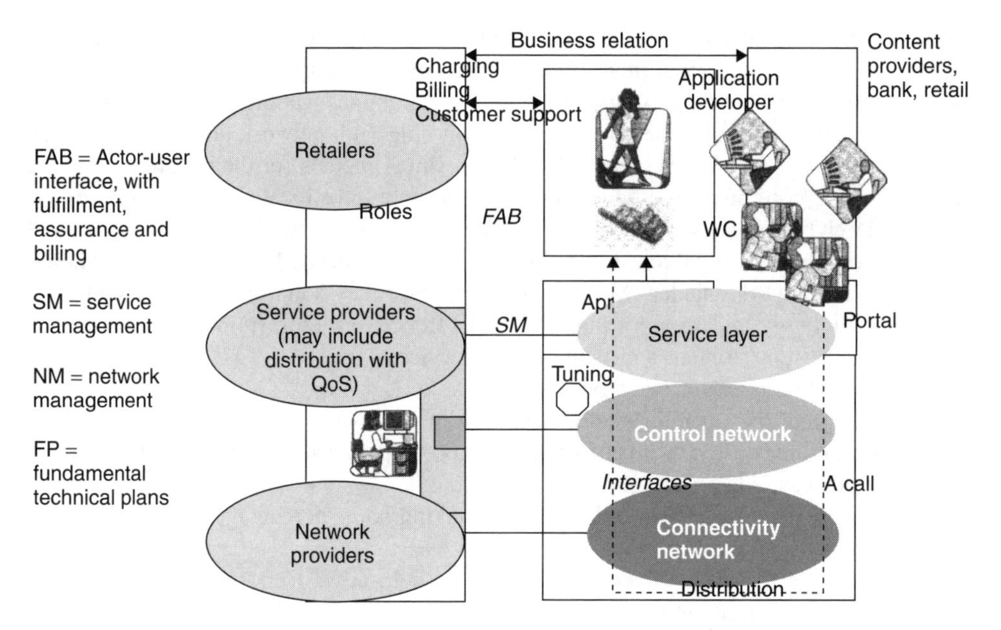

Figure 4.28 Example of a value chain for mobile media distribution

4.5 TOOLS FOR PROFITABILITY CALCULATIONS AND BUSINESS CASES

Since profitability is a key word for telecom business actors, there is a need for tools to support the management staff in decision making and experts in evaluating alternative investments before submitting proposals. The first tool is a broad management tool: the business plan (Figure 4.29). This plan will be briefly dealt with here, whereas the other more narrow tools will be dealt with in Section 4.5.

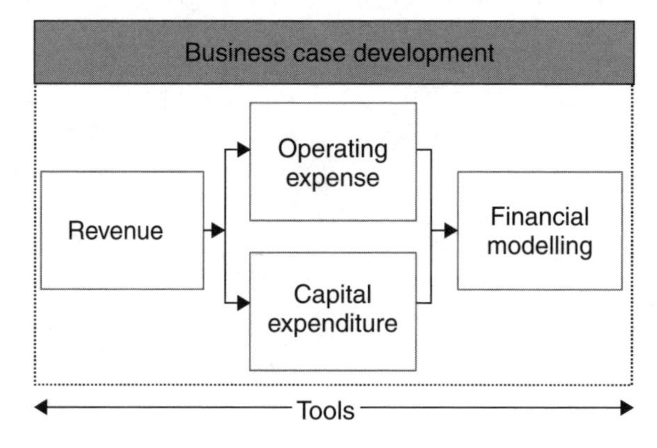

Figure 4.29 The structure of the business case

The following tools are treated in this chapter:

1. The business plan. A shift in business planning.
2. Simplified operator process plans.
3. The business case. A simplified business case work model.
4. A computer-supported business flow model.

4.5.1 A MAIN TOOL: THE BUSINESS PLAN

The *business plan* is a management tool for the operational work towards profitability in a company. The business plan defines the operator strategy, which might include:

- The end-user services they can deliver and will deliver over the forecast period by market segment.
- The position in the value chain and relations to other actors, including roaming.
- Customer segmentation and priorities.
- The market share targeted for end-user services.
- The access services they offer and will offer.
- The market share for access services.
- Development of the 'machine' (if any): O&M (operation and maintenance), capacity, billing, security, QoS, technologies.
- Development of the staff.
- Financial goals.

The business plan always contains a business concept, for example:

'The business concept is to offer telecommunication services and systems which facilitate personal contact between individuals and enables companies to develop and operate more efficiently.'

> 'Services and systems offered should be easy to use, competitively priced and reliable. The network is the nucleus of the operations and infrastructure must support the network services'.

Let us start with a simplified profitability model. See Figure 4.30.

The main elements of this particular model are a telecom network as a service factory, the staff extending, operating and maintaining the factory, the marketing expenses for supporting the service sales on the market, the market battle field and finally the resulting monetary streams and the related financial calculations. Operating expenditure is often called OPEX. Investment expenditure is often called CAPEX (capital expenditure).

A common action plan by the actor when choosing business concept may look like this:

- Examine the market and the competitive situation carefully and choose suitable role(s) in the value chain. The future market share should be sufficient to achieve economy of scale.
- Choose customer segments within the existing regulation framework. The customer segments are normally end users, but there are also internal vendor–customer relationships between actors within network operation and service provision.

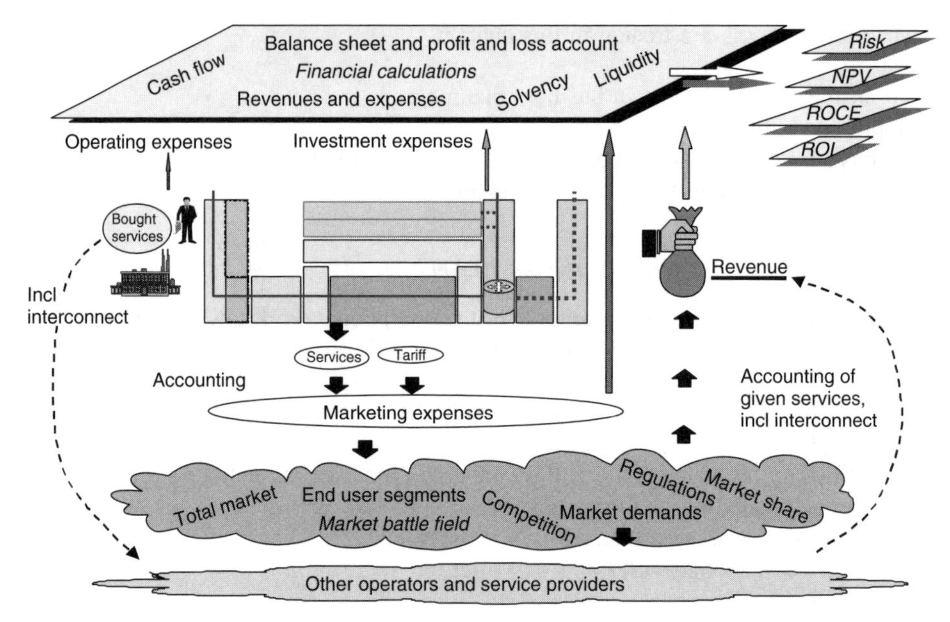

Figure 4.30 A simplified profitability model

- Based on your role(s) and targeted end-user/internal customer segments, choose a suitable range of services for the potential customers, and forecast the market share and income potential, based on a realistic tariff, competitive situation, marketing efforts, market demands, end-user segment potential and market regulations.
- Choose appropriate technologies for investment purposes, based on the service offerings, and estimate expenses for investments, network operation, marketing, leased capacity etc.
- Perform the business analysis by means of financial calculations. ROCE (return on capital employed) is maybe the most important profitability measure used to assess the operating performance of a company. A reason for this is the return on investment employed shows how well the resources of the company have been managed, regardless of how the resources are financed.
- Document it all in a business plan.

The business plan contains guidelines, technology choices and overall targets, such as overall growth estimates. It can be regarded as a choice of direction or consolidation of previously elaborated (open) scenarios and (secret) visions. It is followed by execution.

Such a plan acts as a spider in the overall picture, with the role chosen by the actor/operator, with 'business opportunity arms' to targeted end users and their assumed service profiles, and with guidelines for the corresponding design of the network (if any) and the interconnect requirements.

When available, a relationship between planning/investments and the related effects on the balance sheet, profit and loss account and cash flow is a key tool.

Let us now look at a very simple connection between the business model and the three ROCE components: revenue, expenses and total capital. The equation in Figure 4.31 gives the percentage profit.

Revenue comes from the services in the consumer and enterprise list in the business plan. Specific servers can also generate money internally in the value chain, for example e-commerce servers. (The situation is different for carriers and other actors exclusively working internally in the value chain.)

Expenses are represented by the corresponding operating cost and investments in the network. Below the line is the total network, ideally designed as a tailored platform for the business plan services. *Total capital* is therefore mainly represented by the overall value of the network. (To a much smaller extent there are also investments above the line.) For actors that are internal in the value chain the situation will look different.

Summing up, the business plan is a key tool in the actor business in order to make the company work internally coordinated towards a common target. Today this is more important than ever because of the paradigm shift, illustrated by Figure 4.32.

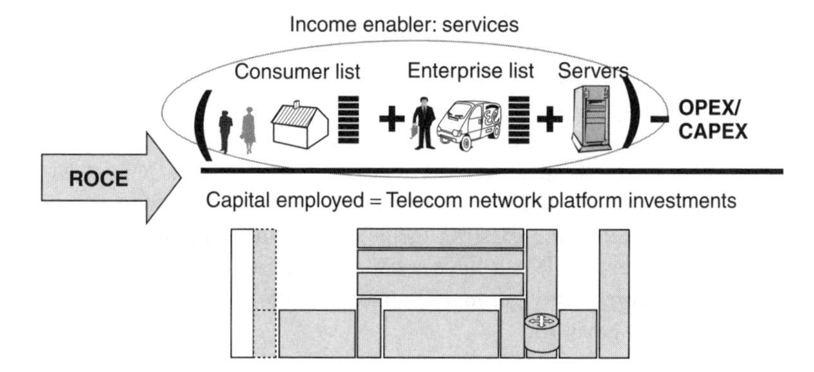

Figure 4.31 Connecting the business plan to ROCE

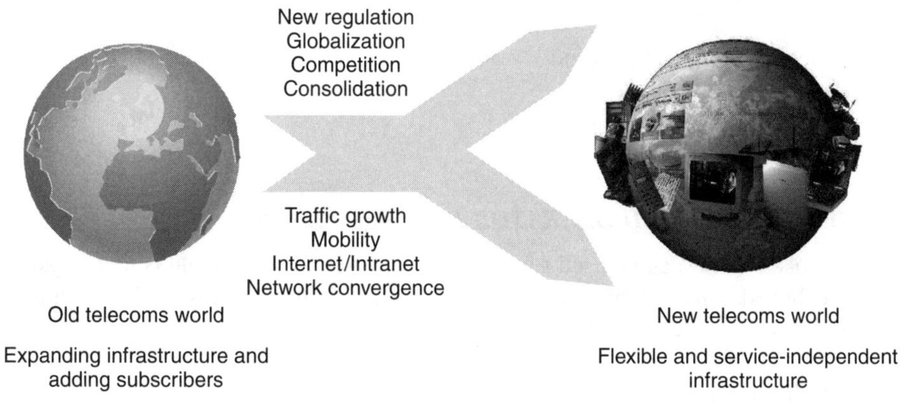

Figure 4.32 The paradigm shift

4.5.2 WHAT IS A BUSINESS CASE?

A business case:

is a tool (with sub-tools as indicated in Figure 4.32)

describes the impact of a decision.

is an analysis that supports decision making by showing the incremental added tangible and intangible value to a business that follows from an investment.

A business case is a tool to facilitate the evaluation of the economical risks. For example, what are the benefits of the introduction of new services and related investments? In principle, both the marginal investment cost required by the new service and the assumed revenues created by the service should be calculated, e.g. using a traditional net present value (NPV) method. The method is explained in Appendix 2.

While the marginal investments are difficult enough to estimate, the revenue calculations are really hard to perform. There are a number of aspects that directly or indirectly impact the revenues. When an end user chooses to subscribe to a particular service delivered by a specific service provider, the choice is not only based on a simple price comparison. Quality and timeliness are equally important factors. There is always a risk of losing the customer to a competitor, which may result in lost revenues from other existing services, which this customer would have used.

The effects in monetary terms when introducing new services are normally: new revenues, increased revenues of existing services, savings, but also new costs, increased costs and investment costs. A specific case of revenue and savings impact would be the introduction of voice over IP (VoIP). Savings would be possible especially in the long-distance network, when replacing existing technologies with a higher operating cost. On the revenue side, however, VoIP can have a cannibalization effect in direct competition with an own higher charged conventional telephony service.

The business case is a basis for:

- achieving a uniform corporate approach to investment analysis
- ensuring formal documentation of assumptions
- comparing alternative investment proposals
- setting investment priorities

The content is implied by the policy and includes:

- key issues
- financial analysis

4.5.3 TOOL: SIMPLIFIED OPERATOR PROCESS PLANS

Operator process plans can be used for business purposes. One example is to compare two or more technical solutions by comparing the impact on the processes. See Figure 4.33. The view is used in Section 4.7.

Among the ingredients of a business plan are a service plan and a technology plan. The services in the service plan are supposed to support the ROCE in the best possible way, and the technology or service implementation plan is supposed to support the service plan.

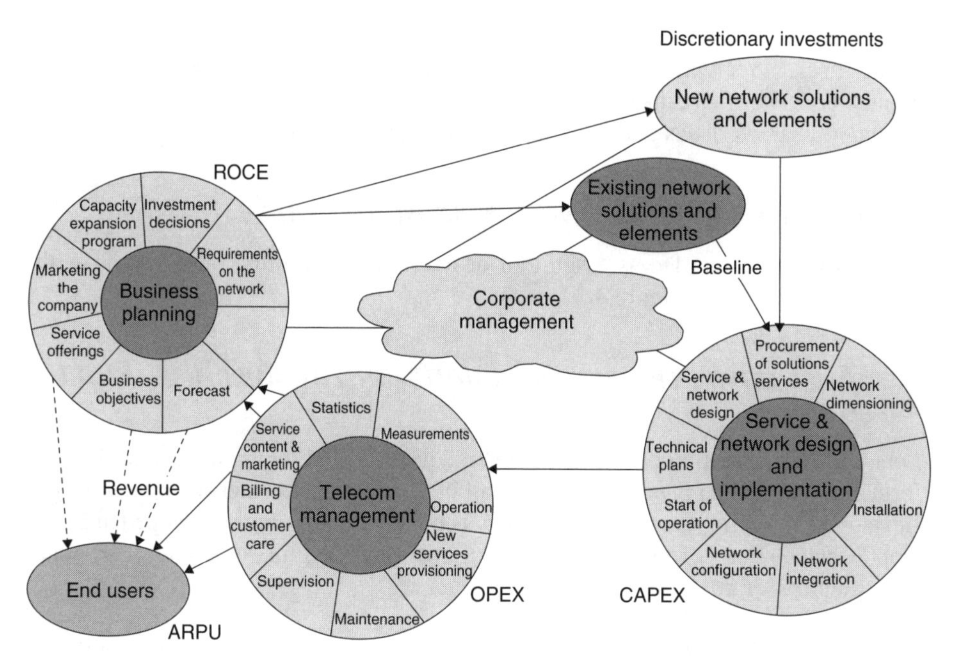

Figure 4.33 The operator process is an excellent tool to analyse the cost impact of various solutions

The service plan is treated in Chapter 5 and the service implementation plan is treated in Chapter 9.

The overall telecom business processes could be described as part of a continuous and dynamic cycle, starting with business development, leading to network design and implementation, going through network operation (which in turn will feed new input data) and coming back to a new business development phase, to start a new cycle.

From this perspective the corporate management takes care of handling this cycle in a way where the main business objectives, including not only short-term financial goals such as profitability or market share (lagging indicators) but also business development leading indicators (customer satisfaction, branding, etc.), are satisfied.

Figure 4.33 shows this dynamic approach to telecom business. It is important to reflect that even when the network, with its continuous upgrades and transformations might be considered the main protagonist of this cycle, it is in reality just a means to satisfy the demands the end users have for attractive services devised and offered by the telecom operator.

It is curious then to realize that most of telecom business literature focuses on network characteristics and technology (or enablers). The cause of it can be found simultaneously in old and new reasons. Old reasons remind us about the characteristics of old telecommunication networks, which were mainly single-service networks. Not having very much to worry or say about the only existing and successful service, the focus was set on how to satisfy or provide it, in the best way.

New reasons refer to the close relationship that can be found between new network technologies (enablers) and new customer services. Frequently enablers open new possibilities on service provisioning and can force a dramatic turn around on telecom business. However, this does not always happen, on the contrary many promising technologies are frequently not able to produce any substantial change on the service menus or service provisioning characteristics and even to cause any attraction for end users.

4.5.4 TOOL: SIMPLIFIED BUSINESS CASE WORK MODEL

This model can be regarded as a sub-tool or a variant of the first one, dealing with all type of investments. See Figure 4.34.

4.5.5 TOOL: A COMPUTER PROGRAM BUILT ON THE BUSINESS MODEL FLOW

The present transformation of telecom networks has a wide-ranging impact. To be able to perform a corresponding business case with all its changing parameters computer programs are developed. The program referred to below is one of the programs developed at Ericsson to support calculations of the impact of the transformation process. Some main goals of this computer program are:

- To demonstrate cash flows as a result of increased and new end-user service revenues for the operator.
- To show the revenue distribution over the value chain and how the operator is positioned to maximize opportunities.
- To show revenue streams for each market segment per end-user service that is offered. Link revenue to capital investment and operating cost for a total business case.
- It is up to the user to assess how suitable the tool is for the business case that is being considered. The tool is quite comprehensive in order to suit many cases. Some examples are: identifying impact of end-user services on the customer network, to sales

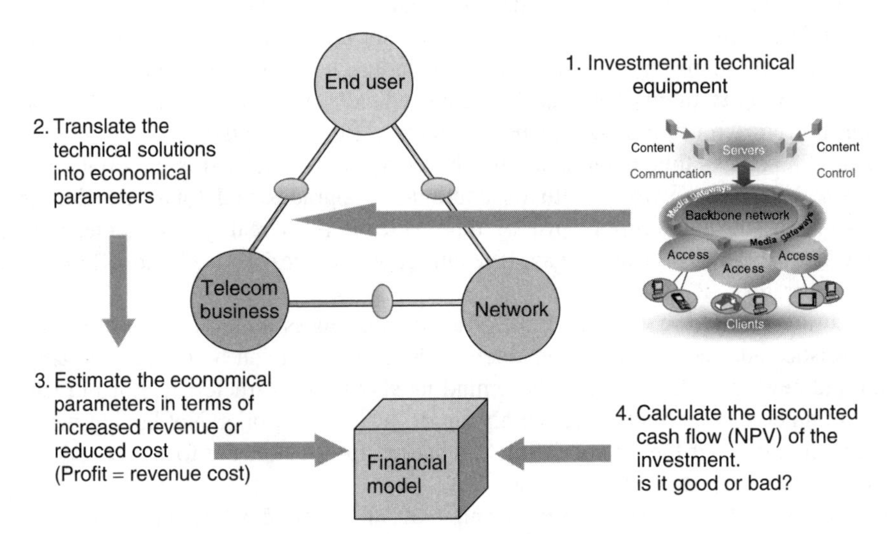

Figure 4.34 Simplified business case model, starting with investments from the network corner

of asymmetric digital subscriber line (ADSL) access services, to sales of a core ATM network, to justification of a move up the value chain and to offering ASP services.
- The tool is sophisticated. Therefore it is detailed and requires a fair amount of input data. If the information is at hand then it is relatively straightforward. Default data in the model reduces the amount of work.

The program contains six functional building blocks:

1. Market model – Details of the overall market in which the actor operates, e.g. size, demographics, average revenue per user (ARPU), market share and penetration.
2. End-user services model – List of existing and future new end-user services (applications) and access services, the timing, barrier to entry, application and service characteristics. Defines the revenue and value chain distribution for each application and access service type.
3. User model – Defines the user profile for each market segment. Identifies the applications they use, the take-up levels, the usage profile, traffic profile. Calculates the per user revenue per application and access service.
4. Business model – Defines the business strategy, based on current operations, current and target infrastructure, applications to offer, market penetration goals etc.
5. Network model – Defines the network structure and capacity requirements based on market growth forecasts and operator network strategy. It estimates the network infrastructure required for a scenario. Determines the cost for each network element applying generic pricing.
6. Financial model – Defines the financial result for the business strategy. Defines the revenues, business and network capital and operational expenditure for the business strategies per scenario.

See Figure 4.35.

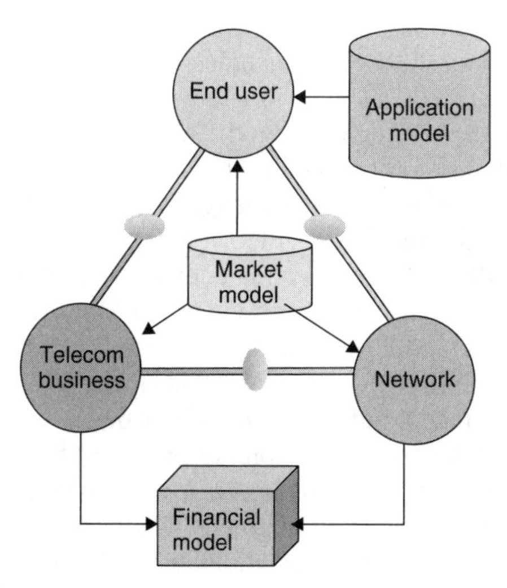

Figure 4.35 Structure of a computer program, which is adding application, market and financial models to the basic triangle

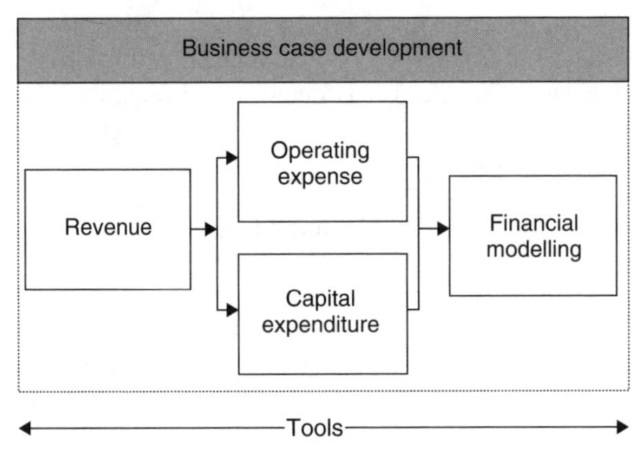

Figure 4.36 Revenue

4.6 REVENUE

Equipped with tools let us now look at the main parts of business case development starting with revenue. See Figure 4.36.The last part on financial modelling is treated in Appendix 2.

4.6.1 WHERE IS THE MONEY?

Since the end user is the main final financing source for new investments etc., the potential revenue from end users is crucial.

A common expression is ARPU (average revenue per user).

Whether the overall portion of money allocated to communication will increase or not is under debate. The Swedish investment bank Carnegie, for example, calculates a doubling of mobile communication services as a percentage of the GDP during the next 10 years. The development in Sweden has shown an increase per household from 2000 SEK a year in 1992 to 5820 SEK in 2001. This includes fixed telephony, mobile telephony and Internet services.

In the UK ARPU decreased during the period 1998–2001 as well as revenue per minute. The data ARPU was not able to counteract this decrease as was expected only a couple of years ago. So, here we see the main challenges, to defend voice ARPU and increase revenues from messaging and mobile Internet.

A key factor is the willingness to pay for content. The ability of mobile devices to receive content such as streaming audio and video will improve continuously. Therefore it is not a very daring assumption that mobile devices will become competitive as one content channel among others, such as books, newspapers, TV and movies, illustrating a true convergence with the media industry (‘mobile media’).

Regarding content, the futuristic generation has proven to be large consumers of pornographic content on the Internet. Many mobile actors are therefore investigating the area of providing erotic content or indeed pornography as a means to increase ARPU with the increased bandwidth of the 3G systems as an enabler. The availability will be restricted by means of codes for interested paying adults. The possible erotic-pornographic services will be of pull type, not push.

A common unit when looking at growth rates is to use the compound annual growth rate (CAGR): the year over year growth rate of an investment etc. over a specified period of time. CAGR is calculated by taking the *n*th root of the total percentage growth rate where *n* is the number of years in the period being considered. This can be written as:

$$(\text{Current value/Base value})^{(1/\text{number of years})} - 1$$

CAGR doesn't represent reality. It's an imaginary number that describes the rate at which an investment grew as though it had grown at a steady rate. Figure 4.37 shows an example of use of CAGR. This figure is a forecast of the UK market growth 1998–2010 (Arthur D. Little).

Mobile voice £5–11 billion CAGR = 7 %

Fixed voice £7.5–8.5 billion CAGR = 1 %

Fixed data £7.6–16.5 billion CAGR = 7 %

Mobile data £0.2–6 billion CAGR = 33 %

For comparison with the real development, mobile data represented 16 % of the revenue for the mobile operator Vodafone for the last quarter of 2002. The SMS service represents most of the data income. One mobile phone per household develops to individual phones per household.

The IP technology for voice services develops with IP PBX, VoIP, IP Centrex and a fast growth in IP-VPN.

Messaging is a forecast favourite, both e-mail, SMS and MMS.

Another example of potential high volume service: The M2M market is untapped. There are 100 million households and homes in Europe alone and over 500 million cars worldwide with 50 million new cars sold every year. There are three million elevators and four million alarm systems in Europe – and 10 million vending machines in the world.

The turn from narrowband to broadband is clear. It can schematically be illustrated by Figure 4.38.

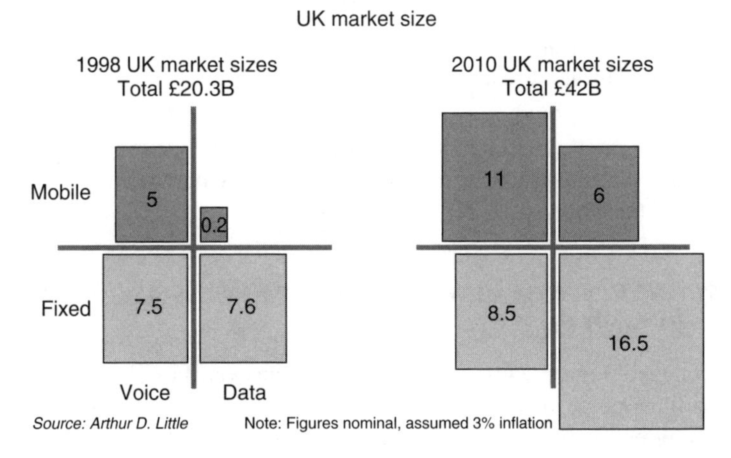

Figure 4.37 A forecast from the UK

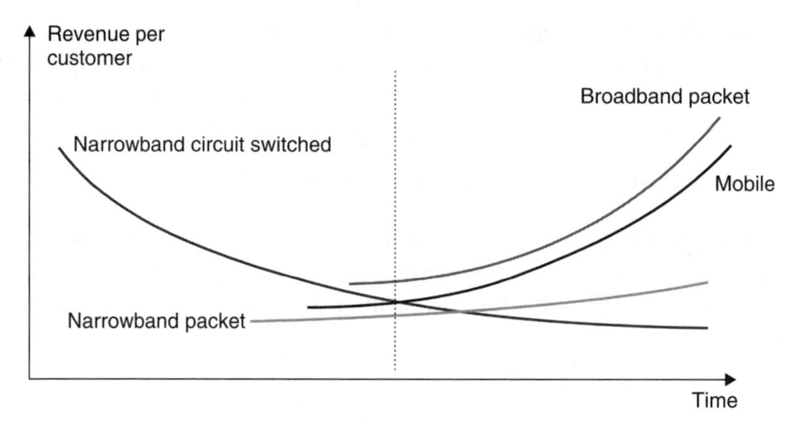

Figure 4.38 Turn from narrowband to broadband – trend in revenue expectations per customer

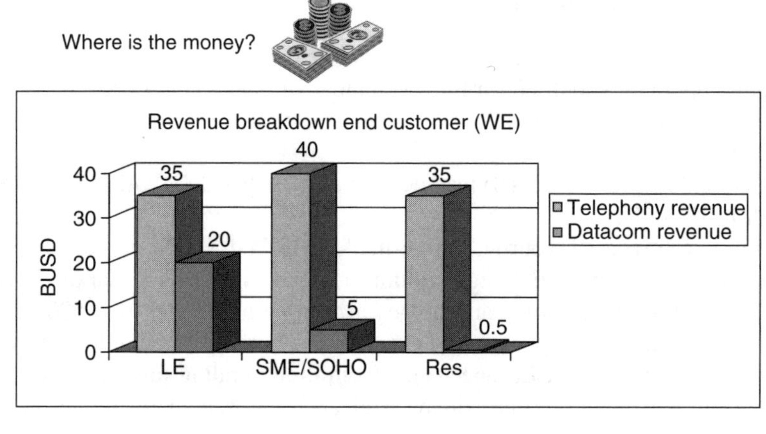

Figure 4.39 Revenue from enterprises (LE in this particular context = large enterprises). *Source*: SCB (Statistiska Centralbyrån), Sweden

The large enterprises are the main source of datacom revenue. See Figure 4.39.

For narrowband circuit-switched services the revenues per customer are decreasing, mainly because of the price pressure on the tariffs.

4.6.2 INTERCONNECT, ROAMING, MULTI-TERMINALS AND LEASING (REVENUE AND COST)

What is interconnect? Figure 3.40 gives a part of the answer. For all types of interconnect there is a need for a point of interconnect (POI). The responsibility for POI creation and maintenance is agreed between the interconnecting actors. But there is a need to define a lot more conditions between the actors such as billing, network management, QoS,

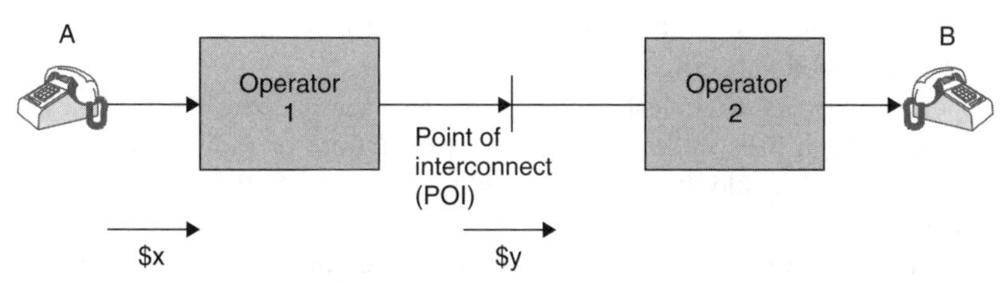

Figure 4.40 Many questions have to be solved between operators

routing, numbering and other fundamental technical plans. The overall terms are usually regulated in a service level agreement (SLA). More on interconnect and SLA is found in Chapter 15, Section 15.5.

Interconnect appears normally both as revenue and as a cost.

Interconnect opens for roaming, where the subscriber can connect to visited networks, belonging to other actors. We can also speak of roaming among a number of technologies provided that the terminals are equipped with such multiple technologies. Then we can also move between different access technologies belonging to different actors. Such possibilities of better use of access investments improve the ROCE value.

An actual term is 'generalized roaming' with a need for appropriate business models for the actors involved.

4.6.3 ACCESS UNBUNDLING (REVENUE AND COST)

For the local COPPER loop there are two possibilities for operators who do not own the access plant:

- To lease the copper line as it is.
- To share the use of the copper line. Example: the incumbent operates PSTN/ISDN. The new operator offers xDSL services.

The copper pair ownership depends on the regulations. If the regulator has enforced 'copper unbundling' the incumbent operator no longer owns the copper access network. In this case the new operator can connect the subscriber in the same way as the incumbent.

Another alternative for the regulator is to go for 'equal access'. In this case the incumbent operator continues to own the copper, but the subscribers can choose their preferred operator and the incumbent must direct all calls to the pre-selected operator. (In Sweden, this has been the case since September 1999.)

A third alternative is that the new operator builds an access network of its own. It is very costly to build a full copper network with pairs to all possible subscribers in an area. This might be profitable only if the result is a market share of 50 % or more in the area. Normally, the incumbent has already connected all the high and medium income people living in the area, and profitability for the new operator would also require churn – an overflow of established subscribers to the new operator.

A cheaper strategy is to invest, initially, in an 'access core', sometimes called a feeder network, which uses multiplexed systems over fibre or radio. If copper for some reason is available it could be equipped with HDSL or SHDSL systems, and later VDSL systems.

'The last mile' or drop to the subscriber is expensive, and a 'pay as you grow' investment strategy should be chosen. Fixed and mobile radio systems fulfil this requirement.

4.6.4 PEERING (REVENUE AND COST)

The traffic conditions in the Internet differ from conditions in telephony networks. The part that initiates a session sends a limited amount of traffic towards the server. The server answers interactively, often with considerable amounts of traffic. This gives an asymmetrical pattern with most traffic generated by the B-side. However for small offices/home offices the situation may be different.

For interconnect purposes the ISPs have 'peering' agreements (peer to peer = 'on equal footing'). Between two top tier peers there is no accounting when using transport resources of the cooperating peering ISP.

The ISPs are ranked in a hierarchical order: primary, secondary etc. Secondary ISPs buy 'transit' from primary ISPs. 'Transit' means that the ISP is allowed to utilize all the peering agreements of the primary ISP. Peering agreements are usually 'free' between operators of approximately the same size. Peering is mutually beneficial to both operators, since they both will increase their performance and connectivity. However, peering is more beneficial to small operators, giving large operators a negotiation advantage. See also Figure 4.41.

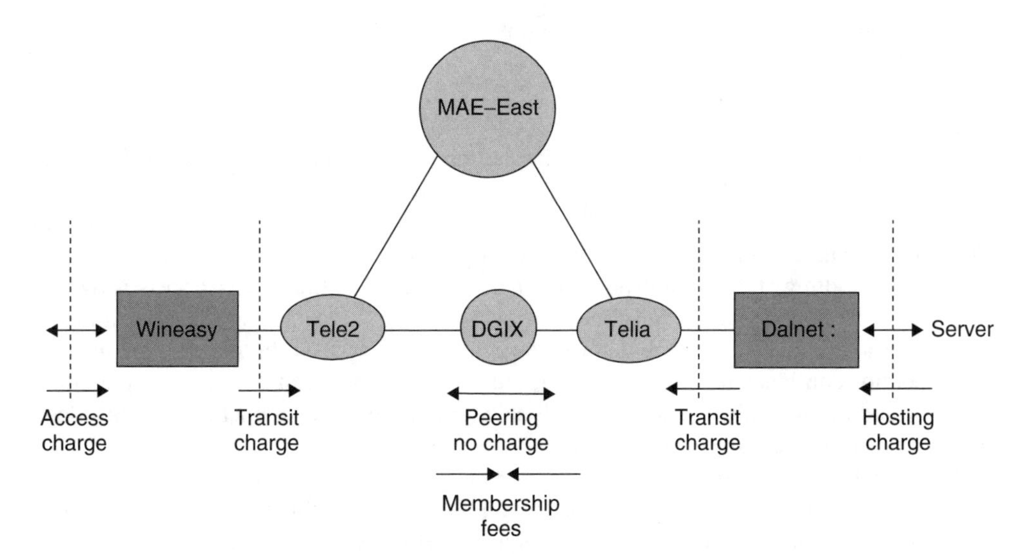

Figure 4.41 Small ISPs pay large ISPs

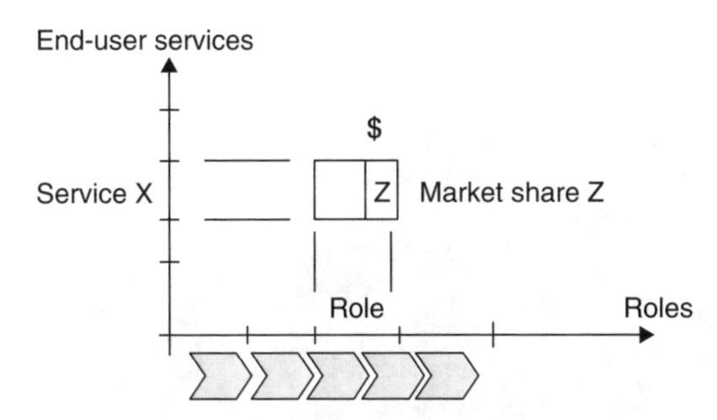

Figure 4.42 Revenue elements from end-user services

4.6.5 IMPACT OF MARKET SHARE

The revenue for a specific actor depends on the role(s), the service offerings, the charging and the market share per service. Both the income per service and the distribution of the revenue between the actors are unstable. Figure 4.42 illustrates the unstable revenue elements from the end users. In the 'good old monopoly days' things were more stable but slower.

The market share for a particular service has a clear relationship to the agility in launching new services, and charging for them.

A common business-oriented approach is to develop business cases for the potential services. See Sections 4.6, 4.7 and 5.2.

4.6.6 IMPACT OF MANAGEMENT

Let us illustrate the importance of a good telecom management with a small 'classical telephony' example. From an operator perspective, the quality depends on anything that affects the number of successful calls, which is the same as the income base. It is easy to make a small business case and see how the answer ratio affects the operator's income. Example:

- A city with 100 000 telephone subscribers
- An average of 12 calls per subscriber and day
- An average income of 1 Swedish crown (SEK) per call

How much will the annual income be affected by an increase of 1 % in the answer ratio? Answer: $100{,}000 \times 12 \times 1/100 \times 365 = 4\,380\,000$. The income increases by more than 4 MSEK per year.

4.7 COST EFFICIENCY

Cost efficiency paired with service quality is a key area of interest for operators to improve competitiveness both in the eyes of customers and financial markets.

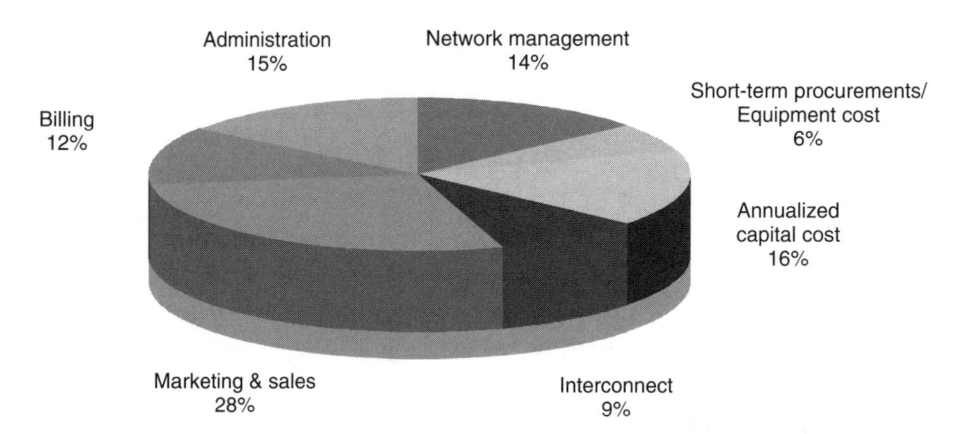

Figure 4.43 Example of costs for an operator. Principal targets for savings: network management and annualized capital cost. *Source*: Federal Communications Commission (FCC)

Figure 4.43 shows an example of the main investment and operation expenses of an operator. Taxes are not included.

It should be noted that the investment expenses are split into annualized capital cost and short-term procurement/equipment cost. Annualized capital costs arise for example when building a telecommunication network. The annualized expenses are related to loans for financing the investment. Short-term procurement is not annualized.

It is crucial to balance the investments in relation to the traffic. As an example, traffic bottlenecks spread and reduce income in other network parts. Over-dimensioned parts increase the capital employed without contributing additional income. In both cases ROCE will decrease.

To better explain CAPEX and OPEX we need a couple of other definitions: baseline cost and discretionary cost.

Baseline costs are required just to keep the network operational and for meeting customer growth. OPEX expenditure is known as *baseline expenditure*, but, as pointed out, investments to meet customer growth are also baseline costs.

Discretionary costs (CAPEX) are investments incurred when expanding the capability of the network to generate new revenue or to modernize, that is reduce the baseline costs.

The objective is to maximize funds available for discretionary investments in order to minimize funds spent on baseline costs, which will drive up revenue and profit.

Any operator has a limited supply of corporate funds to invest in its business and networks. An operator determines a spending level for OPEX, which is usually fixed with a percentage for growth each year, and CAPEX, which can vary depending on how much is available and what is needed. CAPEX is achieved by raising funds externally or using available profit. It is the objective of any operator to minimize the proportion of funds that it spends on maintaining the network in order to meet daily operation and normal growth expectations.

If next generation solutions drive down baseline (OPEX) costs, they will be successful. Figure 4.44 illustrates the concepts: select the new generation network (NGN) to be one of the options in the box ‘Strategy and evaluation of options’. Compare NGN with other options in a business case. Choose the best alternative. Implement the best alternative.

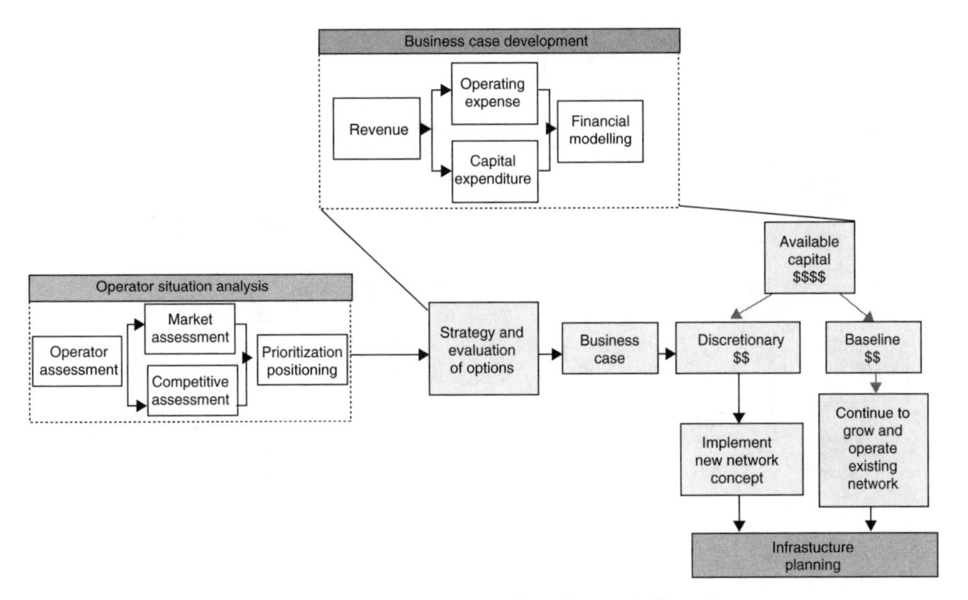

Figure 4.44 The difference between baseline and discretionary costs

Let us start with some baseline costs, illustrated by Figure 4.45 from an expanding cellular operator that continues with a specific technology (GSM). Obviously marketing and customer care are labour intensive parts of the operator processes.

Network or telecom management represents main baseline expenditure, with customer care as an important part. See Figure 4.46.

The Issue of Changing Technology: Uncertainties

When changing technology, or modernizing the network, we consider *discretionary* costs. To evaluate the outcome of a possible change there is a need for some kind of tools.

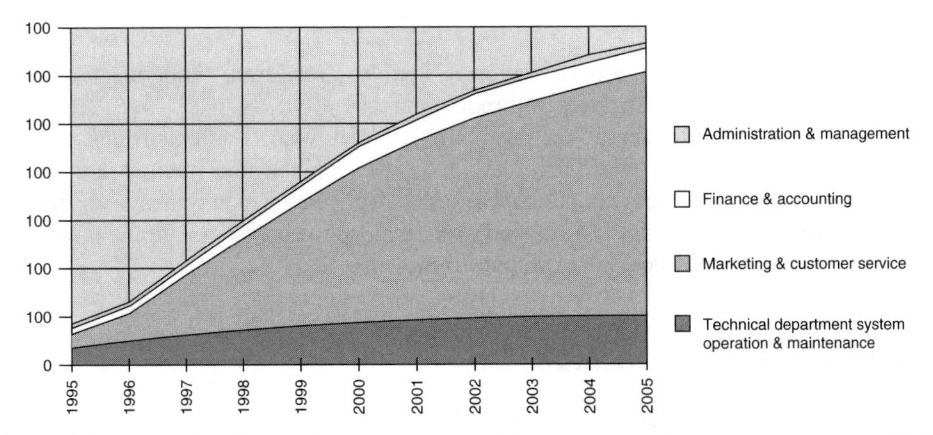

Figure 4.45 Example of staff development for cellular operator with a growing business (GSM technology is used throughout the period)

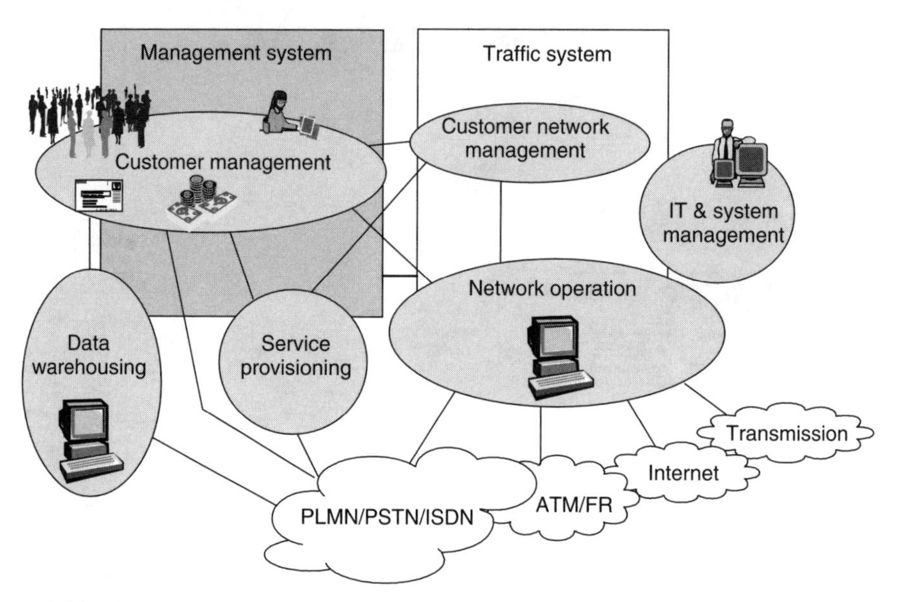

Figure 4.46 Telecom management is a main baseline cost with expenditure for staff, IT and O&M infrastructure

One is the business case, described before. We must in this context also look at the impact of a change on the process chain. Can the processes be performed faster after the change? Or cheaper?

There is a fundamental difference between the introduction of broader accesses, such as EDGE, UMTS and ADSL, and the horizontalization of the network core. The first case is an extension or an upgrade, and the second case is more of a modernization. Extensions have to be paid for by means of the new traffic they are supposed to carry, whereas implementation of a modernized core network design can be compared with a traditional (vertical) way of building and operating a network core for voice, data and multimedia. Of course, it is also possible to use the new 3G accesses as a modernization tool by using them for voice.

In the following, network modernization is mainly considered. Modernization has a limited dependence on new services and regulations.

Upgrades/extensions to access are also briefly dealt with. For extensions like 3G accesses there is a well-known debate going on. Among the uncertainties are regulatory issues, such as operator cooperation rules and coverage requirements of sparsely populated areas, and the length of the end-user 'maturity process' for the new services. Cost efficiency then becomes more difficult to measure.

4.7.1 MODERNIZATION ASPECTS

When modernizing a network it is highly desirable to cut the baseline OPEX costs and increase the funding for discretionary costs. However, in an uncertain world we should carefully select the 'right' cost-cutting technologies, with a high degree of flexibility.

The potential of the next generation solutions contributes with many types of savings: savings resulting from a new organization, new business processes, reduction in network complexity, elimination of network elements, and reduced operational requirements. Let us therefore in the following chapters choose a number of significant processes and compare the traditional and the new solutions.

Seamless migration solutions should be used when introducing new technology, to retain existing customers. Through a successive and business-driven transition, services and applications can be shifted to inherently future-proof and scalable solutions.

Discretionary Core Investments

This chapter has a business-oriented focus. However, in order not to lose the firm connection between telecom business and technology, the chapter also contains some technical elements. For a deeper study of *core and backbone techniques* refer to Chapters 3, 12 and 14, and the core part of the CD 'IP internetworking' mentioned in Chapter 1.

Keywords for the new core are multi-service and multimedia. The transformation affects both mobile and fixed networks.

The core network is divided into three layers: the connectivity layer, control layer and application layer. User applications reside at the edge of the network, accessible through the network services. Communication and control servers comprise the control layer, providing call control and session management, and control service provisioning and usage as well as network infrastructure resources. The connectivity layer including the backbone network takes care of the packet transport. End users connect to the multi-service network through access points.

Process Differences between the Traditional Vertical Networks and the New Horizontal Structure

Figure 4.47 exemplifies what process activities to analyse when evaluating alternative solutions.

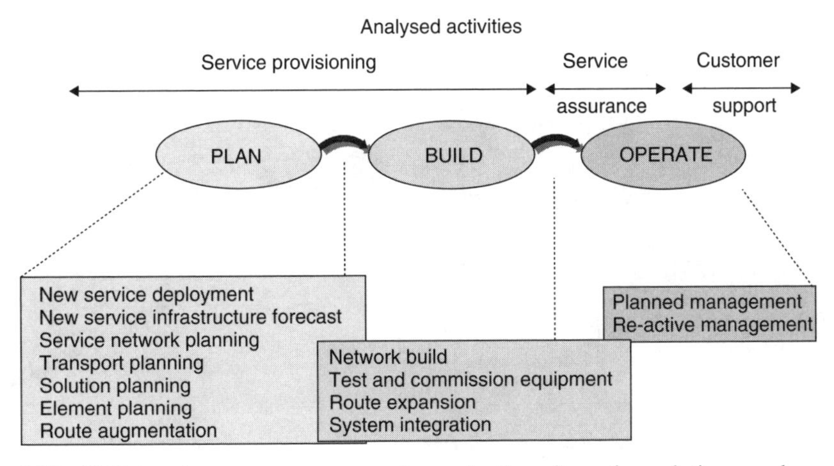

Figure 4.47 Utilizing the operator processes for evaluating alternative solutions, such as vertical versus horizontal

Overall Result of the Cost Comparison: Cost of Ownership for the Network Core

Cost of ownership combines CAPEX and OPEX (Figure 4.48). According to Figure 4.46, the new architecture and the old one has approximately the same CAPEX, whereas OPEX differs significantly. However, other sources claim a significantly lower CAPEX for the new architecture as well.

From Figure 4.48 the conclusion cannot be drawn that CAPEX is not decisive, since this is just an example. Anyway, the large impact of the next generation on OPEX is clear. In addition, the operator need for agility and flexibility is strongly supported.

4.7.2 RESULTS OF PARTICULAR COMPARISONS

Traffic Planning

Let us start with the planning area. Traffic planning becomes much easier with the new network. Compare the route reduction in Figure 4.49. Going from many networks to one or a few simplifies the work, saving resources and time.

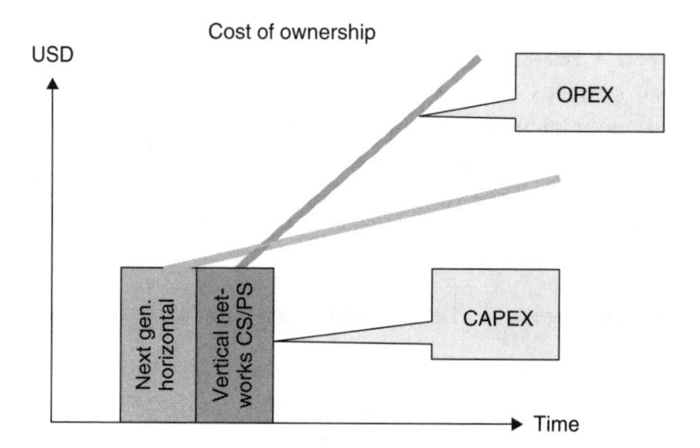

Figure 4.48 Cost of ownership includes CAPEX and OPEX

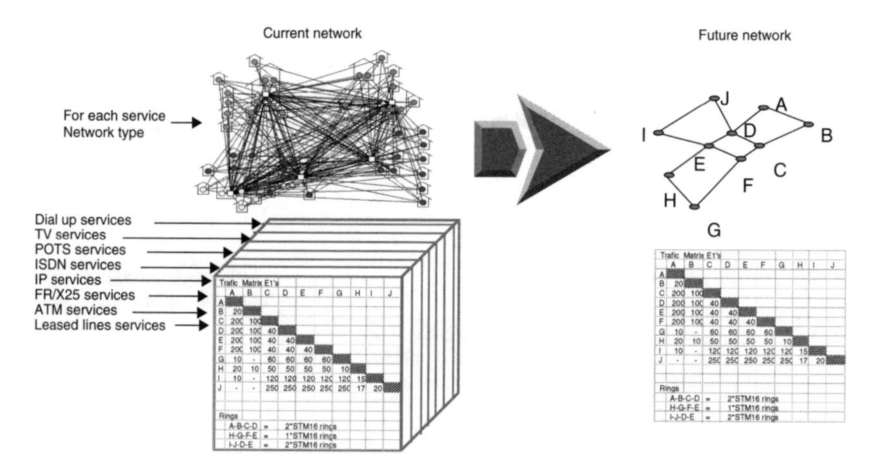

Figure 4.49 Traffic planning comparison

The traffic engineering process is used to determine the volumes of traffic flowing between different points on the network in order to dimension the infrastructure required at the:

- access interfaces
- switch nodes
- transport networks, cross connects, protection etc.

In order to do this the process must be repeated for each traffic type on each network and these traffic flows aggregated where common transport exists.

This is further complicated by the fact that the input forecasts are quoted in the number of services and must be converted to traffic.

A converged network eliminates the parallel traffic engineering processes and collapses it into only one aggregated matrix, which is further simplified by the non-meshed or partially meshed ATM network topology. The example in Figure 4.49 illustrates how the process of traffic planning can be simplified if all products and services are provided on one common platform.

The network diagram on the left is a telephony network taken from a real example but simplified, since only transit routes are shown. In practice this would be even more complex with hundreds more routes. Similar networks of differing complexity would exist for each product or service type.

Service Provisioning and Customer Care

When all services are provided on one common platform it is easier to migrate customers from one service to another, both in the procedures required and the physical changes required. No longer are you disconnecting the customer from one network and reconnecting them to another. This is especially important in keeping operations costs down and in particular where exchanges are remote or no longer manned. See Figure 4.50.

- Single activation and configuration systems can be adopted.
- All the activation systems that are needed for 'vertical networks' require training of operator staff. Deficiencies in competence imply errors and delays in getting services to the market or connected to the customer.
- The same user interfaces can be made available for all services. This maintains the familiarity with the configuration procedures.
- This familiarity can dramatically reduce the cost of processing churn in competitive markets
- There are fewer elements to update, interact with, upgrade and activate services on.

The importance of customer care for churn and retention rate can hardly be overestimated. Since it is crucial for the business, it is wise to be cautious to cut here:

- It costs five times more to acquire new customers than to retain old ones.
- One customer lost will inform seven others why!

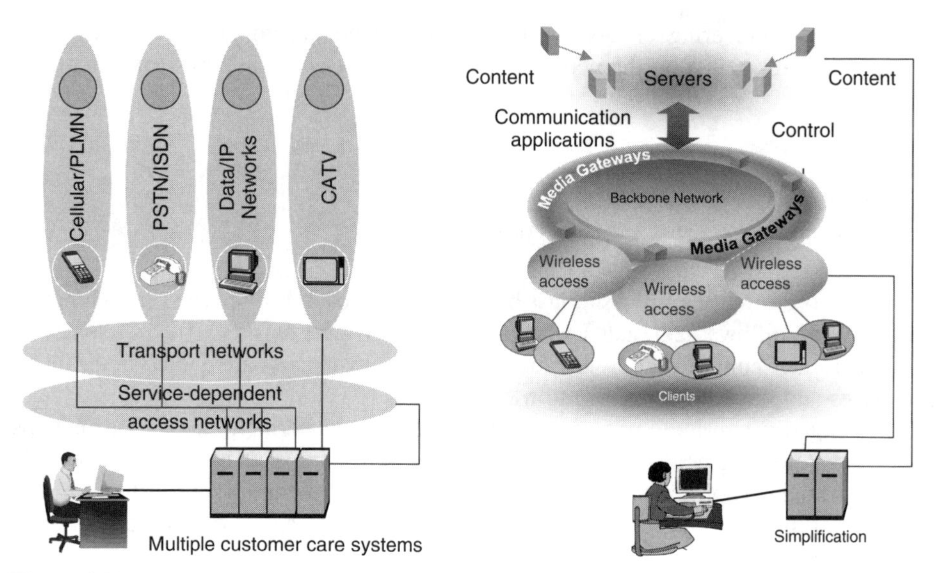

Figure 4.50 Handling a horizontal network is simpler than handling many vertical networks

Monitoring and Capacity Management

- Operational complexity of monitoring and interpreting network statistics and parameters is directly proportional to the number of elements and number of service networks.
- The horizontal network combines multiple networks and considerably reduces the number of network elements, resulting in less effort and more streamlined processes needed to undertake capacity management.
- There are fewer interfaces and only one single core technology to monitor.
- Traffic is monitored only in its bulk aggregated form.

Much of the traditional traffic engineering can be disposed of, as detailed trends are no longer required

4.7.3 COST EFFICIENCY IN THE ACCESS NETWORK

For a more in-depth technical study of *access techniques* refer to Chapter 13 and to the CD 'IP Internetworking', referred to in Chapter 1.

The access network is the main contributor to capital expenditure. As an average more than 50 % of all network investments are related to the access part. It is, however, also a door from the subscriber to the core network and the possible services and applications, or, with another view, a door from the network operator to the 'money', the paying subscriber. The size of the door corresponds to the maximum bandwidth of the services. A larger bandwidth implies an increased income potential.

Pay as You Grow

'Pay as you grow' is a good rule that is applicable on the expensive access network. If the investments are much ahead of the related revenue much equipment (and capital) is poorly utilized. See the example on GSM radio access investments in Figures 4.51 and 4.52. A fairly modest over-investment may result in a prolongation of the payback time of half a year.

Reusing Existing Investments

For discretionary investments targeting an increased bandwidth the cost-efficient operator might adopt this approach: 'A successful communications provider needs a stepwise approach in building cost-efficient broadband multi-service access networks that cater for

Cumulative no. of RBS sites
1000
900
800
700
600
500
400
300
200
100
0
Over-investments
Balanced investments
Case 2
Case 1
0
1
2
3
4
5
6
7
8
9
10
Year

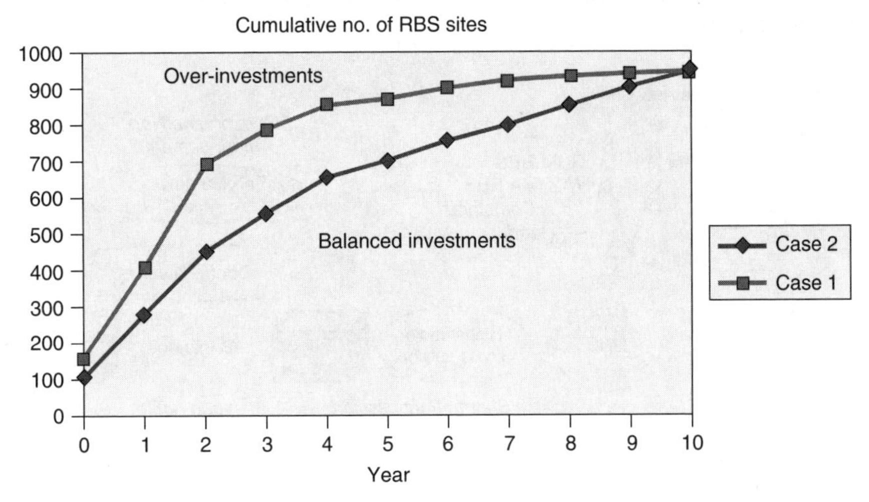

Figure 4.51 Example of various strategies for access network investments

The cellular race requiring a well-balanced investment plan: an example including two scenarios
Accumulated operating income KUSD
700000
600000
500000
400000
300000
200000
100000
0
-100000
-200000
Over-investment
Balanced investmentse
Case 2
Case 1
0
1
2
3
4
5
6
7
8
9
10
Years of payback time (crossing zero)

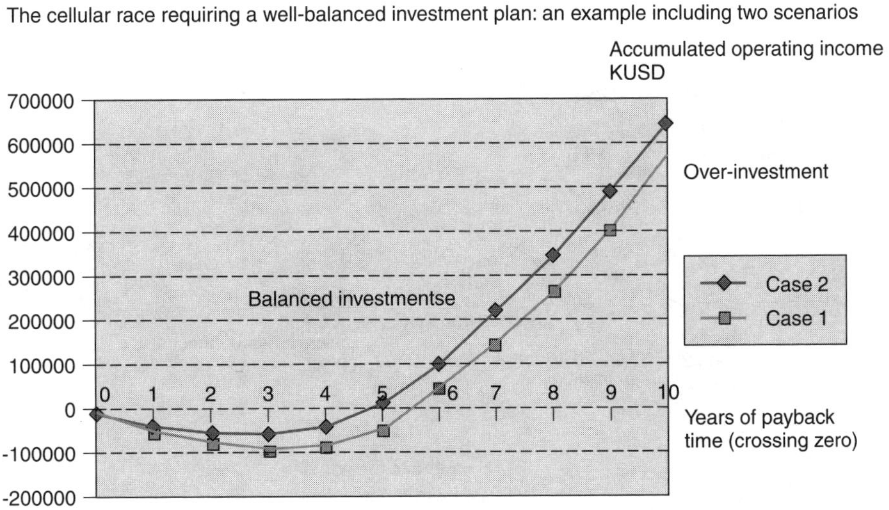

Figure 4.52 Example of various strategies for access network investments

the new demands of end users while at the same time reusing existing investments and securing new revenues'.

This approach fits extremely well in the case of established access operators with investments in copper, masts etc. The best example today is perhaps implementing 3G mobile systems with maximum reuse of 2G investments. See Figure 4.53.

Convergence Fixed Mobile in the Access Network: Network Sharing

Cost-efficiency optimization includes expenditure from leasing or renting capacity from other operators. A new operator relationship is called 'network sharing'. This is a way to reduce CAPEX by means of radio access sharing or geographical area sharing, introduced by UMTS operators.

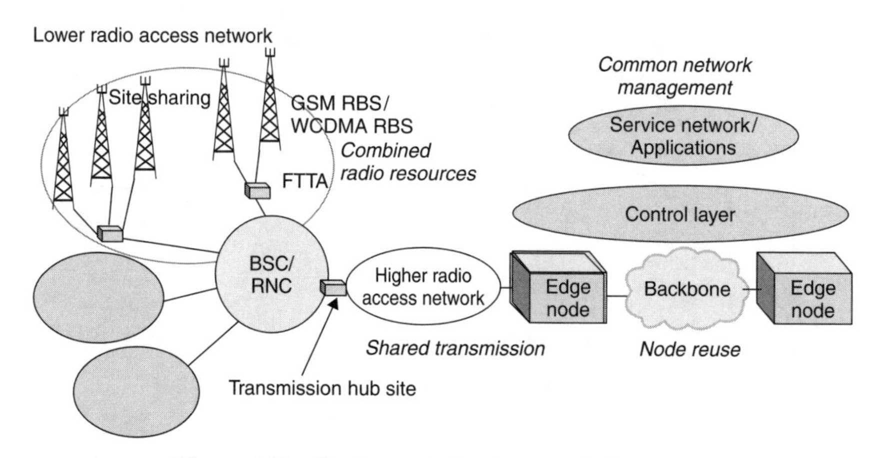

Figure 4.53 Sharing resources between 2 G and 3 G

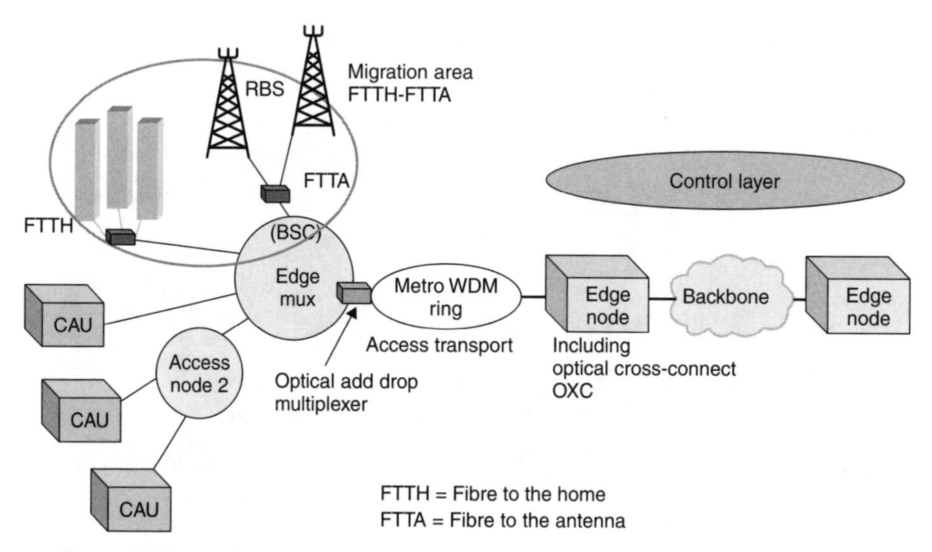

Figure 4.54 Sharing investments in fixed and mobile access in a metropolitan area

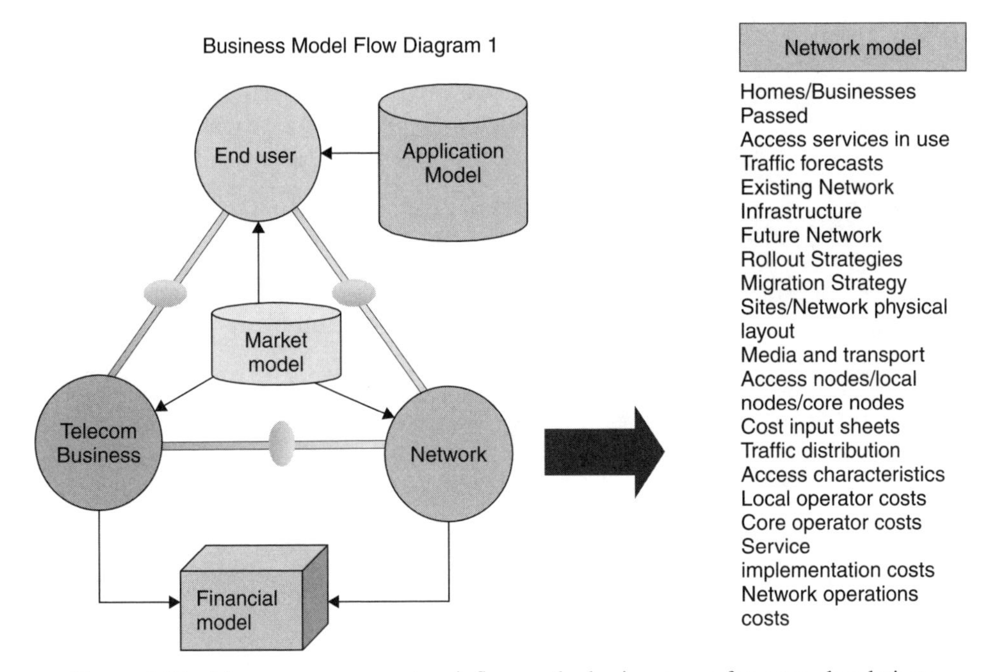

Figure 4.55 Many access parameters influence the business case for network solutions

If permitted by regulators fixed and mobile accesses will probably converge. Going the UMTS way also makes network sharing a possibility, in order to decrease overall investments and improve ROCE. Another interesting feature would be to enhance bundled fixed and mobile offering. See Figure 4.54.

Introducing IP/Ethernet on Fibre Systems

According to source information from Yipes, Dell 'Oro, Yankee Group, Extreme Networks, Juniper Networks, compiled at Ericsson, the cost components equipment, bandwidth management and annual upgrades are much cheaper for IP/Ethernet than for IP/ATM/SONET or IP/SONET.

For a regional network with five hubs and 10 rings the cost might just be about 20 % of the other alternatives.

Using Computer Programs for Access Optimization

In order to achieve optimal cost efficiency a computer program might be useful, or even necessary. The reason is the very large number of parameters that are involved in network optimization. See the computer program model in Figure 4.55.

The calculations are made according to a financial model and expressed for example in net present value (NPV) or internal rate of return (IRR). A typical go/no-go value of IRR could be 15 %.

5

Services

5.1 INTRODUCTION

5.1.1 GENERAL

This chapter is more task and target oriented than the previous general chapters. It integrates the three corners dealt with in Chapters 2, 3 and 4.

Services are the merchandise of the telecom market. The end users consume and pay, the actors provide and sell and the network is the tool.

The main goal of this chapter is to support the revenue side of the ROCE (return on capital employed) equation. Return is specified as revenue minus cost. This chapter is linked to Chapter 4 by means of Figure 4.31 (see also Figure 5.2), where revenue is represented by a consumer service list, an enterprise service list and possible additional

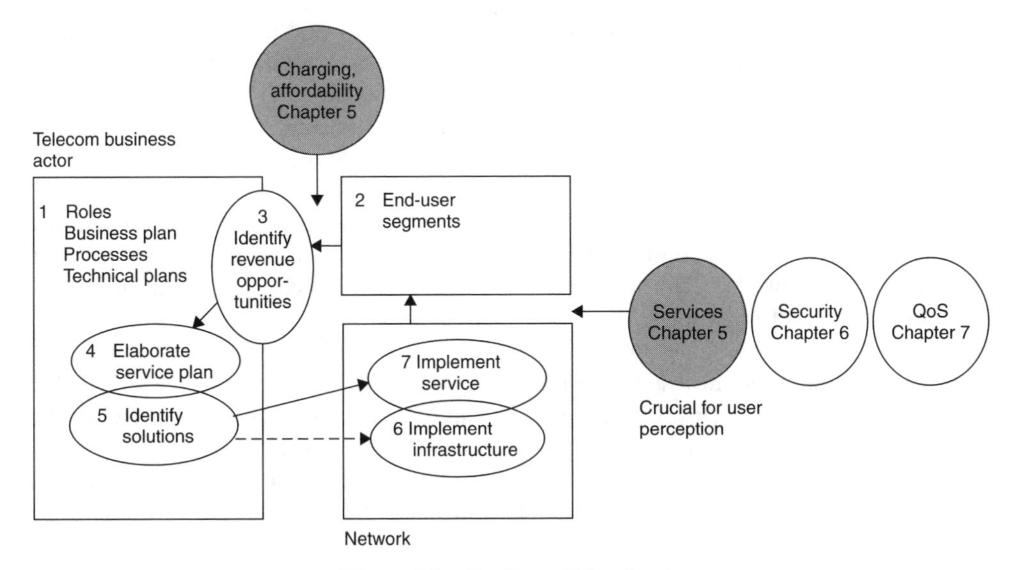

Figure 5.1 Position of the chapter

Understanding Changing Telecommunications – Building a Successful Telecom Business. Edited by A. Olsson
 ISBN: 0-470-86851-1

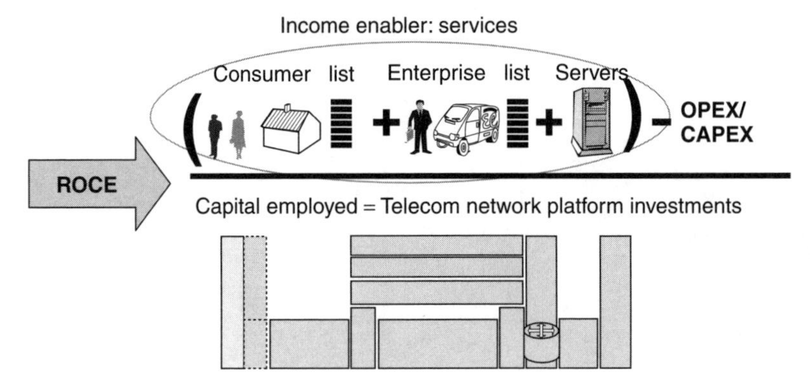

Figure 5.2 The importance of services as revenue generators in the ROCE equation

income from servers to other actors. At the end of the chapter we will also connect the service plan to the network, since services may be prohibitively expensive when the network does not easily match the service requirements. Service-related cost efficiency is included, various ways of reducing the bandwidth requirement are explored. This type of feature, such as coding standards, can be regarded as 'economy of service' (EoS). (The overall EoS includes for example aggregation and concentrating devices as well.) EoS functions can often be regarded as a kind of middleware. However, whereas EoS functions have a clear cost-saving goal, middleware in general has a taste of prefabrication and re-use to shorten development time. Finally the chapter discusses a number of charging methods.

The service plan with its lists of services should be marketed in the user's language, whereas the implementation of the service plan could use another more technical internal language. If the technical language is used towards the market the merchandise becomes termed 3G services, WAP services, ISDN services, GPRS services etc. This terminology may make it easier to define which investments are suitable, but may be confusing for many ordinary potential customers.

If the customer language is used there is a need for a kind of translation from this language down to the detailed implementation and service provisioning. Such a translation is illustrated in Figure 5.3. The service implementation mentioned in the figure is treated at an overview level in Chapter 8.

The importance of customer orientation can be illustrated by means of a comparison. Imagine an operator, but instead of running a telecom service, he runs a travel service. He offers holidays (the service) to people (users) and to achieve this he needs some means of getting people to the holiday, i.e. a plane (transport mechanism).

The professional operator does not sell holidays by promoting the features of the plane. He sells the holiday. The plane is only there to make sure that people arrive at their holidays in an efficient manner.

The business model implies that the money an operator (whether telecom or travel) earns comes from the users of the service. Chartering a plane costs the operator money. If enough people buy holidays, the operator will be able to pay the plane company and make a profit.

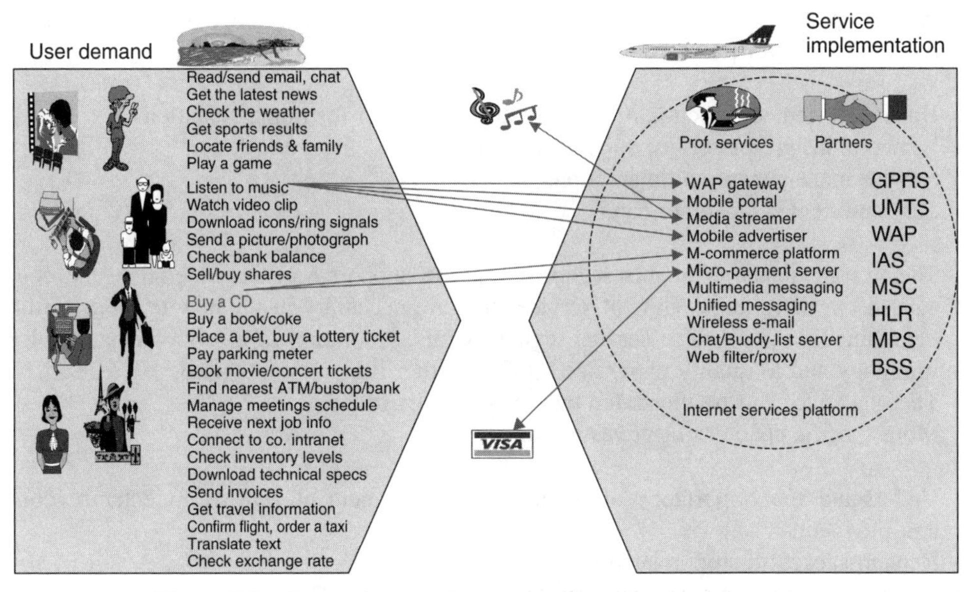

Figure 5.3 Connecting user language with service implementation

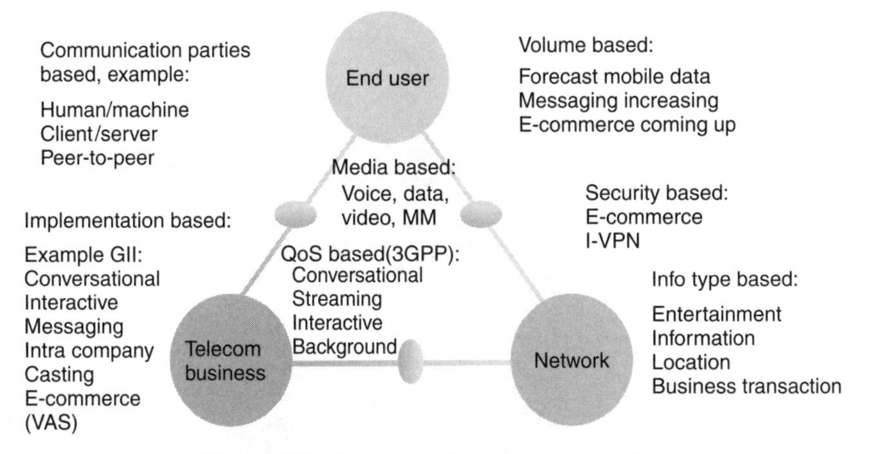

Figure 5.4 Examples of service segmentation

Among the translation tools there is service segmentation or grouping. Some segmentation methods are close to the user and others closer to implementation. A number of alternatives are presented in Figure 5.4.

5.1.2 OBJECTIVES

Understanding service attractiveness and potential, and the success factors are crucial conditions to make the business cases realistic and a good base for decisions. When summing up the required need for service-oriented competence, it covers all the main parts of the operator processes:

- Business-oriented selection of services and partnering for implementation.
- Network design for service support, including service network design.
- Service management and marketing.
- Service-oriented customer care.

The overall objective of this section is thus to support revenue generation and contribute to the multi-angle view of services that are necessary in the new telecom world. Special attention is given to service segmentation, elaboration of a service plan, service terminology and to quality of service (QoS), another important integrator.

The objectives can be illustrated by means of Figure 5.5.

More specific chapter objectives are to:

- Understand the importance of digital rights management (DRM) to enhance content provision.
- Explain success factors related to services.
- Know two main ways of grouping services.
- Explain web services and multimedia services.
- Know the main possible principles for charging.

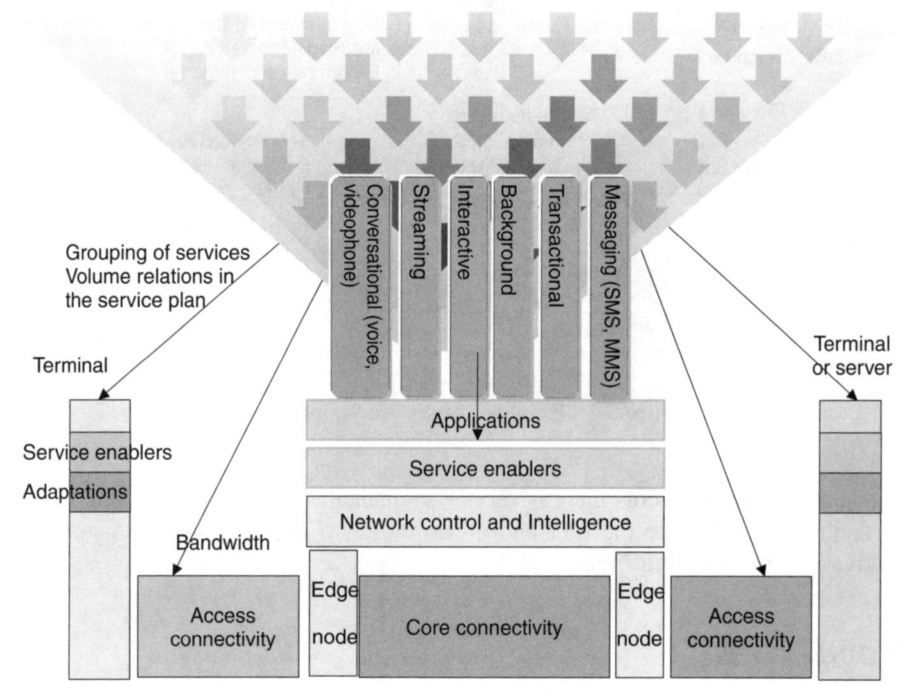

Figure 5.5 Hundreds of services are grouped into a number of segments, with defined requirements on the network and terminals

- Explain what a service plan is and the criteria for service selection.
- Understand service QoS and related network requirements.
- Describe three methods to decrease service bandwidth needs.
- Understand the relation between service requirements and design of various network parts.

5.1.3 TERMINOLOGY

The service area is dynamic. Today we talk about multimedia services, web services, positioning services, video services and so on, proving that 'services' can have many different flavours. Other important terms are media stream and digital rights management (DRM).

What is DRM?

DRM poses one of the greatest challenges for content communities in this digital age. A main reason is that digital files can be easily copied and transmitted, breaking copyright law.

DRM covers the description, identification, trading, protection, monitoring and tracking of all forms of rights usages. Since content is a key to success for the integration between the telecom and the media industries, a solution is crucial.

What is a Media Stream?

See Chapter 1, terminology.

What is Multimedia?

Multimedia is a key target for ongoing development. Multimedia is any combination of voice, video and data, used to enhance the end-user experience to something above traditional telephony. It can include both conversational and streaming services (explained below).

Multimedia can be described as a combination of two or more media elements synchronized in time and/or context. 'Conversational multimedia' is a subset of multimedia, offering real-time, multidirectional communication between two or more parties, using multiple media concurrently. Multimedia services that are combined with non-media related applications, such as directory information or location, can be referred to as 'combinational services'.

Multimedia services can be:

- Narrowband Internet access
  - Enhanced VoIP (web-based calling, service invocation etc. (voice + web))
  - Web browsing and e-mail
  - Unified messaging
  - Appliance control (home, office control and surveillance etc.)
  - Presence services (voice, text)

- Broadband Internet access
  - Streaming services
    Video and audio downloads
    Instant messaging (video/audio etc.)
    Presence services (video, voice, text)
  - Conversational services
    Multiparty video/audio conferencing
    Gaming
    Gambling
  - Real-time interactive digital video/audio (live TV etc.)

Mobile multimedia using the Internet will probably be a significant, perhaps dominating communication form in the future.

Advanced 'virtual reality' based on multimedia using sound, images, M2M data and feelings will be there if we are prepared to pay for it. In the meantime 'augmented reality', AR, will probably be a practical and popular form of communication. The real and virtual reality is mixed to help or enjoy people in certain situations. One example: you put on your special AR glasses through which the real-world surroundings can be seen. Additional information as text or pictures is projected on the glasses depending on the actual context. You get the sound via the earplugs. When you for instance walk in a new place you can get relevant information of where you are or about buildings or landmarks that you pass.

Even if spectacular new applications can be foreseen, a more immediate and important application area is that of support for disabled persons to compensate for a lost or degraded sense or ability to move.

Conversational Multimedia

Conversational multimedia is bi-directional (interactive) communication (e.g. for video telephony) by means of real-time end-to-end information transfer from user to user. For conversational multimedia calls the following functions require state and functions in the call session server:

- Support for enhanced addressing (Enhanced because of multimedia)
- Enhanced security (Enhanced because of multimedia)
- Enhanced charging (Enhanced because of multimedia)
- Supplementary services

Note that the streaming multimedia (e.g. video streaming) can be transparent to the core network, i.e. no call session server is needed.

What are 'Web Services'?

Web services are briefly explained in Section 3.13. A more detailed explanation is given in Appendix 1.

5.1.4 HISTORY AND SUCCESS FACTORS

History

Historically voice services have been the main source of telecom income. Even if the data traffic is larger today, globally, voice income is still at least equal because of the different charging methods between voice and data. Since fixed and mobile voice is maturing in many countries, the development of the telecom sector will depend more and more on non-voice services or enhanced voice services including multimedia.

Long-term forecasting of service (phone calls) development is another historical feature. It goes without saying that this is more or less impossible in a multiple service environment with considerable uncertainty about the market for individual services.

Success Factors: Network

In this less predictable situation the main tool, that is to say the network, must permit flexibility and agility to respond to the actual development. This is a main reason for the 'horizontal network'. In addition there is an increased need for flexibility in the connected terminals, which leads to a development where services are downloaded to the terminals.

This does not mean that the actors can ignore their services once they can provide a large flexibility. With a good 'service feeling' and professional marketing the actors can increase their market share. And the network is by no means self-adaptive to the actual service load. On the contrary, service control becomes a matter for terminals as well as the network. This is because the services entail a number of attributes. Service implementation is treated in Chapter 8.

A specific success factor in the network is the introduction of a service layer.

Success Factors: Services 1

Östen Mäkitalo in Sweden has proposed the following criteria for successful services (also mentioned in Chapter 1):

- Sufficient security
- Sufficient QoS
- Sufficient bandwidth (sometimes included in QoS)
- Availability of terminals
- Sufficient coverage
- Sufficiently cheap and easy to use

These requirements are considered later in this chapter.

Success Factors: Services 2

Quoting Wireless World Research Forum services have to be attractive, intuitive and easy to use, with personalization and ubiquitous access. The services must also be developed easily, deployed quickly and, if necessary, altered efficiently.

Regarding the easy-to-use goal there is a concept called 0–1–2–3 that is used by Ericsson. It can be explained as follows:

0. Manuals: no manual should be needed for a user to get started with mobile Internet services. Configuration parameters are automatically sent to the user's phone over the air.
1. One-button access to your personalized homepage: each user has a personalized portal home page that is automatically set up from the start. The user can access the home page through a single push of a button and then personalize the home page to suit individual needs.
2. Number of seconds to service activation: it should not take more than two seconds after pushing the button for a user to access the personalized home page.
3. Number of clicks to services: the design of the solution ensures that no more than three button clicks are required to reach the desired service on the mobile Internet.

5.2 THE SERVICE PLAN

5.2.1 INTRODUCTION

For public operators and service providers the service plan develops, when implemented, into the content of the service pages in the public directory. Some promising sources of revenues and promising individual services are found in Chapter 4. Establishing a good service plan requires knowledge about all the triangle corners:

- **Telecom business:** the selected role. Actor competence. Operator processes. Business case impact by the service. Business model. See also Chapter 4.
- **End users:** who are the target groups? What do they like? ARPU? Is it possible to make a forecast? Marketing strategy? Pricing strategy? Customer care. See also Chapters 2 and 4.
- **Network:** how to achieve the success factors of the telecom business (QoS, security, bandwidth, coverage, easy-to-use terminals). See primarily Chapters 6, 7 and 8.

5.2.2 ELABORATION OF A SERVICE PLAN

Methodology

Figure 5.6 refers. First, there is a need for service selection strategy. As an example, the following questions could be raised as input for decision about possible new services:

- What is the significance for the user? For the provider?
- How soon can the service be in operation?
- What about the value chain to produce the service? Is there a need for cooperating parties/partners to make the chain complete? Competence?
- What is the introduction cost?
- What is the profitability?
- Is there an impact on other services and, if so, to what extent?
- Are there any risks?

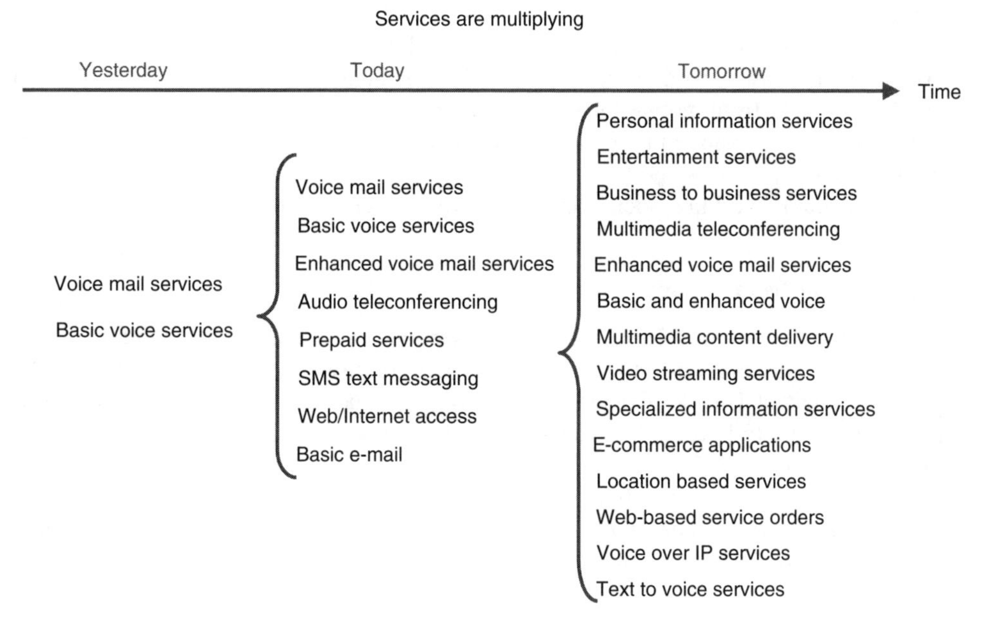

Figure 5.6 Elaborating a service plan involves a steadily increasing amount of work. *Source*: Telemanagement Forum

- What is the expected service penetration over time? Growth potential?
- What is the average user traffic volume?
- What tariff is reasonable?
- Are there any applicable regulations? (For example full deregulation)
- What are the service attributes?
- What about technical level? (Need for consultants and partnering roles for this reason?)

Based on the outcome of this investigation an introduction and promotion strategy might be elaborated. A strategy could include the following elements:

- **Overview:** service description, roles and actors
- **Financial:** revenue and costs
- **Marketing strategy:** segments, pricing, marketing channels
- **Business model:** flow of money between roles. Success factors: End user and operator benefits, etc.

Some Trends as an Input for Selection of Services

- Within a few years a majority of mobile and fixed users are expected to use some sort of advanced communications solutions. MMS services are being introduced as 'standard' features in both fixed and mobile subscriptions. Corporate business users are shifting from leased lines to voice and data IP-VPNs supporting new services, e.g. video conferencing etc.

- Human-centric services have been overlooked on behalf of technology centric. Services must be easy to use. Mobile telephony could fuel 3G.
- Some people draw a parallel between the development from radio to television with the development from mobile telephony to mobile data/Internet/multimedia with video sequences, games, photography and multimedia mail. The change is considered about the same but will take some time.
- Instead of home working, socializing is becoming more popular.

Lessons Learnt from I-mode

See also section 5.3.6 on SMS messaging and section 5.8 on bandwidth requirements

- I-mode shows us that information exchange between users is important. E-mail is the most used part of I-mode. Content created by other end users is very valuable. MMS has the potential to become big, as well as instant messaging. Keyboards are important to enter more information.
- It is important to learn more about end users, for example mobile Internet users and how they use the services. The youth segment is very important. They have been leading the way for mobile Internet so far (SMS, I-mode).

  Interactive communication is preferred rather than retrieval of information.

Application Platforms for Large Groups

Instead of targeting users with a number of single services, application platforms can be designed for various purposes and target groups. An example could be a virtual society platform. The supported activities could for example be *purchasing* (e-commerce, B2C, retail fashion, retail grocery store, car dealer, food on line, mail order clothing store), *social services* (childcare, school, public healthcare, social security), *transport services* (public transportation, booking office), *housing rental agency* (Yellow Pages), *leisure* (golf course, sports club) and *enjoyment* (cinemas, gambling, games and betting, TV on demand).

The applications are reached through a portal for the selection of a group of services. Evidently, the growing range of services needs more and more application support servers, such as bank servers, education servers, advertising servers, game servers etc.

Service Packages

Service or application providers can package services focused on specific profitable user segments. The same service, e.g. e-mail, can of course be part of several packages. Examples are packages for ‘Youth’, ‘Executives’ and general ‘Life enhancement’. The Youths are expected to be the first to try the new services followed by Executives and ‘Life enhancers’.

Packages consisting of low-cost voice and data services will be possible and probably attractive for 3G operators.

5.3 A COMMON SEGMENTATION OF SERVICES FOR MOBILE INTERNET

Mobile Internet content can be loosely categorized into five areas. Such services are either now available or under development. It is also common to use a more coarse segmentation into two types, mentioned in Section 3.4.1:

- **Wireless Internet (WI).** Traditional Internet 'desktop services' are available in the mobile device ('cutting the cords').
- **Mobile media (MM).** The mobile device becomes a new 'media channel' in addition to TV, radio, newspapers, with direct and 'on-the-spot' interaction and consumption.

5.3.1 BUSINESS SERVICES

The most widely used service may be access to e-mail, office faxes, intranets and extranets, directories and other business services over a wireless device, be it a phone, a communicator or a WAP-enabled laptop computer. Many operators already offer a range of these services. Content isn't much of an issue for these services, but operators do need to figure out how best to configure the portals that subscribers use to access such services.

5.3.2 TRANSACTIONAL SERVICES

The kind of services that are already big business on fixed-line Internet – online purchasing and banking – are also becoming more commonplace on the mobile Internet. One content issue is how to develop an easy-to-navigate interface for subscribers who want to check on their stocks.

5.3.3 LOCATION SERVICES

A rather different sort of applications takes advantage of the mobility of the user: location-based services. These types of services typically help the user with directions, whether it's a travel route that will avoid a traffic jam, or the location of the nearest three-star restaurant via the Michelin Guide. The content issue here is as much about text.

5.3.4 INFORMATION

Information is what the Web is all about. Information services for the mobile Internet tend to focus on the kind of information that users need when they are on the go. These services range from newspaper portals and business directories to cinema guides and weather forecasts.

Probably the biggest content issue here is whittling down existing information into something useable on a wireless device. Subscribers are going to need and want just the bare essentials. Processes for collecting personal preferences and customization software will become very important.

5.3.5 ENTERTAINMENT

Last but by no means least are those more light-hearted services that some analysts say will drive the bulk of the first subscribers to the next generation of mobile Internet.

Today's WAP services offer a host of entertainment features, including horoscopes, lottery results and sports scores. On a more interactive level there are games that can be played alone or in competition with other users. The development of new games and adaptation of existing games is a huge opportunity for content providers. In addition, electronic music distribution (EMD) took off before digital rights management technologies were fully developed.

5.3.6 MESSAGING

Messaging is a transfer technique rather than a content type
The general simple procedure for the sender is to:

1. Write or fetch the desired phone number from the internal directory
2. Write the text message or fetch a ready-made message
3. Add sound and graphics (EMS) or music and a pointer to a photo or moving pictures in an MMS database
4. Send!

Let us now have a look at the three types of messaging.

Short Message Service (SMS)

SMS has been on the market for several years but became a success, especially among young people just a few years ago. Text messages, a maximum of 160 characters long are sent between mobile phones. During the first half of 2000, 161 million messages were sent in Sweden. One year later, during the first half of 2001, this figure was almost three-fold, i.e. 463 million SMS were crossing the air. In Japan the major operator NTT experienced a great success with 'I-mode' that includes SMS as one of the services.

So, what is the reason for the SMS success? One answer is that it is interactive and relationship oriented. In some sense it is also cheap, simple and direct. It could be regarded as cheap since the service is normally included in the mobile subscription and the user only pays when it is used. The user can also reduce the number of transmitted characters by replacing 'I am glad/happy!' with ☺, i.e. sentences are replaced by simple figures, so-called smileys.

SMS is also quite fast. It is sent directly to the desired mobile phone, with minor delay and saves the user from calling any message database.

Enhanced Messaging Service (EMS)

EMS includes SMS and adds simple graphics and sound. It is based on the SMS standard so no new functions are needed in the operators' networks. It was generally introduced during 2001 and it goes under the same subscription as SMS. New, slightly modified mobile phones are required.

Multimedia Messaging Services (MMS)

MMS makes it possible to send text, music, photos and moving pictures between mobile phones. The size of the memory in the receiver's phone sets the limit of the message size. MMS cannot be used in circuit-switched 2G networks such as GSM or CDMA but was introduced in 2.5G, i.e. GPRS. MMS can be fully utilized in 3G networks as W-CDMA within UMTS. In 3G networks the quality of the services will improve and video streaming will be added.

MMS requires new functions in the operators' networks and new mobile phones.

Use of SMS

The magazine *Mobile Choice* carried out a poll during August and September 2001 of the preferences of 1100 British mobile phone users. The survey results illustrate to some extent the relationship between telecommunication and the previously described human needs, and make the use of SMS clearer:

- To start a relationship – 17 %
- To break a relationship – 9 %
- While driving a car – 21 %
- During work hours – 86 %
- While having sex – 14 %

The conclusion was drawn that mobile phones and text messaging are the ubiquitous communication phenomena of our time.

5.4 SERVICE SEGMENTATION FOR PLANNING

Service segmentation is based on a pragmatic selection. Since QoS and security are success factors used in this book, the majority of classes come from these two areas. In UMTS four traffic classes have been identified, which can be used as a definition of service requirements:

- conversational
- streaming
- interactive
- background

The services are characterized by the QoS parameter, which is related to different radio access bearer attributes. The main QoS attributes used to define a service are bit rate, transfer delay, transfer delay variations/jitter and bit error rate (BER).

The UMTS network designer must fulfil QoS requests from the UMTS application or user.

In addition to these four classes the global information infrastructure (GII) presented another two classes according to a taxonomy described in the *Telecommunication Journal of Australia*, 47(2), 1997.

GII considers *transactional and messaging* services as separate from other conventional groups. A particular advantage is that both these service types were represented among the list for mobile Internet. See Sections 5.3.2 (Transactional services) and 5.3.6 (Messaging services). Messaging is in addition a very important service type, today it is the biggest non-voice service used over mobile networks. Some analysts expect transactional mobile services to successively become one of the biggest non-voice services. In addition, transactional services represent the important security area.

This book thus uses a set of six service classes as an input to service implementation. See Figure 5.7. All the services are covered, but some services need one or more service enablers, such as positioning services.

5.4.1 CONVERSATIONAL CLASS

1. Video telephony – Video telephony is a real-time service, which is the most demanding type. Due to the nature of video compression, the BER requirement for video is even more stringent than that of speech. It is indeed very important to distinguish between conversational real-time services and other services, because of the development towards multimedia paired with the original basic mismatch between the conversational services and the packet-based technologies.
2. Voice telephony – The integrity of the existing telephony service is a key concern for service providers. The following distinction is useful, recognizing that classical telephony and IP telephony are really two different services:
   - Classical telephony, i.e. the existing PSTN telephony service with its characteristics and service features, irrespective of whether it is run over circuit or packet-switched (ATM and/or IP) networks.

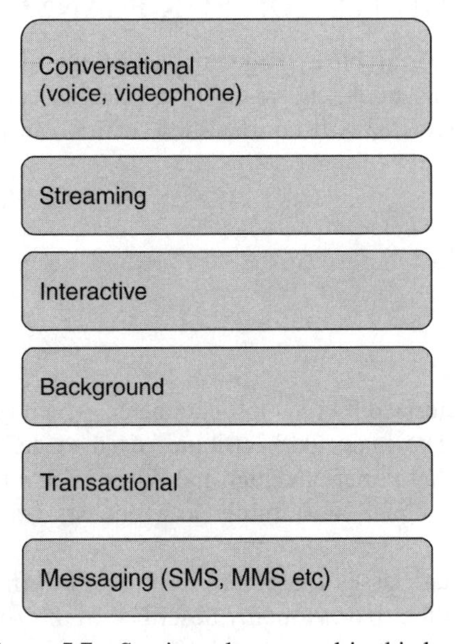

Figure 5.7 Services classes used in this book

- IP telephony, i.e. voice services defined as part of multimedia standards, run over IP and provided separately or as a component of a multimedia service.

Conversational Mode Requirements

Conversational mode is used when two or more persons are involved in a conversation demanding short response times and a certain and stable quality. Some main examples are voice conversation (e.g. telephony), audio conferencing and video conferencing. The main requirements are:

- **Voice quality –** low, medium or high ('hi-fi') depending on the application and usage cost. The technical correspondence is the frequency range of the transferred voice. Traditionally 300–4000 Hz has been a requirement among network operators, specified by the ITU.
- **Picture quality –** low, medium or high depending on the application and usage cost. Quality can be described primarily by the following parameters:
  - black/white including different grey scale granularities or
  - colour with different fidelity levels, for example normal colour (8-bit representation), high colour (16 bit), true colour (32 bit)
  - resolution represented by pixels/square inch. Normal alternatives are 640×480 (lowest quality), 800×600, 1024×768, 1280×1024 and 1600×1200 (highest quality).
- **Delay –** minimal/low. The delay consists primarily of network propagation delay (which can never be zero), also by buffering, format conversions (e.g. A/D and D/A conversions) and any other type of handling of the information stream. The ITU has in recommendation G.114 specified the following values for one-way delay for voice (telephony):
  - 0–150 ms – acceptable for most user applications
  - 150–400 ms – acceptable provided that the operators/administrations are aware of the transmission time impact on the transmission quality of the user application
  - 400+ – unacceptable for general network planning purposes.
- Voice and pictures must be fully synchronized in video conversations. This means that the above values are also valid for the picture component.
- **Delay variation –** none (as perceived by the user). Buffer functions at the receiver side can handle smaller delay variations, but the delay, as described above, increases.
- **Noise –** none/low.
- **Service accessibility –** no or low congestion is the adequate measure in circuit-switched networks. In packet-oriented networks the effective throughput, measured in bit/s is a more adequate measure (to avoid 'slow' connections).
- **Service availability –** high, meaning no or low probability of interruptions of the service.
- **Bit error ratio (BER) –** low or medium depending on application.

5.4.2 STREAMING CLASS

Multimedia streaming is a technique used for transferring information such that it can be processed as a steady and continuous stream. Streaming technologies are becoming increasingly important with the growth of the Internet because most users do not have fast enough access to download large multimedia files quickly.

Typical users of the streaming class are streaming real-time applications, for example audio/video retrieval:

- streaming audio (news, music etc.)
- streaming video (news, music, traffic information etc.)

With streaming, the client browser or plug-in can start displaying the data before the entire file has been transmitted. For streaming to work, the client side receiving the data must be able to collect the data and send it as a steady stream to the application that is processing the data and converting it to sound or pictures. Streaming applications are very asymmetric and therefore they withstand more delay than the more symmetric conversational services. This also means that they tolerate more jitter in the transmission. Jitter can easily be smoothed out by buffering.

Streaming and downloading are compared in Figure 5.8. Streaming offers obviously a shorter waiting time before viewing.

Internet video products and the accompanying media industry are clearly divided into two different target areas:

- Web broadcast (for example on-demand lectures)
- Video streaming on demand

5.4.3 INTERACTIVE CLASS

Interactive class and background class are mainly intended for traditional Internet applications like WWW, e-mail, Telnet, FTP and News. In both classes the resources are reserved

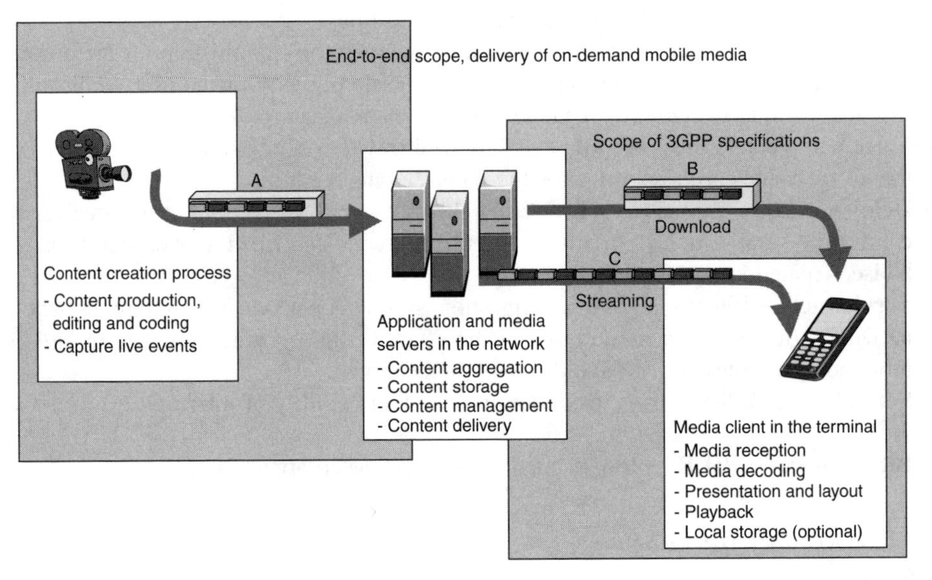

Figure 5.8 Distribution of mobile media content. The C alternative is much more demanding than the B alternative, which can use time to reduce the bandwidth

dynamically. Responsiveness of the interactive applications is ensured by the separation of interactive and background applications.

When the end user, either a machine or a human, is online requesting data from remote equipment (e.g. a server), the interactive scheme applies. In a normal case the user is requesting information. During a common packet call a bursty sequence of datagrams is exchanged.

This phenomenon is very important when developing a traffic model. Burstiness is an inherent feature of this class.

Examples of human interaction with remote equipment are web browsing, e-learning, interactive e-mail and database retrieval. Server access is another application in this class, for example when several persons at different geographical locations play a game together supported by a game server. A third example is collaborative working, i.e. when several persons work together on the same object and exchange information on that object directly between their respective terminals.

Examples of machine interaction with remote equipment are polling for measurement records and automatic database enquiries (tele-machines)

Location-based services belong often to the interactive class. See below.

Interactive traffic is dependent on the request response pattern of the end user. At the message destination there is an entity expecting the message (response) within a certain time. Round-trip delay time is therefore one of the key attributes. Another characteristic is that the content of the packets must be transparently transferred (with low bit error rate).

For mobile systems two types of users are included, namely laptop users and mini browser users. Users referred to as mini-browser users are assumed to be running their applications over efficient protocols and are commonly known as 'thin clients'. A palmtop or pocket phone type is an example of such clients.

Service session characteristics include:

- Duration of one session
- Byte volume up and down per session
- Packet size up and down link
- File size down
- Inter request times
- Number of sessions per day
- User access (rate, QoS)
- Distribution of sessions during the day

Location-based Services

It is easy to predict that location-based services and applications will become one of the key dimensions in UMTS. A location-based service is provided either by a teleoperator or by a third-party provider that utilizes available information regarding the terminal location.

Computer Games

Playing a computer game interactively across a network is one example of an application that can be part of the interactive class. However, depending on the nature of the game, i.e. how intensive data transfer is, it may belong to the conversational class with high requirements for the maximum end-to-end delay.

5.4.4 BACKGROUND CLASS

Examples of background communication are status information, remote measurement and M2M communication. M2M communication is used for example for file transfer between operational systems and between operational and supervisory systems.

Remote supervision including alarms is shown in Figure 5.9. A terminal module (without screen and buttons) sends information about status and position. The module can be placed in different machines and vehicles. The information will be presented in a mobile terminal or computer. A service technician takes care of appropriate actions.

Data traffic applications such as e-mail delivery (including attachments), SMS, downloading information from databases and reception of measurement records can be delivered background since such applications do not require immediate action. The delay may be seconds, tens of seconds or even minutes.

Among background services are also:

- Telemetry.
- Corporate database access. This application is used for field services etc.
- Fleet management (couriers, trucking, taxi) including positioning. This service might be very common and bound to proliferate, for example in combination with vehicle navigation systems.
- Information push (news, financial data, sports, traffic, weather and so on, possibly with positioning).
- Other non-real time services, for example DNS traffic, push/pull services, other mobile information services which are both commercial and society related and software downloads of terminal properties.

Background traffic is one of the classical data communication schemes that is broadly characterized by the fact that the destination is not expecting the data within a certain

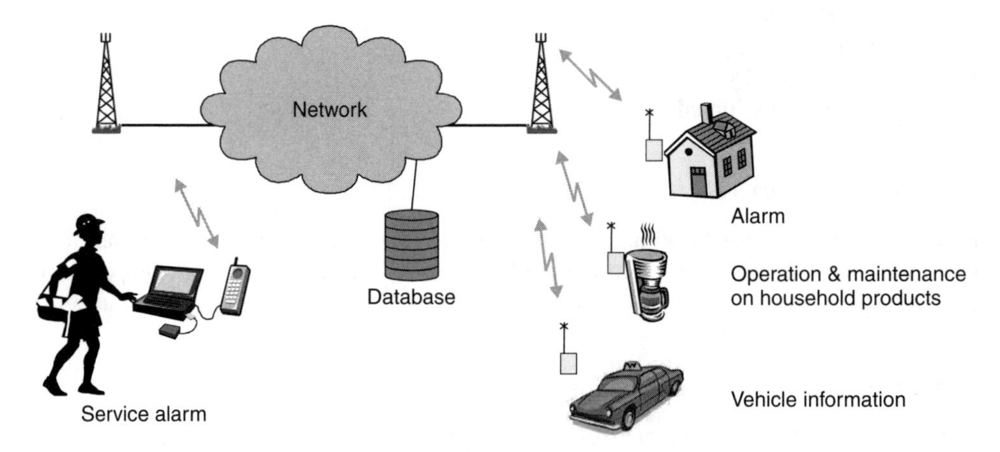

Figure 5.9 Example of remote supervision

time. It is thus more or less insensitive to delivery time. Another characteristic is that the content of the packets does not need to be transparently (isochronously) transferred. Data to be transmitted have to be received error free.

The electronic postcard is an example of a new application that is gradually becoming more and more common. The terminals have built-in cameras. Transfer of digital photos has started to increase.

Traffic in the interactive class has higher priority in scheduling than background class traffic, so background applications use transmission resources only when interactive applications do not need them.

5.4.5 TRANSACTIONAL CLASS

Transactional means financial transactions B2B and B2C where security and information accuracy are crucial. 'Micro payments' for parking and cinema tickets must be possible.

Transaction-based services have caught the interest of stakeholders who are eyeing business opportunities. The development depends a lot on trust and security. The mobile phone becomes a personal trusted device (PTD).

Typical transaction-based services include supporting infrastructure services throughout the various stages of the service life-cycle, such as issuing the WAP identity modules (WIM), certificate authority (CA) services and payment-clearing services. Banks are strong players here.

5.4.6 MESSAGING CLASS

Messaging can be integrated into the background class (or interactive class when interactive). However, it is a revenue-wise important type of service and it requires special equipment. This could justify its own service class for the network designer. Messaging offers user-to-user communication via storage units with a store-and-forward mailbox, and/or message handling, such as information editing, processing and conversion functions. See Section 5.3.6.

5.5 VALUE-ADDED SERVICES

The meaning of the term value-added service (VAS) is not very clear. In *Understanding Telecommunications*, book 1, a distinction was proposed between IN and VAS services, where IN was network oriented and labelled advanced fundamental technical plans, and VAS was in principle implemented outside the network and more equal to information services. In any case VAS is interpreted as something more than ordinary telephony.

The examples below show that service enablers play a key role for VAS. Service enablers are further described in Chapter 8.

Let us take *location* as a VAS example. By using information about the subscriber that is available in the home location register (HLR) or mobile positioning centre (MPC) the information can be tailored to the subscriber position. The HLR knows the cell where the subscriber is located. The MPC can give even more exact positioning when equipped with GPS, and can also provide geo-navigational services.

Messaging functions for MMS and other messaging services is another added value.

Video playback can also be regarded as a value-added content service.

For a business user who can access the same information from many devices a *synchronization service* is appreciated.

Imagine a business user who owns a mobile phone, a PDA, a laptop computer, and a desktop PC. Assume that it is possible to perform an update to a contact's phone number stored on the mobile phone, and that this update would be transmitted to other devices and databases. This broad update could include a contact list on the PDA, a contact list within Outlook on the laptop and desktop PC, a corporate CRM database, and potentially an Internet-based calendar and contact list that the user might occasionally access. The need for keeping this data in sync with the other data sources is an absolute requirement in order for the devices to be of their greatest value, and for the user to work efficiently.

5.6 ECONOMY OF SERVICE BY MEANS OF CACHING

A new dimension in network design is introduced by content-based networks. In order to save capacity in the network modern cache systems can be used. Caching is a type of middleware function.

When surfing you often don't know the location of the 'B-subscriber' (the source). The content can be web pages, e-business, bank services, streaming video, broadcasting etc. Considering the traffic load the source should be as close as possible to the surfer. Such source could be an intermediate cache-type storage in a mirror server at the edge of the network. In order to perform intermediate storing, the various services must be discriminated. This means that you have to enter layer 4 in the Internet. Content-based 'layer 4 switches' can direct traffic to the cache to be stored there for a time, which could vary from seconds to hours. News has a fairly long storage time, despite the fast ageing process. Information with high requirements on security, such as a bank account balance is not stored. An ordinary web page can contain about 50 objects, text, figures etc.

The content view opens up for content networks with different QoS and content server nodes. Figure 5.10 illustrates a content-based view of a small network.

5.7 ECONOMY OF SERVICE BY MEANS OF SAVING BANDWIDTH

The Internet explosion and the intra-company business traffic have been instrumental in a process towards wideband and broadband services offerings. See Figure 5.11 which relates to the fixed networks. However, there is normally no linear correlation between bandwidth and revenue.

It should be noted that there are several ways to fulfil bandwidth requirements of the original source. This means that it is difficult to define a bandwidth need as it appears in the transmission system.

Transfer of high quality pictures can for example be solved either by *high bandwidth* (bits/s) or by using efficient codecs with *compression algorithms* (see Figure 5.12). Moreover, if the user data is not transferred in conversational or streaming modes, *time* is a third parameter to play with.

WAP provides faster communication through compressed and encoded data and headers. A very efficient header compression is offered by the standard robust header compression

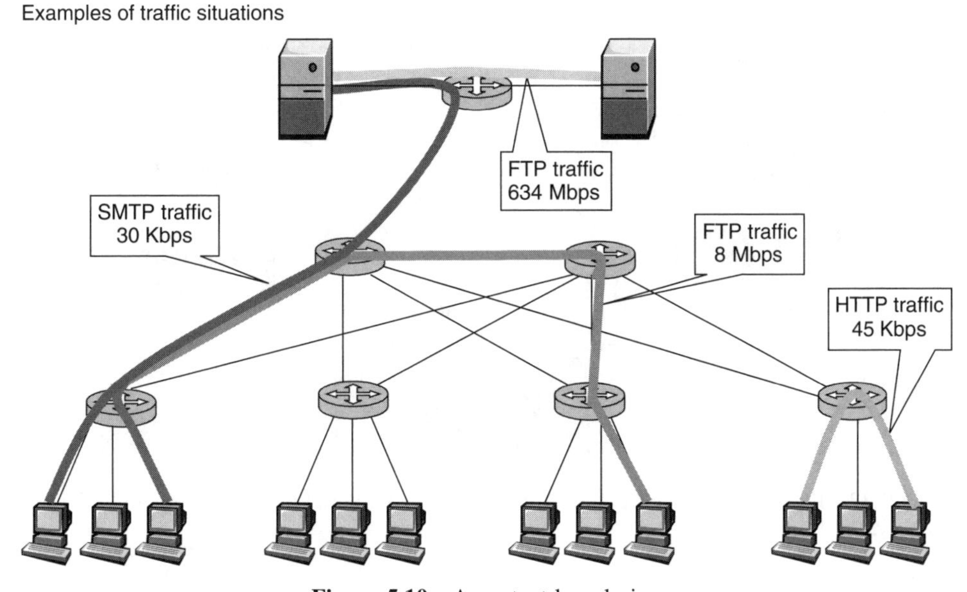

Figure 5.10 A content-based view

1. Voice (fixed PSTN developing towards cellular PLMN)
2. Dialled up Internet
3. Broadband Internet 'Always on', etc.

Figure 5.11 Driving applications in the fixed network

(ROHC). See Figure 5.12. The ROHC can be used for services such as VoIP, video over IP, SIP signalling, WWW and e-mail.

- ROHC is included in 3GPP specifications from release 4. It is defined in TS 25.323, Packet Data Convergence Protocol (PDCP) specification.
- ROHC is also included in 3GPP2 (CDMA2000).

Combinations of these tools are used to get the most economical transmission to fulfil the service requirements.

5.7.1 3GPP MEDIA CODECS

Every media codec developed and standardized in, or adopted by, the 3GPP is suitable for use in wireless applications. Important factors affecting codecs are bit rates, complexity

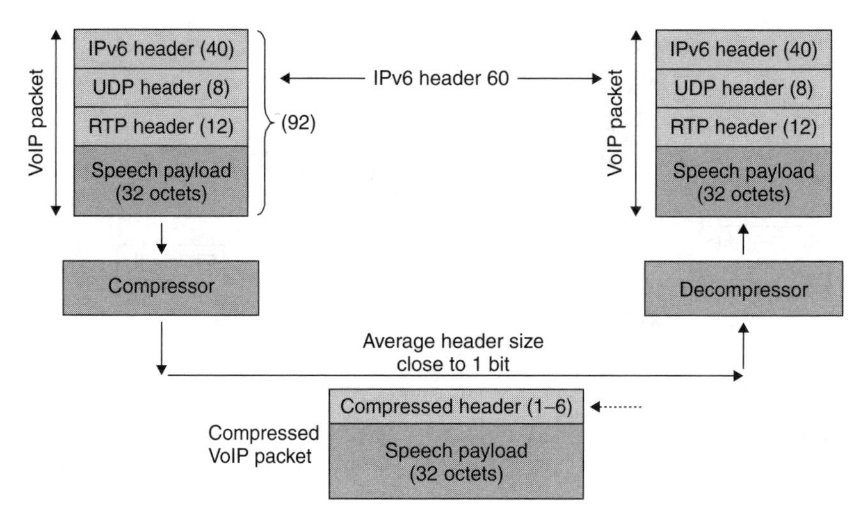

Figure 5.12 Example of header compression from 60 bits down to about 1 bit. ROHC operations – VoIP example

(in terms of millions of instructions per second, MIPS), memory (in terms of RAM and ROM), quality in error-free conditions and in typical wireless environments, and cost of implementation. A number of standardized media types for text, voice, audio, images and video are listed below:

- XHTML basic: XHTML Basic is designed for web clients that do not support the full set of XHTML features, for example mobile phones, PDAs, pagers and set-top boxes.
- AMR: the AMR voice codec (3GPP TS 26.090) encodes narrow-band (4 kHz 'telephony' bandwidth) voice using one out of eight rates between 4.75 kbit/s and 12.2 kbit/s. The media type is well suited to all voice and news clips. AMR was standardized by ETSI in 1998. AMR will be further described in Chapter 8.
- AMR wideband: the AMR wideband voice codec (3GPP TS 26.190) encodes wideband (8 kHz audio bandwidth) signals using one out of nine rates between 6.60 kbit/s and 23.85 kbit/s. The media type is well suited to all kinds of voice and audio clips, and provides high-quality sound. AMR wideband was standardized by 3GPP in 2001.
- MPEG-4 AAC-LC: the low-complexity version of the Advanced Audio Codec of MPEG-4 provides a range of bit rates and quality levels. The media type is well suited to high-quality audio at bit rates of 64 kbit/s and above. MPEG-4 AAC was standardized by ISO/IEC in 1999.
- JPEG: the JPEG compression provides lossy compression of typically photographic images with a compression ratio of 4:1 to 30:1 leading to typical mobile compressed file sizes of 5–20 kbytes. JPEG was standardized by ISO/IEC in 1992.
- GIF: A format for colour raster images that is among the most widely used formats for graphics compression. While GIF87a supports only still images, GIF89a supports simple animation sequences.
- H.263: the H.263 standard supports video compression for applications, such as video conferencing, video telephony, and video streaming. H.263 aims particularly at video coding for low bit rates (typically 20–30 kbit/s and above). The H.263 baseline was

approved in 1995, with a new version including the profile and level specifications from 2000.
- MPEG-4 visual: the simple profile is suited for mobile application scenarios since it is optimized for low bit rate coding and includes coding tools for increased error resilience.

There are also other classical standards and standards for VoIP, for example:

- PCM G.711: the well-known codec used in PSTN networks is the PCM codec G.711. It samples at 8 kHz and codes every sample using 8 bits resulting in a 64 kbps data rate. G.711 is an example of a Waveform codec.
- GSM codecs: in the European cellular GSM system Vocoder type codecs are used, these are also suitable for voice over IP applications. The data rate is 13 kbps for the GSM Full Rate (FR) codec and 12.2 kbps for the GSM Enhanced Full Rate (EFR) codec.
- G.723.1 Algebraic CELP. 6.3 and 5.3 kbit/s. 30 ms frames. Algorithmic delay 37.5 msec. (CELP = Code Excited Linear Prediction).
- G.726 ADPCM (Adaptive Differential PCM). 40, 32, 24, 16 kbit/s.
- G.728 Low delay CELP. 16 kbit/s. Very low algorithmic delay (0.625 ms).
- G.729 Conjugate structure CELP. 8 kbit/s (and 6.4 and 11.8 kbit/s). 10 ms frames. Algorithmic delay 15 ms.

5.7.2 WIRELESS APPLICATION PROTOCOL (WAP)

The presentation below is very short. A detailed description can be found in the book *GPRS and 3G Applications* and in *Ericsson Review Nr 1 2000.*

Purpose

- To bring Internet content and advanced services to digital cellular phone and other wireless terminals.
- To create a global wireless protocol specification that will work across different wireless network technologies.
- To enable the creation of content and applications that scale across a very wide range of bearer networks and device types.

Features

- Global standard.
- Optimized for narrow band bearers.
- Optimized for hand-held devices with limited capabilities.
- Integrates telephony services with micro-browsing.
- The WAP browser should do for mobile Internet what Netscape did for the Internet.

The main advantages of the WAP standard are:

- Faster communication (compressed and encoded data). The wireless session protocol (WSP) increases the efficiency of transmission over wireless. Comparing with HTTP,

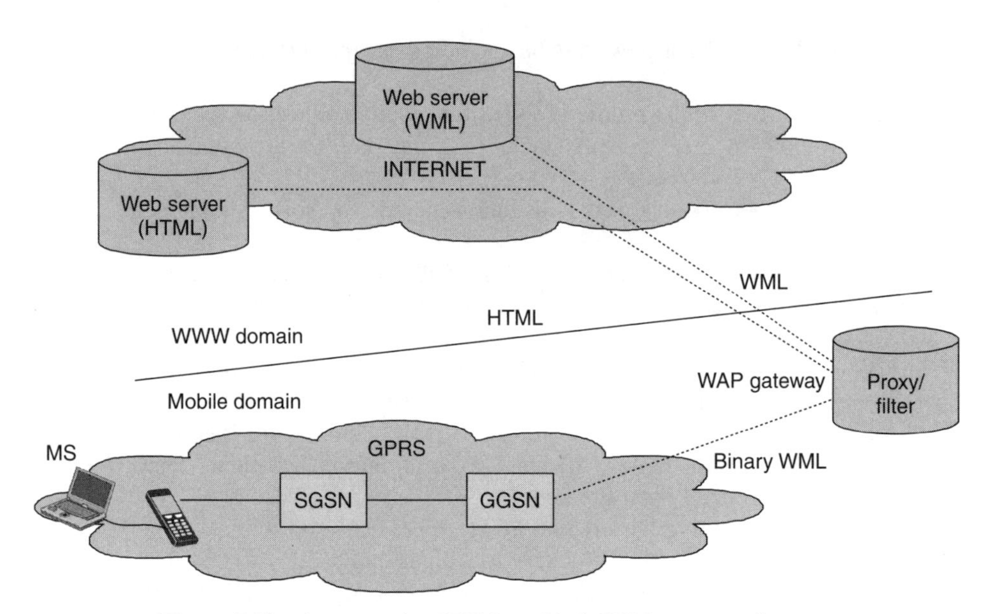

Figure 5.13 An example of WAP-enabled GPRS access to Internet

content is sent as binary ones and zeroes instead of plain text. The headers are encoded into short, binary sequences. These measures increase the efficiency of WAP significantly. As an example, an HTTP header of 32 bytes in plain text can correspond to 2 bytes of encoded representation.

- In the example in Figure 5.13, the WAP client (roaming MS) communicates with two web servers via GPRS. The WAP proxy gateway translates WAP requests to WWW requests thereby allowing the WAP client to submit requests to the web server. The proxy also encodes the responses from the web server into the compact binary format understood by the client.

If the web server provides WAP content (e.g. WML), the WAP proxy retrieves it directly from the web server. However, if the web server provides WWW content (such as HTML), a filter is used to translate the WWW content into WAP content. For example, the HTML filter would translate HTML into WML.

5.7.3 SUMMARY OF BANDWIDTH SAVING METHODS

See Figure 5.14.

5.8 BANDWIDTH REQUIREMENTS

In line with what has been said above, services may need a large span of bandwidths on the transmission line. See Figure 5.15, which shows a large area for streaming and download services.

Some further indications may be of interest. The following list is collected from a number of sources (even if is delicate to indicate figures at all when the range is so wide):

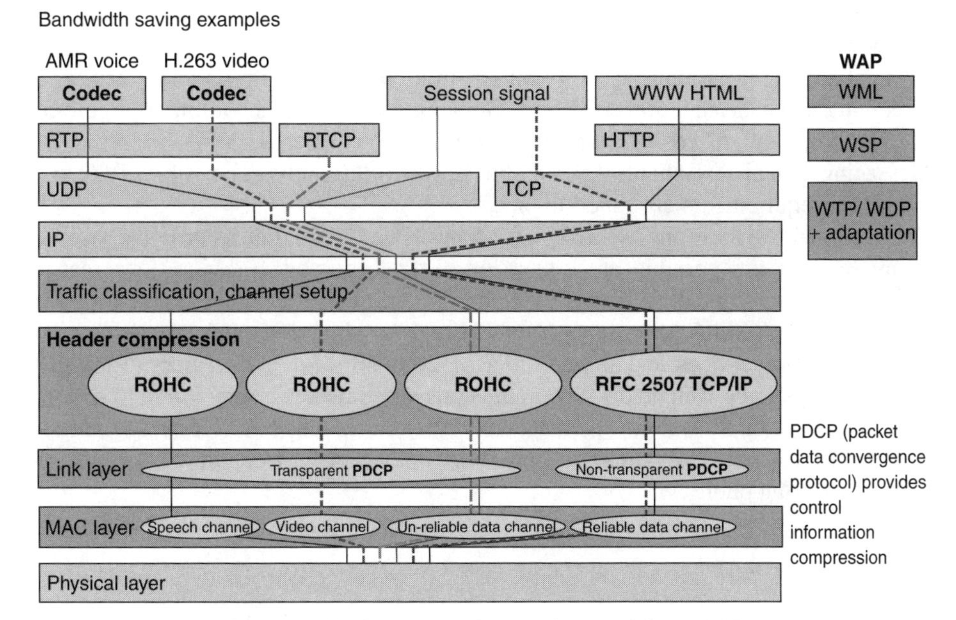

Figure 5.14 Summary of bandwidth-saving methods

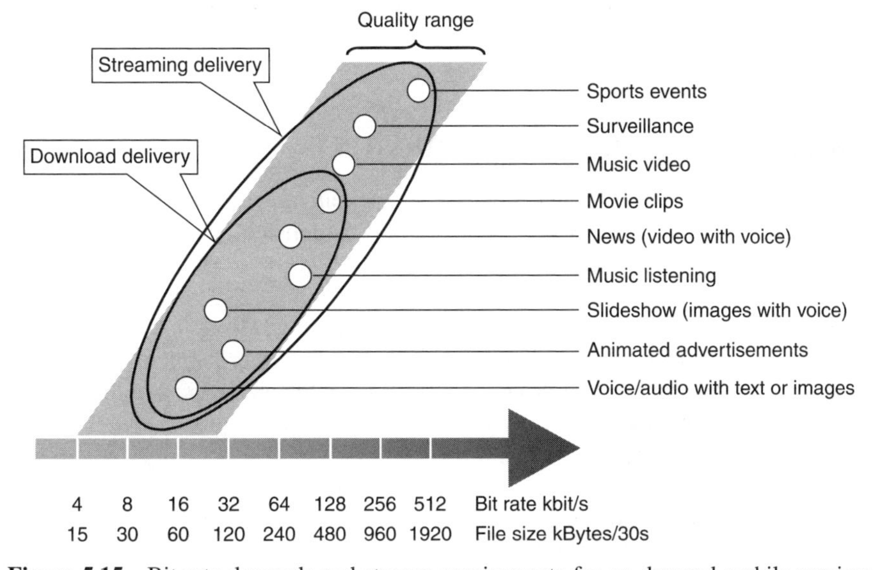

Figure 5.15 Bit rate demands and storage requirements for on-demand mobile services

- Conversational voice services including IP telephony at around 10 kbit/s.
- Conversational video telephony from around 64 kbit/s to more than 1 Mbit/s.
- Interactive browsing/downloads around say 100 kbit/s with a broad spectrum up and down. WAP browsing contains significantly fewer bytes and can therefore use a narrower bandwidth.

- Interactive multimedia browsing around say 300 kbit/s with a broad spectrum up and down.
- Interactive gaming around say 300 kbit/s with a broad spectrum up and down.
- Streaming music around 50–200 kbit/s. Streaming in general according to Figure 5.15.
- Background 'any' speeds. Background accepts what is offered.
- Messaging e-mail text about 30 kbit/s but down to 9.6 kbit/s is possible. With multimedia the requirements are much higher.
- Transactional services 'any' speeds, since focus is not speed, but security and accuracy.
- Many services are possible at quite a low bit rate around 10 kbit/s. One reason is WAP, another one is the Japanese I-mode system. The maximum speed for I-mode download is 28.8 kbit/s for top range models, and 9.6 kbit/s for standard handset models for 2G I-mode services and in the order of 200 kbit/s for 3G services. 9.6 kbit/s is sufficient for simple I-mode data. Of course this speed makes it impossible to download live movies through I-mode. I-mode on 3G is much quicker: it permits 384 kbit/s download (typically 200 kbit/s) and 64 kbit/s upload in best conditions and therefore allows video telephony.

5.9 SECURITY

Security has developed to become a vast area. A lot of security improvement measures can be done outside the network, for example in the operating system of the terminals Authorities for encryption key handling and authorization certificates reside outside the network. Still there are a lot of measures that can be taken inside the networks. See Chapter 6.

5.10 FUTURE SERVICE DEVELOPMENT

Initially it should be pointed out that WLAN systems can already cope with broader bands than 3G, as indicated in Figure 3.45. The evolution of application areas in mobile systems after 3G can be described as three main streams. See also Figure 5.16.

1. More of the same – This evolution stream contains mobile applications in 3G that benefit primarily from higher bit rates in 4G. Quality improvements relative to 3G are a main benefit together with new applications demanding very high transmission bandwidth. We can expect these benefits in optional combinations:
   - Very low, almost no response times.
   - Improved picture quality (very high resolution, 3D photos with no movements delay at zooming and turns).
   - Virtual and augmented reality.

   One usage example is new generations of mobile e-commerce where the user inspects an item before making a purchasing decision. Another example is new generations of mobile group communication for collaborative working, video conferencing and interactive games, all these with virtually unlimited number of attendees.

   Group communications used for dispersed work groups or game groups with anywhere participants can get much better resolution pictures, faster response times and improved 3D virtual reality capabilities.

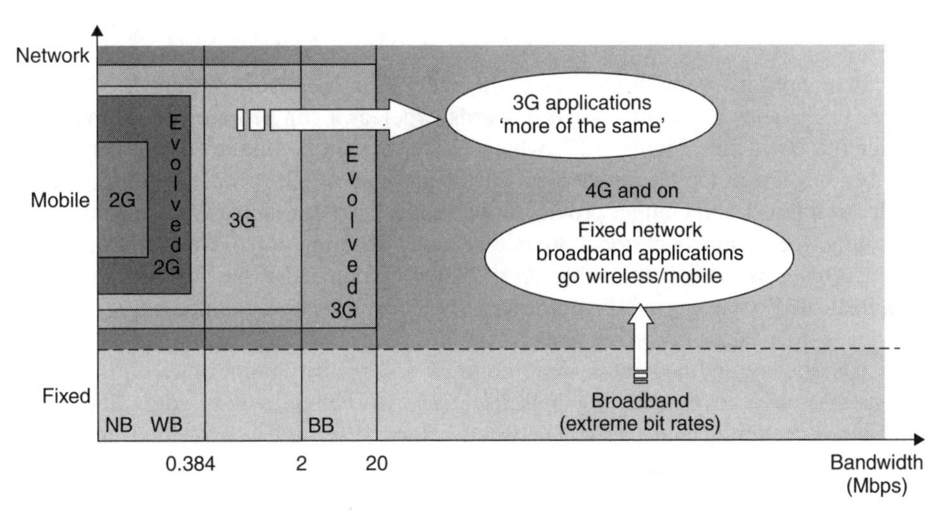

Figure 5.16 New possibilities will be offered by '4G' wireless systems

2. 'Wired network broadband applications go wireless' – Here high ('extreme') bit rate applications that exist in the wired systems can become mobile. For example, distributed communities/workgroups, video conferencing and digital TV for mobile devices with large screens. Seamless mobility between wireless and wired networks becomes possible for these applications.
3. 'Wireless specific broadband applications – true mobile and broadband' – This evolution stream enables applications that are dependent on true real-time mobility and high bit rates. This means for instance high-quality mobile conversational multimedia with fast response times. Another example is augmented/virtual reality applications where 3D is one obvious area that is expected to take off when the technical prerequisites become generally available. Customized applications based on moving 3D pictures of high quality in the mobile communicator can be foreseen in the future. The used applications will also extend from today's hear and sight to involve all other human senses. The vast majority of applications are still to be defined. But let us try to give some short examples:
   'Consultation of a medical expert on the move'. A surgeon is at the airport when contacted for an urgent consultancy regarding a patient. High-quality moving 3D pictures, interleaved with voice and data, transfer the necessary information for the discussion with the expert. High bit rate, i.e. more than 2 Mbps can be used and no delays are experienced during the discussion. This is an example of broadband mobile conversational multimedia.
   'Tourist application'. A tourist visits the Old Town in Stockholm. He puts on his AR glasses and starts the tourist application. He sees the real surroundings but additional sight information is superimposed on his glasses. His personal virtual assistant Lisa guides him when he wants. The name and history of the buildings he looks at is shown. A map with walking instructions to the 'Old Cyber Café' are shown when he thinks of (or in early version says) the name of this café. (Augmented reality (AR) mixes real and virtual realities to get relevant and continuously updated information on the surrounding environment (buildings, events etc.) when walking around.)

5.11 PRICING: CHARGING IN THE NEW TELECOM WORLD

Let us first look at a simple case that illustrates how to set a charging level. A high end-user value tends to push the tariff upwards, whereas a competitive environment tends to lower the tariff. See Figure 5.17, which shows a typical mixture of elements in the subscriber bill. From the figure we also can get an idea of the revenue streams.

With the Internet and content provision the traffic becomes increasingly asymmetric as a consequence of heavy occasional downloads and streaming video. Normally the content is directed towards the 'calling subscriber'. Video on demand is an extreme case.

'Sophisticated' charging of IP -traffic is possible by checking the content at the application level by means of a proxy server. The drawbacks are that such a server will give delays, cost money and become a weak point from a security point of view.

To satisfactorily cover charging it is necessary to include a few words about interconnection, accounting and peering between actors. 'Peering' is explained later in this chapter. With such a complex value chain it is not only the end users that are charged.

5.11.1 CHARGING END USERS

'Distance is dead and time is running out. How communications and services will be paid for in the future will change the Telecom business radically.' In the traditional telecom world charging has been based on subscription fee, time and distance-based periodical charging.

In the new telecom world probably only the *subscription fee* will survive in the long run. Subscribers want a simple charging structure and preferably just one service retailer in order to receive only one bill. (The other fees are likely to be replaced by *access fees* (similar to access provider fees), transaction fees for e-commerce, advertising fees

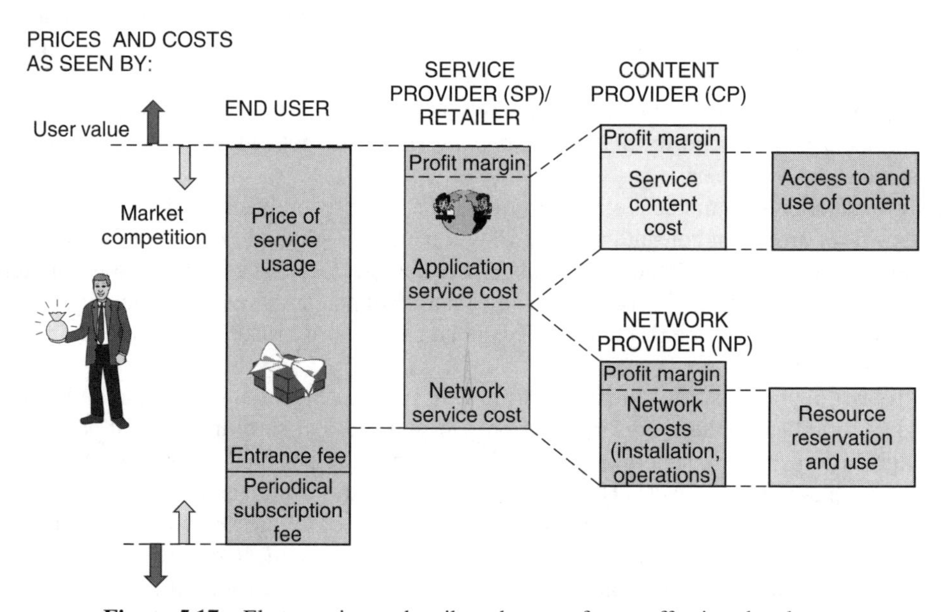

Figure 5.17 Elements in a subscriber charge – forces affecting the charge

for advertisers, and possibly usage/content fees measured in for example Mbytes for *applications*.)

The imminent forms of traffic transfer (access, 'find and connect', network service) charging can be based on the following parameters:

- Volume – the amount of data transferred, measured in for example Mbytes.
- Duration – for example the duration of a session over the packet path in GPRS (a packet data protocol (PDP) context session), or duration of voice, video conference or video telephony calls.
- Time – date, time of the day, day of the week.
- Final destination – the destination address, such as an address in a specific network.
- Location – the location of the subscriber. Perhaps different charging in 'home zones' and 'office zones' and 'hot spots'.
- QoS – the defined QoS classes (conversational, streaming, interactive and background) may be replaced by something more precise, which can be based on priority in the network.
- Reverse charging – the sending party or a third party pays.
- Free of charge
- Flat rate – a fixed monthly fee.

Prepaid will be used also in the new telecom world. Probably prepaid and flat rate will dominate during the next few years. For prepaid the operator can check the account in real time, to make sure that the money is available, before the end user can use a service.

For UMTS operators the access fees may contain:

- Airtime fees, as a function of transferred Mbytes, QoS level and traffic class.
- Monthly subscription fees.
- QoS level per traffic class.
- Connection fees.

Obviously QoS level and traffic class are getting much more into focus.

In the long run usage/traffic volume may disappear (compare TV) as subscribers might leave the operator. The service content/application part could be charged by the access operator or the service provider/content provider. See Figures 5.18 and 5.19.

For IP traffic the Call Data Record in PSTN might become the Internet Protocol Data Record (IPDR). It is possible to charge for individual IP-based services by looking at the application at application level. However, IPDR requires processing and adds cost to the traffic handling. Application level charging is not possible for the encrypted and tunnelled IP-VPNs. The main service provision fee is the application service revenue per application and segment. The pricing should reflect the end-user value of the application content. The charging may consist of:

- activation fees
- monthly fee per pay service
- set-up tariff per pay 'session'
- per-minute tariff of pay 'session'

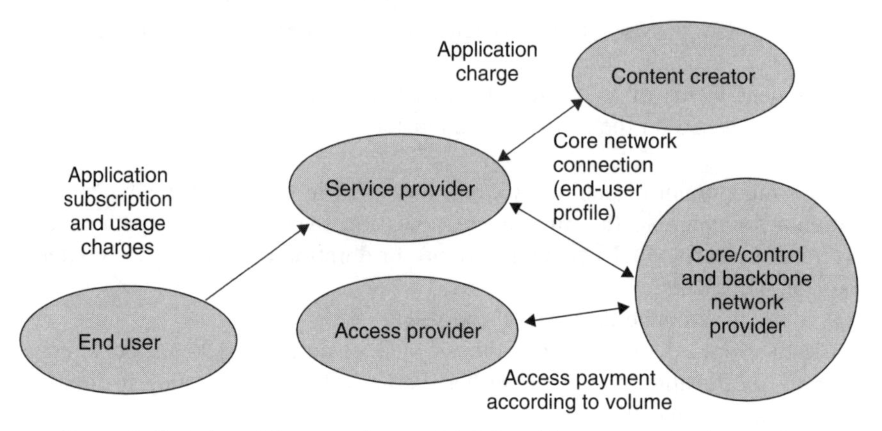

Figure 5.18 Financial transaction model 1 for video communication services

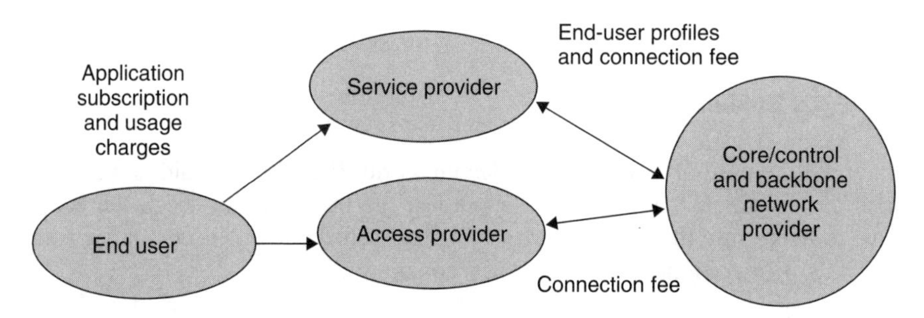

Figure 5.19 Financial transaction model 2 for video communication services

- per-charge unit tariff of pay ‘session’
- operator commission

(Note: ‘Session’ above is not equal to a TCP session.)

Flexible Charging, Example

It is not advisable to use the same type of charging for all services. A security service for example does not generate traffic (or very little traffic). Here is an example of flexible charging:

Chat	20 pence/session
Photos	50 pence/picture
Music	50 pence/song
Weather	Free
Emergency	£1/month
Monthly cost:	**£10–20**

Micro payments

The e-commerce and m-commerce 'micro payments' (such as parking fees, cinema tickets, bus ticket) are an income source for the payment transaction handler(s). The handling can be split into payment support and charging. For the end user it is convenient to have micro payments included in the telecom bill.

5.12 THE SERVICE PLAN VERSUS THE NEW ARCHITECTURE

We have elaborated a promising (end user) service plan and want to build the network accordingly, based on a development plan and technology plan for the network. The end user services reside in peer-to-peer terminals, clients or servers. A terminal can also take client and server roles. A client–server scenario can be described according to Figure 5.20. Between the servers and end users are the various service provider roles.

It should be noted that services used to be offered without a service network. The advantages with such a network are explained in the Chapter 9. The services have to rely on the underlying network for the 'find and connect functions', security and end-to-end transfer quality.

The implementation could at least partly be illustrated by means of a game called ABC. See Figure 5.21.

Obviously the game uses WAP and UMTS or GPRS. The clients need colour displays.

5.12.1 THE OVERALL BUSINESS VIEW OF THE HORIZONTAL NETWORK

See Figure 5.22 for a business view of the horizontal network. Every architecture needs a competitive service-oriented edge to become implemented and survive. That makes it possible to explain the new network architecture from a business point of view. The various features indicated in the figure will be explained in the following chapters. The various network parts will be described in more depth in other chapters.

5.13 THE CORE NETWORK AND THE SERVICE PLAN

The ideal core should satisfy both tele-centric and data-centric operators. For established voice operators, the most debated area when entering the horizontal way of building networks is IP versus ATM as a bearer in the packet backbone. See for example Figure 5.23. A third possibility is gaining ground: multi protocol label switching (MPLS). MPLS employs both IP and ATM techniques (or IP and other layer 2 technologies). The bearer issue including the MPLS solution will be handled in more detail in Chapters 8 and 12.

ROCE-based business objectives applied on the core may look like Figure 5.24, which shows some relevant considerations when selecting transport technology to support the business plan.

In principle the core/backbone network must be able to offer a high QoS that is the same for short and long distances. Most of the permitted bit error rates and other degradations should therefore be allocated to the shorter, but expensive and exposed access part. A sufficiently low bit error rate in the backbone is achieved by means of fibre-based systems. Delays and delay variations should also be kept low. The fibre contributes

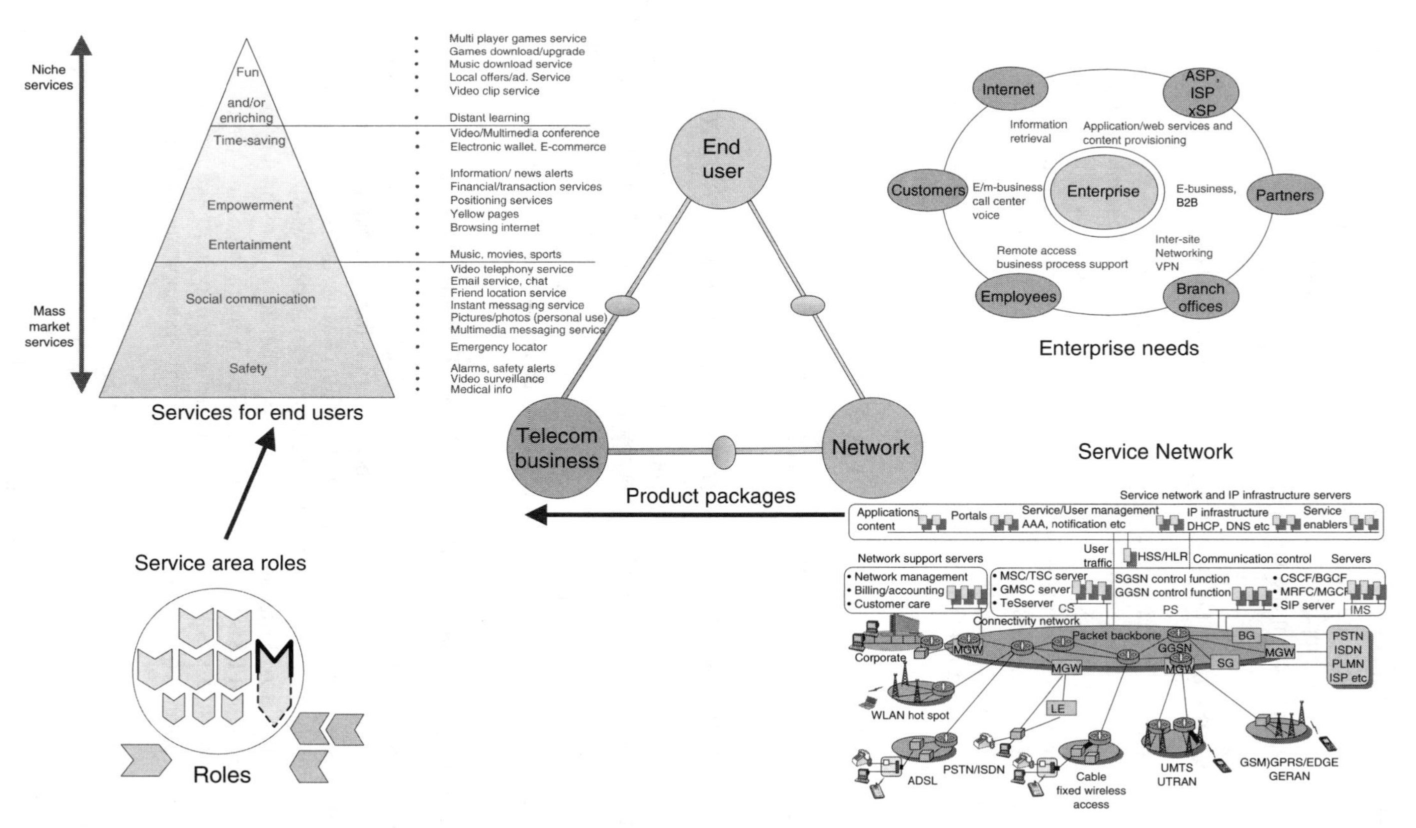

Figure 5.20 Integrating the corners at service network revenue level

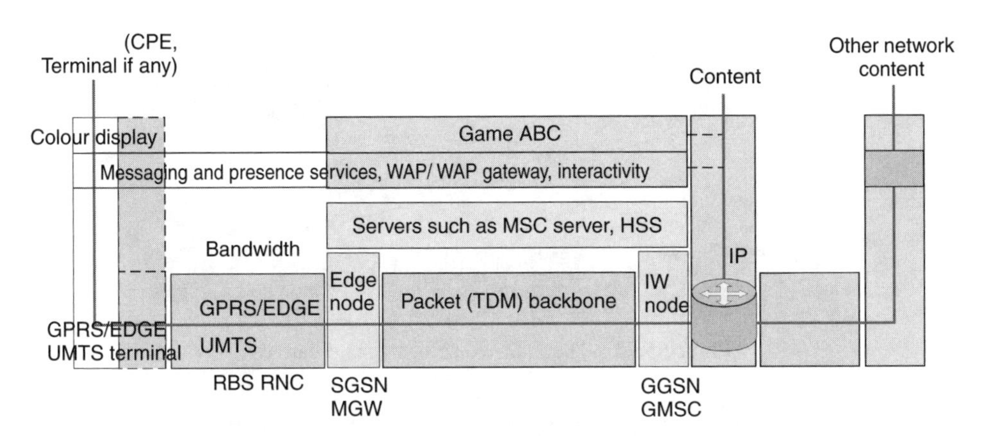

Figure 5.21 A network designed for a game called ABC, made for mobile terminals

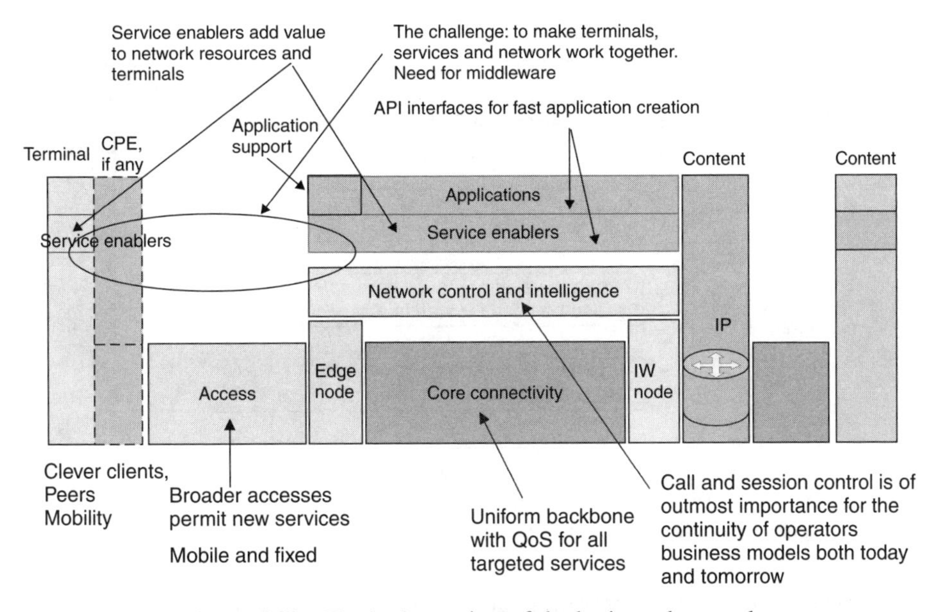

Figure 5.22 The business view of the horizontal network

0.005 ms/km propagation delay to the overall delay. For packetized voice other delays (coding, packetization, dejitterization) are together normally larger.

The main issue from a business point of view becomes the delay variations for real-time services. IP was not designed for real-time services, but is improving. ATM AAL1 and/or ATM AAL2 are initial candidates in layer 2 in order to guarantee multi-service capabilities.

- ATM can carry multi-services with different QoS requirements.
- The ATM core functions like a single large group switch, enhancing engineering and dimensioning.

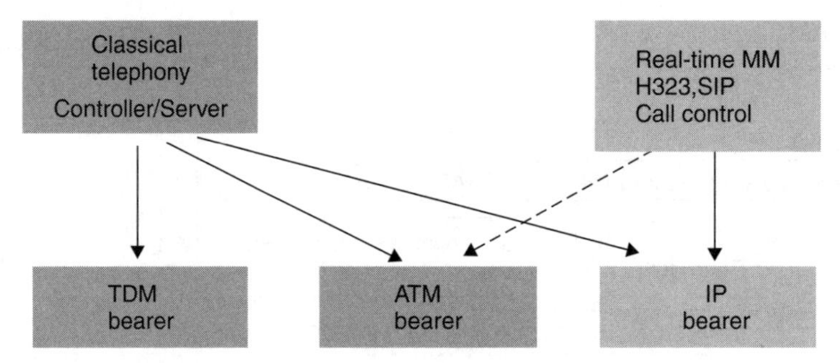

Figure 5.23 The main voice bearer alternatives

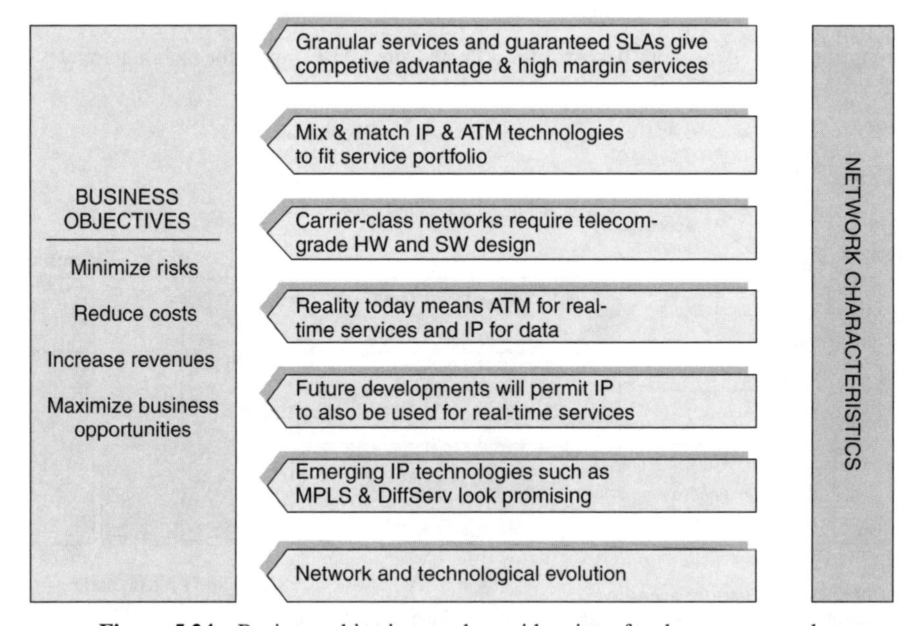

Figure 5.24 Business objectives and considerations for the core network

- High capacity in backbone networks (up to hundreds of Gbit/s).
- Efficient forwarding of IP packets (MPLS).
- Cost benefits in equipment.
- Savings in operations cost and transmission cost.

The critical question is when is IP ready to provide carrier class quality for voice and other real-time services. Figure 5.25 gives an indication.

5.14 THE ACCESS NETWORK AND THE SERVICE PLAN

For a more in-depth study of access techniques refer to Chapter 13.

The access network is the main contributor to capital expenditure. As an average more than 50 % of all network investments are related to the access part. It is, however, also a

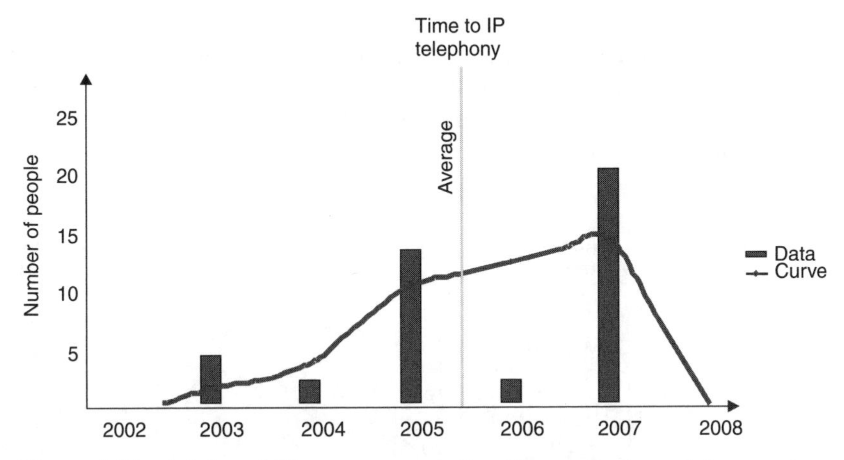

Figure 5.25 Opinion of 50 operators regarding IP maturity for real-time services

door from the subscriber to the core network and the possible services and applications, or, in another view, a door from the network operator to the 'money', the paying subscriber. The size of the door corresponds to the maximum bandwidth of the services. A larger bandwidth implies an increased income potential. However, there is no linearity between bandwidth and revenue. It is also a matter of quality. High-quality narrow band services such as telephony continue to be a main income source for tele-centric operators.

Some ideas of the connection of services to systems is shown in Figure 5.26.

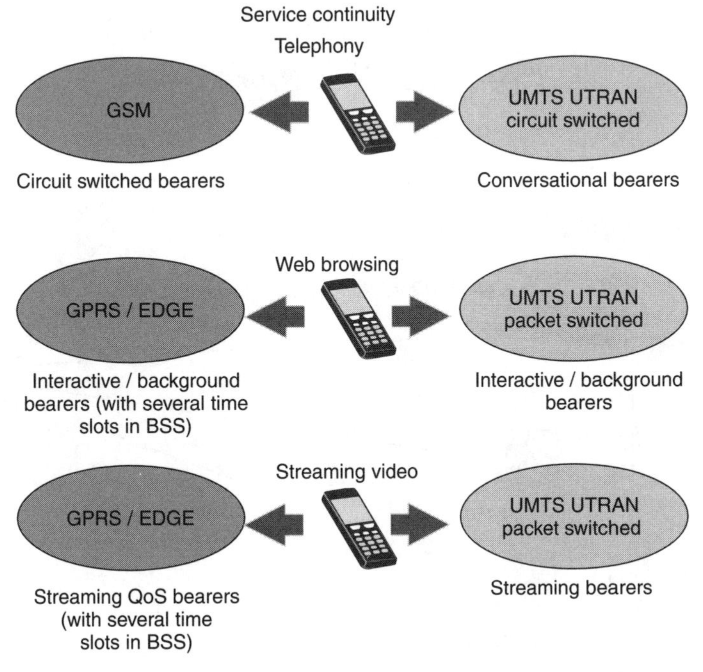

Figure 5.26 Example of matching services and networks

Business cases are suitable inputs to take decisions. For the expensive access network stepwise extensions give good results. Examples of stepwise approaches are GPRS and EDGE on the mobile side and the xDSL systems on the fixed side. UMTS also represents a step, since it complements GSM for established GSM operators, but the step is more expensive with new base stations. See Figure 5.27.

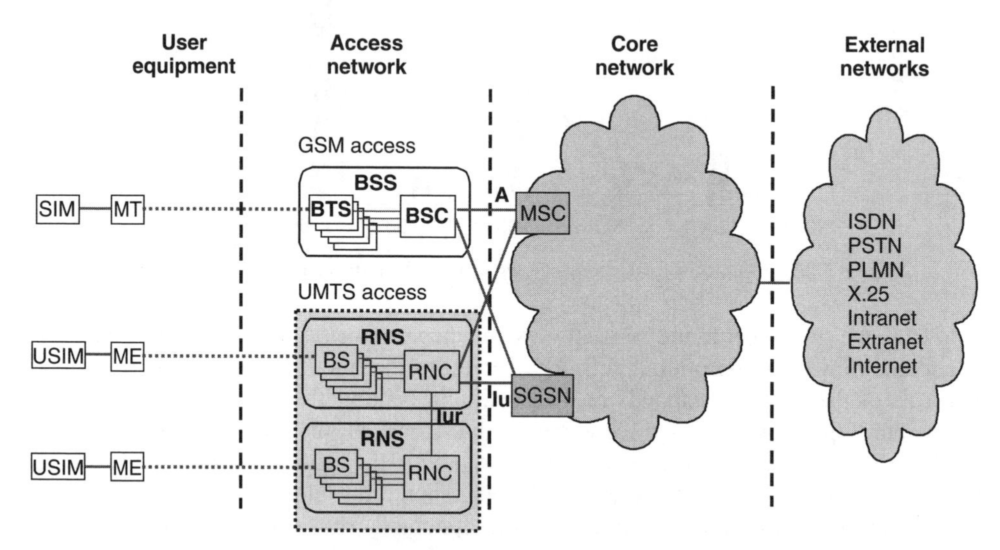

Figure 5.27 UMTS can be regarded as a complement to GSM

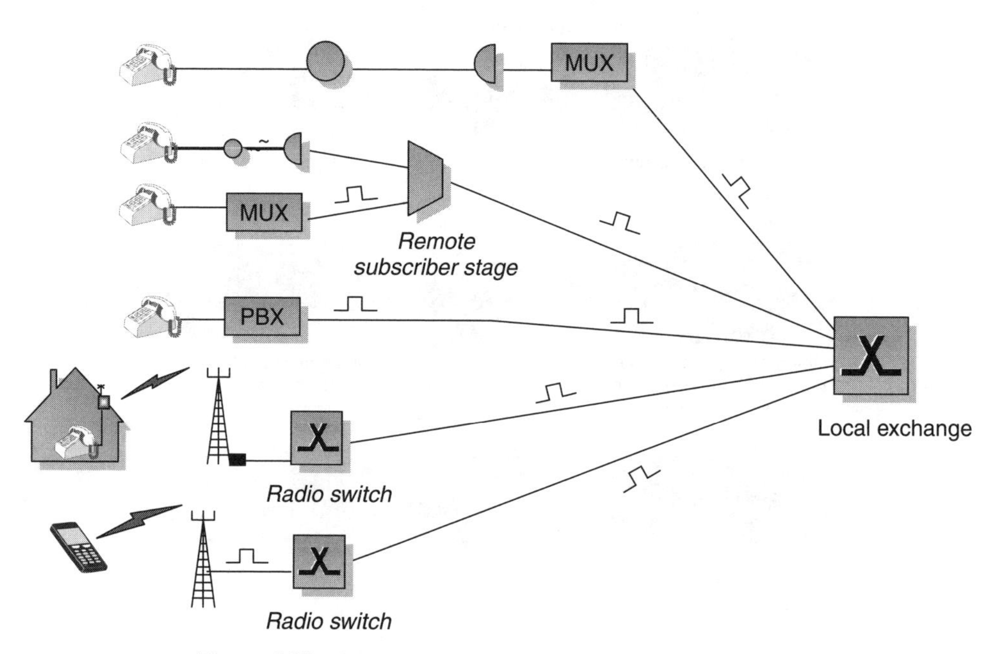

Figure 5.28 Access network shape for baseline investments

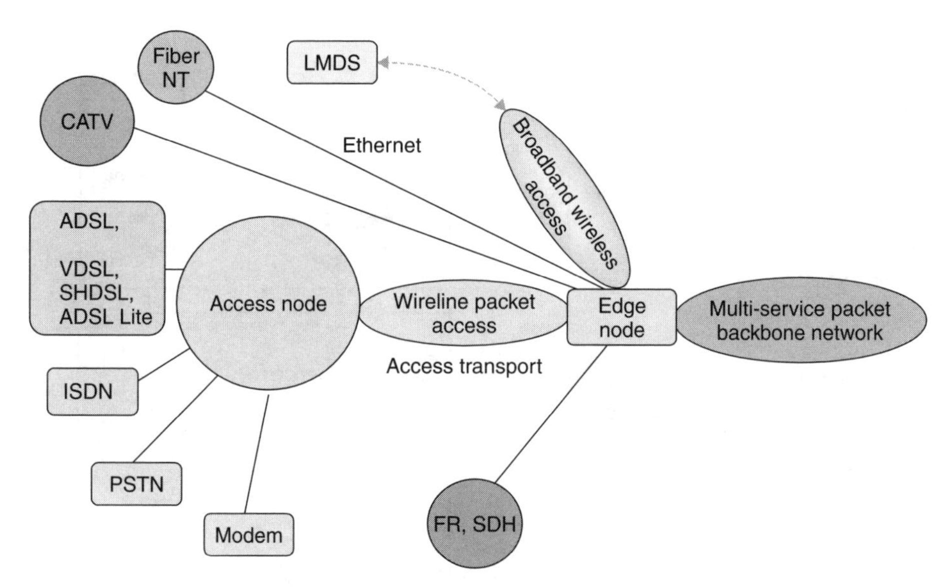

Figure 5.29 Transport in the access network

In 3G, the enhanced bandwidth capabilities and advanced terminals enable video and audio, either real time, near real time or download. Examples are two-way video conferencing with audio, video streaming, hi-fi music streaming and interactive games. The speed and quality of the service will enable new presentation formats.

The need for new base stations is substantial when extending mobile systems to 3G, in view of the maximum coverage area radius: GSM 900 about 35 km, GSM 1800 about 8 km and UMTS cells about 5 km radius.

About 10 years ago the fixed access network would have had the look of Figure 5.28. To be fair it must be pointed out that the figure is voice oriented. Thus, there were also dedicated and switched (X.25, frame relay, ATM) data accesses to the enterprise sector. It is easy to see that the network was not designed for Internet traffic.

Today the corresponding access network might look like Figure 5.29. The difference is striking, proving the strong impact of the Internet paradigm. The financial expressions are explained in Appendix 2.

Figure 5.29 shows an access network shape for discretionary investments, especially for broadband. For fibre systems IP/Ethernet is gaining ground with advantages regarding price and bandwidth management.

5.15 TELECOM MANAGEMENT AND THE SERVICE PLAN

Because of the increased revenue focus in the present competitive environment the management system must be a money-maker as well as save money. It has to adapt to the convergence and horizontalization of the networks. It should also support operator/service provider moves along the value chain, especially the new service layer.

This requires a layered management, at least with connectivity (access, core), control and service layer management. See Figure 5.30.

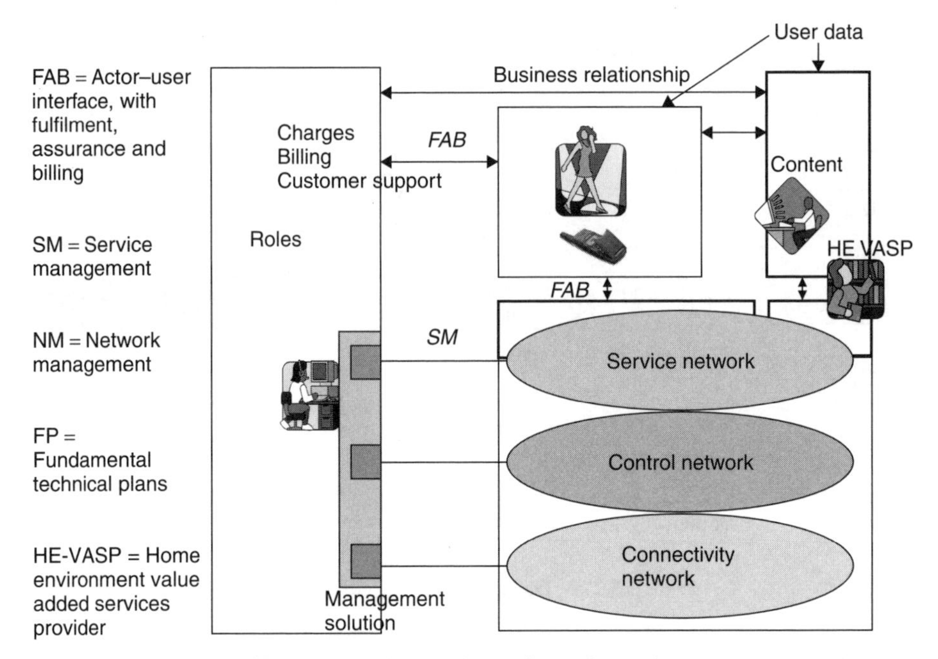

Figure 5.30 A layered network requires a layered management

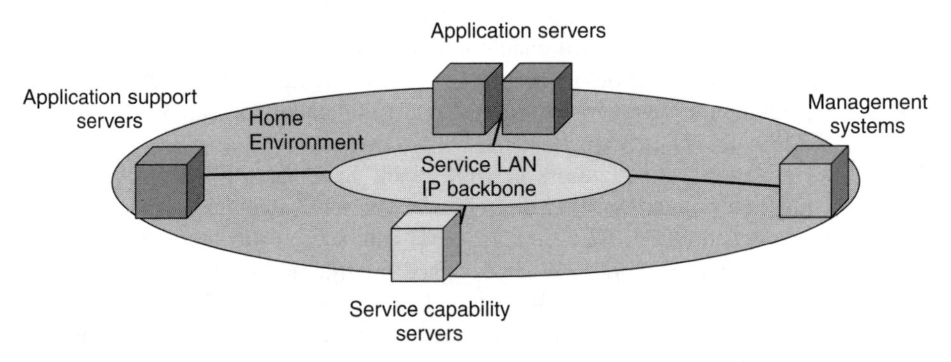

Figure 5.31 The new service layer requires management

Typically, the service explosion requires a new management focus: the service layer management. The service networks have their own management systems. The service capability servers and application support servers (together called service enablers) in the service network are potential key revenue enablers. Applications can be created by third party or the network operator/service provider. See Figure 5.31. Service management is treated in Chapter 16.

6

Security

'The trick is not to hide, but to remain secure in the open.'
'On the net, nobody knows that you are a dog.'

6.1 OBJECTIVES

A main objective of this chapter is to integrate security and telecom. Some specific goals of the chapter are to:

- Define what is meant by security in the telecom world
- Describe the main threats against security
- Explain the main methods for authentication
- Explain the function of a firewall and a DMZ (demilitarized zone)
- Explain encryption and tunnelling

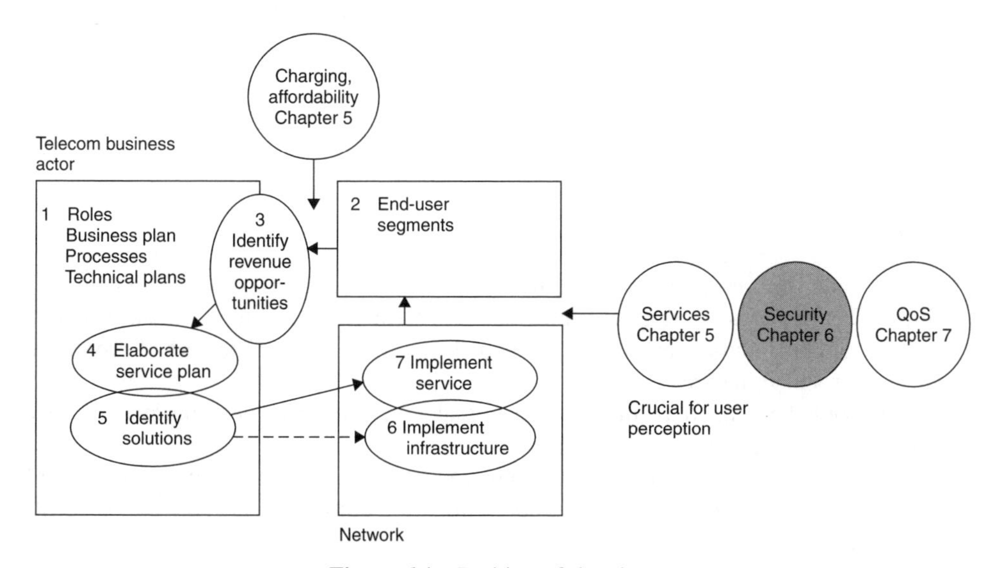

Figure 6.1 Position of the chapter

Understanding Changing Telecommunications – Building a Successful Telecom Business. Edited by A. Olsson
 ISBN: 0-470-86851-1

- Explain IPsec
- Explain the terms AAA, RADIUS, LDAP, TLS
- Explain cryptography and encryption
- Describe what security protocols are found at different layers
- Understand what is important for designing the security policy and the problems that appear
- Discuss security in radio networks

6.2 THE GOALS OF THE USER AND ACTOR. TERMINOLOGY

Security is classified as a fundamental plan in this book. Moreover it is classified as one of the success factors. Security is a *process*, not a product. Security can never be guaranteed to 100 %, because of the significant human impact. Therefore *trust* is important for users and actors.

The fundamental security-related requirements of the user are:

- Integrity – data cannot be manipulated by unauthorized users.
- Confidentiality – data cannot be accessed by unauthorized users.
- Availability – data is accessible to authorized users.

Often *authenticity* is mentioned as a fourth fundamental requirement. Otherwise it has been regarded as part of integrity. Authenticity means that we know with whom we are communicating: we authenticate ourselves and form an integrity-protected secure session between each other.

For a company, the needs might be somewhat developed:

- Which systems are most critical to the business?
- What information needs protection?
- In what way should the systems and information be protected?

Integrity

Integrity means that information being transported or stored cannot be manipulated or modified. This includes altering, changing status, deleting and creating. As an example, an e-mail could be manipulated while it is queued on a mail-server. One way of protection is using electronic signatures.

Integrity could be defined as 'the assurance of accuracy, completeness and performance according to specifications'.

There are three aspects of integrity: authorized actions, separation and protection of resources, and finally error detection and correction. Sometimes another two requirements are included:

- **Authentication:** identity should be proved, for example by means of certificates. Authentication addresses the simple question 'Who are you?'
- **Non-repudiation:** What is sent/received is sent/received and cannot be denied.

Confidentiality

Confidentiality means that information being transported or stored is prevented from being revealed to unauthorized parties. Differentiation is necessary for access to different sorts of information. The type of access could be reading, viewing, printing or even just knowing the existence of an object. Confidentiality is sometimes called secrecy, privacy or prevention of intrusion. The most common confidentiality measure is encrypting the information.

Confidentiality could be defined as 'the ability to avoid disclosing information to anyone who is not authorized to use it'.

Availability

Availability has turned out to be a complex term. In general it means that assets are accessible to authorized parties. It is then supposed that the assets are working properly. Thus non-availability can have other reasons than security-based ones (in a narrow sense), such as technical failures. Therefore, one paragraph in this chapter will deal with this type of non-availability as well.

Availability could be defined as 'the state of being able to ensure that users can use any information resource whenever and wherever it is needed in accordance with applicable privileges'.

An authorized party should not be prevented from accessing objects to which he/she/it has legitimate access. For example, a security system could preserve perfect confidentiality by preventing everyone from reading a particular object. However, this system does not meet the requirement of availability for proper access. Availability is sometimes known by its opposite, denial of service. The user wants, with maybe the highest priority, a very high availability from a telecommunication system, sometimes expressed with a series of nines, such as 99.99 %. This requires a working and accessible system, one that is not severely disturbed by physical damage, viruses etc.

A particularly serious threat is offered by unauthorized staff taking control of management systems for the operation of a network. A thorough authentication is therefore required before allowing commands from the management system to operate on a traffic system. Despite the complexity and importance of the management system most of the content of this section will deal with traffic-related protection.

Goal for the Actor

The purpose of the security plan and measures is to provide reasonable assurance to the user of the integrity and confidentiality of information, and availability of network resources, servers and services.

6.3 THE PROBLEM

The public Internet is increasingly used to access security-critical services like banks and e-commerce. Technically, the public Internet is, however, an unreliable network with many possibilities for an intruder. And on top of an insecure IP network, we usually run insecure applications.

Potential threats are illegal access to information, inadequate identification of users and resources, tampering with information, masquerading, eavesdropping, subscription fraud, non-payment for services, bypassing security controls and service disruption.

Malicious users can easily tap (eavesdrop) IP traffic, redirect traffic, insert false packets, modify packets, mount denial of service attacks and introduce harmful software into systems. Future wars might well be fought in cyberspace.

One of the greatest threats in all computer systems is the company's own staff. There are not so many investigations done in this area but most of them show that the staff inside the company is one of the most frequent intruders in computer systems.

6.3.1 SECURITY: A FUNDAMENTAL PLAN?

For the actor the security area can be seen as a huge fundamental technical plan with relationships and implications to most other plans. This statement will be developed later (e.g. see Figure 6.11). Whereas security measures protect the service and the functioning of other plans, they have often an adverse effect on QoS. One reason is that many security measures increase the overhead of the traffic, simultaneously hiding header information in encapsulated packets. Sometimes there is a need to 'undress' the packet, especially in gateways, when changing between technical solutions like in a WAP gateway or a proxy server. This is not attractive for the staff responsible for security.

This also means indirectly that the service is affected in a negative way. At the very least extra cost has been incurred. Another negative impact is that the availability is reduced by the number of precautions, such as passwords and other types of access restrictions.

The overall immediate result becomes a significant cost and possibly a reduced QoS. This cost has to be paid by the users in one way or another. But this is not the end of the story: as mentioned the user has some fundamental requirements on the information transfer: *integrity* and *confidentiality*. Once these requirements are jeopardized, it has a severe impact on the utilization of the networks. The first type of service to be affected is transactional services, typically B2B and B2C. Consequently e-commerce services are rejected by many people, because they don't trust the confidentiality and integrity of the service offered. An enquiry in the USA in 2001 asked: 'What puts people off buying on the Internet?' The result was clear: 86 % answered: 'Fear about security of sensitive information'.

Let us at state, therefore, that security is one of the most critical areas for success within the telecom area. It has now even become a national concern. At a time when the risk for cyber war has become a reality there are countries that are considering development of their own 'counter virus'.

The Internet publication 'News.com' has published an article with the heading 'Will the Web ever be secure?' The answer that is given as: 'Based on where we have been and where we are heading, we have every right to expect the Web to become increasingly secure as security continues to enable, rather than strangle business performance'. This is just one voice (Steve Mills) out of a huge number of views, with divergent opinions. The gradually increasing market share of the more vulnerable wireless access on behalf of wired access must for example be considered.

The Wireless world research forum stresses the need for systems that provide wireless users affordable broadband mobile access solutions for the applications of secured wireless mobile Internet services with value-added quality-of-service (QoS) through application layer all the way to the media access control (MAC) layer.

When looking back to the 'good old days' in telecommunication there was no or very little talk about security. This is another aspect of changing telecommunications, which obviously calls for an explanation:

- Has the telecom system become more vulnerable?
- Why all these hackers now?

Let us start with *vulnerability*, which has a strong correlation with the IP technology success.

The Internet protocol (IP) has no inherent security function. It is relatively easy to forge the addresses of IP networks, modify the contents of the IP packets, replay old IP packets and read the contents of the IP packet in transit.

Figure 6.2 shows terms or network elements that are related to security: firewall, VPN, AAA and PKI (public key infrastructure).

The new connected world with always on and communication between things offers new possibilities to enhance life, but also large vulnerability risks: fridges and washing machines can stop working because of a virus. Since mobile Internet will become powerful (it will for example manage e-commerce) and ubiquitous, it will require a particular security. The powerful mobile units, some of them with sensitive information, will no doubt frequently be lost or stolen. Each year about 4 million mobiles are stolen in Western Europe. The wireless interface can be a leak of information. The companies need rules for the use of portable computing equipment.

Many IP problems come from the protocol stack itself. TCP/IP was designed to operate in a tough environment, but the creators focused on reliability rather than security. The

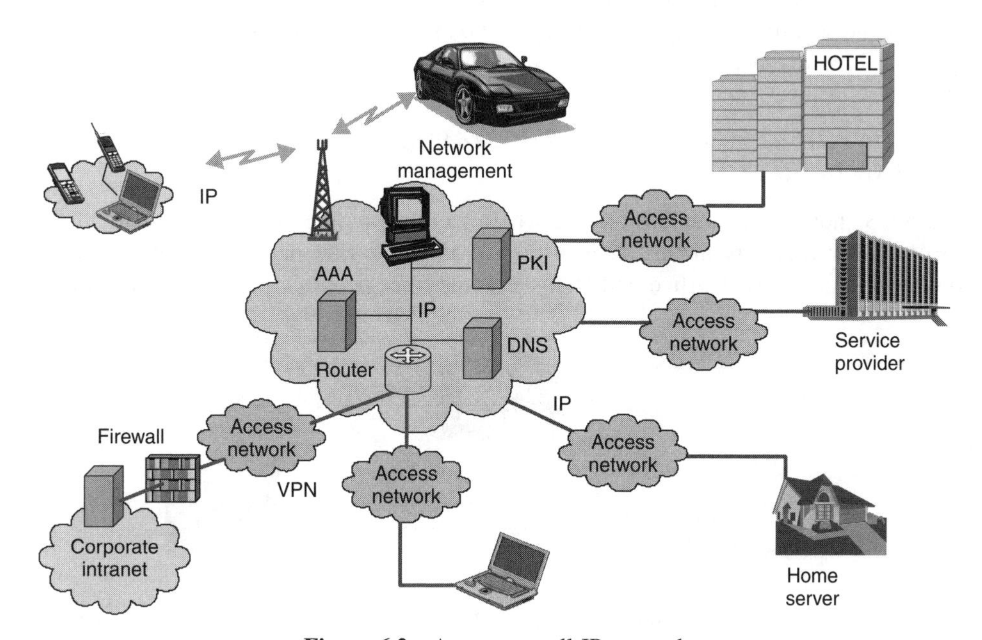

Figure 6.2 A common all-IP network

initially targeted users worked within the technology sectors of an academic world. This was a pretty safe environment and the risk of security violations was very limited. Very few people expected that the protocols would be misused by attackers trying to break into systems using these protocols.

Sending an e-mail over the Internet resembles sending a postcard rather than sending an enveloped letter. It takes a hacker 10 minutes to spoof an e-mail (send an e-mail with a false sender). E-mail is used extensively today for business communication, although it uses plain text messages that almost anyone can read and/or manipulate. This vulnerability makes it an *untrusted* network, which must be borne in mind when planning for services directed to the public. Untrusted means that it is fairly easy to wiretap and/or alter information.

Web services offer another type of mobility: service mobility. When applications are composed by service elements from many sources, a reliable authentication and authorization of the sources is crucial.

Turning to the *crimes*, it seems there are Internet users for which bypassing security barriers, either for fraud, sabotage or pure fun purposes, is a reason of proud and increased self-esteem (hackers' phenomenon). Violating security systems has become more a challenge than an action that is socially condemned.

Obviously there are also people using the Internet for child pornography and extremist propaganda.

The overall cost for Internet criminality is hard or even impossible to calculate. What was the cost of Sobig F? Klez? Love bug? We are talking about millions of infected computers and billions of US dollars per virus. A hacker can invest a year in development and launching a specific virus attack.

The conclusion is that a cloud of uncertainty and threat will remain for a long time over Internet-based communication. In a way it is a race between development of security enablers and security 'disablers'. Security is one of the fastest growing IT areas, measured in annual investments.

6.3.2 CONFIDENTIALITY AND AUTHORIZATION PROBLEMS AND INTRUSION

Figure 6.3 shows an example of a case where we have problems with confidentiality.

Eve, the intruder, is in this case a passive intruder. She just listens or records the communication between Alice and Bob. Reading other people's e-mail is an example of

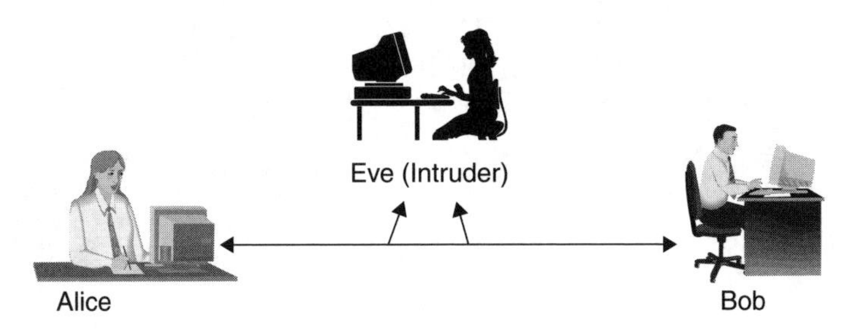

Figure 6.3 Eve has unauthorized access to the communication between Alice and Bob

a passive intruder. Depending on what is written in the e-mail, the harm could be minor or major for either Alice or Bob or both.

Ethernet has the weakness that all messages are broadcast on the LAN to all connected stations. It is thus fairly easy for an intruder to pick up the information and record the traffic. Figure 6.4 shows an example of monitoring on a LAN.

The simplest case is of course when A and B communicate without any encryption. This makes life very easy for the intruder. Note that the monitoring can lead to a second attack by the intruder: he may monitor user identities and passwords and then access the system with the stolen user identity.

6.3.3 AUTHENTICATION PROBLEMS

Figure 6.5 shows a simple example where Alice thinks she is communicating with Bob but she is not. An intruder (Eve) is acting like Bob without Alice knowing that. This means that Eve is taking a rather active part as she is acting like another person or company.

The problem of authentication also exists in real life. Try to empty your bank account without proper identification papers or via a telephone line. You will not succeed as the bank has some rules regarding authentication.

The simplest way of getting around an authentication system is to steal someone's user identity and password. The way to do this can of course differ but monitoring, as described above, is one way. Figure 6.6 shows how it can be done.

Once the intruder has the false identity and the password, it is almost impossible for the system to detect this. One single access with the stolen user identity might never be detected.

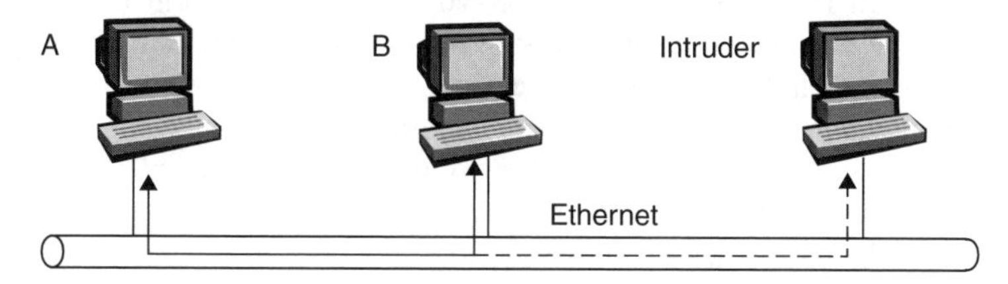

Figure 6.4 An intruder using monitoring

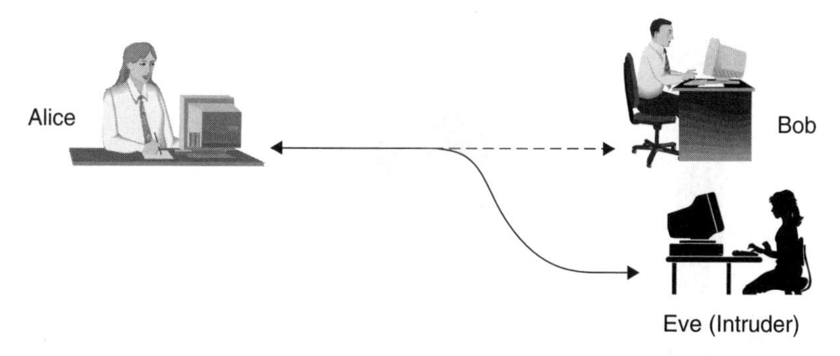

Figure 6.5 You never know who you are communicating with

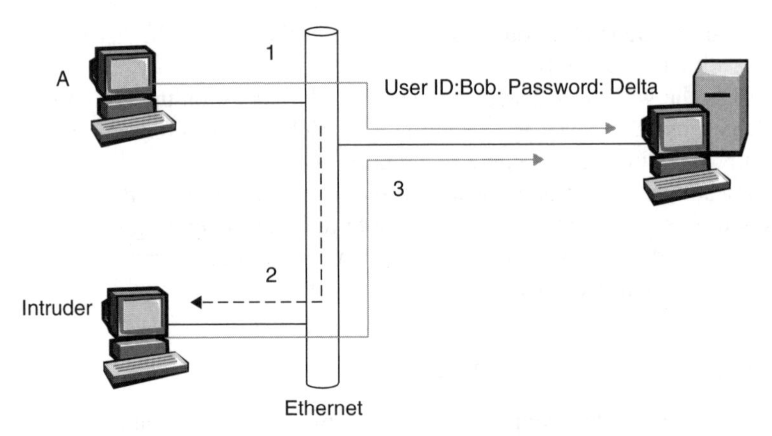

Figure 6.6 Access by means of stolen identity

6.3.4 INTEGRITY PROBLEMS

The area of integrity addresses the problem of altered information. An intruder may not only listen to a communication between two other parties, he or she may also alter information sent between A and B. Imagine an economical transaction between a person and a bank. The intruder may alter the message sent to the bank so money is transferred to his bank account instead. This can be done without the knowledge of the bank or the person requesting the change. Figure 6.7 shows an example where the intruder (Eve) alters some information in the message sent from Alice to Bob.

The problem with integrity is that Bob probably trusts the information coming from Alice as some type of authentication has been done. The intruder can in this case cause great damage to Alice and Bob. A good security system prevents this from happening.

When you receive an e-mail from someone, you think that what you see on the screen is what the sender wrote. Unfortunately, that might not be the case if an active intruder has changed the mail. Without encryption, it is very simple and can be done in any editor. Figure 6.8 shows the case when an active intruder modifies an e-mail.

Exactly how the intruder gets the mail is not interesting, the main point is that it is possible in a number of ways and that you should not trust e-mail for critical information (business or private).

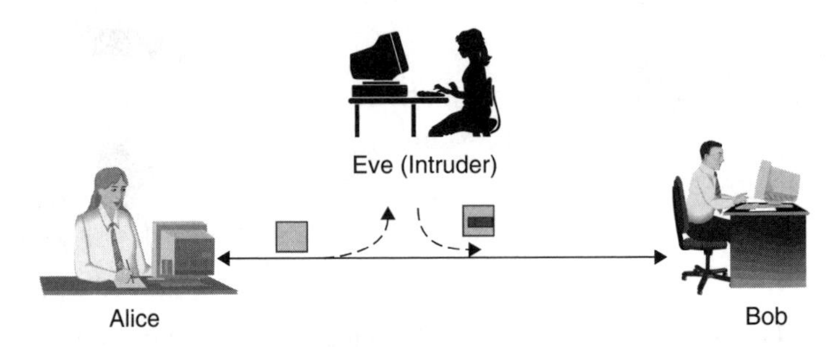

Figure 6.7 Integrity problems

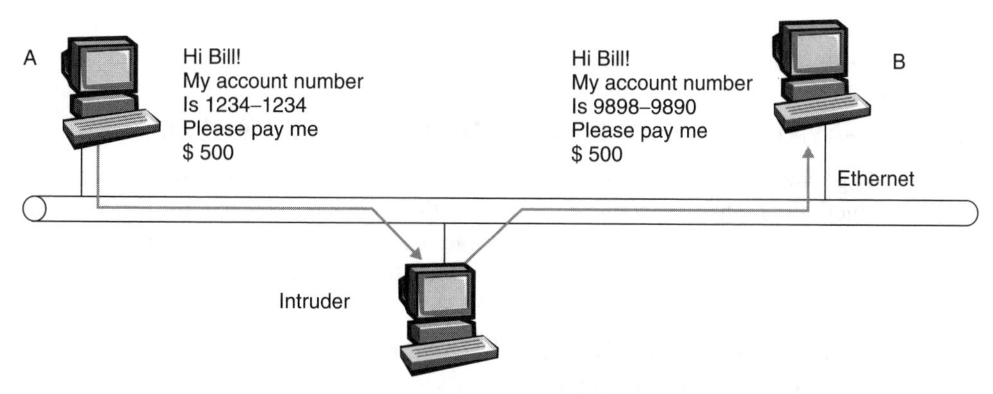

Figure 6.8 Modification of a mail

Non-Repudiation

Non-repudiation means that you have sent some information or document to another party, and a number of weeks later you deny having sent it. It could, for example, be some legal document regulating a deal between you and the other person. You say that it must have been some other person that sent the document. To say who is right could be very difficult and time consuming. Please note that the function may be affected by laws and regulations in a country. For example, a digitally signed document may be valid as a proof while another country does not accept it. Figure 6.9 shows an example of the problem.

If an offer is sent between Alice and Bob in traditional format (paper), the offer is written on paper and a person signs it. The signature as well as a date can be used as proof of the sent offer. In a world with only ones and zeros, this has to be solved in another way by the security system.

Denial of Service

Denial of service is a threat against the availability of a computer system. By sending storms of messages (for example, web page requests or e-mails), a server becomes busy handling these faked messages and no 'useful' tasks can be done by the system. Ordinary users cannot use the system in a normal way. There are several ways of doing this:

- One user lets many computers send heavy traffic to one and the same destination. The origin can thus be difficult to trace as the actual traffic is generated by other computers.

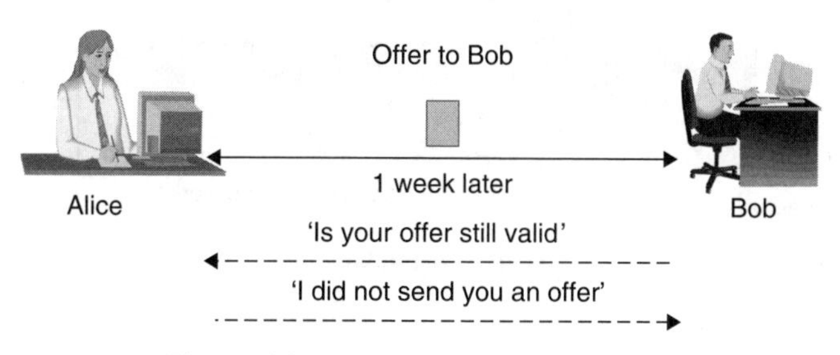

Figure 6.9 Example of non-repudiation

- An application sends thousands of e-mails to a specific user. Bugs in e-mail software can be utilized in a way that e-mails multiply themselves and overload the system/server.

6.4 NON-AVAILABILITY FOR NON-SECURITY REASONS

This is a large area. All the traditional reasons for non-availability remain, but the successive reduction of physical volume decreases the number of elements, which is an advantage for reliability and availability. Other factors make the network more vulnerable, with the introduction of web services as one contributing factor.

6.5 CONNECTING SECURITY TERMS INTO TELECOMMUNICATION

So far the vocabulary has mainly been security oriented. It has also been service oriented. What is needed is to apply or even translate the security requirements into the telecommunication field. This could be done in various ways. One approach chosen for this book is to use the same tool as in Chapter 3: the fundamental technical plans (Figure 6.10). Another one is to use the 'stack'.

The first measure is explained in Chapter 3: to integrate security within the fundamental plans. To make a parallel with QoS, it would have been logical to call the security area SoS, security of service (for the user), or security of system (for the actor).

The next measure is to look at the nature of attacks in a broader way. What is really being attacked?

Figure 6.10 can be developed into Figure 6.11. Let us use this FP figure to illustrate the attacks.

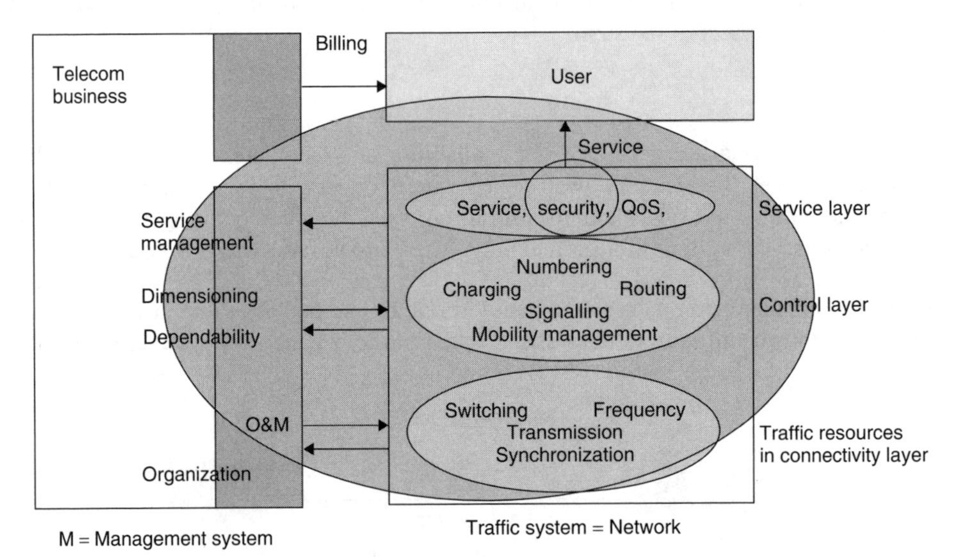

Figure 6.10 Security is in this book seen as a fundamental technical plan (among other views)

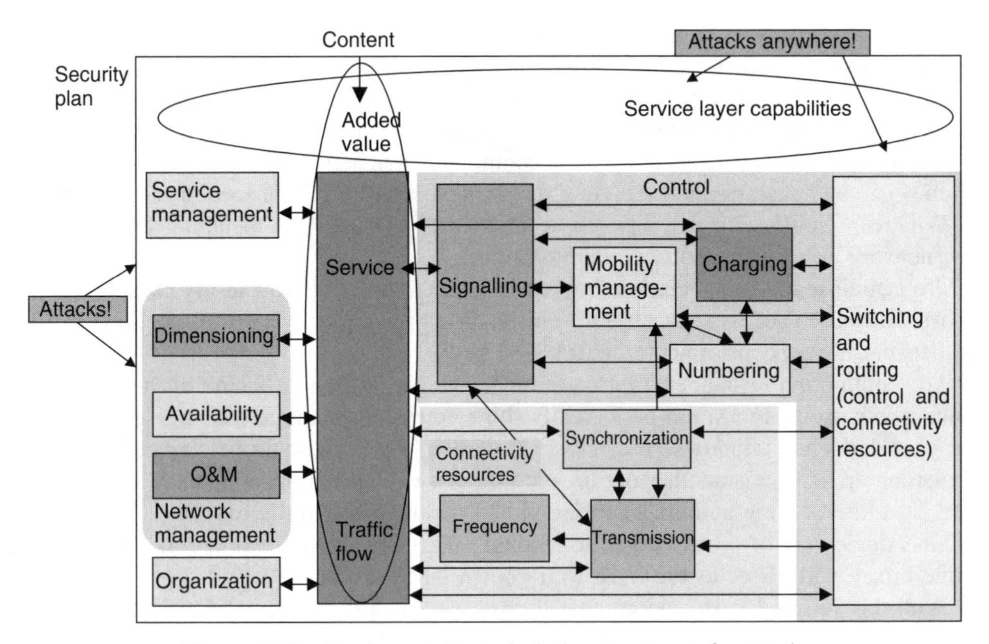

Figure 6.11 Fundamental technical plans are target for attacks

The IP *control plane*: the attacks can hit the addressing functions. The domain name system (DNS) is the directory service inside the Internet. It is the DNS that translates a WWW or IP address (for example 10.30.100.45). Suppose that an intruder wants to manipulate the DNS database. If possible, then a user may think that he is communicating with his bank, while in fact he is communicating with the intruder system (which could be a copy of the bank's home page). In this way the intruder can get hold of, for example, bank account numbers and passwords.

Other attacks can hit the *routers*, whether seen as control or connectivity elements. The routing plan uses routing tables in the routers for controlling traffic flows and making decisions based on packet content and configuration.

The routers are accessible through the network and are therefore often targeted by attackers trying to hack their way in. If an attacker can control the traffic flow in the network, he can get his own packets to any destination (misrouting).

The *charging* system is of course interesting for hackers, as well.

The *signalling* systems are the nerve systems of the communication. Protection of signalling (SIP, SDP) traffic has priority.

Regarding the *connectivity* layer, transmission is a weak part for eavesdropping in radio systems. At low level, hardware such as switches and relays might be destroyed. Electronic attacks can make the network stop working. Wires could be cut. Fire and water are other risks.

Attacks on the *management* part may be more severe than others. Management security is of particular concern regarding both network protection against intentional damage (sabotage from hackers and intruders) and fraud detection and prevention. A very strict authorization and authentication is therefore necessary.

At the *service and content* layer the network can be flooded, leading to denial of service. Virus attacks and other malicious programs hit the contents. Data get corrupted. Unauthorized users gain access to the communication network. False data is inserted. Money can be transferred to the wrong receiver. B2B and B2C transactions are disturbed.

What can we do with the network, to protect it? Let us use a metaphor and consider the network as a house, that we want to protect.

We can close the doors and windows and lock them. Then we can give a key only to those we want to access to our house. This is an organizational way of protection, similar to network access control and an encryption key.

You could improve data encryption through rotation of the key. Think of this as having your own personal locksmith periodically show up at your home and change all the locks on your doors and windows. You dictate how often the key should be changed.

Setting up a fence is another option, with the firewall as a correspondence. The firewall acts as a filter, letting authorized traffic in and locking other traffic out.

One single way of protection is not enough, but there are protection concepts that are quite broad, with IPsec as the most well-known one.

With this in mind it is obvious that the security plan in reality is distributed all over the network. Therefore, one way is to attach a security part to all other fundamental plans to ensure that they are all functioning properly. However, there would still be a need to have a chapter with no correspondence in the other plans, with the overall security policy and the translation from security language to concrete measures in the network. A common complementing or alternative way is to use layered protection. Both methods will be used.

6.6 MAIN WAYS TO IMPLEMENT SECURITY

6.6.1 SECURITY POLICY IN A COMPANY

Security policies are in conflict with user friendliness. The tighter the security, the less is the convenience to users. Users will therefore find a way around a policy if it's too difficult to comply with.

Some general security advice:

- Understand how your system normally functions. Notice unusual events that can help you to catch intruders before they damage the system. Use auditing tools to detect unusual events. You should know exactly which software you rely on, and your security system should not have to rely upon the assumptions that all software is bug-free or that your firewall can prevent all attacks. Create appropriate barriers inside your system so that if intruders access one part of the system, they do not automatically have access to the rest of the system. The security of a system is only as good as the weakest security level of any single host in the system.
- Passwords and encryption keys are secrets. The more secrets, the harder it will be to keep all of them. Security systems should be designed so that only a limited number of secrets need to be kept.

- Physical access to a network link usually allows a person to tap that link or inject traffic into it. It makes no sense to install complicated software security measures when access to the hardware is not controlled.
- Administrators, programmers and users should consider the security implications of every change they make. Understanding security implications requires lateral thinking and a willingness to explore every way in which a service could potentially be manipulated. Users must have access to qualified support to ensure accurate use of security systems.
- The security management and technical staff must work with the security solution on an everyday basis. Solutions for secure networks are not static.
- Education is important.

6.6.2 THE LAYERED VIEW

Let us now look at the security of the network and services in a layered way. The layering does not normally cater for the distribution dimension, but the view is very common. Security measures can be introduced at all levels in an IP stack. In Figure 6.12 a simple two-dimensional security view is provided.

It is not possible to manage with protection at one single layer. A further protection measure, apart from access protection, is the partitioning of networks into zones.

Figures 6.12 and 6.13 give examples of protection enablers at various layers. In Figure 6.12 the measures could be:

- For the application layer: encryption of transmitted user data, proxy filtering.
- For the network user: all types of access control to the network. See AAA below.
- For the control layer: control of individual and general security measures at the connectivity level, such as tunnelling.
- For the connectivity layer: tunnelling, filtering.

Among the missing standards in Figure 6.13 are PGP (pretty good privacy, see also Section 6.9.1) at the application level and link layer encryption. Link layer encryption could for example be used for leased lines and frame relay or ATM WANs.

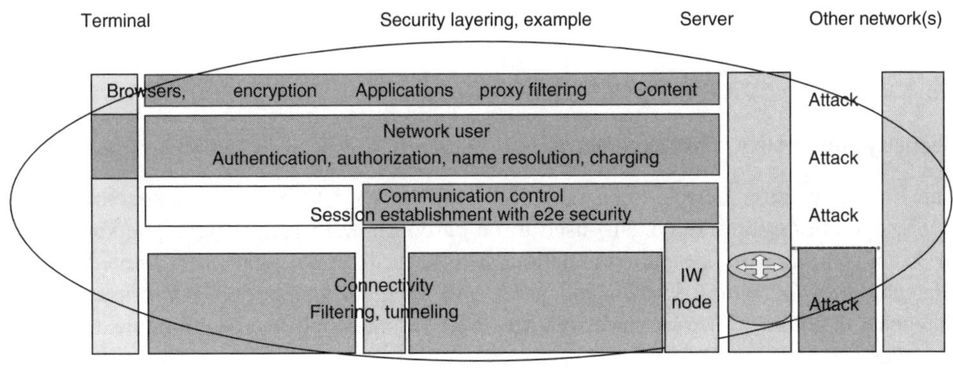

Figure 6.12 A two-dimensional view

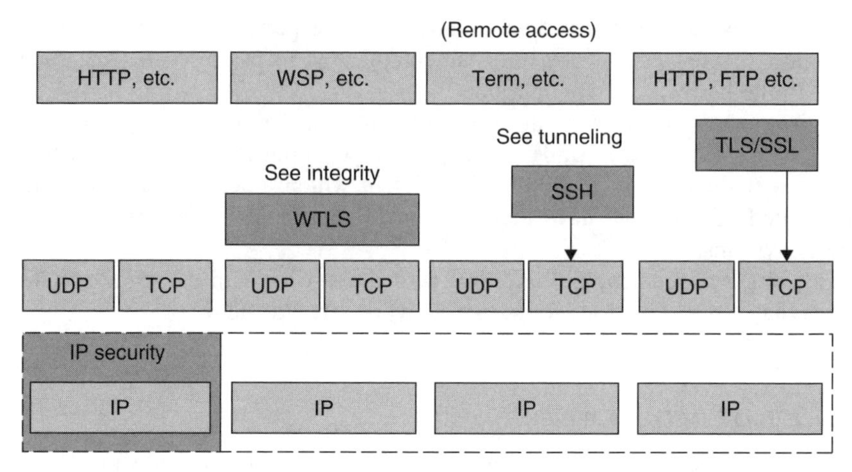

Figure 6.13 A layered view with four protection methods indicated

Encryption, in order to achieve confidentiality protection, can be used at all layers. Regarding protection at the link and physical layers, solutions are normally not continuous end to end, because of the link-by-link architecture.

In GSM, for example, the physical layer is encrypted between the mobile and the base station. In WLAN according to IEEE 802.11 layers 1 and 2 are used for encryption in a standard called wired equivalent privacy (WEP).

Since the IP protocol is normally used end to end an IP layer encryption can/should be maintained all the way. A common layer 3 encryption is part of IPsec, commonly used for IP-VPN.

Encryption at higher layers increases the security even further, but has the drawback that the applications need individual encryption. An example is secure socket layer (SSL) at layer 4 that can protect HTTP web traffic.

The layered view is applicable to many standards in the new network. 'Ordinary' standards for content to user communication are covered in Chapter 3. Service and business-oriented standards are explained in Chapter 5, security standards in Chapter 6, QoS standards in Chapter 7 and specific enabler standards in Chapter 8.

6.6.3 THE FUNCTIONAL (FP) VIEW

Security for Network Management

The communication between the network elements (NEs) and the management entity (OSS, operations support system) used to be based on X.25 technology. The knowledge of X.25 is limited to a small population. The security risks were then limited. Today IP technology is widely known and used, and even its standards (RFCs) are open to the general public. CORBA and Java are used for the high-level communication with the nodes.

Despite considerable risks of security violations when working in an IP environment, there are not many other options available. The strong technological shift towards the

Internet for data communication, the increasing maturity of Internet security alternatives and the fact that the connectivity backbone will be based on IP or IP/ATM, force us more or less to use IP for data communication (NE to/from manager) within the management system.

The main communication protocol is simple network management protocol (SNMP). Main security features in SNMPv3 are encryption to assure privacy, and authentication of management–agent communication to guarantee sender identity and message integrity.

A solution could look like this:

- The communication system is built in three layers. A transport IP part, a secure data communication network and an application layer with management support applications.
- Access is possible only by means of specific logon servers. A public key server is used to generate certificates. The CORBA traffic is secured by means of security modules at application level in the nodes, verifying authority and authenticity. Telnet and FTP traffic are protected by means of SSH/TLS.
  Security information is stored in a LDAP server.

Security for Network Resources

Network resources are normally controlled by the management system. As long as the management system is run by authorized staff there is only the human factor that could cause harm to the network through the management system. Making sure that only authorized staff can access the management system can be achieved by means of digital certificate based on secure encryption.

Then there are threats from physical damage.

A third threat comes from the service network, which could control network resources via the API interfaces. Therefore an authorization/authentication procedure protecting the network is provided in the service network (application)–control interface. A similar protection is required in the user–service network (application) interface.

As mentioned in the introduction, integrity targets are supported by the separation and protection of resources. In order to control the security, network security partitioning into zones should be used for large networks. The borders might be points where protection method changes, for example from L2 tunnelling to L3 tunnelling. Possible border control elements for security inspection are firewalls, filtering routers or IPsec end points. Packet filtering towards backbone and ISP/corporate, as well as downlink/uplink address screening, is found in GPRS edge nodes. When applied to a large company the main idea looks like Figure 6.14.

Numbering – DNS Protection

DNS is, as explained earlier in this chapter, a very vital part of an IP network. DNS Security sees to it that the client gets an answer from the correct DNS server, that the answer is related to the query and that no changes have been made to the answer by an intruder.

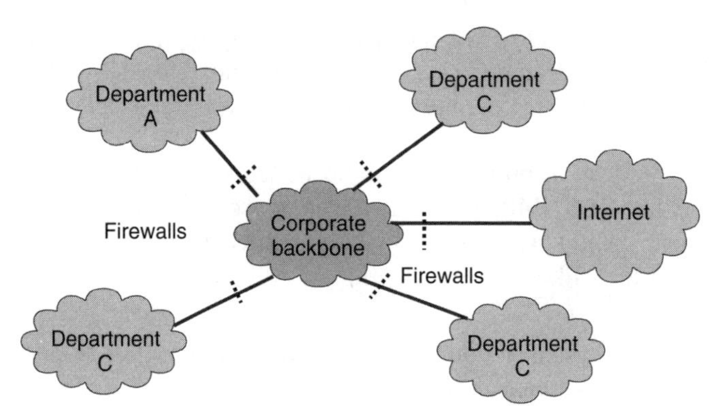

Figure 6.14 Network partitioning into zones

Security Needs and Solutions for the Business Sector

For economical reasons the protection level is different between different services. Services with the highest security requirements are e-commerce and computer networking, such as remote access, LAN interconnect, extranets between enterprises and links between enterprises and application service providers.

Ultimately, Internet VPNs will be the global means of business communication just as the voice network is today.

Casual e-mail is considered less risky, but at least business e-mail should be protected.

Voice over IP may introduce risks for other services and needs protection against eavesdropping/wiretapping.

In today's workplace there is a need to provide access to employees even when they are outside the office. This need arises first for the 'road-warrior' employees such as a sales persons or installation teams, but also for accommodating home workers and 'day-extenders'. Traditionally, such access has been provided with dial-in connections to a corporate network access server (NAS). See for example Figures 6.15 and 6.16. This approach has, however, several drawbacks:

- The corporation has to invest in the installation and management of the NAS.
- While easy and inexpensive from the same geographical area as the corporate offices dial-in access from overseas locations may be prohibitively expensive.
- The introduction of new access technologies such as digital subscriber line (DSL or xDSL, such as ADSL) with higher bandwidth but shorter distances is making the corporate-owned NAS approach no longer possible or necessary.

For these reasons corporations are moving towards the model where local access is provided by for example the DSL provider and connectivity to the corporate Intranet is provided through IP networks. This provides substantial benefits to corporations in terms of how fast, easily and flexibly they can provide remote access.

The solution for secure applications has initially been building a private network, but with the wide-spread use and availability of the Internet it is of large economic interest

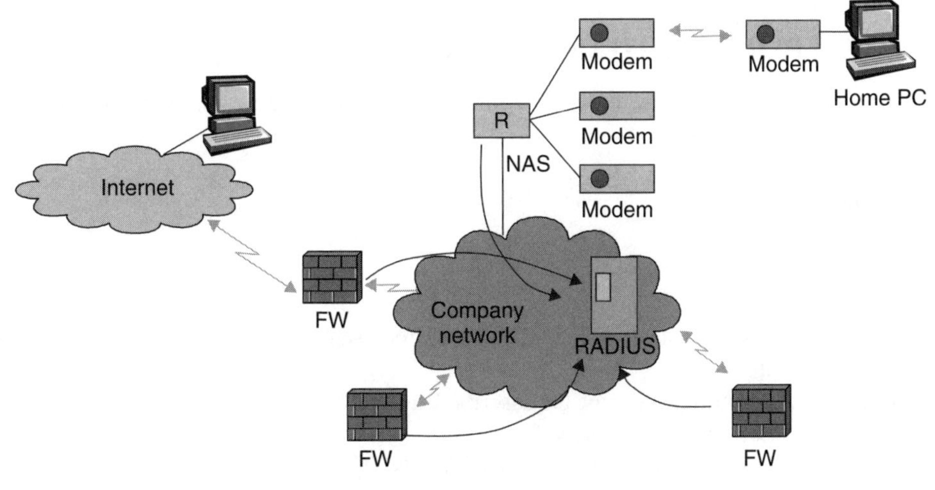

Figure 6.15 RADIUS dial-in

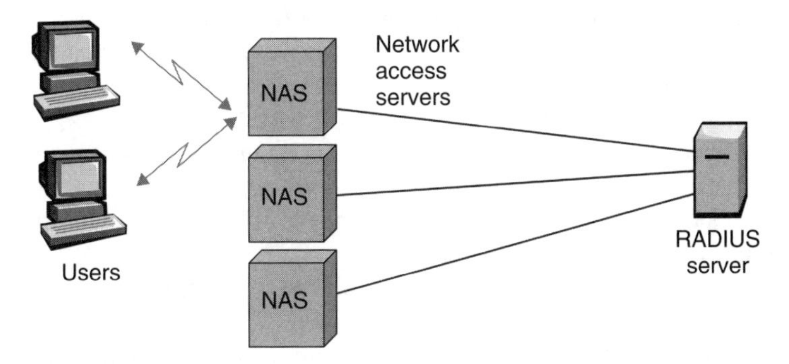

Figure 6.16 A RADIUS connection. The RADIUS position can be seen in Figure 6.16

to introduce security there. The main secured Internet service is called virtual private network (VPN). It is tailored for wide area enterprise networking and remote access to enterprises. No precise definition of VPN exists. A rough definition is: 'Private (mostly data) traffic utilizing a public network'. Another alternative is 'An encrypted tunnel over unreliable networks'. A third definition is: 'A Virtual Private Network is a private network that makes use of the public telecommunications infrastructure, maintaining privacy through the use of tunnelling protocols and security procedures'.

The main idea is to create a secure path through the insecure Internet environment by using tunnelling, encryption, keys and other security enablers.

Also company–customer information flow should be secure, based on suitable routines. Certainly the customer does not wish that anyone could read his conversation with the company. Yet, high security levels with cumbersome routines might scare off a company's customers. The solution is informing people of the consequences of intruders. By knowing risks, people may better accept security routines.

6.7 INTEGRITY AND CONFIDENTIALITY BY ACCESS CONTROL – AUTHENTICATION

Authentication is one of the three As commonly found in figures showing IP-oriented networks. AAA stands for:

- Authentication
- Authorization
- Accounting

Authentication is a control of the identity of the user, whereas authorization means allocation of rights.

The last A, accounting, will not be described in this chapter as the function is related to charging and not network security.

Authentication means controlling the user identity. The fundamental mechanism is that another party has to show that he knows or is able to do something that nobody else can do. Often the one who wants to verify the authenticity of somebody must challenge him and the challenged party must respond with the right answer. Such a response is to encrypt the challenge with the right key so that the party can prove that he/she possess the key. It is very important that the challenge is truly random. If the intruder can predict the next challenge, he can ask the right response in advance.

Access restrictions by means of authentication is nothing new. An interesting possibility deployed by the old Egyptians and Greeks, who did not have identity cards, is to use biometrical methods. 'Biometry' means 'measurements of humans' with original measures such as length, eye colour, skin colour and scars.

A true personal identification can only be achieved using unique personal properties. A key, a personal code or a photo can all be misused.

Two fairly cheap alternatives are voice recognition and personal signatures. Voice recognition is tailored to voice communication, and therefore easy to introduce for voice supporting media. This type of verification requires storage of voice prints. The problem with voice recognition is that any type of background noise, or voice changes because of a cold, can cause legitimate calls to be blocked. A more futuristic alternative is scanning of the iris. See biometrics below.

Even simpler is a system that is used for personal verification of roamers, who are routed to a call centre, where they verify their identities by answering one or more short questions. However, some people consider the system inconvenient, which may lead to churn.

Some common authentication methods use RADIUS (remote authentication dial-in user service) and LDAP (lightweight directory access protocol). See below.

It is important to understand that the right response is not a guarantee that we are directly communicating with the right party. There is a very serious attack called 'man in the middle'. Let us take an example from the real world. Mallory claims that he is a master chess player although he is not. If some chess master challenges him to play, Mallory arranges the game to happen in the computer network. Then Mallory challenges still another chess master. When the first master makes a move, Mallory mediates it to another master. They both think that they are playing with Mallory – but actually they are playing against each other. The same can be easily done in the Internet where a challenged party can quickly challenge the right party and give the right answer.

The solution to the man in the middle attack is to carefully combine authentication and key exchange.

6.7.1 BIOMETRICS

Biometric recognition uses fingerprints, faces, the iris and other parts of the body for identification. If successful and affordable, such methods can fully replace identity cards, codes and passwords. This will no doubt make life easier, since we won't have to carry identification cards or keys or remember a lot of digits any more.

Biometrics is nothing new. The old Egyptians used it to handle 'secure transactions' when the society became more organized and anonymous. The biggest advantage is that nobody can steal your biometric identity as opposed to keys and codes.

Computer-supported biometry came into use in criminology for comparing a fingerprint with fingerprints in a catalogue. One of the safest methods is scanning of the iris. The technique was initially developed in Cambridge. Even twins have different iris. Scanning is possible also with glasses. A half-year trial in the USA gave encouraging but not perfect results.

Less reliable are voice recognition and (handwritten) signatures. Mobile phones could use biometry to support e-commerce, most likely using fingerprints. The fact that we cannot change our biometrical identity might be a drawback if the stored identity is stolen.

6.7.2 RADIUS

RADIUS is a flexible system, which can be used for many types of authentication services. Its original use was for dial-up connections to companies or to the Internet. The main idea is to protect and/or charge dial-up connections. The building blocks of a RADIUS system can be seen in Figure 6.15.

Radius consists basically of a client and a server. The client, who is situated between the modem pool and the server, asks the user for identity and a password. A typical client location is a NAS, which is the access point to the Internet/intranet. The RADIUS server is nothing but a server somewhere in the network. Users are fixed or mobile users accessing the network. See Figure 6.16 and 6.17.

End-user authentication in a voice network (PSTN/ISDN/PLMN) interacts with a RADIUS server located at the ISP service network or in a corporate network. For this purpose, the voice network includes a RADIUS client.

6.7.3 DIAMETER

Investigations have shown that RADIUS is ill-suited for inter-domain purposes in a roaming network Diameter is a new AAA security protocol used with mobile IP.

6.7.4 TRANSPORT LAYER SECURITY (TLS)–SECURE SOCKET LAYER (SSL)

The TLS–SSL protocol was developed by Netscape Communications to provide security and privacy over the TCP/IP protocol. Figure 6.13 gives the layer position. The protocol supports server and client authentication.

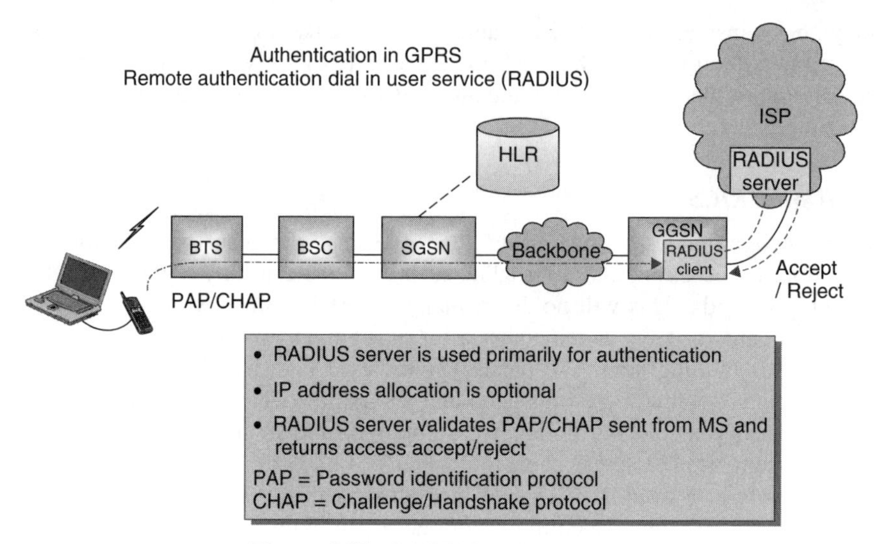

Figure 6.17 RADIUS authentication

TLS is application independent, allowing protocols like HTTP, FTP and Telnet to be layered on top of it transparently. The TLS protocol is able to negotiate encryption keys as well as authenticate the server before data is exchanged by the higher level application.

The TLS protocol maintains the security and integrity of the transmission channel by using encryption, authentication and message authentication codes.

A variety of cryptographic algorithms are supported by TLS–SSL.

6.7.5 WIRELESS TRANSPORT LAYER SECURITY (WTLS)

WTLS is based on TLS. As opposed to TLS, WTLS can operate over UDP. The WTLS layer is modular and the application decides what security level and features it will use. Figure 6.13 indicates the layer position.

6.7.6 LIGHTWEIGHT DIRECTORY ACCESS PROTOCOL (LDAP)

LDAP is a protocol that is used by many applications to retrieve information in a library in a standardized way. The information could deal with:

- Persons and enterprises
- Configuration of software and equipment
- Certificates for different security services
- Authorization rules
- Support for QoS including traffic priorities in the network. Directory enabled networking (DEN) uses for example LDAP to store attributes and traffic classes.

The LDAP protocol is an IETF standard (RFC1779). LDAP exists in a number of versions. LDAPv3 has an improved functionality for national characters and cross-references between different servers. Contrary to many other TCP-based protocols, like HTTP, FTP and Telnet, LDAP uses coded commands over the network. The transaction coding is called binary encoding rules (BER).

An alternative directory service with less success is called Whois++.

E-mail programs get LDAP support to search e-mail addresses. LDAP is an important part of Windows 2000 Active Directory and Netware NDS (Netware Directory Services).

All versions of web browsers have support for LDAP.

LDAP is also a base for public key infrastructure (PKI). Almost all PKI systems use LDAP servers to search, verify and revoke certificates.

The LDAP transfer must be safe, in particular when handling public keys. What if someone else pretends to be the LDAP directory? LDAP can be encapsulated in SSL or TLS.

6.8 INTEGRITY BY ACCESS CONTROL – AUTHORIZATION IN ENTERPRISES

Data that is available in an internal company system must be protected. Only authorized staff must be able to access, read and change such data. Such security functions are built into most network operating systems. There are in principle two common ways to solve the problem:

- Users must log on with user identification and a password to get access to the network. In addition, different resources can be allocated to different users.
- The network is open to all, but when you want access to a specific resource you have to use a password which is related to that resource.

It is possible to have many levels of security within a company or system. A good idea is to avoid using more security than necessary to avoid unnecessary inconveniences. What is experienced as complications can lead to confusion and less efficiency. However, one of the greatest threats in all computer systems is the company's own staff. Most of the investigations done in this area show that the staff inside the company is one of the most frequent intruders in computer systems.

6.9 INTEGRITY BY ACCESS CONTROL – FIREWALLS

6.9.1 WHAT IS A FIREWALL?

In general, a firewall is a node (computer and/or router) often located at the border between the public Internet and servers or a private intranet. See Figure 6.18. The purpose of a

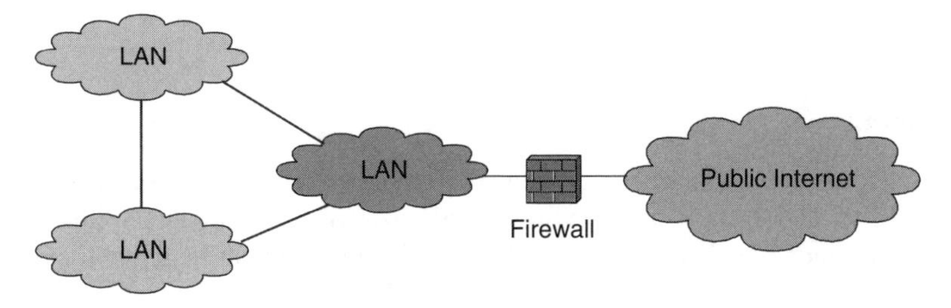

Figure 6.18 Location of a firewall

firewall is to block access to the intranet from the public Internet. People outside the intranet must not be able to access information inside the intranet or perform actions such as destroying information.

At the heart of all firewalls is a packet filtering or screening mechanism, based on an interpretation of packet type. The simple firewall cannot check the application level, but identifies IP, TCP, UDP and ICMP. The filtering is controlled by a table. By reading the TCP header, especially the TCP port, it is also possible to determine the type of application. Port 80 in the header means web traffic. Thus the firewall can for example decide that web traffic can pass whereas FTP traffic is blocked.

The firewall translates the IP addresses by means of network address translation (NAT). With NAT the internal IP addresses in a LAN or Intranet are not visible from the outside. This increases the security for the enterprise and reduces the necessary number of IP addresses.

6.9.2 TERMINOLOGY FOR PROTECTED AREAS

Protected Network (Inside)

All internal systems are placed on the protected network and cannot be reached from any external source. However, this is no reason to reduce the level of system security used internally. If an attacker gets inside, we must also have good host security.

High Risk Area – Demilitarized Zone (DMZ)

It is common to connect a screened subnet to the firewall. To this subnet, often called a demilitarized zone, public servers such as web servers can be located. Such servers are thus available to the outside and therefore possible targets for attacks.

Only computers with a strategic function in the company should be on the DMZ segment.

Thorough logging should be applied to all DMZ systems. The log files should be transferred in a secure manner to the inside network for analysis and storage.

Alarms should be triggered if an attempted attack occurs.

Bastion Host

A bastion host is a computer that is continuously monitored, logged and upgraded, and which serves the purpose of being the point of entry to all traffic from outside to inside networks.

Proxy Firewall

A more advanced firewall is the proxy firewall. It can perform virus control of documents, traffic logging to track intrusions, blocking of JAVA/Active X components. Since the proxy functionality includes analysis of the application layers, it must have separate programs for each application, such as SMTP (e-mail), HTTP (web applications), FTP, Real Audio and Telnet.

VoIP

With increasing voice over IP traffic there is a need to mix VoIP and ordinary IP data traffic. Since the data traffic should pass firewalls, there is a need for firewalls supporting the SIP protocol used for mixed traffic.

6.10 CONFIDENTIALITY: ENCRYPTION AND KEY MANAGEMENT

6.10.1 BASIC CRYPTOGRAPHY

The first solution to protect IP packets is to simply encrypt the packet. This guarantees that no one can read the content. However, we must remember that encryption does not prevent modification of the packet. It means that it provides confidentiality but not integrity.

For example, if a packet consists of just one bit of data: 0 or 1, then it is 0 or 1 in encrypted form and someone can change the bit. In this example, the modifier can be quite sure that he can change the message by changing the bit. In most cases this is not that trivial, but it is still possible to garble the message so that the receiver does not notice. Another problem with plain encryption is that you cannot encrypt headers of the packet for communication reasons, however, the modifier can still often successfully attack by modifying headers.

The two main purposes of cryptography are:

- to maintain the confidentiality of messages; and
- to guarantee the integrity of messages.

Confidentiality is provided by encryption, whereas integrity can be provided by authentication codes or digital signatures. A fairly common standard for cryptography is called *pretty good privacy* (PGP).

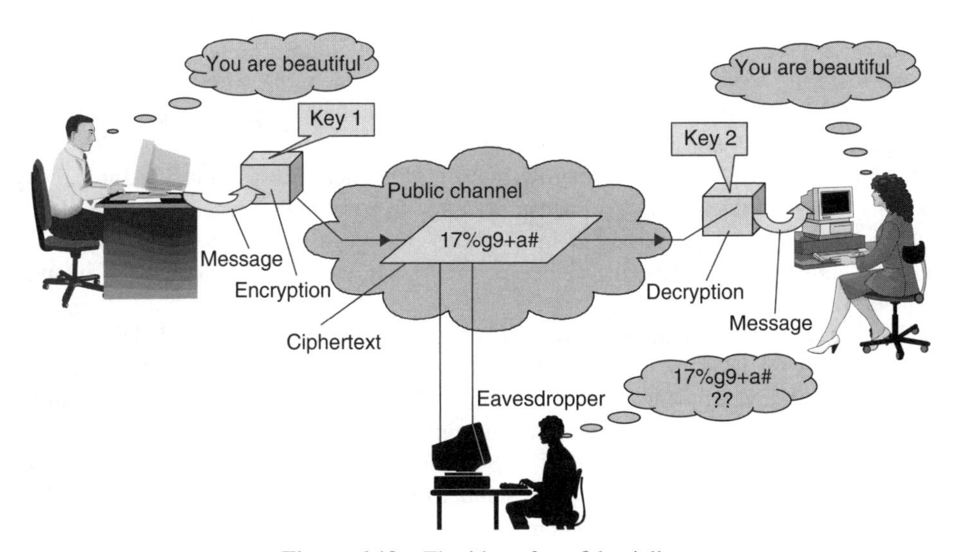

Figure 6.19 The idea of confidentiality

PGP was initially created by one person. It has become widespread because it is a freeware. PGP provides a data integrity for messages and data files by using existing technologies for security:

- Electronic signatures
- Both symmetric and public key (asymmetric) encryption
- Compression
- Key management

When a user encrypts plain text with PGP, PGP first compresses the plain text. Data compression saves transmission time and disk space and, more importantly, strengthens cryptographic security. Files that are too short to compress or which don't compress well aren't compressed.

6.10.2 ENCRYPTION

The encryption of packets protects IP traffic. Before sending a message the sender uses a key (k1) to encrypt it. The cipher text is then sent over a public channel that is open to eavesdroppers.

To read the message, the recipient uses a key (k2) to decrypt the cipher text, thereby retrieving the message. Although active adversaries can insert packets and modify communication, for this discussion we assume only passive listening.

If k1 and k2 are equal, the system is symmetric. Otherwise, it is said to be asymmetric.

To guarantee security, k2 must always be kept secret, whereas k1 can be made public, provided it is infeasible to derive k2 from k1. If indeed this is the case, then the system is called a public key system.

Public key systems offer many interesting possibilities. For instance, anyone can send an encrypted credit card number to an online shop using the shop's public k1. Since only the shop possesses k2, no one but the shop can determine the number. If a symmetric system were used, the shop would have to exchange unique keys – privately and in advance – with every potential customer. The security of public key systems is always based on the difficulty of solving certain mathematical problems, whereas symmetric schemes are more *ad hoc* in nature.

The main drawback of public key systems is that their mathematical nature always makes them less efficient than symmetric systems; in particular, because the size of keys in public key systems is measured in kilobits – the keys of symmetrical systems are only one-tenth as large. Thus, the choice of encryption method depends on the intended application.

The most widespread symmetric system is the data encryption standard (DES), which was developed by IBM in the mid-1970s. The DES used a key size of 56 bits. With recent hardware developments, however, a 56-bit key size space can be entirely searched by advanced machines within hours or even minutes.

For this reason, the National Institute of Standards and Technology (NIST) initiated the development of the advanced encryption standard (AES), which supports 128- to 256-bit keys. Moreover, unlike the development of DES, the AES design process is open to the public. A key length of 128 bits is considered secure. IPsec applies AES encryption.

6.10.3 ELECTRONIC SIGNATURES

Electronic signatures enable the recipient of information to verify the authenticity of the information's origin, and also verify the integrity of the information. Thus, public key electronic signatures provide authentication and data integrity. An electronic signature also provides non-repudiation, which means that it prevents the sender from claiming that he or she did not actually send the information.

6.10.4 WHY PUBLIC KEY INFRASTRUCTURE?

How can you prove who you are over the Internet? In 'real life' you use some type of identification card (for example a driving license or a passport) to prove who you are. In the world of only ones and zeros, the digital certificate is your ID card. The ID cards in real life are distributed according to strict rules to avoid someone else getting hold of your ID card. The way of handling digital certificates is what PKI is all about.

PKI is a way to solve the problem of the management of keys. In a large system, with thousands of users, there must be some automatic way of managing keys for encryption and authentication. With key management, we usually mean:

- generation of keys and digital certificates
- distribution of keys and digital certificates
- storage of keys and digital certificates
- validation of keys and digital certificates
- revocation of keys and digital certificates
- expiry of keys and digital certificates

A consequence of the encryption is the need for keys in order to understand the message. Key handling requires what is called 'secure associations'.

The PKI standard allows a single login to a system with the aid of a 'smart card'. This card contains a private key, and the public key is stored in a directory. This requires a trusted party to act as Certification Authority (CA), issuing necessary certificates (Figure 6.20).

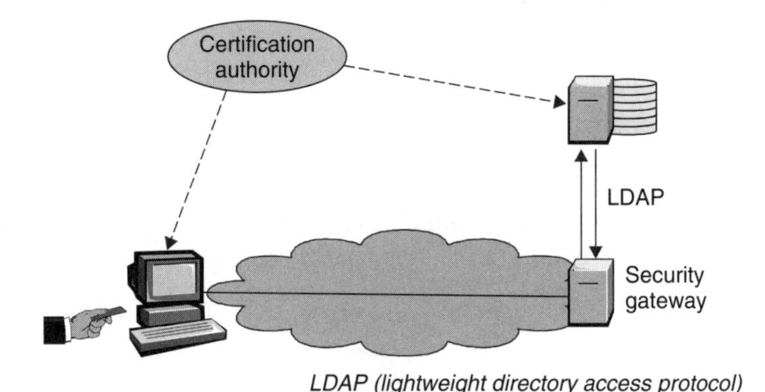

Figure 6.20 The need for a Certification Authority

The protocol LDAP is used to get the public key from the directory. Once again, it is important to know you are talking to the right person, and that nobody pretends to be the LDAP directory.

6.11 CONFIDENTIALITY BY TUNNELLING

There are many definitions of a tunnel: A tunnel is a safe point-to-point connection over an unsecure network, but can also be seen as an emulated connection-oriented connection over a connectionless network. Furthermore, tunnelling is an enabler for encapsulating packets (of different types such as IP, IPX) inside a common infrastructure.

Even better is to regard the tunnel as an overlay network creating for example a corporate wide area network within a public IP network.

Tunnelling simplifies life for the routers along the path, since they do not have to understand all tunnelled encapsulated protocols. Layers 2 and 3 are employed for tunnelling. Tunnelling is used for:

- designing and setting up a VPN
- remote access to companies
- interworking between IPv4 and IPv6 numbering
- supporting host mobility, such as mobile IP (RFC 2002)
- private numbering plan (compare VPN in PSTN/ISDN with public and private numbers)
- multiplexing several protocols into one link
- carrying other protocols than IP across the IP network
- creating a dedicated overlay backbone (e.g. MBONE and other tunnels for multicast).

The use of tunnels ensures that the tunnelled traffic may not exit from the tunnel. However, it is still possible for malicious users on the path between the tunnel end points to listen and inspect all the tunnelled payload traffic. It is also possible for an attacker to insert packets into a tunnelled stream. Since the packet content is normally encrypted it is, however, difficult to understand what is transferred.

The lower part of Figure 6.21 shows layer 3 tunnelling with basic IP and transport protocols encapsulated by 'IPsec' and a new IP header, existing between IPsec gateways. The upper part of the figure shows an alternative to tunnelling, defined in the IPsec standard. See below.

There are two main parts of a tunnelling protocol:

- setting up the tunnel (and tearing it down)
- encapsulating one protocol in another

The tunnel technique has proven reliable. To provide a secure communication path from the employee to the intranet, a secure tunnel is created.

6.11.1 L2TP

One of the most frequently used protocols today is the layer 2 tunnelling protocol (L2TP). Another tunnelling protocol is MIP (mobile IP).

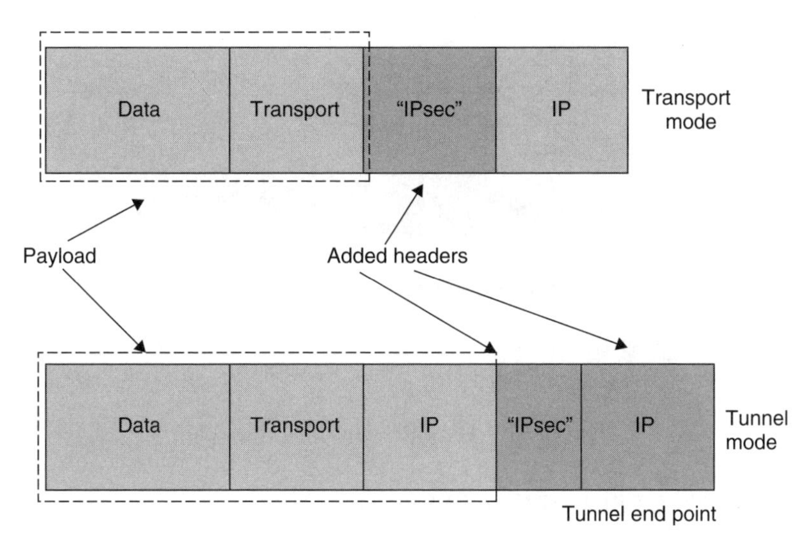

Figure 6.21 Layer 3 tunnelling

L2TP is a secure protocol used for connecting VPNs over public lines such as the Internet. It is essentially a combination of two other secure communications protocols: PPTP and Cisco Systems' L2F.

L2TP is a popular mechanism for accessing corporate networks. It provides a tunnelling facility that can be used both in a voluntary and in mandatory manner. In the voluntary case the user's PC establishes an L2TP connection to the corporate tunnel server over any ISP or Internet access method.

In the mandatory case the ISP's remote access concentrator creates a tunnel from a specific user and to a specific destination automatically as the user logs in. This destination could be a particular kind of ISP or service, a server in a corporation that the ISP has an agreement with, or ISP equipment in a location that is closer to the served corporate than the access concentrator.

Both models are useful in their own situations. The main advantage of the voluntary tunnelling model is that there is no need for agreements with the ISP. The method is mainly used for road-warrior access from around the world. The main advantages of the mandatory tunnelling method are that there is no need for extra configuration of software on the PCs, and the omission of the tunnelling and security headers from the packets on the typical network bottleneck, the access link. The latter is particularly useful in the case of mobile communication.

6.11.2 GENERIC ROUTE ENCAPSULATION (GRE) PROTOCOL

The GRE protocol is used in conjunction with the point-to-point tunnelling protocol (PPTP) to create VPNs between clients or between clients and servers.

GRE encapsulation is achieved as shown in Figure 6.22.

The original IP packet is encapsulated into a GRE packet. The GRE packet itself is encapsulated into an IP packet. This means an IP packet is transported inside an IP packet. The outer IP packet is forwarded like an ordinary IP packet within an IP network.

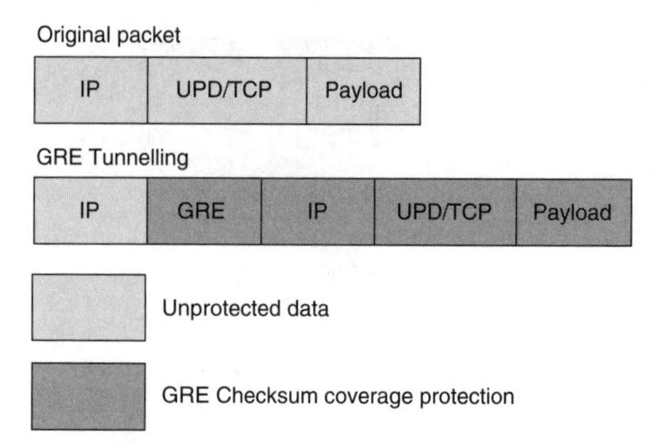

Figure 6.22 The GRE tunnelling principle

6.11.3 SECURITY IN REMOTE ACCESS

Secure shell (SSH) is a protocol that permits secure remote access over a network between two computers. The protocol negotiates and establishes an encrypted connection between what is called an SSH client and an SSH server. The SSH protocol consists of three major components:

- Transport layer protocol. This protocol provides server authentication, confidentiality and integrity.
- User authentication protocol. This protocol authenticates the client to the server.
- Connection control. This protocol multiplexes the encrypted tunnel onto several logical channels. SSH can handle several data streams over the same TCP connection.

6.12 CONFIDENTIALITY AND INTEGRITY BY IPsec

The Internet protocol security (IPsec) provides a set of protocols and a robust mechanism to provide security to the Internet protocol (IP). The protocol is a cornerstone when building VPNs. IPsec forms a secure layer from one network node to another, protecting all IP packets concerned at the IP layer (it does not protect individual applications from attacks). It is rather an extensive standard, which has become part of the IPv6 standard. For IPv4 it is optional. IPsec provides:

- Confidentiality
- Integrity
- Authenticity
- Non-replay protection

Different IPsec security levels are available through various modes and security protocols. IPsec can be used in two modes:

- Transport mode (not common)
- Tunnel mode

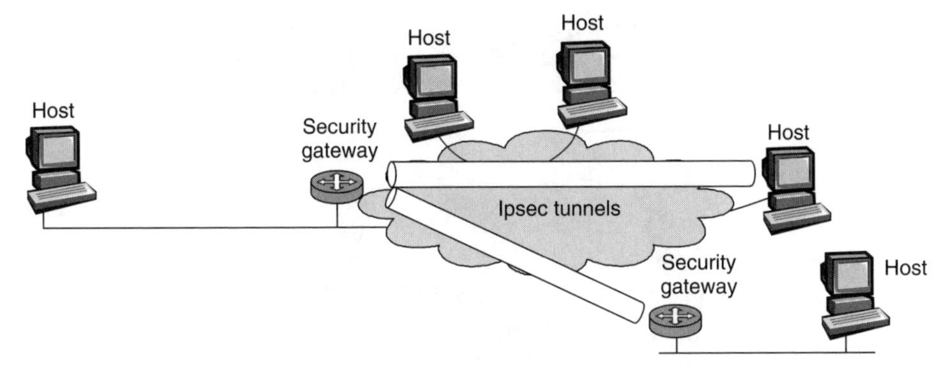

Figure 6.23 IPsec tunnels starts and ends at security gateways or hosts

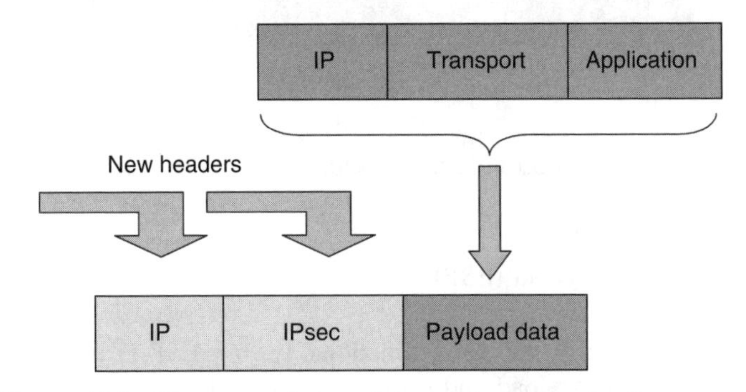

Figure 6.24 The encapsulation of the original IP packet into a tunnel

The tunnel mode allows an IPsec entity to encapsulate, encrypt and forward IP packets through a logical point-to-point tunnel to another IPsec entity. The tunnel mode encapsulates the entire IP header and the payload. See Figures 6.23 and 6.24.

The tunnel mode is normally used between two machines when at least one of the machines is not an end point of the connection. For example, IPsec in the tunnel mode can be deployed between two firewalls located between a client and a server.

The major advantage of the tunnel mode is that an attacker can only determine the tunnel end points and not the true source and destination of the tunnelled packets. Thus the tunnel mode protects against traffic analysis. Another advantage of the tunnel mode is that if offers tunnelling of IP packets with private IP addresses in IP packets with public IP addresses. Thus, the IPsec tunnel mode offers some solutions to the addressing problem.

6.12.1 TRANSPORT MODE

The transport mode (Figure 6.25) should primarily be used between the end points of a connection (hosts). It encapsulates and encrypts only the data portion (payload) of each IP packet, but leaves the header untouched.

The transport mode is less secure than the tunnel mode, since it does not conceal or encapsulate the IP control information and does not provide traffic flow confidentiality by hiding the source and destination of the IP packet.

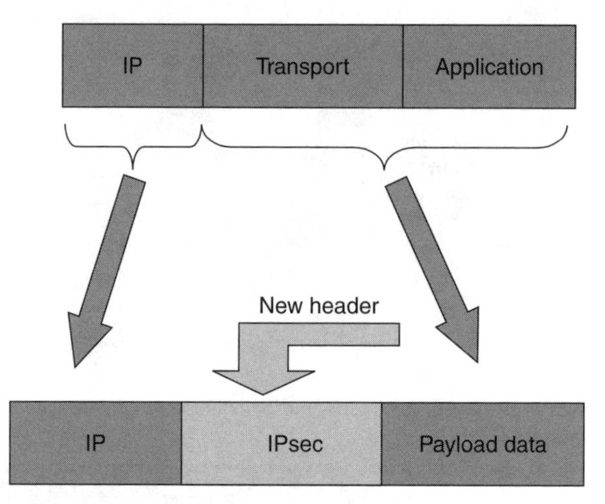

Figure 6.25 The transport mode

IPsec uses two protocols to provide traffic security:

- Authentication header (AH)
- Encapsulating security payload (ESP)

AH and ESP supply partly the same functionality, but ESP is broader and includes encryption methods for IP payload and (in the tunnel mode) for a part of the IP header. AH is described in RFC 2402 and ESP is described in RFC 2406.

The AH header is 24 bytes and includes a security parameter index (SPI) field, a sequence number field and an authentication data with variable length. The SPI identifies a security association (SA) while the sequence number field is counter for anti-replay service.

AH can be used in the transport mode and the tunnel mode.

6.13 CONFIDENTIALITY AND INTEGRITY FOR MAIL BY S/MIME

E-mail has been used within the Internet community for a very long time. From the beginning, all e-mails were text-based, and text was the only transferred information. Later on, the mail standards were extended to embrace enhanced text, graphics and audio in the body part of the mail. The new standard was multipurpose Internet mail extensions (MIME). The Internet standard secure MIME or S/MIME was developed to add security by providing authentication, integrity, confidentiality and non-repudiation for Internet e-mail. S/MIME is also attractive for HTTP information retrieval and for corporate mails, now that more and more mobile users need to communicate securely with colleagues on corporate intranets. See Figure 6.26.

S/MIME relies on the existence of a PKI under the control of a trusted CA. S/MIME is fundamentally used for point-to-point security.

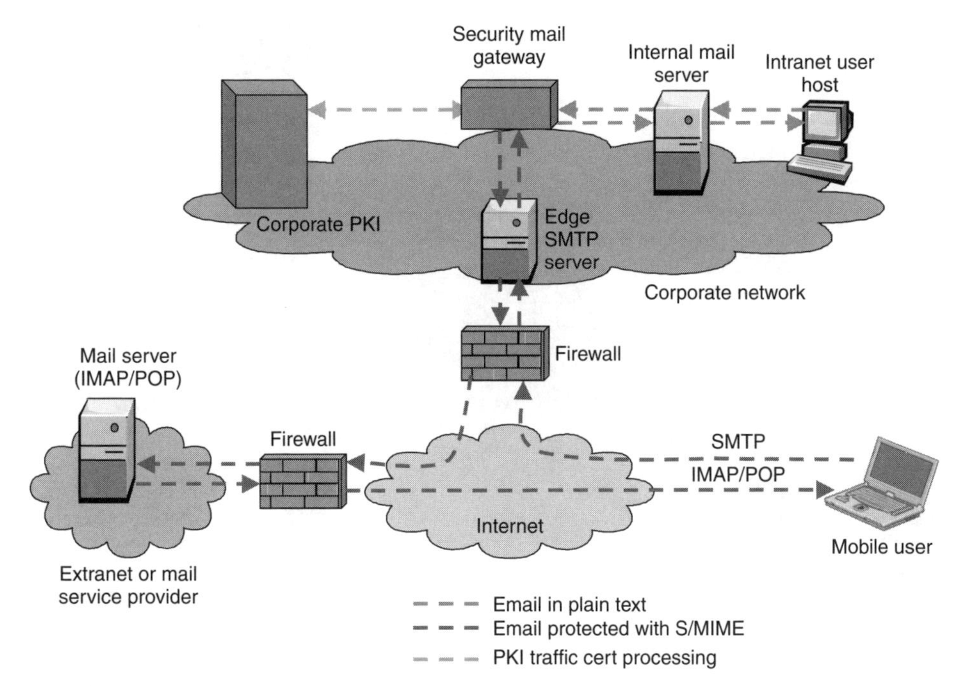

Figure 6.26 Corporate e-mail architecture example – The Internet and corporate sections firewall–security mail gateway protected by S/MIME

6.14 APPLICATIONS AND SOLUTIONS

6.14.1 VPN

As mentioned before, the main VPN idea is to create a secure path through the insecure Internet environment by using tunnelling, encryption, keys and other security enablers.

VPN technology allows an organization to securely extend its networking services over the Internet to remote users, branch offices and partner companies, allowing access e.g. through user ID and passwords. VPNs always include encryption of some type, typically IPsec is employed.

The IP-VPN tunnel can be regarded as a competitor to a frame relay or leased line connection for creation of wide area corporate networks. The tariff is based on IP with no distance-dependent fee, as opposed to frame relay and leased lines. The solution is cheaper, global and flexible. A leased line, a PSTN/ISDN VPN or maybe a frame relay/ATM connection can be used for intra-company voice, apart from public voice networks. See Figure 6.27.

The tunnels form VPNs. (Note: there is another VPN type for voice, mentioned above.) A VPN is a private data network that makes use of the public telecommunications infrastructure. Typically, a VPN uses a private numbering plan.

There are three types of VPN:

- Corporate-built VPN, totally handled by the company. Telecommuters can be connected by means of L2TP. (layer 2 tunnelling protocol). The tunnel is set up by NAS at an ISP.

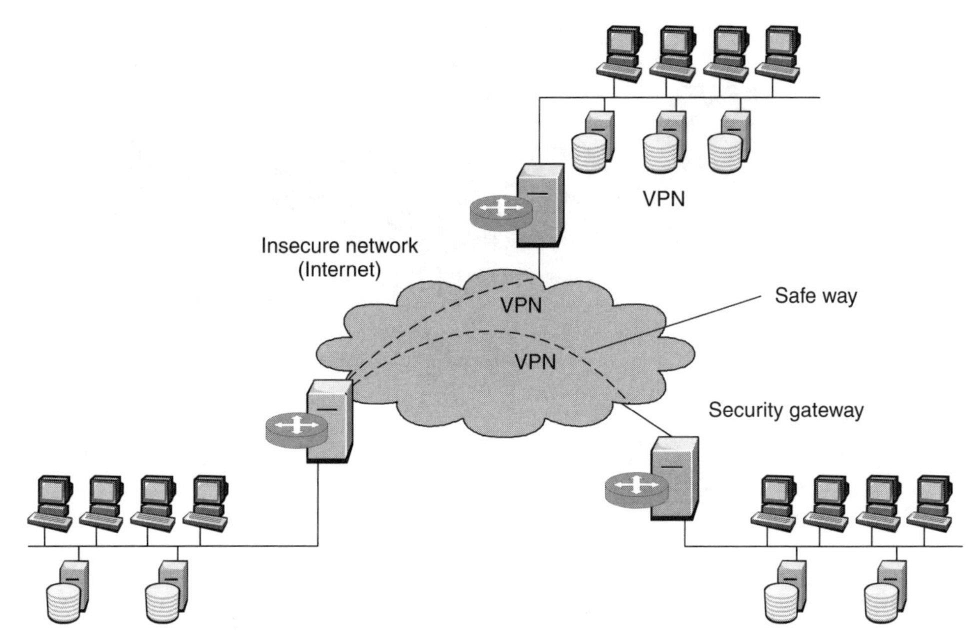

Figure 6.27 A possible VPN architecture

- Customer premises based IP-VPN. About the same solution but with central VPN management.
- Network-based IP-VPN. Everything is done between edge routers.

One of the driving factors in this development of enterprise networks is the minimization of costs for both equipment and maintenance. At the same time adequate security measures must be employed. The use of VPN-based tunnelled remote access sessions substantially decreases the operational costs for providing access to home workers and road warriors. This is due to the reduced need to rely on long-distance modem calls. Flexible configuration options in enterprise NASs make it possible to allow both PC-to-enterprise and operator-to-enterprise tunnels, thereby allowing the enterprises to outsource and contract remote access services from operators.

VPNs can also provide gaps in security when the end node is connected to secondary networks. For example, say someone is connected to their corporate network via VPN dial-up or the Internet from their home. In their home they have a wireless network with no security in place. A hacker outside that house could connect to the PC via an unsecured connection, and share the VPN tunnel without the worker's knowledge. This is referred to as 'split tunnel' vulnerability.

6.14.2 MOBILE SYSTEMS

SIM Application Toolkit

The SIM Application Toolkit (SAT) is a technology that lets the SIM card issue commands to the telephone. These commands range from displaying menus and getting user input to sending and receiving SMS messages. The SIM toolkit is excellent for implementing security critical applications, since it allows for custom encryption. One potential drawback is that the operator owns the SIM card, and to place applications on it you need the operator's permission.

What is USIM?

USIM (universal subscriber identity module) is the smart card for third generation (3G/UMTS) mobile terminals. The role of SIM cards in 3G concerns the traditional benefits of portability, security and individuality.

Mobile commerce is a big driver for security functions in the terminals for authentication and cryptography.

Tunnelling

In GPRS and WCDMA/UMTS one of the main protection methods is to send end user data in tunnels.

GPRS and WCDMA/UMTS use tunnels for signalling and user data traffic. The protocols are called GPRS tunnelling protocol – user (GTP-U) and GPRS tunnelling protocol – control (GTP-C).

GPRS tunnelling protocol (GTP) is used in the internal backbone. If required, IPSec can be used internally as well. See Figure 6.28.

Tunnelling towards external networks will only be performed if the interworking case so requires.

The media gateway and the GGSN are the tunnel end points for the GTP-U for packet mode communication. The GGSN tunnel end point is situated on the border between the basic WCDMA/UMTS core network and the ISP's point-of-presence (POP).

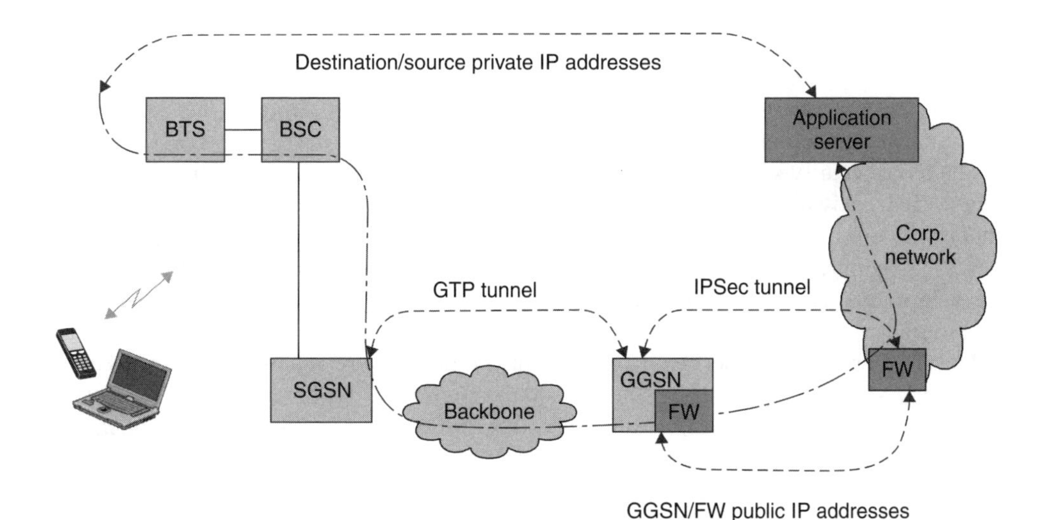

Figure 6.28 Packet routing and tunnelling

The SGSN (serving GPRS support node) supports the GTP-C protocol for control signalling between SGSN servers and GGSN. GTP-C is transported by UDP/IP and contains functionality for SGSN media gateway function – GGSN tunnel management and control.

GTP-C protocol for control signalling between SGSN servers and GGSN supports end-user authentication interacting with a RADIUS server located at the ISP service network or in a corporate network. For this purpose, the GGSN includes a RADIUS client.

The optional IPsec function makes it possible to have secure transfer between SGSN and GGSN nodes and between GGSN and external networks. The IPsec may also be used to secure the link between the GGSN node and the network management terminal. The security function also includes the packet filtering options in order to protect against intrusion or denial of service attacks.

6.14.3 E-COMMERCE AND M-COMMERCE SOLUTIONS

Micro-payments and protection of m-commerce are important subjects.

The security measures include access control, user authentication, signing, data integrity, non-repudiation.

6.14.4 WLAN SOLUTIONS

A WLAN security solution should provide transparent mobility and security for mobile users who want to access their company networks.

WLANS have basically an inferior protection compared with wired networks. The reason is that radio waves propagate outside the intended area. In non-controlled areas there is a risk of false base stations. The idea could be to attract clients and get hold of their passwords.

Thus both clients and infrastructure are unreliable and must be authenticated. On the other hand wireless networks can have a higher availability if the access points have secure positions, and the coverage between access points overlap. Then a lost access point can just result in a handover.

To protect for unauthorized access-to-access points many WLAN operators use the function access control list (ACL) in the access points. ACL defines what clients can access the base stations.

The basic security solution in WLAN according to IEEE 802.11/802.11b is called WEP 128 (wired equivalent privacy 128), with the ambition to protect against unauthorized connection or eavesdropping.

WEP generates secret shared encryption keys that both the information source and destination stations can use to alter frame bits (pieces of data) to avoid disclosure to eavesdroppers.

The solution includes encryption. WLAN should provide firewall protection, preventing unauthorized access. A WLAN solution should preferably be based on IPsec. Encryption/decryption and authorization keys could also be employed to provide authentication, automatic security association management, and to protect wireless traffic.

Network access control can be implemented by using a service set identifier (SSID) associated with an access point (AP) or group of APs. The SSID acts as a simple password for network access and provides minimal security since wireless clients can share it.

802.1x is related to the 802.11 standards and covers two distinct areas: network access restriction through the use of authentication, and data integrity through WEP key rotation.

The 802.1x standard recommends the use of a RADIUS server in conjunction with two data communication protocols: extensible authentication protocol (EAP) and transport layer security (TLS).

802.1x implementations also improve data encryption through rotation of the WEP 128 key.

6.15 SUMMARY WITH IPsec AND FP FOCUS

With the proliferation of access methods to the Internet in mind, and a development towards more sophisticated attacks from hackers, the conclusion is that protection must be extended beyond access restrictions. In particular, the traffic itself must be protected to a large extent. There are a number of protection methods for different applications, such as TLS/SSL for WWW traffic, FTP and Telnet, Wireless TLS for WAP WWW, and S/MIME for E-mail.

A more general protection is related to the common IP layer.

The biggest Internet layer security protocol family is IPsec, which is a fairly complex umbrella IETF standard, developed since 1995. IPsec is an open, vendor-independent architecture and a kind of VPN standard. It can also be used for leased lines and frame relay traffic. The main aim of IPsec, however, is to provide privacy, integrity

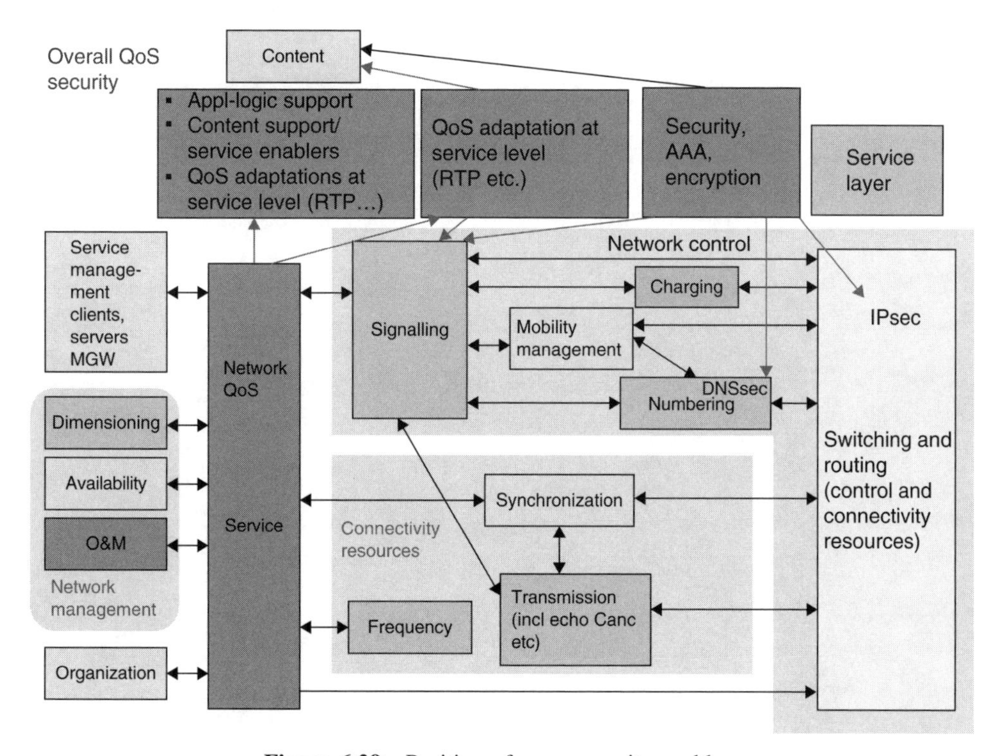

Figure 6.29 Position of some security enablers

and authenticity to information transferred across IP networks. IPsec addresses all of the threats to Internet (such as loss of privacy and data integrity, identity spoofing and denial of service) in the network infrastructure itself.

IPsec provides IP network layer encryption. Normally tunnelling is used, but there is an alternative, called transport mode.

The customers to IPsec are all IP-based protocols, such as TCP, UDP and ICMP.

For staff working with security a layered view and an 'FP view' are useful, for example when positioning the security enablers. See Figure 6.29.

7

Quality of Service

'Quality of service is not what it used to be.'

7.1 OBJECTIVE

The overall objective of this chapter is to support the elaboration of a Quality of Service plan and the implementation of a service plan as far as Quality of Service is concerned.

7.2 INTRODUCTION

7.2.1 WHAT IS QUALITY OF SERVICE?

Quality of Service (QoS) is the collective measure of the level of service to a subscriber (ITU-T E.800). The service in question used to be voice, but now many types of services must be considered. Chapter 5, for example, dealt with six service classes. Yet, voice is

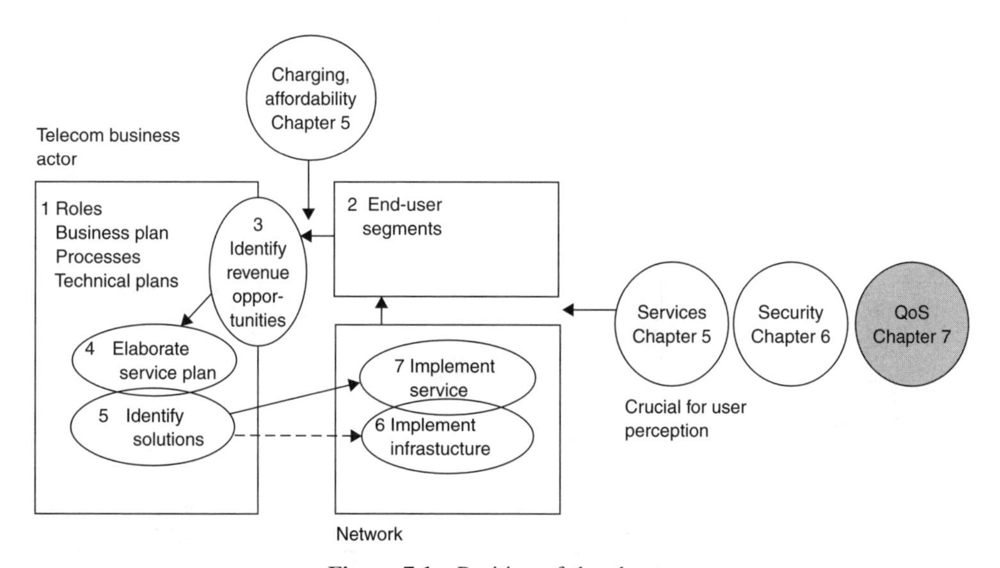

Figure 7.1 Position of the chapter

Understanding Changing Telecommunications – Building a Successful Telecom Business. Edited by A. Olsson
 ISBN: 0-470-86851-1

still one of the main services, especially when it comes to generating revenue. Also, voice is still a main focus of QoS discussions. A main reason is that voice is an isochronous service, which means that the decoded signal ideally should have the same audio frequency and frequency stability as the sent signal. With the present technology convergence the subject of voice and multimedia in packet networks comes into focus.

Against this background a lot of the performance criteria in a packet network are included in this chapter:

- Packet delay (or delay in general)
- Jitter (delay variations)
- Packet loss or bit error rate
- Bandwidth or throughput
- Echo
- Silence encoding, if background noise if suppressed

The above criteria are network oriented. Let us call it bearer level QoS. Nowadays the user may also perceive an application-related QoS performance. At least for mobile network terminals, various perceptions occur within the possible terminal capability range, depending on service enablers, application support enablers and service layer middleware. See Figure 7.2.

QoS is one of the critical success areas for a telecom actor, since the perceived end-to-end QoS has a major impact on user perception and satisfaction and consequently on churn.

Compensation for a poor QoS might be to offer the service in question for free during a limited time period, or at a lower price. However, such solutions might need financial perseverance and are not considered is this chapter. The opposite solution, at least at bearer level, is to over-dimension the network, which would reduce ROCE.

Regarding the QoS business model, let us return to the holiday sales at the beginning of Chapter 5. It was argued that the travel operator does not sell holidays by promoting the features of the plane. He sells the holiday. However, it would certainly be possible to have different travel classes for charter trips, similar to the differentiation in hotel standards.

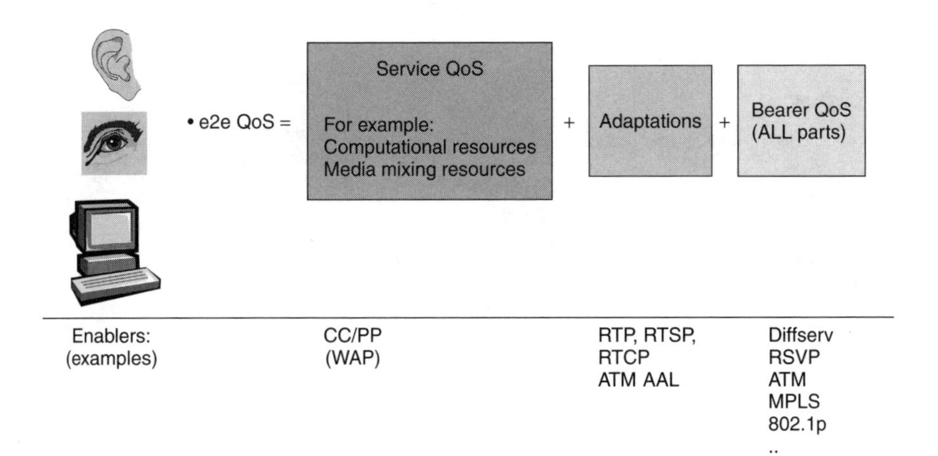

Figure 7.2 Perceiving QoS at service terminating points by sets of QoS enablers

By paying a different overall price to the travel operator including flying standard class, the airline company gets paid for the provision of different classes of transport.

7.2.2 HISTORY

QoS has always been an important part of telecommunication. Various measures have been used. The probability of congestion for traditional voice trunk calls is still indicated in the number of congested calls per hundred call attempts. Analogue circuits were particularly expensive and investments were expressed in number of circuits. Other measured circuit characteristics related to service quality in telephony networks are echo, cross talk, signal-to-noise ratio and lately bit error rate (BER). The characteristics were directly associated to the selected circuit, as well as the bandwidth, which was fixed and standard for all calls. A fairly common current perception of QoS is wider than, but not too far from, what has traditionally been called 'The transmission plan'. This plan caters for *end-to-end perceived quality of voice* in circuit-mode networks, originally in analogue networks and later in digital networks. The broad introduction and success of mobile systems brought echo aspects into focus, since efficient voice coding, as applied for example in GSM (global system for mobile communication), takes time (about 80 to 90 milliseconds). Echoes with such time delays are very annoying.

7.2.3 THE COMBINATION OF VOICE AND IP

The current importance of QoS is very much associated with the transition from circuit switching to packet switching, with the success of the IP technology, with the development towards multimedia and finally with the success of mobile systems. One consequence is the appearance of voice, video and multimedia over IP, basically a best-effort technology. Going one step further we get voice/multimedia over IP over wireless, 'VoIPoWL', the real challenge. IP was not designed for voice traffic, and wireless was not designed to carry IP traffic. Therefore the most critical QoS terms have become delay, delay variations/jitter and packet loss. This has indeed led to a semantic change of the meaning of the term QoS. Today, some people tend to use the expression in a narrow sense, such as: 'QoS is a necessity for (packetized) real-time services' or 'A specific QoS can be described as a set of parameters that describe quality requirements of a stream of data'.

The QoS of IP telephony (or VoIP, voice over IP) and associated charging, in relation to subscriber perception has been a hot question ever since the birth of the 'IP paradigm'.

What is then so attractive about real time services over IP?

- IP technology enables unification of the private or public network by allowing voice to be treated as a data application.
- Once achieved, the unification enables us to decrease network operating cost, and opens up the door for innovation and fast service creation.
- For rational reasons VoIP is already a reality, especially in the local area network of many enterprises. For residents it could for example be combined with ADSL subscriptions (VoIPoADSL).
- The main advantage of running IP all the way over the air interface is service flexibility. To date, cellular-access networks have been optimized in a two-dimensional space

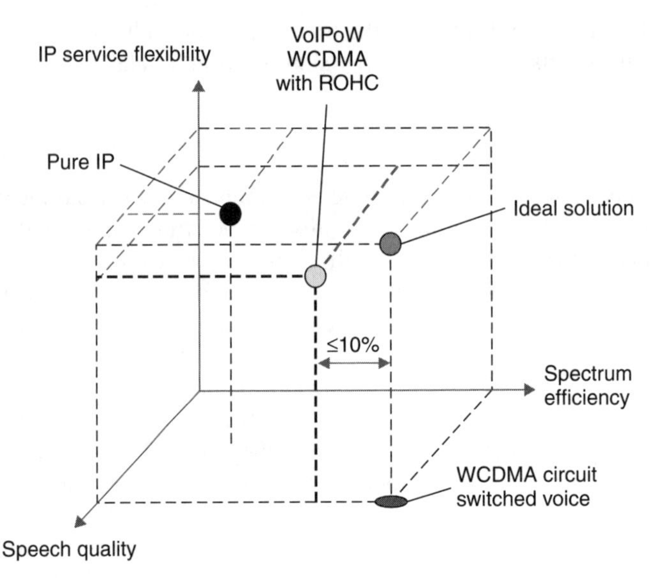

Figure 7.3 A main target is QoS, also spectrum efficiency and service flexibility

whose axes are voice quality and spectrum efficiency. IP introduces a third dimension in terms of service flexibility, but adds protocol overhead, contradicting the goal of spectrum efficiency. See Figure 7.3.

The major challenges ahead for VoIP are QoS and interworking with legacy networks and systems. In fact, very few issues regarding telecommunications, if any, have raised such strong and passionate polemic as QoS. Few telecom areas have been targeted by such intense research and frenetic development since the mid-1990s.

7.3 PERCEPTION OF QOS

7.3.1 SERVICE REQUIREMENTS

The eye and ear are sensitive to delays in interactive conversation, but maybe even more sensitive to delay variations. They are less sensitive to bit errors, especially the ear, since one bit represents very little information, and the ear/eye/brain will fill in small missing information. For multimedia, lip synchronization (= limited skew) is important. The computer on the other hand isn't very sensitive to delay, but even one bit error can be devastating. See Figure 7.4.

As already mentioned, an overall goal of this chapter is to support the implementation of a service plan as far as QoS is concerned. Considering the present power of the user, let us start with the user perception of service quality. See Figure 7.5.

7.3.2 TELEPHONY

Perceived Quality

Perceived quality end-to-end at application level is what matters for the user. What is it that determines whether we perceive the quality of a phone call as good or bad?

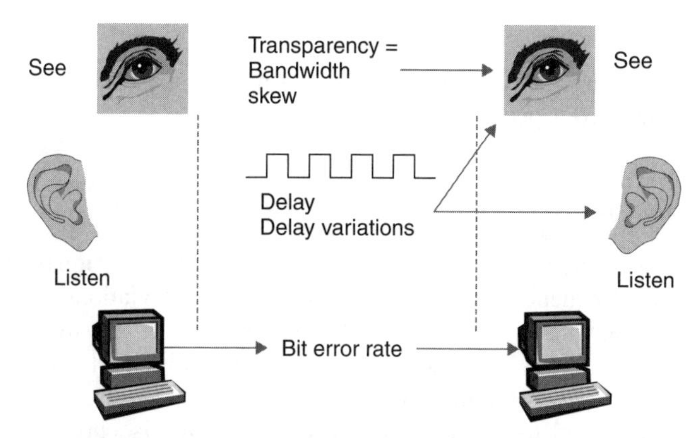

Figure 7.4 Different demands from different terminating points

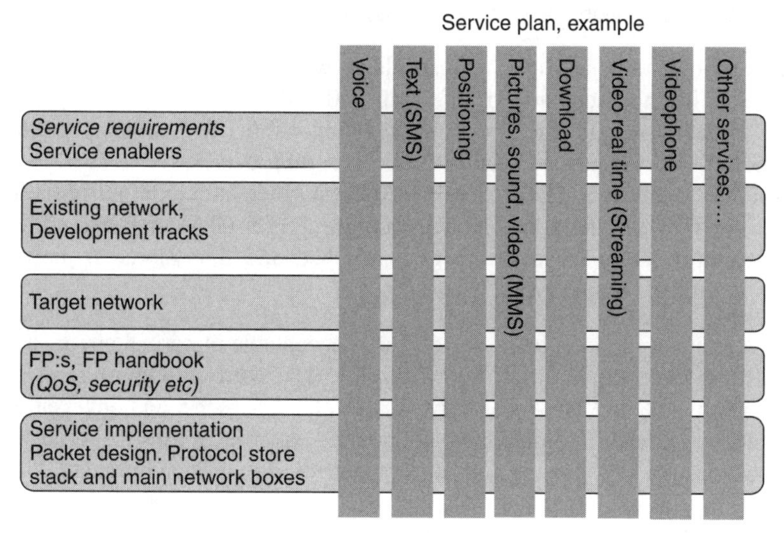

Figure 7.5 The service plan and some (horizontal) implementation steps, starting with the service requirements covered in this chapter

- You here what is said.
- You here who is talking.

This requires a sufficient signal level without too much noise or echo.

The quality is often measured by letting a test panel compare the sound between telephony from different sources and rating the perceived quality. The average rating is called the MOS (mean objective score) value. ITU-T Recommendation P.800 provides the following scale: 5 = excellent; 4 = good; 3 = fair; 2 = poor; and 1 = bad. The current target for VoIP is class 4, which is the same as ordinary circuit-mode voice systems.

Another current target for VoIP is called carrier grade. It means extremely high reliability with a down time of about 5 minutes per year, known as five-nines reliability (99.999 %).

Delay and Jitter

Delay is an important area for conversational real-time services like telephony and video telephony. There are two aspects regarding delay:

- Total delay.
- Delay variation or jitter.

Voice is an isochronous service, which means that the decoded signal should have the same frequency and frequency stability as the sent signal. Deviations depend mainly on impact from active elements in the path, but also on rerouting. If the stream of information is uneven (also called 'jittery') or the values arrive in disorder, it becomes difficult to hear what is said and who is talking. What is called adaptation protocols in the enabler part are intended to cancel the deviations, and to restore the isochronous relation. Jitter is eliminated by means of delaying packets individually to restore the original intervals. The price to pay is an increased overall delay. The process is called 'dejitterization'.

The total delay gets worse at certain occasions:

- Connections over a long distance (especially when one or more satellite link(s) is used – a satellite hop adds a delay of about 240 ms).
- Voice is coded using advanced compression algorithms to save bandwidth (as in a GSM phone). See Chapter 5, Section 5.7.1 on coding delays.
- Connection is made over networks using different voice coding, requiring a translation in a gateway between the networks (for example PSTN-GSM or Internet-PSTN).
- The voice information is packet or cell switched and thus stored in buffers in the switches during the transport.

Typically, a maximum total delay of 150 ms is acceptable in normal circumstances, and 400 ms in more exceptional cases. See Table 7.1. The corresponding figures for packet loss are typically 5 % and 10 % respectively. See also Chapter 5 on conversational class.

For IP telephony the TCP protocol cannot be used, since TCP asks for retransmission in case of erroneous packets. This would cause a varying delay when retransmission occurs.

Secondary Problems with Delay

Echo and talker overlap are problems that arise from high end-to-end delays in a voice network. Echo is noticeable by the ear/brain at about 25 ms distance between talk and echo. Talker overlap is the problem of one caller stepping on the other talker's speech.

Table 7.1 Acceptable delays

One-way delay (ms)	Description
0–150	Acceptable for most user applications
150–400	Acceptable provided that administrators are aware of the transmission time impact on the transmission quality of user applications
400+	Unacceptable for general network planning purposes, however, it is recognized that in some exceptional cases this limit will be exceeded

Noise

The signal-to-noise ratio should be as low as possible. The digitization of the trunk network, followed by the digital GSM and ISDN accesses, led to a significant improvement of the quality. One result was drastically reduced intelligible cross-talk.

Comfort Noise

In a normal case noise has a negative impact on QoS. However, if there is no background noise at all, you feel uncomfortable, because you do not know if the connection is broken or not. To take care of this, a so-called 'comfort noise' is added at the receiver's end in applications where everything is coded as silence below a certain threshold. This goes for both mobile telephony and IP telephony.

7.3.3 DATA COMMUNICATION

So far we have discussed the quality of telephony. Some of this is applicable for other services as well, some of it is not. The ability to successfully connect to other subscribers is important for voice services, since this is the basis for charging. For Internet service providers (ISP) often a flat rate applies. Anything that makes the subscribers select you instead of your competitor as their ISP, will feel like 'good quality'. So what affects the perceived quality when we for example connect to the Internet from home?

- **Short set-up time**
  Bandwidth
  Fast data delivery without excessive queuing
  Low bit error rate (BER)
  Accessibility to many fellow subscribers
  Differentiated priority
  Some services, or some users, need a higher priority and are prepared to pay for it. This is true for real-time services. Or you can imagine a business user at an airport who wants to download mail with attachments before entering the aircraft. Differentiated service classes, differentiated billing and resource reservation are methods used to improve this aspect of quality.
- **Security**
  Business information, money transactions and data of a private nature all demand some kind of protection against 'eavesdropping'. This can be solved with encryption, or by using tunnelling. There are also methods for putting an electronic signature on a data message, to ensure the receiver that the sender is who he claims to be (compare to the desire to hear who is talking in a telephone conversation). Public key ciphering is a method that solves both encryption and the signature. More security issues are included in Chapter 6.
- **'Future-proofness'**
  If you use a certain data communication (for example an e-mail program), you do not want to lose information when you upgrade to a newer version (for example, have to rewrite your address book). This has more to do with the terminal software than the

actual communication network. For the operator, the choice of technology for implementing a service should be future-proof, which means it must be *scalable* and possible to upgrade to new versions. Otherwise the operator might be in a situation where the original investments have to be totally replaced by new investments. Compare ROCE.

- **Reliability**
  In telecommunications we have been used very reliable networks that are up and running almost all the time. This is in contrast to the local enterprise networks, where stop-time during maintenance or upgrades have been common.

7.3.4 MOVING IMAGE (VIDEO)

There are a few quality areas specific to video:

- **Delay variation**
  Video is an isochronous service, just as telephony. The information is sent in a stream (streaming video) that has to be consistent and steady, not jumpy and jittery. Otherwise the picture will freeze and jump forward.
- **Total delay**
  There is a difference between one-way broadcast services and two-way interactive services. When broadcasting rental films over a network, the delay will not affect the quality. In a video conference, or when using net-meeting on a PC, the delay is as important for the quality as it is for telephony.
- **Bit error rate**
  Depending on the compression technique that is used, bit errors might cause problems. The more the information is compressed (to save bandwidth), the more is lost if a packet is lost, or the more is misinterpreted if a bit error occurs in a packet that arrives. So normal TV broadcasting is not sensitive to bit errors, whereas net-meeting on the PC is.

7.3.5 MULTIMEDIA (VIDEO + AUDIO)

The same quality aspects as for moving image apply to multimedia communication, with one addition:

- The synchronization between the sound and the picture (lip–voice synchronization) has to be good. The gap in time between them is called skew. The skew risks increase if voice and picture are sent over different channels, or if they are coded separately.

For all these services, quality also depends on factors like short set-up times, being able to communicate with many others (having many film titles to select from), being able to find them (catalogue or enquiry services), mobility and clear billing information.

7.3.6 QOS CLASSES

Service Classification

In this part of the chapter a brief summary of the connection between services and QoS classes is presented.

The real behaviour depends a lot on the traffic situation in the network. In calculating a traffic model, the end-user behaviour with regard to particular applications needs to

Table 7.2 UMTS traffic classes and some related requirements

QoS class	Transfer delay requirement	Transfer delay variation	Low bit error rate	Guaranteed bit rate	Example
Conversational	Stringent	Stringent	No	Yes	VoIP, video conferencing, audio conferencing
Streaming	Constrained	Constrained	No	Yes	Broadcast services (audio, video), news, sport
Interactive	Looser	No	Yes	No	Web browsing, interactive chat, games, m-commerce
Background	No	No	Yes	No	E-mail, SMS, database downloads, transfer of measurements

be determined. The mix of applications and the daily traffic distributions must also be characterized in order to carry out analyses and calculations.

All types of packet data traffic will not have the same requirements on delays, packet loss etc. As a result traffic may be divided into four different classes: streaming, conversational, background and interactive. See section 5.4.

Real-Time Applications

Streaming The fundamental characteristics for QoS are to preserve time variation between information entities of the stream, e.g. video streaming.

Conversational Here the fundamental QoS characteristics are to preserve time variation between information entities of the stream and to have a low delay, e.g. voice.

Non Real-Time Applications

Background In the background class, the destination is not expecting the data within a certain time but is expecting a preserved payload content, e.g. e-mail.

Interactive A request/response pattern is important in the interactive class and the payload content must be preserved.

UMTS QoS Classes

It is foreseen that UMTS should be designed for the case where the bottleneck is *not* the core network, but the UMTS bearer access speed. A summary of the UMTS QoS classification is shown in Table 7.2.

7.4 THREATS TO QOS

Multi-operator connections, extensive layering and interconnected technologies create many more interfaces than in the monopoly era. See Figure 7.6.

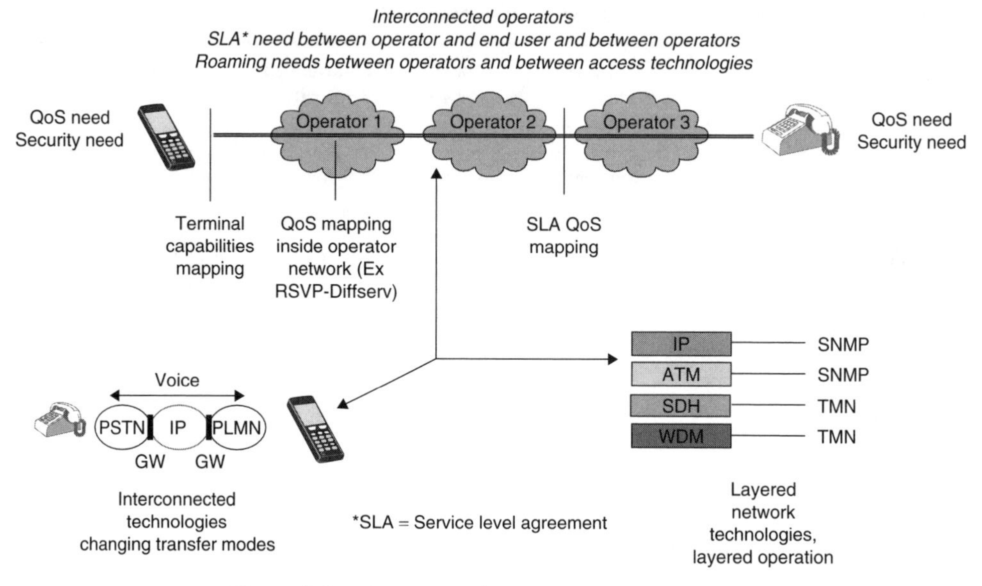

Figure 7.6 The new environment is a threat to QoS

Deficiencies in fundamental plans (other than QoS) create a number of threats. See Figure 7.7.

7.4.1 PASSING SEVERAL OPERATORS' NETWORKS

When nobody has the overall responsibility for the total delay, the question is how long it will become. There is also a need for mapping between different QoS technologies. See also Chapters 15 and 16.

Service level agreements (SLAs) are necessary in order to regulate the QoS between different operators.

7.4.2 CHANGE OF TRANSFER MODE

A telephone connection from a PSTN network, via an IP backbone trunk to a mobile phone in a GSM network is an example of this type of connection. The voice is (probably) coded differently in the three different networks. This means that a translation has to be done at interconnecting gateway points between the networks. The translations take time (in this case it could add to more than 100 ms), which adds to the total delay. This means a reduced quality for the interactive real-time services, such as telephony and interactive multimedia (like video conferences). Delay is not a problem for data communication, where a gateway might translate from one mail format to another one or from one ASCII table to another. Instead the readability might be affected.

7.4.3 LAYERED NETWORKS

The horizontal network gives us increased freedom of choosing technology in the various layers. It is cheaper to manage one network than many. But there are two sides of the

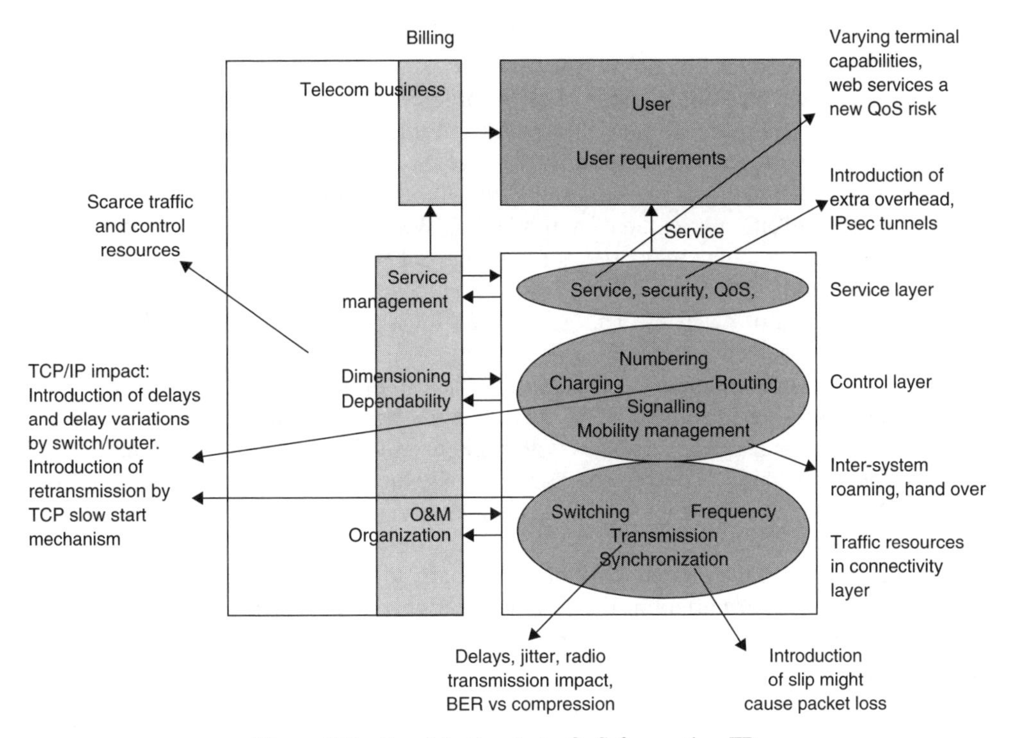

Figure 7.7 Possible threats to QoS from other FPs

coin. The introduction of a new, widely layered approach gives initially an increased complexity and a poorer performance compared with tailored solutions.

When combining 'ordinary' layering with for example an IPsec tunnel, carried over IP, carried over frame relay, over ATM, on SDH, using WDM on optical fibres, what will the overall quality and efficiency become? What about QoS control? Is a kind of feedback possible from underlying layers to the application level? SLA between layers? In Chapter 12 the standard GMPLS is presented, with a kind of common overall approach to a number of layers. Another trend is to reduce the number of layers to a minimum.

7.4.4 TCP/IP AND ATM PACKETIZATION EFFECTS

Delays and Delay Variations in Packet Systems

Packetized systems are queuing systems. Traditionally all packets were treated equally: first in, first out. Having all packets share the same queue has much the same effect as having all highway traffic share a single lane: congestion and delays at the intersections, depending on the distribution of packet arrivals. The burstiness (peak rate to average rate) of many interactive applications should be borne in mind. It might be 200 times as much as voice, with a burstiness set to 1 for a reserved path.

Bandwidth sharing led to a situation where bandwidth started to be related for the first time to QoS, and delay handling became an important issue. IP networks have basically very little sensitivity to delay as data packets do not need synchronous reception and eventually can be retransmitted on error detection, some time after the original dispatch.

The packetized voice applications (VoIP, VoFR, VoATM) use short packet/cell sizes (typically below 100 bytes) to minimize delay and jitter.

In a packet-handling device at least three processes can create queues: the arrival process; the serving (switching) process; and especially the output serialization process (see Figure 7.8). Serialization is a kind of multiplexing. When long data packets are occupying the output, excessive delay may arise for waiting packets used for real-time traffic.

The packetization delay for an ATM cell is 6 ms, when used for a single voice channel and with a payload of 48 bytes. This is the time for a 64 kbit/s or 8 kbytes/s stream to fill the payload space.

In IP data networks the best-effort data traffic has an average packet size of around 1.5 kbytes and a possible maximum of 64 kbytes. Now let us consider the case when this traffic competes for network resources with packetized voice traffic. Provided the overall traffic volume is within the design conditions, no significant queuing will exist in network buffers. Delay will then be kept within reasonable values, and voice quality will be good. See Figure 7.9.

Nevertheless when the total amount of traffic starts to rise, there will be queuing in the network buffers (router buffers in IP) and then the long packets of IP data (more than 15 times longer than voice packets) will be included in the queue causing long delays for the voice traffic.

The amount of the overall delay in buffers might then become high enough to be annoying for the people involved in a voice conversation. The main advantage ATM has in terms of QoS compared with IP voice in data networks is derived from the fact that all the packets (cells) in ATM are short (53 bytes) and have the same size. Therefore there are no chances that long packets introduce unacceptable delays in the VoATM traffic.

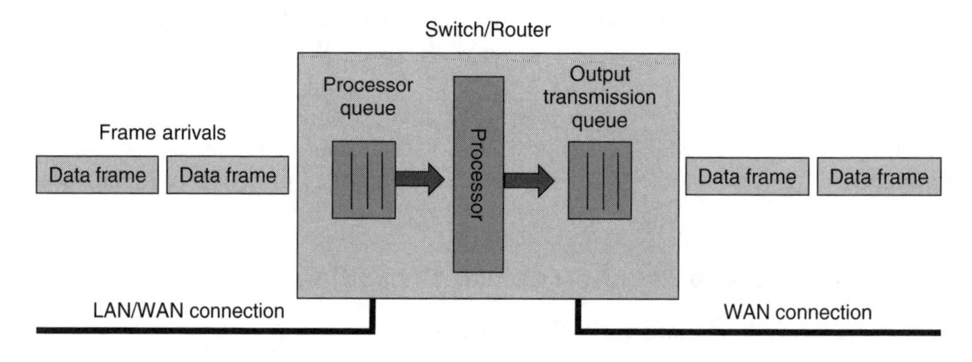

- In this example there are 2 single server queues in the switch/router
- The processor is the server for the processor queue
- The transmission line connecting the switch/router to the WAN is the server for the output transmission queue
- To analyse the queuing delays and queue sizes it is necessary to describe:
  - the arrival process
  - the serving process
  - the serialization process at the output.

Figure 7.8 Critical processes in a packet-handling device

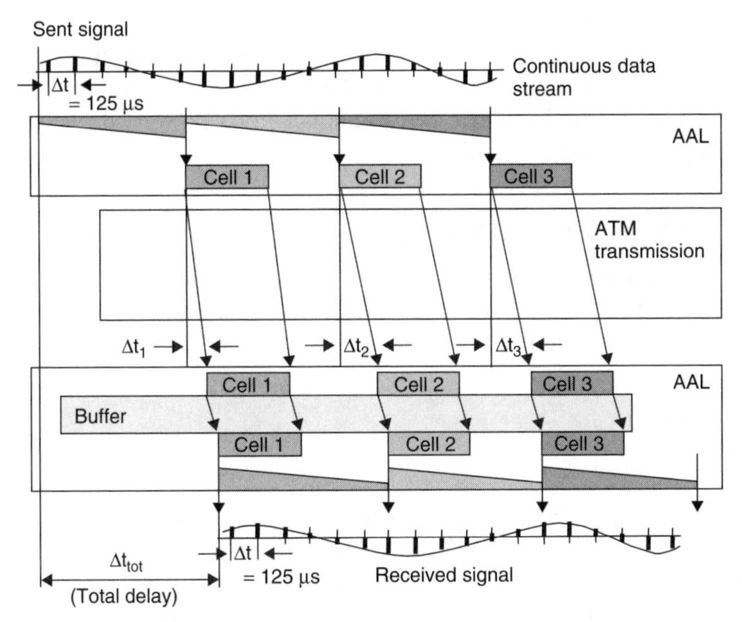

Figure 7.9 Packetization delay when converting PCM to ATM

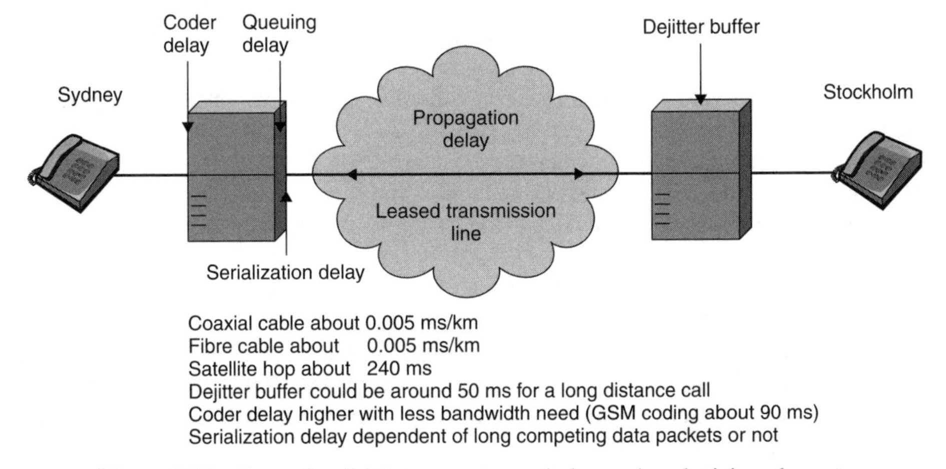

Figure 7.10 Example of delays over transmission and packetizing elements

In IP networks the delays occur mainly at routers throughout the network. Congestion has an additional effect in a packet switching network: When the queues fill up, excess packets are simply discarded. See Figure 7.10.

Mobile Environment

Among degradation sources for IP/wireless QoS are:

- Strongly varying transmission conditions – Fading, shadowing, interference, packet loss, etc. From bandwidth to power dependence.

- Limited unpredictable bandwidth.
- Handover – e.g. from GPRS to UMTS or between different capability cells in UMTS.
- Hierarchical cells – high bandwidth in small cells, low bandwidth in large cells.
- Varying terminal capabilities with respect to computational power and display size – PDA, mobile laptop, PC in wireless LAN.

7.4.5 CONNECTIONLESS TRANSFER

In connectionless transfer packets can arrive at the receiver in the wrong order, for example when the path has changed. This and other displacements in time are reasons for buffering, adaptation and 'dejitterization'. See Figure 7.12.

7.4.6 TCP EFFECTS

Slow Start Mechanism

TCP has a major responsibility for the QoS received by the application on top of TCP/IP, with an error-free transfer as the main goal, at least for best-effort data traffic. To achieve this goal the transmitting end must adapt send speed and possible necessary retransmission to information at suitable speed limits from the receiver and congestion information from the network.

Congestion control is a distributed algorithm that is used to share network resources among competing users. It consists of two components: a network algorithm that updates and feeds back, implicitly or explicitly, congestion information to sources; and a source algorithm that dynamically adjusts rate (or window size) in response to congestion in its path.

In the current Internet, the source algorithm is carried out by TCP. In case of congestion TCP reduces the send speed. This speed reduction and the subsequent speed recovery has been a subject for much interest during the last few years. Why? Let us first look at Figure 7.11.

A reasonable QoS goal is that the inherent bit rate of a particular system should be possible to use for traffic, and thus generate corresponding income. The traffic types using TCP such as e-mail, file transfer or web browsing will be affected by a reduced transfer speed caused by TCP. Real-time applications such as IP telephony do not normally use TCP (but UDP). This traffic cannot slow down or retransmit, and tends to both cause congestion and suffer from loss.

TCP behaviour can force the carried traffic to become much lower than system speed would permit. The TCP algorithm uses window size, threshold and acknowledgement as the main parameters. A window is the maximum traffic volume that is not yet acknowledged by the receiver. It can be expressed as the maximum number of sent packets that are not acknowledged. In Figure 7.11 the traffic is segmented and the sender is waiting for six segment acknowledgements travelling between receiver and sender. Figure 7.11 also shows the round trip time, RTT (or round trip delay). RTT is equal to the time it takes for a packet to travel to the receiver plus the time it would take to travel in the opposite direction. A sending window is the window that is agreed with the receiver. A threshold is a speed limit set by congestion indications in an iterative way, in order to operate below congestion. It is therefore also called congestion avoidance.

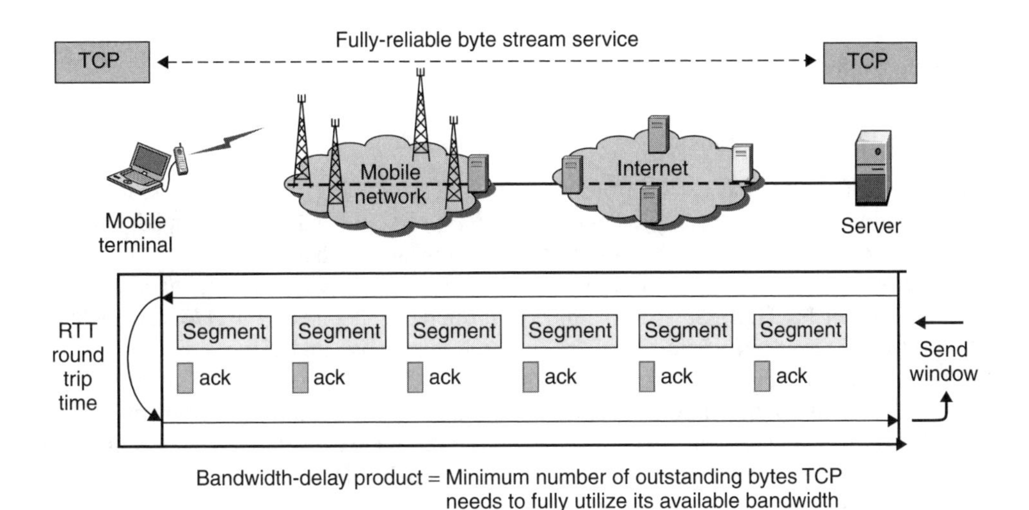

Figure 7.11 Important factors in TCP/IP over radio are round trip delay (RTD) and bandwidth-delay product

The applied speed is determined by the lowest value of the sending window and congestion avoidance. It should be added that the threshold varies in time, and that the TCP algorithm forces the sender to reduce the speed (to a small window) and threshold (to 50 % of the previous one) significantly after congestion indications. This heavy speed reduction and the subsequent speed recovery, although quite fast, are called the slow start algorithm.

TCP and Radio

TCP is designed for high quality lines. As indicated before, however, the radio path might suffer from a low, unstable bandwidth, handovers, short periods of no connection, sometimes high BER and long, varying round trip delays (RTD). This does not fit TCP very well. BER causes packet loss. RTD is affected by buffering and serialization of data, propagation time and necessary retransmissions.

The possible ways to retain a higher speed include cutting the RTD by using different TCP connections for the radio path and the rest of the path, replacing the retransmission mechanisms with a faster one, or making retransmissions only over the radio path. Alternative retransmission methods require a mechanism at a lower level in the stack below TCP. Such a solution uses the radio link control (RLC) protocol at the link layer. However, in the last case, TCP will experience delays when RLC retransmits, which might cause a congestion indication to TCP, invoking the slow start mechanism.

Using different TCP connections between the terminating points has other drawbacks such as time for protocol translation and an impact on security (TCP is above IPsec in the stack, for example).

With a delay from sending to reception of acknowledgement of up to 450 ms, which is not unrealistic with significant retransmissions at the radio path, the bandwidth-delay product for WCDMA operating at 384 kbit/s is 20 kbytes. At 64 kbit/s it is around

4 kbytes. This is the required number of outstanding bytes TCP needs to fully utilize the available bandwidth. A low bandwidth-delay product is the goal.

TCP and High Speed Links

As a result of research at the California Institute of Technology and other research units an alternative or at least a complement to the TCP protocol has been developed. The aim is to reach robust and stable networking at 100 Gbps and higher speeds.

The protocol is a source algorithm which is called 'fast active queue management scalable TCP/IP' or FAST/IP. See for example http://netlab.caltech.edu/FAST/overview.html. The protocol is especially interesting for research, enterprise and ISP needs.

The largest difference compared with ordinary TCP is the smooth behaviour of the speed adaptation to the network capacity limitations, based on measurements of the delay between sent packet and acknowledgement. Preliminary results indicate a significant increase in transfer speed and utilization factor of existing Internet capacity.

To summarize the TCP aspects: TCP/IP is used much more widely than it was designed for. There are obvious risks that the user can perceive the problems that may arise. The network designer must therefore look for evolving improvements, in order to maintain or improve user satisfaction.

Security

Securing connections with IPsec should be done end to end if possible. As mentioned above a split of a TCP path into more path sections might give a shorter bandwidth-delay product. The split could be done by means of a proxy server, but such solution creates a non-wanted split in the IPsec path as well. See Figure 7.12.

7.4.7 COMPRESSION OF IP VOICE PACKETS BY ROHC

It could be expected that a strong header compression should lead to a decreased QoS, at least in environments with degradations. For ROHC, however, there are also positive effects, according to Figure 7.13.

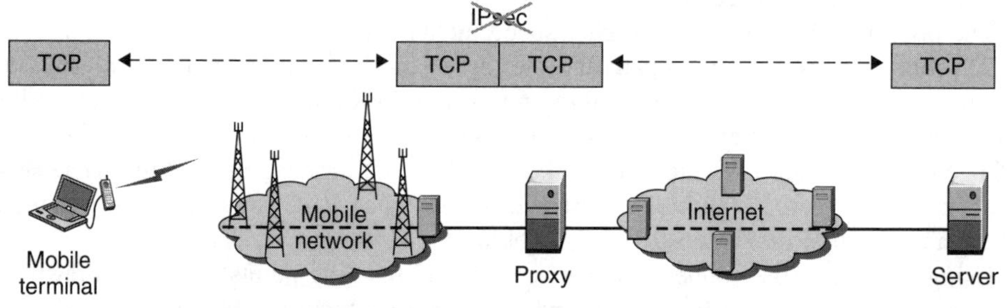

Figure 7.12 QoS and security in conflict

- Compression results
  - The **40 octet** IPv4/UDP/RTP or **60 octet** IPv6/UDP/RTP headers can be compressed down to a minimal size of **1 octet**
- Average size is just above minimum
- Consequences
  - Reduced bandwidth demand
  - Headers less error sensitive (smaller)
  - Reduced delay, since fewer frames can be sent in each packet

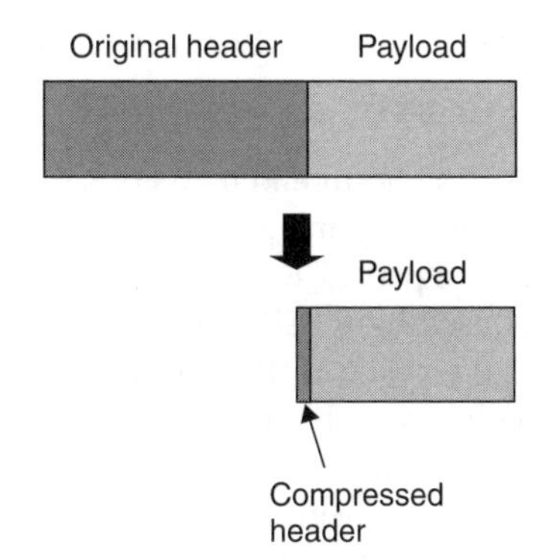

Figure 7.13 Compression of voice packet headers can have positive effects as well

7.5 QOS ENABLERS

In order to enable real-time services over IP three protocols will be briefly presented: RTP, RTCP and RTSP.

For QoS adaptation of ATM to various traffic types the three adaptation layers AAL 1, AAL 2 and AAL 5 are provided.

Differentiated services and integrated services are used for resource allocation, as well as MPLS, which is also a traffic-engineering tool.

All these enablers will be briefly treated. See also Figure 7.14.

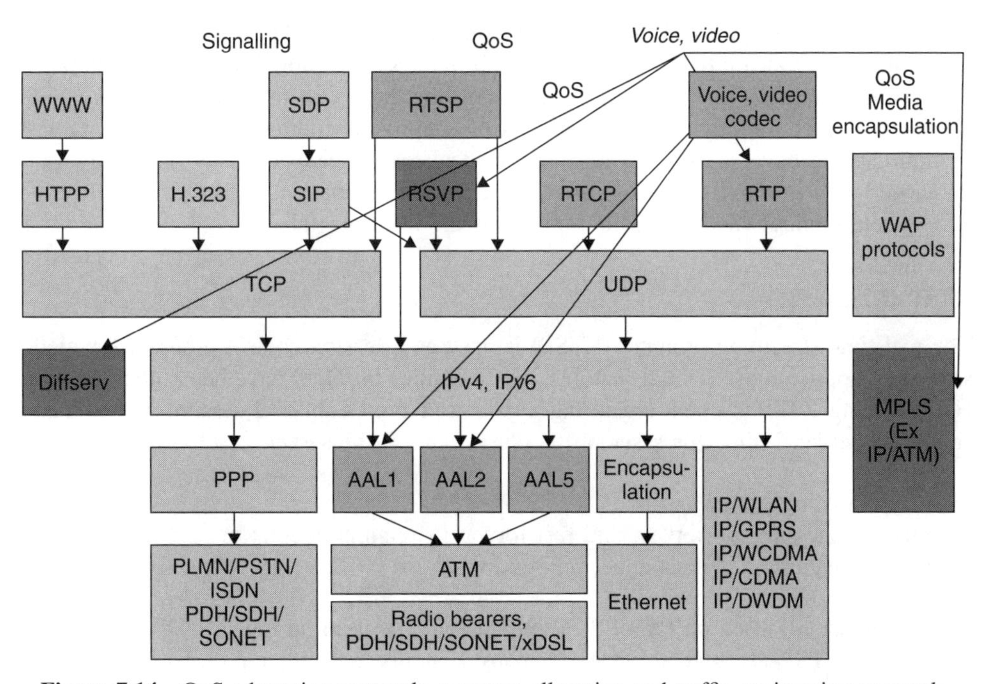

Figure 7.14 QoS adaptation protocols, resource allocation and traffic engineering protocols

7.5.1 ADAPTATION-TYPE ENABLERS

RTP (RFC 1889)

RTP provides end-to-end delivery services for data with real-time characteristics, such as interactive and streaming audio and video, including IP telephony. RTCP (real-time control protocol) is a part of RTP and helps with lip synchronization and QoS management.

RTP is used by RTSP, SIP and H.323 for the user data portion of these protocols.

The RTP services include payload type identification, sequence numbering, time stamping and delivery monitoring. The time stamp is related to the sampling or the presentation or composition time of the media carried in the payload of the RTP packet. It is used for playing back media at the correct speed, and together with RTCP, it is used for synchronizing the presentation of other streaming media. Applications typically run RTP on top of UDP to make use of its multiplexing and checksum services. Both protocols contribute parts of the transport protocol functionality. RTP supports data transfer to multiple destinations using multicast distribution if provided by the underlying network.

RTP does not provide any mechanism to ensure timely delivery or provide other QoS guarantees, but relies on lower layer services to do so. It does not guarantee delivery or prevent out-of-order delivery, nor does it assume that the underlying network is reliable and delivers packets in sequence. The sequence numbers included in RTP allow the receiver to reconstruct the sender's packet sequence, but sequence numbers might also be used to determine the proper location of a packet, for example in video decoding, without necessarily decoding packets in sequence.

RTCP

RTCP is the control protocol that works in conjunction with RTP. RTCP control packets are periodically transmitted by each participant in an RTP session to all other participants. The primary function is to provide feedback on the quality of the data distribution. The feedback may be directly useful for control of adaptive encodings but experiments with IP multicasting have shown that it is also critical to get feedback from the receivers to diagnose faults in the distribution. Sending reception feedback reports to all participants allows determining whether possible problems are local or global.

RTSP (RFC 2326)

The real-time streaming protocol (RTSP) is used as a session control protocol for media streaming applications. Several features and semantics in RTSP have been inherited from HTTP. Using RTSP, a client streaming application can establish a session with a media streaming server. Using this session, the client can ask the server:

- to start streaming media;
- to pause, back-up and replay, and fast forward streaming media; or
- to stop streaming and disconnect the session.

RTSP is usually used on top of TCP but can also be used on top of UDP.

RTP and RTSP are used together in many systems, but either protocol can be used without the other.

RTSP establishes and controls either a single or several time-synchronized streams of continuous media such as audio and video.

ATM Adaptation

See also the parts on ATM in section 7.4.

In order to take care of all kinds of services ATM has its own protocol stack. It consists of three layers:

1. The ATM adaptation layer (AAL). It is here that 48-byte data units with AAL and payload are created.
2. In the ATM layer the five-byte header is added.
3. The physical layer takes care of the mapping of the bits onto the physical medium.

ATM is designed for all types of media. Different services demand different kinds of adaptation support and therefore different kinds of AAL protocols. AAL1 takes care of constant bit rates, especially voice coded with 64 kbit/s. AAL2 takes care of variable bit rates like compressed voice and video. AAL 2 is a 'small cell' transfer mode of its own with dedicated AAL 2 switching. AAL5 takes care of data. See Figure 7.15.

7.5.2 NETWORK-DRIVEN QOS COMPONENTS

802.1p, Ethernet QoS

The idea with 802.1p is the use of flags in the media access control (MAC) header to establish packet priority (not reservation!) in shared-media 802 networks. 802.1p awareness is requested for devices, hubs and switches.

The issue of differentiating between network packets, and perhaps treating them differently according to the applicable QoS differentiation, is done by means of the MAC

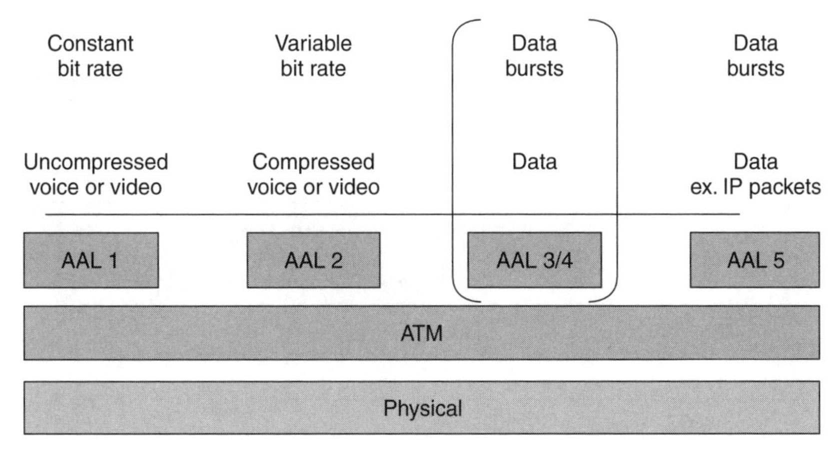

Figure 7.15 Adaptation protocols in ATM

header. The MAC header (the lower half of Layer 2 in the ISO OSI Model) is the only part of a packet that hubs or switches investigate in their scope of work.

802.1p provides prioritization of packets traversing a subnet by the setting of a three-bit value in the MAC header. Thus, when the local segment becomes congested and the hub/switch workload results in the delay (dropping) of packets, those packets with flags that correspond to higher priorities will receive preferential treatment, and will be serviced before packets with lower priorities.

7.5.3 QOS IN IP NETWORKS

In IP networks the routing is normally based on hop count and delay. This often leads to an uneven traffic load in the network. Some links become heavily congested while others remain almost idle. Resource allocation should therefore be complemented by traffic engineering and performance optimization.

The initiatives on QoS in IP networks are conceptually similar to those in ATM networks and consist of:

- Admission policy (possible shaping).
- Assignation of specific paths for synchronous (real-time) traffic preventing the competition from long IP data packets.
- Prioritization (certain packets can be dropped in case of congestion).
- All QoS methods have the same goal which is to open a differentiated lane for real-time communications out of the sometimes congested heavy traffic highway assigned to best-effort traffic.

The main enablers are RSVP through resource reservation, MPLS through label switching and differentiated services with a defined per hop behaviour in each router based on the traffic identification. See Figure 7.16.

RSVP is normally used for reserving bandwidth in the access. RSVP offers guaranteed quality. It is heavy work to maintain RSVP in large backbones.

At the edge nodes, when entering the Internet from the access network, a policy decision is made. Depending on who the user is, what type of traffic he/she generates, and other factors, for example time of the day, or month, the data is assigned a certain traffic class. Traffic shaping and policing might also occur at this point. The purpose of shaping is

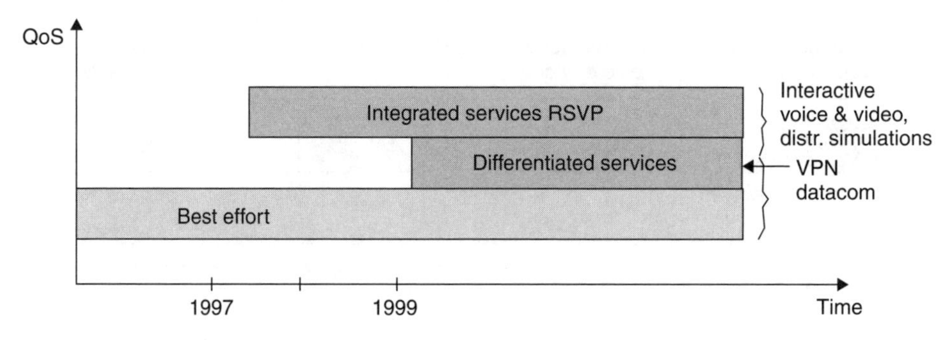

Figure 7.16 QoS methods and their main applications

to even out peaks in the traffic. This means that shaping introduces a delay and should preferably strike the non real-time traffic.

Policing looks at the agreed service level and prevents traffic above the agreed levels in cases of high load. This means that a packet loss is introduced, so this should not strike loss-sensitive data. See Figure 7.17.

7.5.4 INTEGRATED SERVICES

Resource Reservation Protocol

RSVP carries and disseminates QoS information to QoS-aware network devices along the path between a sender and one or more receivers for a given flow. See Figure 7.18.

Integrated services supports real-time transfer and guaranteed bandwidth for specific 'flows'. A flow is a distinguishable stream of related datagrams from a unique sender to a unique receiver that results from a single user activity and requires the same QoS. The flow/connection must then be identified and treated in the same manner through all intermediate nodes. The flow is unidirectional and it is identified by the destination IP address and the port number.

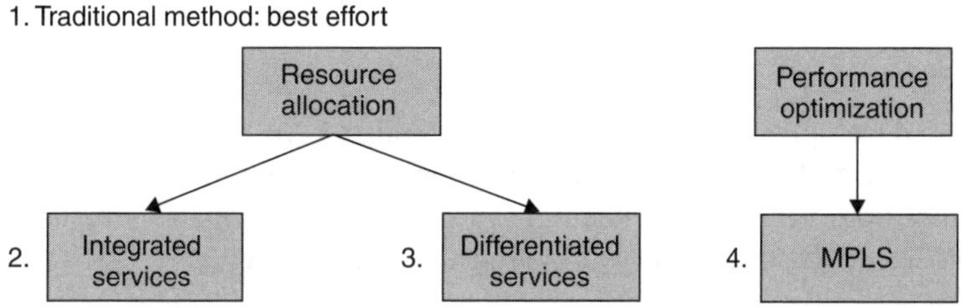

1. Traditional method: best effort
Resource allocation
Performance optimization
2. Integrated services
3. Differentiated services
4. MPLS

Figure 7.17 Best QoS is achieved with resource allocation, traffic engineering and performance optimization

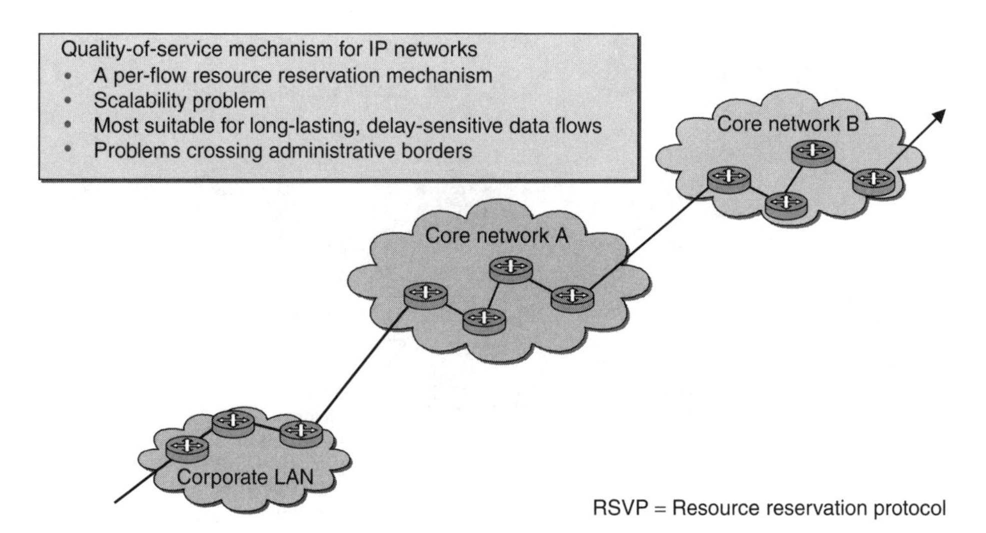

Figure 7.18 A reserved RSVP path for a specific flow

A path and a reservation are set up through signalling, and the signaling protocol is called the resource reservation protocol (RSVP). The receiver actually performs the reservation, and it must be updated continuously during the duration of the connection. This makes it possible to reroute if necessary, and it prevents reservations from 'hanging'.

7.5.5 DIFFERENTIATED SERVICES

More text about differentiated services (Diffserv) can be found in Chapter 12. What follows here is just a brief introduction.

Essentially, differentiated services specify the transfer priority of a packet as it passes through each network device on its journey through the network.

In order to allocate this priority, differentiated services marks the packets with a code point value, called the DiffServ code point, which is used by network devices such as routers to determine the per-hop behaviour (PHB) treatment.

Diffserv does not give any guarantees. It is like travelling by train without knowing when you will arrive. The standard is a compromise between reservation and prioritization. (you travel faster than packets with a lower priority). Diffserv is considered a scalable technology that can offer differentiated services for various needs in large networks.

7.5.6 MPLS

MPLS and a development of MPLS called generalized MPLS are further dealt with in Chapter 12. What follows here is an introduction. See Figure 7.19.

MPLS (multi-protocol label switching) introduces connection-oriented label switching mechanisms inside the otherwise connectionless IP technology.

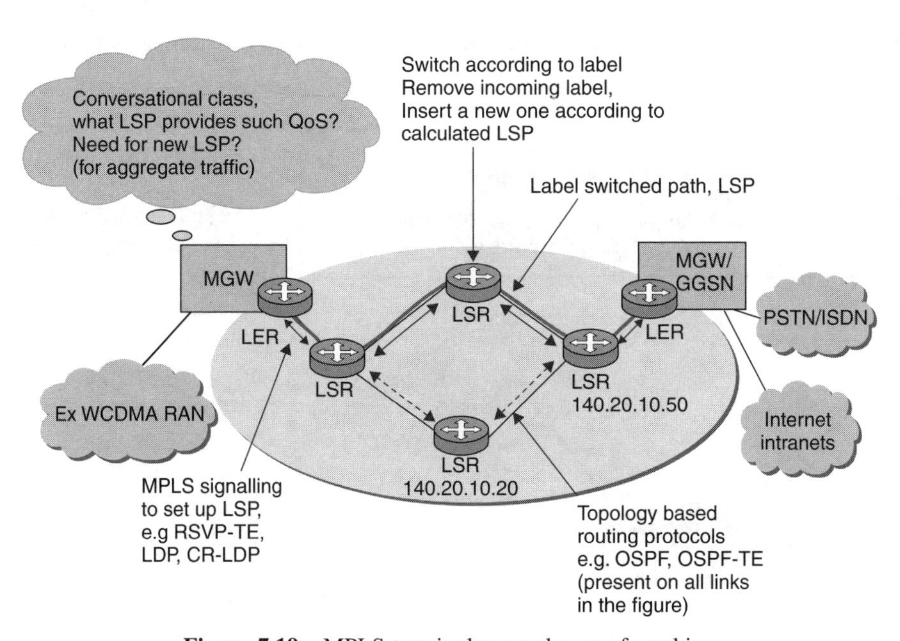

Figure 7.19 MPLS terminology and way of working

MPLS offers dynamic bandwidth allocation, traffic engineering, scalable IP over ATM support and support for VPNs.

MPLS introduces the concept of label-switched paths (LSP) for both individual and aggregated traffic streams. Separate LSPs are used to serve flows with different QoS requirements. LSPs are set up between media gateways based on topology information provided by routing protocols, such as OSPF (open shortest path first). Each LSP carries traffic from only one specific service class. The LSPs are created with a specific set of characteristic parameters (carried in MPLS signalling) according to the type of traffic (voice, signalling, etc.) it is intended to carry. LSPs may be set up either statically or dynamically according to the actual bandwidth needed between the media gateways. The end-points of an LSP on the border of the MPLS enabled core network are defined as label edge routers (LER). At these points traffic from surrounding networks will be labelled and switching will be carried out based upon the label which is exchanged on a link-by-link basis. The LER function is implemented as part of the media gateway function.

Each LSP carries traffic for a number of users. The number of LSPs as well as the number of calls a LSP is intended to carry is a trade-off between available resources and additional load produced by the MPLS signalling, which is required to maintain the LSPs. Not all types of traffic carried through the network require the use of LSP paths.

MPLS trunks are used to apply adequate traffic engineering, security and resource management on the core network.

Traffic engineering facilitates performance optimization while utilizing network resources economically, efficiently and reliably. MPLS encompasses techniques for reliability so that network service outages arising from errors, faults and failures can be minimized.

7.6 QOS AT THE APPLICATION LEVEL

7.6.1 COMPOSITE CAPABILITIES/PREFERENCES PROFILES

The present transfer of web information is normally based on recognizing a particular browser type. An alternative is to use composite capabilities/preferences profiles (CC/PP). In order to use CC/PP the terminal is provided with a user agent with a CC/PP profile.

CC/PP is a language for describing what the user agent can (currently) do by means of content negotiation. This information would then be conveyed to the originating server as part of an HTTP (or other protocol) request, and it is up to the server to decide how to use the user agent profile to best meet the needs of the user agent client. The two primary ways in which a profile might be used are selection and transformation. Selection is the process by which the originating server chooses an appropriate representation of requested web content from a finite set of existing representations.

Transformation, on the other hand, assumes that there is no finite set of representations, but that content is flexibly created based on the properties expressed by the user agent profile. The content would be stored in an XML-compatible format and then transformed into an appropriate language (or modules thereof) that could be understood and optimized for the user agent, such as XHTML or WML.

Negotiation can include information on the user agent's capabilities (physical and programmatic); the user's specified preferences within the user agent's set of options; and

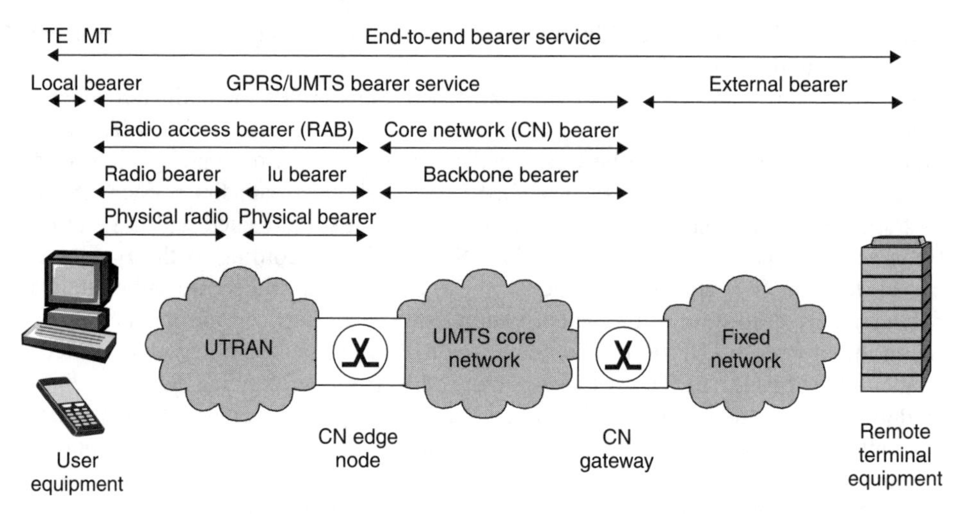

Figure 7.20 Many bearer services are part of the UMTS system, provided there are choices, suitable bearers are negotiated, especially the RAB bearer across the radio access

specific qualities about the user agent that can affect content processing and display, such as physical location. See Figure 7.20.

CC/PP is designed to work with a wide variety of web-enabled devices, such as PDAs, desktop machines, laptops, WAP phones, phone browsers, web television units and specialized browsers for users with disabilities. Proxies may also be used to provide markup transformation, transmission or caching services for CC/PP-enabled clients and servers.

7.7 IMPLEMENTATION OF QOS IN UMTS

UMTS is targeted for multimedia traffic with different QoS requirements. Attributes that define the characteristics of the transfer may include throughput, transfer delay and data error rate. This section covers two main ways to manage QoS in UMTS: RAB (radio access bearer) mapping for the access network and QoS handling.

7.7.1 RAB MAPPING

The composite system contains many layers of bearers. Many layers are stratified on top of each other. From a QoS point of view the RAB plays an important role, since UMTS allows negotiation of the choice and attributes of the radio bearer. An RAB offers circuit (initially) or packet mode and a number of bandwidths. The bearer class, bearer parameters and parameter values are directly related to an application as well as to the networks that lie between the sender and the receiver.

Bearer negotiation is normally initiated by an application, while renegotiation may be initiated either by the network (for example in a handover situation) or by an application. An application-initiated negotiation is basically similar to a negotiation that occurs in the bearer establishment phase: the application requests a bearer depending on its needs and

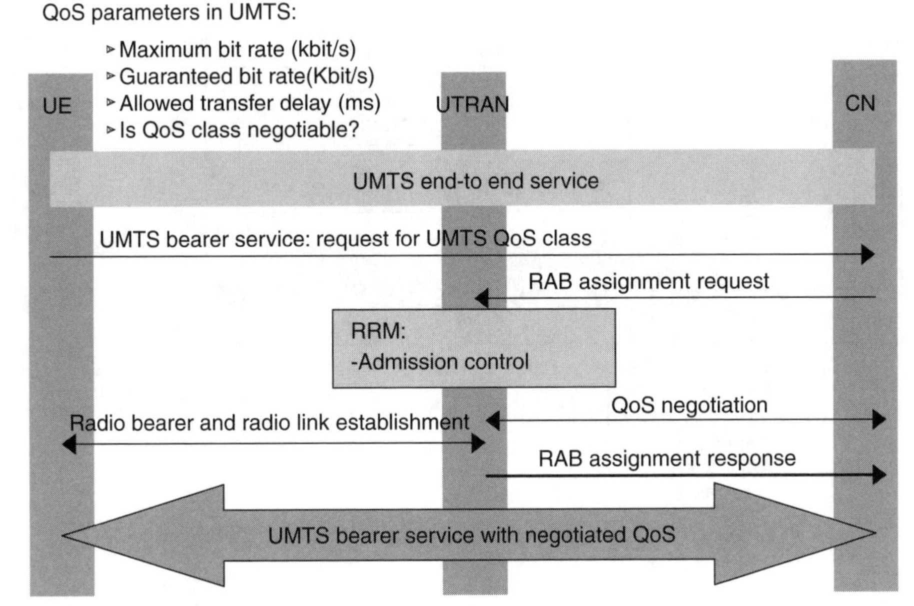

Figure 7.21 RAB negotiation for the UMTS access network

the network checks the available resources and the user type of subscription and then responds. The user either accepts or rejects the offer. The properties of a bearer service affect the price of a service. See Figure 7.21.

At the start of UMTS service not all of the QoS functions will be implemented, and therefore delay-critical applications such as speech and video telephony will be carried on (more expensive) circuit-switched bearers, Later, it will be possible to support delay-critical services transported as IP packet data with QoS functions.

7.7.2 QOS HANDLING

Another UMTS QoS issue is the choice of QoS enabler for IP traffic (RSVP, MPLS etc.), ATM service classes or frame relay.

The applications and services are divided into four different QoS classes according to Chapter 5 and beginning of this chapter.

The subscriber QoS profile, if any, is stored in the HLR. It is possible to store a differentiated QoS. The SGSN server, the MGW and the GGSN are responsible for QoS negotiation and handling.

For a specific call the QoS that is requested from the application or the user is mapped to the QoS profile that is negotiated at PDP context activation. The PDP QoS is mapped to the QoS mechanisms of the concerned networks, such as: DSCP values in Diffserv networks, MPLS labels in MPLS networks, ATM service classes, frame relay traffic parameters, etc. See Figure 7.22.

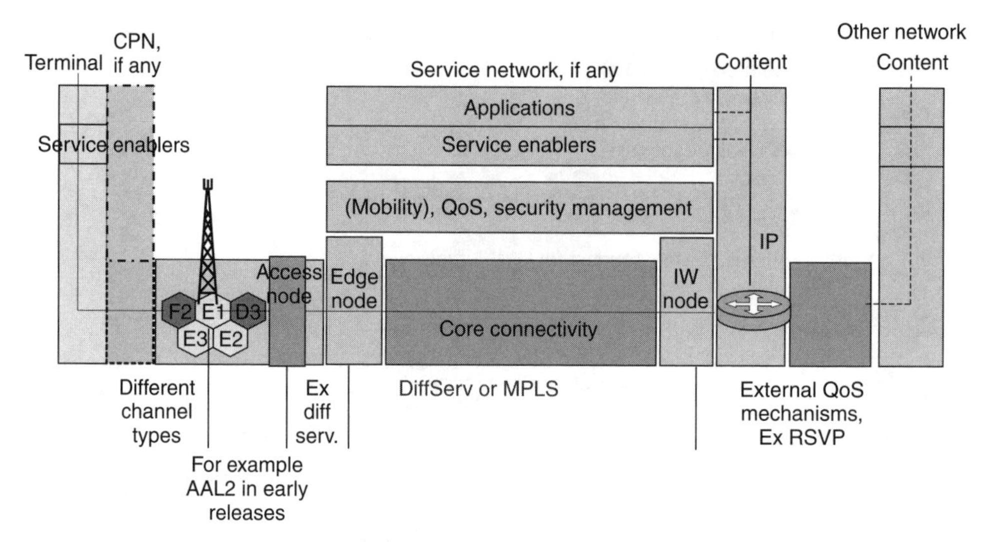

Figure 7.22 Example of chosen QoS enablers in 3GPP R99

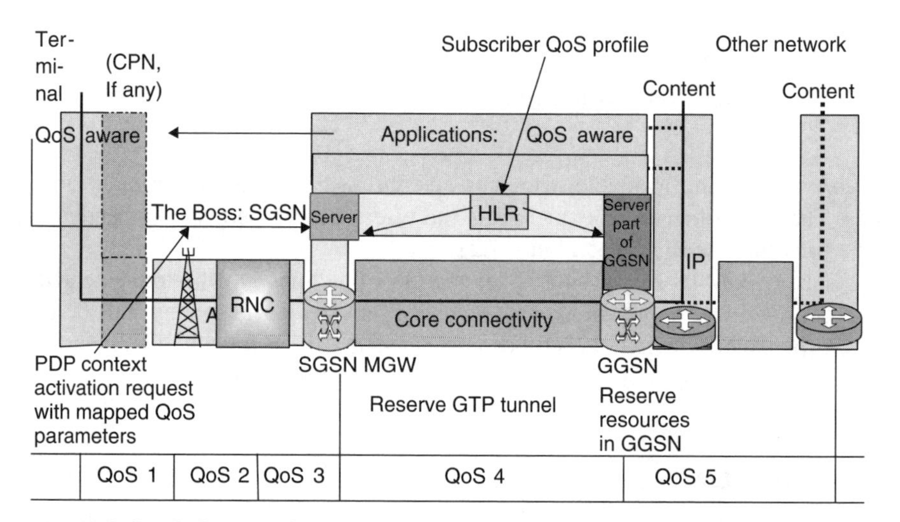

Figure 7.23 QoS communication example – the SGSN server is still collocated with the SGSN MGW

It is important that the fixed path has sufficient capacity for the wireless traffic. Otherwise the scarce resources of the wireless path might not be used to its potential and the QoS might be affected. The SGSN MGW implements QoS handling and RAB mapping. Fallback of QoS level is possible when roaming between UMTS and other mobile networks. See Figure 7.23.

8

Service Implementation

'The most difficult part is not to migrate the technology but to migrate the minds of the users.'

8.1 OBJECTIVES

The ideal vast and ambitious goal of this chapter is presented in the upper part of Figure 8.2. Please observe the word 'competitive'. Since this book is broad and context based, Chapter 8 can only cover a fraction of the overall answer to such a challenging goal. However, most other chapters are directly or indirectly supporting this chapter.

Thus, Chapter 8 integrates findings in many of the previous chapters, exemplified by the following list:

- **Chapter 1 – success-related inputs –** main success factors are user related: application QoS (e2e, end-to-end), bandwidth, security (e2e), coverage (including roaming!), suitable devices, easy to use, affordable.

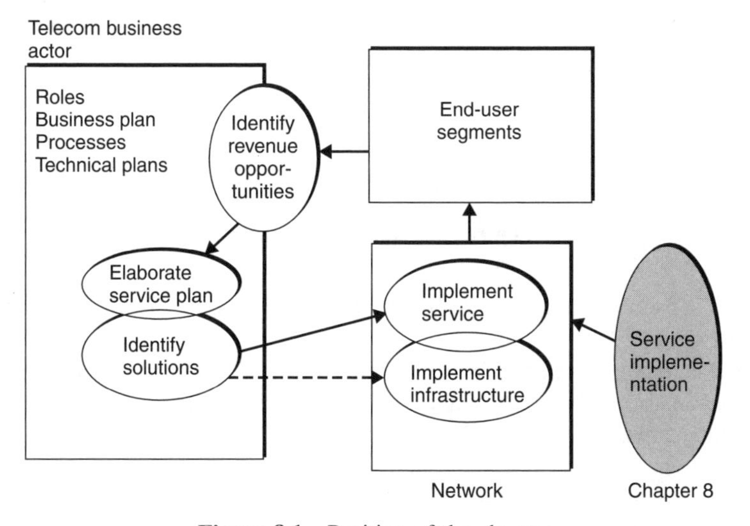

Figure 8.1 Position of the chapter

Understanding Changing Telecommunications – Building a Successful Telecom Business. Edited by A. Olsson
 ISBN: 0-470-86851-1

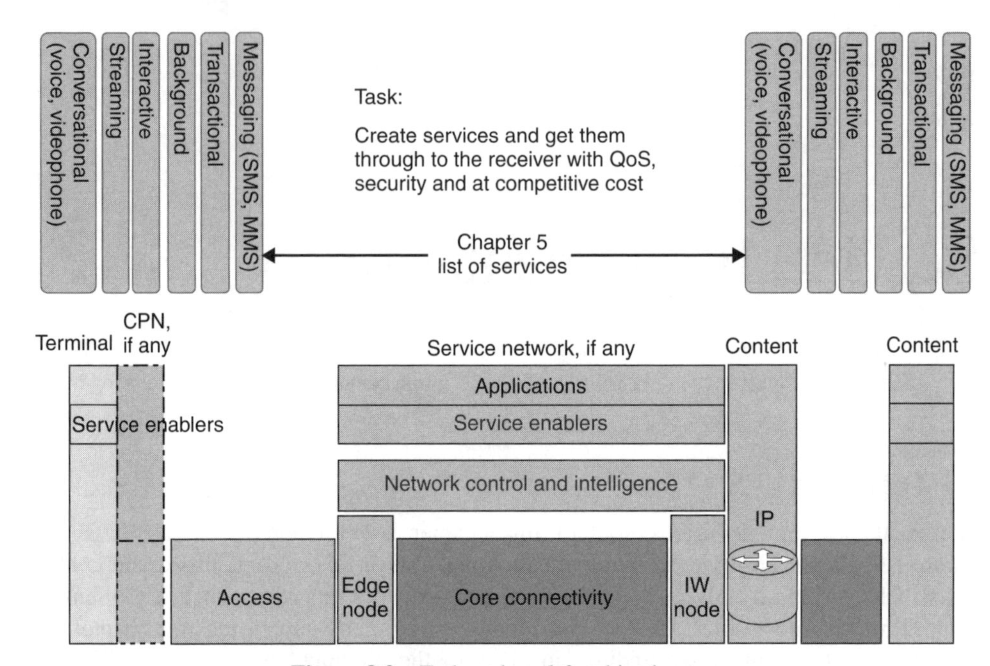

Figure 8.2 Task and goal for this chapter

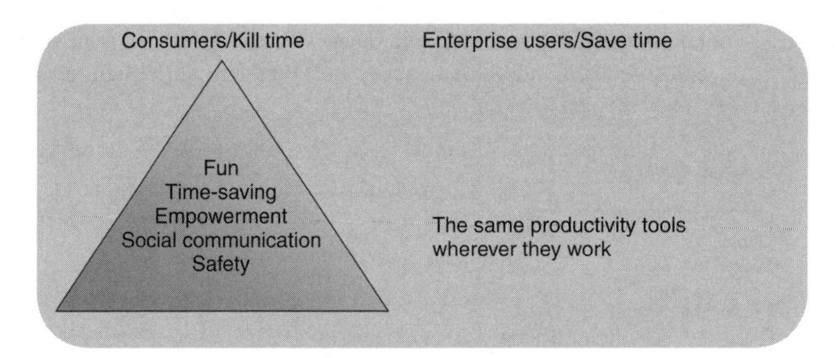

Figure 8.3 Main user demands

- **Chapter 2 – user-related inputs from chapter 2 –** some main user demands are shown in Figure 8.3.
- **Chapter 3 – network-related inputs –** the converged structure with media industry input ('person to content'). Architecture: traffic system with control plane and user plane. Management system. H 248. IPv6. API. FP: numbering, routing, switching, synchronization and other plans are necessary. Dimensioning aspects.
- **Chapter 4 – business-related inputs –** role(s), business rules, ROCE, OPEX, CAPEX, revenue generation, cost efficiency. Agility in time to market increases market share.
- **Chapter 5 – service plan related inputs –** service plan, service classes, business-based features: compression of header, ROHC, compression of payload, charging, need for middleware, impact on core and access network.

- **Chapter 6 – security-related inputs –** security goal, enablers and implementation.
- **Chapter 7 – QoS-related inputs –** QoS goal, enablers and implementation

8.2 CHAPTER STRUCTURE

This chapter examines the support for a vast service plan from a number of angles, starting with the target network (Figure 8.4). In reality, the approach of describing a target network is like shooting at a moving target, with a need for a continuous update. However, the alternative of managing without any kind of target makes focusing almost impossible. In fact the target network used in this book has been partly described in the previous chapters, in particular in Chapter 3. It could be called the 'horizontal, all-access network'.

There is also a need for a systematic approach. Figure 8.5 provides support. The structure of this chapter is mapped onto parts of the ITU diagram. However, products and price lists are not treated in this book. The main modifications/additions are:

- 'Service plan' and 'service enablers' have been added.
- 'GoS' (grade of service) has been replaced by 'QoS' and 'Digitalization process + ISDN (alternative scenarios)' have been replaced by 'Target network + Development track'.

Implementation of particular service types is covered in the following way:

- Messaging and transactional services are 'service-enabler oriented' and treated in Section 8.6.5.
- Conversational and streaming services (and partly interactive services) are QoS oriented and treated in Section 8.7.3.
- Background services are not very demanding, and are not treated at all in this chapter. Ordinary IP networks are suitable bearers.

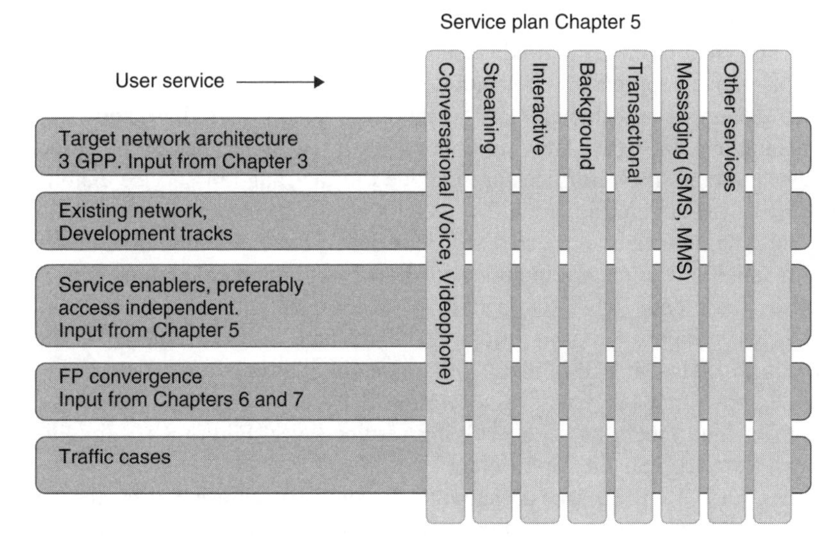

Figure 8.4 The service plan is examined from a number of angles, starting with the target network

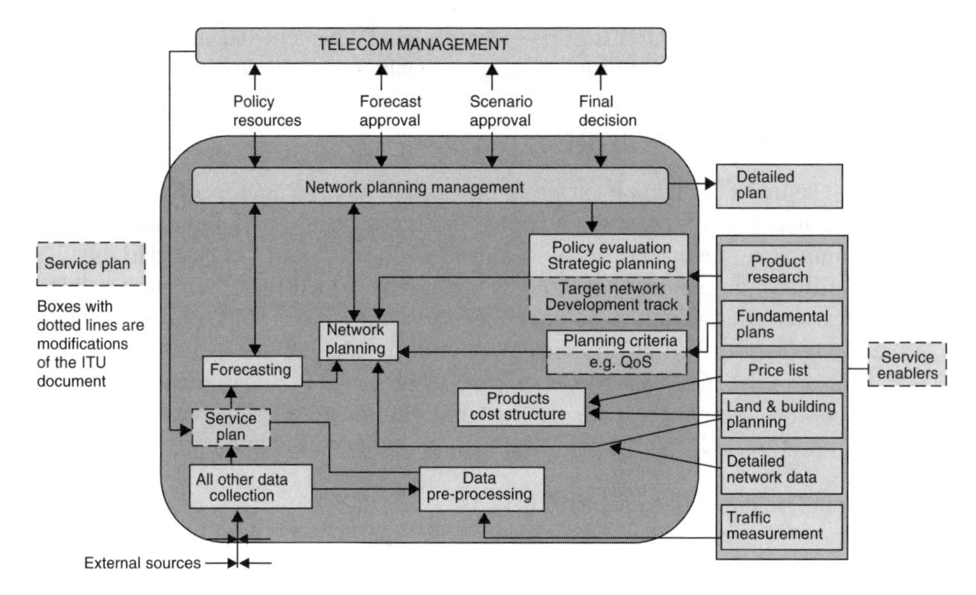

Figure 8.5 This ITU Figure with additions supports Chapter 8

8.3 TARGET NETWORK

The target network should support the services in the service plan in a competitive way. It should fit the main roles mentioned in Chapter 4.

3GPP has presented network architectures for mobile systems such as Figure 8.6. Our idea is to briefly cover most of the building stones shown in the figure. Let us immediately take a look at the IN type of services, represented by SCP (service control point) and extensively covered in a previous book (Book 1).

Many IN functions can be classified as advanced fundamental plans, for example advanced routing (IN-VPN, virtual private network), advanced numbering (number portability, universal number) and advanced charging (prepaid, televoting, freephone) In CAMEL (customized applications for mobile network enhanced logic) there are functions for changed B-number and changed tariff. Figure 8.7 (also Figure 3.39) shows alternative IN implementations. In Figure 8.8 a number of 'VAS' (value-added services) services are indicated. In principle these services are included in the range of service enablers in this book. CAMEL was covered in Book 2.

Figure 8.6 is somewhat complicated and does not represent the initial and target stages. Therefore a series of figures are included to show the gradual development.

This also represents what is called a development track in the book. See Figures 8.8–8.10. Figure 8.8 is a bridge to GSM and similar systems of the second generation, with UTRAN, SGSN and GGSN as main additions.

Figure 8.8 shows release 4 with a separation between control and connectivity layer for CS domain multimedia and the evolving service layer. Figure 8.10, finally shows a simple target network version, based on IP.

Let us then turn to an 'all-access' network. It could be represented by the conceptual Figure 8.11, taking care of the requirement for a competitive solution. Observe that this figure makes no differentiation in the access layer, which is shown in Figure 8.12.

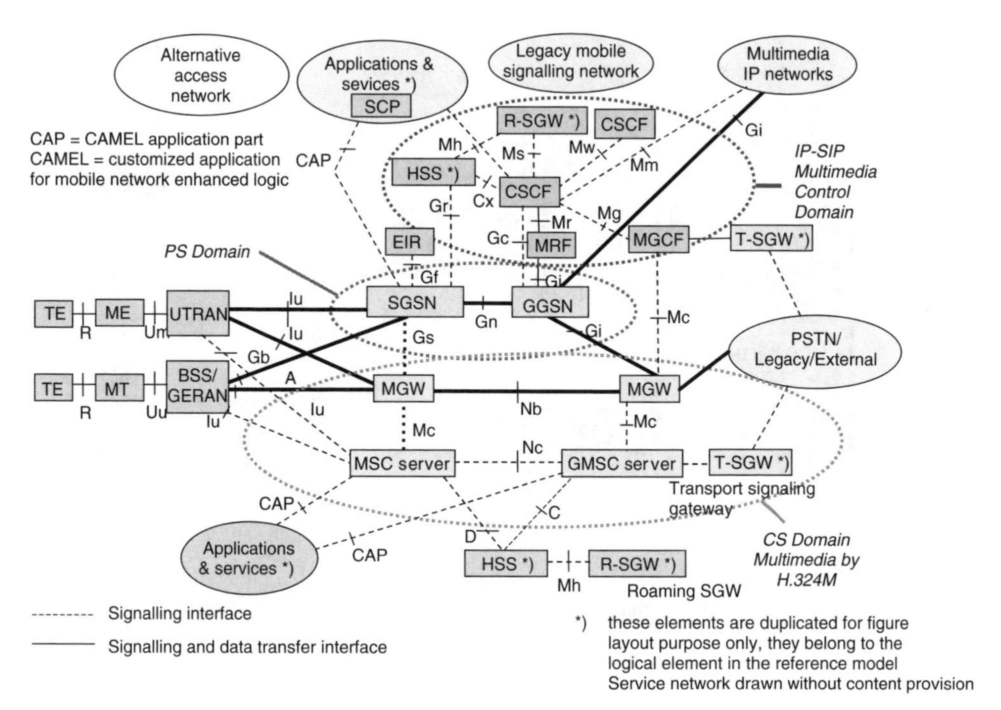

Figure 8.6 3GPP network architecture, example

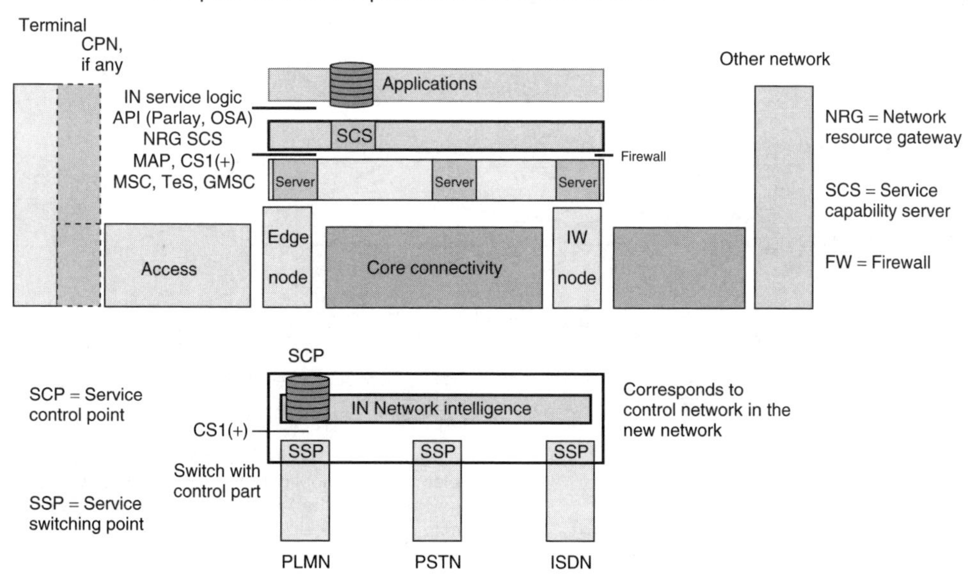

Figure 8.7 Conceptual view of IN services

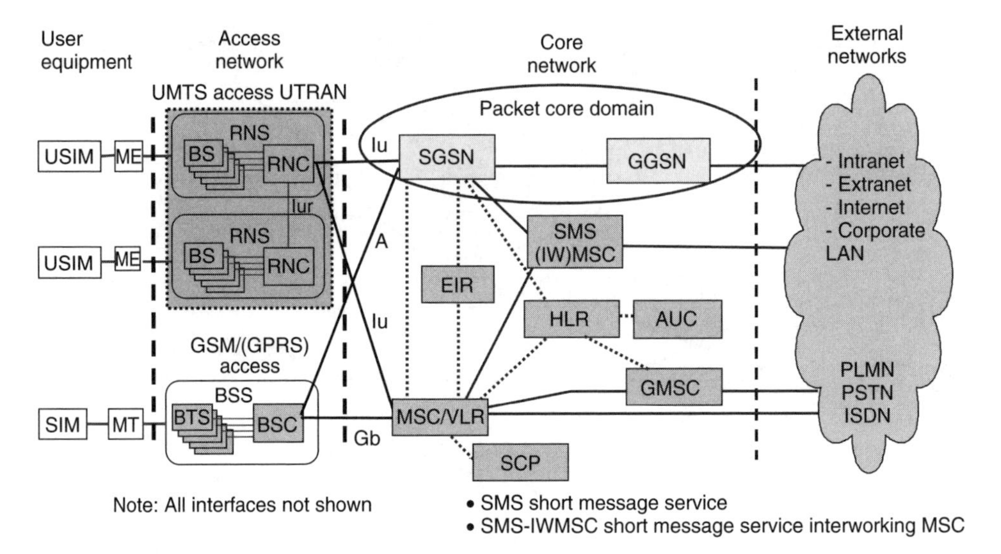

Figure 8.8 The UMTS network before physical separation of connectivity and control layers – instead there is a clear separation between the classical circuit-switched domain and the packet domain

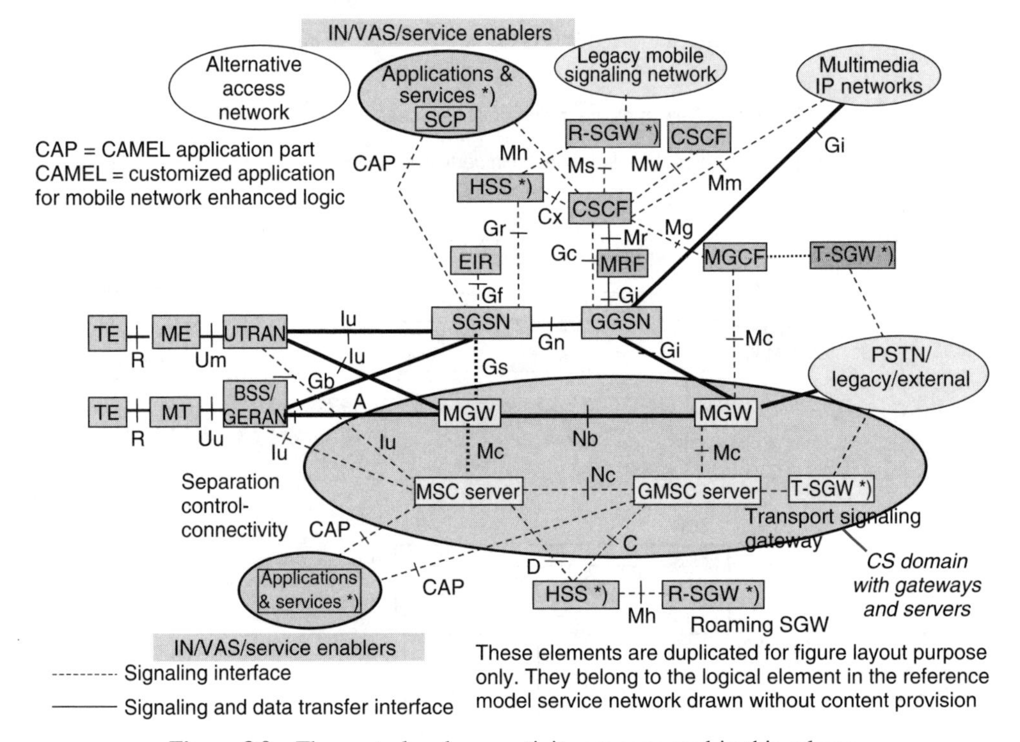

Figure 8.9 The control and connectivity are separated in this release

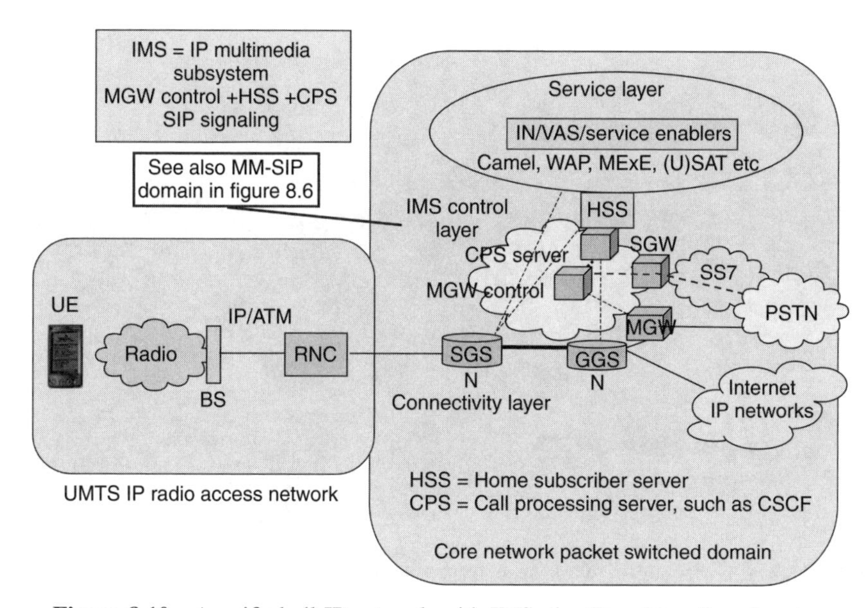

Figure 8.10 A unified all-IP network with IMS, the IP multimedia subsystem

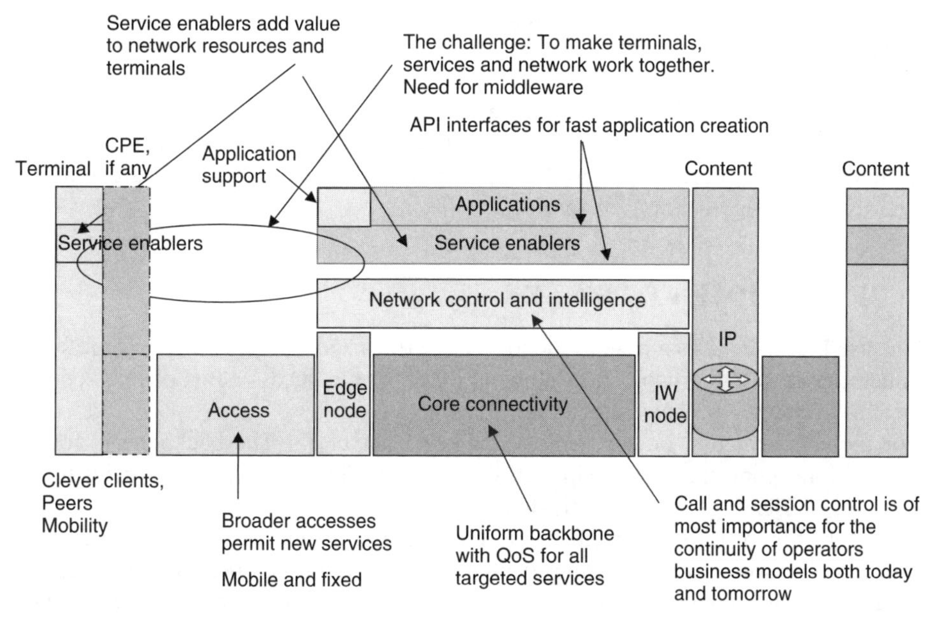

Figure 8.11 The horizontal network from a distributed business angle – all parts contribute to the competitiveness and cope with a broad range of services

Figure 8.12 gives a good idea of the complexity in the overall picture. The access transport parts are mainly used in metropolitan and urban areas and are justified by the distance between terminal and gateway/edge nodes. They consist of self-healing optical rings. See for example Figure 4.54.

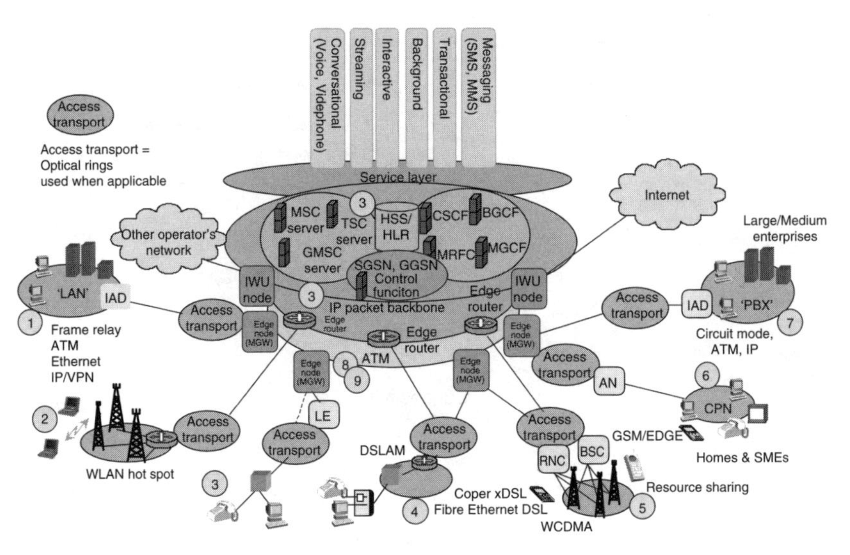

Figure 8.12 The horizontal network seen from the access side, with the demanded services at the top

The figure represents a fairly unified backbone, control layer (two main parts) and service layer. Many people suppose that we are heading towards a final, hopefully simple network with SIP and IP as the main technologies for signalling and the network layer.

The service layer provides application supporting VAS services (e.g. SAT/USAT, Java, positioning, messaging and WAP), network FP supporting IN services such as CAMEL, and 'middleware features' such as XML, and MPEG-4 and other coding/compression standards opting for economy of service.

8.4 DEVELOPMENT TRACKS

Let us use Figure 8.12 as a base to identify some roles and tracks, and also for references to other chapters of the book. It is obvious that it is the access network that represents the main differentiation between actors.

For more information on the access area see Chapter 13. The backbone is found in Chapter 12, the control network with SIP signalling in Chapter 14 and the service network in Chapter 9. The edge nodes are treated in Chapter 11.

Case 1 in Figure 8.12 is an enterprise data access, which could be operated by the incumbent operator, other competing fixed network operators or operators targeting enterprise traffic, often operating globally. The technical solutions for wide area corporate networks develop from leased lines, X.25, frame relay and ATM to IP-VPN tunnels.

The IP-VPN tariff is based on IP with no distance dependent fee, as opposed to frame relay and leased lines. The solution will be cheaper, global and flexible. VPN is mainly presented in Chapter 7.

The FR/ATM/IP-VPN operators could expand their business and incorporate telephony. A traditional PBX could be replaced by an IP-PBX or an IP-Centrex solution (Case 7). See Figure 8.13.

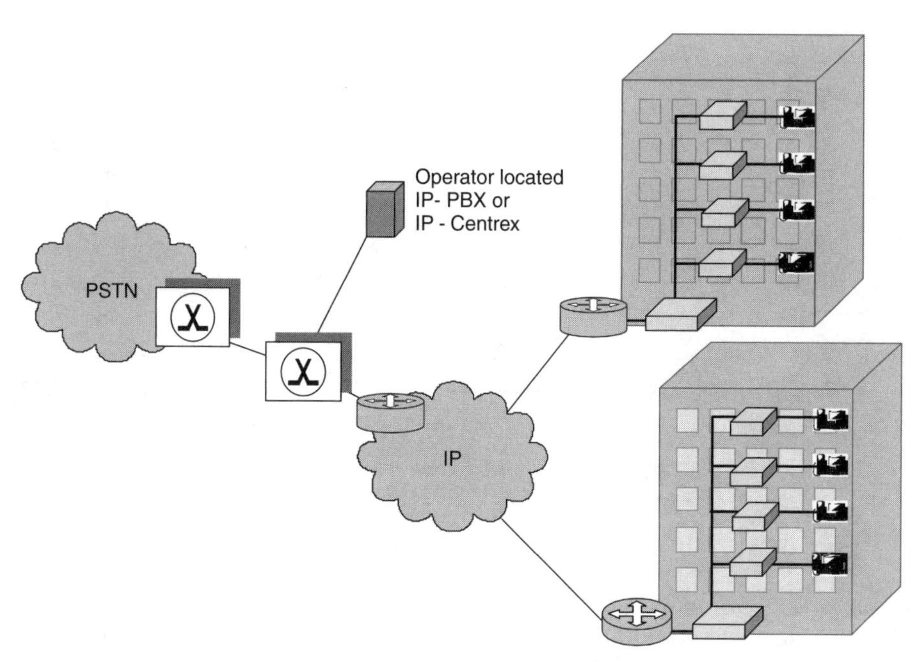

Figure 8.13 IP-PBX is located at the CPE (not shown) or outsourced as operator located and managed, or as an IP-Centrex

Another IP voice solution is related to the backbone network: voice trunking, defined in Appendix 3, Objectives.

The operator might also take care of modem traffic in the PLMN/PSTN/ISDN by means of NAS connections (Case 8). This is called Internet offload or data offload and is also defined in Appendix 3, Objectives. WLAN traffic (Case 2) and xDSL IP traffic (Case 4 and 6) might be connected with or without IP telephony. The expansion into voice traffic requires equipment such as gatekeepers and voice gateways.

In fact, the development could end up in the target network with IP multimedia. See Figure 8.14.

The WLAN operation for Internet and intranet access (Case 2) can also expand or integrate with other networks, such as 3 G. See Chapter 15.

The incumbent (fixed) operator (Case 3) has clear ambitions and possibilities to offer all types of services in the list. It is true that for example MMS is seen as a mobile service, but the fixed network has a long tradition in messaging. The incumbent operator has for some time been in the forefront regarding migration of the backbone to packet mode (IP, ATM, MPLS). Considering that voice is the main service, a telephony server (TeS in Figure 8.14) is a key component in the network.

An example of a migration scenario is given in Appendix 3.

The xDSL solutions (Case 4) may be based on ATM or IP as transfer mode. They will be connected to the packet backbone accordingly.

The mobile development (Case 5) has already been touched on in Figures 8.8–8.10. The focus is cost efficiency with the introduction of 3 G networks.

Case 6 shows a home network. An example is shown in Chapter 10.

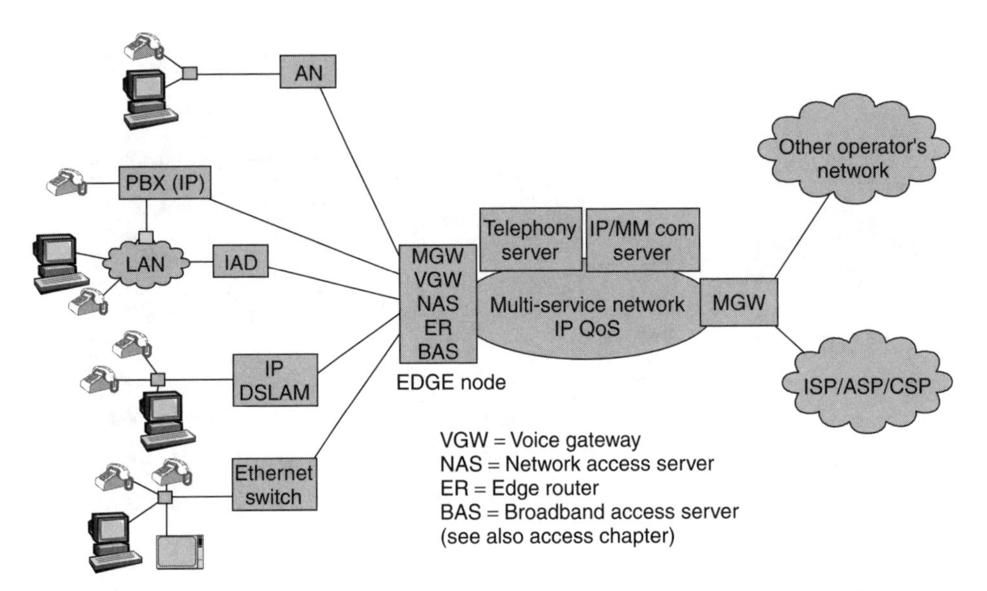

Figure 8.14 Example of a convergence that has started from the datacom side (LAN, IP) and reached the multimedia stage (MM server), and that has incorporated voice (telephony server, PBX, access node) in between. Both voice and multimedia require servers at the control layer

8.5 INTRODUCTION TO PACKET DESIGN

So far a heterogeneous situation with a number of transfer modes in a number of development tracks has been described. The heterogeneous situation demands gateways for the user plane and the control plane. However, the situation is converging towards a more unified packet mode.

Packet design becomes a crucial task for all actors. In a wide sense the task also embraces the content of the payload, apart from all signalling and encapsulating protocols. In fact, the design and content of the packets travelling in the network tells us a lot about end-user needs and the telecom business. See Figure 8.15 that for example could initiate a discussion on VoATM versus VoIP. Thus, a particular service might be differently implemented by different actors, resulting in different packets. SMS has so far been carried by circuit mode, but this will change to packet mode (Figure 8.16).

From Figure 8.15 the conclusion is clearly that the basic packets are far from enough, when for example using stratification and tunnelling. We suspect that quite a number of inputs are involved in the almost instantaneous packet generation. This is confirmed when checking the infrastructure support in IP best effort networks (Figure 8.17).

It is also confirmed when looking at the need for information at the start of a packet communication after attaching, let us say a mobile terminal in GPRS to a packet network. The information referred to is called PDP (packet data protocol) context. It might be a logical association between a mobile station and a public data network running across a GPRS network. The context may define aspects such as routing, QoS, security, billing, type of network, network address, access point name and radio priority. Largely, we are again back to the fundamental plans.

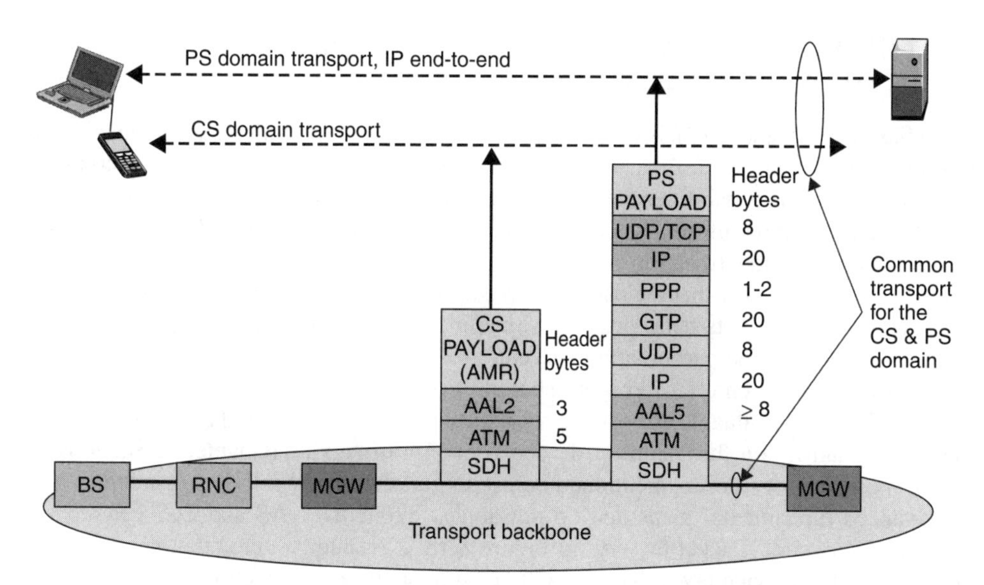

Figure 8.15 Packets travelling in the network, example

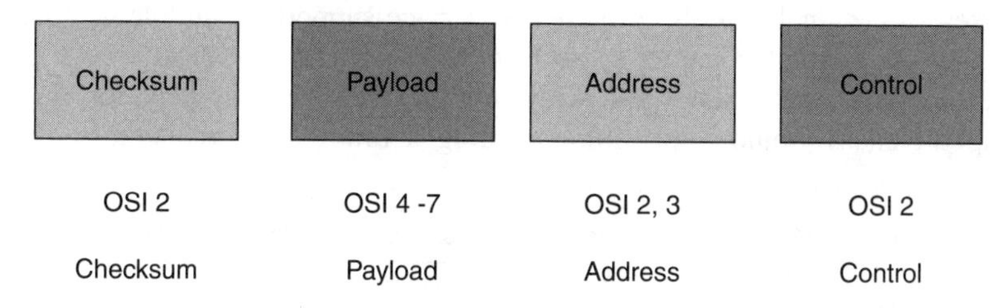

Figure 8.16 Basic packet composition

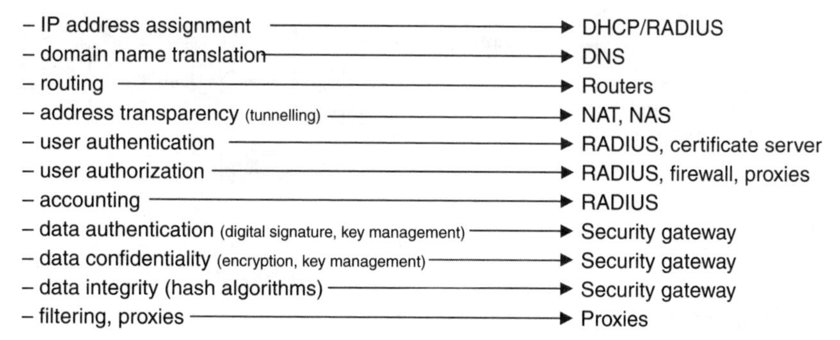

Figure 8.17 IP network support functions

8.6 THE ROLE OF FUNDAMENTAL TECHNICAL PLANS IN PACKET DESIGN

From Figures 8.15 and 8.17 it is easy to recognize the following fundamental technical plans: routing, QoS (= checksum), numbering/addressing, security, charging and accounting. When extending the capabilities the QoS plan is even more important. From the PDP (packet data protocol) context information interconnection needs may obviously be involved, such as foreign operator identification (numbering) and a gateway.

One might simply say that the packet propagation in the network is a teamwork between the packet design and its data, possible signalling, and the programming of the network elements, including the participating servers.

The packet end-to-end transfer work should be as cheap as possible for the actors in the value chain, and as valuable as possible for the users. Therefore we add to the fundamental plans VAS and IN on the income side and economy of service (Chapter 5, Sections 5.6 and 5.7) on the cost side. As mentioned before IN functions are basically regarded as parts of advanced fundamental plans and do not appear explicitly. VAS and EoS get separate boxes in the overall 'Packet factory' in Figure 8.18. It should be noted that many FPs are necessary at the service layer, with or without support from IN functions implemented in the service layer.

EoS includes such cost-efficiency measures as payload compression, e.g. MPEG-4, header compression (especially ROHC), caching and content-based routing. In addition to the boxes shown in Figure 8.18 there are service-supporting middleware/framework functions (see for example Chapter 9, Section 9.3).

Other cost-efficiency measures deal with multiplexing and IP aggregation. In this context, competitiveness would imply implementing a future-safe architecture, composing

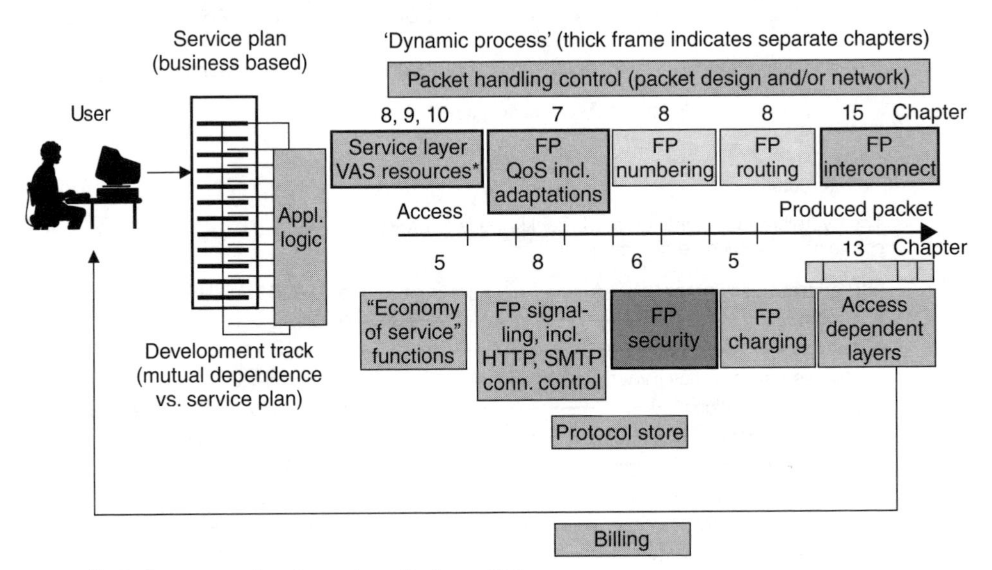

Figure 8.18 Pedagogical support for understanding packet creation and propagation – many other FPs are possible, as mentioned in Chapter 3

an attractive service plan, choosing appropriate service layer enablers and 'Economy of service' functions, and finally tuning the FP 'knobs' for best security, quality and cost efficiency.

The service-layer enablers and more about FPs will be covered later. As indicated in Figure 8.17 QoS, security and interconnect FPs are covered in separate chapters. Later in this chapter there are separate overviews on the numbering and signalling plans. Some routing features are inherently covered in the MPLS standard. Section 8.10 on traffic cases illustrates 'packet handling control'.

Let us now apply a top-down approach. In fact the service layer part should ultimately be independent of the underlying 'find and connect machine'.

8.7 TOP-DOWN APPROACH TO PACKET DESIGN

8.7.1 THE COMPONENTS OF THE WHOLE SYSTEM

Figure 8.19 shows one way of segmenting a communication system: content, application logic, service enablers, adaptations to the bearer technology, and finally the bearer network. The left side represents the terminal device with corresponding functions. Adaptations are considered part of the QoS plan

8.7.2 THE APPLICATION LOGIC

Applications are sometimes called end-user services. The complex terminology relationship between applications and services, including application logic, is found in Chapter 1.

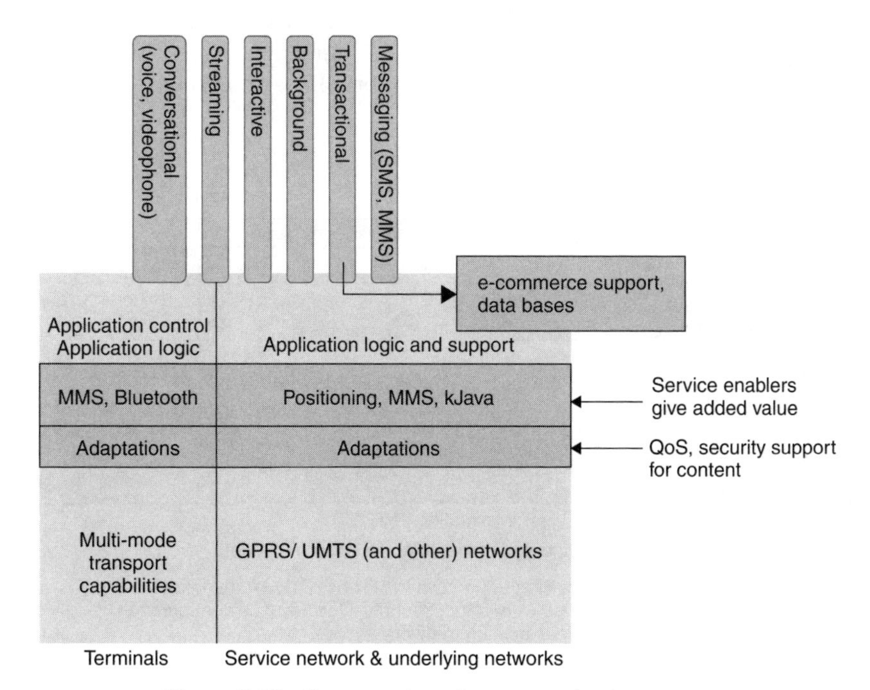

Figure 8.19 Segmentation of a communication system

8.7.3 SERVICE ENABLERS

There are two semantically different service enabler groups. Both are applied in the service network and/or terminals.

- The original Parlay/OSA service enablers consist of service capability servers (SCS) and application support servers (ASuS – an Ericsson term). Service capability servers interface the network, acting like bridges between the applications and the network resources. Application support servers perform other tasks like security and charging.
- The OMA service enablers have a much wider scope. OMA service enablers are the service supporting elements of the terminals and the service network. There is a substantial overlap between Parlay SCS and OMA service enablers.

In the following service enablers means the sum of Parlay SCS and OMA service enablers.

The service enablers have a crucial task: to tie terminals, networks and applications together into an efficient service production unit.

The text below takes mobile terminals as an example, but this is no restriction.

Service enablers will play a crucial role in future application development. The possibilities are almost endless, particularly when it becomes possible to combine two or more enablers. An example is combining location-based services with messaging. An overview of the position of the service enablers in the network is found in Figure 8.20.

The following view below has been presented and developed by Christoffer Andersson. For each and every mobile Internet device, there is a set of enablers that decide what could be done with it. The basic GSM/GPRS handset only supports voice, SMS and WAP, therefore these are the kind of applications available. As handsets get more and more advanced we see more and more enablers added into the devices and networks. One new enabler is positioning support, enabling a report of what the actual position is. This

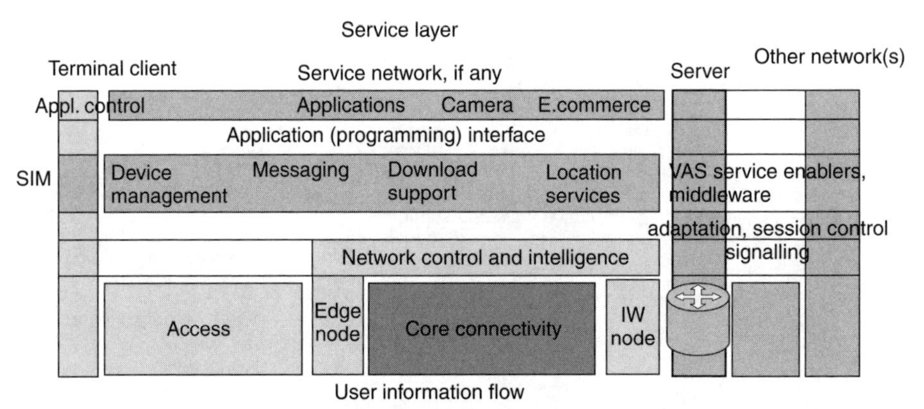

Figure 8.20 Service enabler position in the network

opens up a new range of applications based on this feature, where you might for example have a taxi ordering application that knows in which city you are.

Other examples of enabler support in terminals are multimedia messaging service (MMS), video playback and Bluetooth. All of these enablers open up a new set of applications for the user. Just as the networks themselves have migrated by adding first GPRS and then 3 G, these enablers add more and more applications capabilities step by step. WAP and SMS were followed by enhanced message service (EMS, see Chapter 5) and SyncML (see below). Even more recent are Java and MMS and video playback. See Table 8.1. From the application point of view this is what decides what the GPRS and 3 G applications will be, not primarily the networks themselves. This means that you can actually know what categories of applications there will be in 3 G networks by knowing what enablers the 3 G terminals will support.

The goal of SyncML is that the devices will be entirely plug-and-play with any data stored on any network, enabling a device to synchronize and update with any database, in any location, at any time. Without the ability to access and manipulate accurate data at any time and from any place, mobile devices lose their value. The challenge is to develop mobile applications that are accessible on multiple handheld devices in areas of intermittent wireless coverage, which will still allow users to access data with a 'change once – update everywhere' feature. See also Chapter 5.

Table 8.1 Four sample enablers

Enabler	Handset/Network support needed	Main added capability	Example applications
WAP 1.1/2.0	Handset and network	Transactions and viewing information via a browser	Buy cinema tickets, check sports results
MMS	Handset and network	Sending and receiving messages of text, sound, pictures and (later) video	Take a picture with a (sometimes built-in) camera and send it to a friend. Get the daily Garfield cartoon
Positioning	Network mainly (GPS in handsets for added accuracy)	Uses network to fetch user position when allowed (used together with another enabler like WAP)	Find the closest gas station. What will the weather be like?
Java (primarily J2ME: Java 2 Micro Edition)	Terminal primarily (network for provisioning and charging)	Standardized execution environment that allows programs to be downloaded to the handset	Games with moving objects. Stock trading where tickets are updated in real time

You can view an enabler as a catalyst that, when available, opens up a new range of applications. For example, the positioning enabler removes the need for the user to type in his current address and makes a new range of position-aware applications possible, like 'The weather near me', 'Buddy locator'. (A Buddy locator service is a friend-finder service. When used at a conference it enables users to receive alerts when their colleagues arrive, and tells them where their colleagues are within a conference venue.)

Although some of these applications will be completely new, the majority are known applications that now work a lot better. This is a key success factor for the introduction of new technology – migrate step by step into more advanced features without quantum leaps. It has been shown many times that the most difficult part is not to migrate the technology but to migrate the minds of the users – it is hard to change habits!

Some enablers primarily need support in the terminal while others require some kind of server in the network (WAP, MMS etc.). Generally, a *service enabler is a client component plus a server component that adds a certain feature to the network.* As an example, a weather application might use WAP to present the content efficiently on the screen. Then an operator might introduce positioning support in the network and now the applications developer can present the weather report differently depending on where the user is currently located.

MMS is one of the key enablers. Content is created on top of MMS. Some content will be downloaded via WAP or created on the handset. With a camera built into the phone or attached to it, the user can more easily create appealing content on top of MMS.

We need to view MMS as an enabler in order to create focus on the dearly needed content. Then there will be sites that not only host ring tones and logos showing Britney Spears but also advanced MMS shows with animations, pictures and sound clips that create early multimedia experiences on the nice colour screen handsets.

8.7.4 ADAPTATION AND BEARERS

When combining service groups, service enablers and the stack from the QoS chapter, the peer-to-peer or client–server communication system in Figure 8.21 can be drawn. Because of space problems MPLS and RSVP are left out in this particular figure. Since the client and the server (terminal device and service network) cooperate in enabler creation, it is fair to draw the enabler as an end-to-end function. What now remains are the 'Economy of service' functions mentioned in Figure 8.18. As seen from Chapter 6 there are security protocols as well (IPsec, WEP 128, WTLS etc.) at different layers in the stack. See for example Figure 6.13. Tunnelling further adds to the complexity.

To make the picture even more complete let us take another look at the service layer. OMA and others are working with web services, in particular for the service layer. Introduction of such technology makes service enablers accessible to any service provider. Among the web service protocols are WSFL, UDDI, WSDL, SOAP, XML and others. See Appendix 1.

Every protocol belongs normally to a fundamental plan.

8.7.5 SERVICE IMPLEMENTATION SUPPORT IN THE SERVICE LAYER

Among the six segments in the service grouping, transactional services and messaging services are very much dependent on the service layer for their implementation.

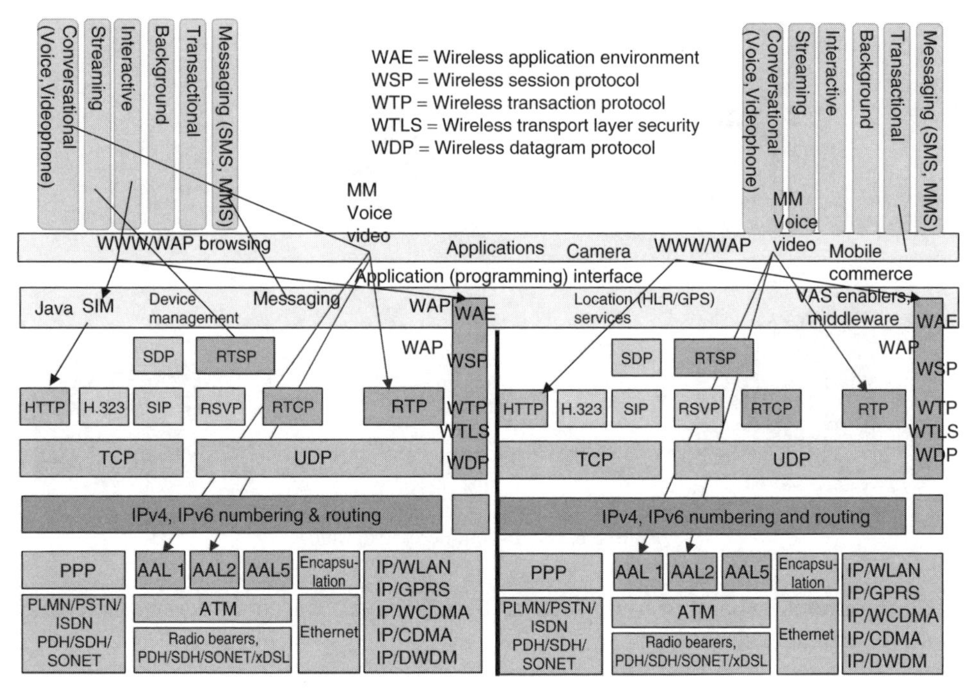

Figure 8.21 An all-access architecture where the service groups are carried by IP or ATM transfer mode. The figure also shows some stratification possibilities, for example IP over ATM over SDH

Transactional Services

Transactional services are considered to have a very large potential. Standardization, accuracy and security are key enablers. Transactional services are very dependent on application support servers at the service layer for e-commerce, micro-payments for parking, cinema tickets etc. The services are not time critical, but the accuracy in the transaction must be very high.

Transactional services should be protected as close to the application as possible. IPsec has many advantages, but it is located at the network level. Two closer protocols are secure socket layer (SSL) and S-HTTP, which complement each other. Whereas SSL is designed to establish a secure connection between two computers, S-HTTP is designed to send individual messages securely.

An e-commerce service could work like this (see also Figure 8.22).

1. You want to buy something from the Internet and give the shop your card number.
2. The shop connects to the bank to verify that the card is valid and that there is enough money/credit on your account.
3. The bank responds with a Yes or No.
4. You get confirmation on your purchase or a message that you don't have enough money.
5. The bank will bill your account for the purchase.
6. The goods will be delivered. (This is really easy when you buy software, somewhat worse for hardware.) As mentioned in Chapter 5 there is a need for a large number

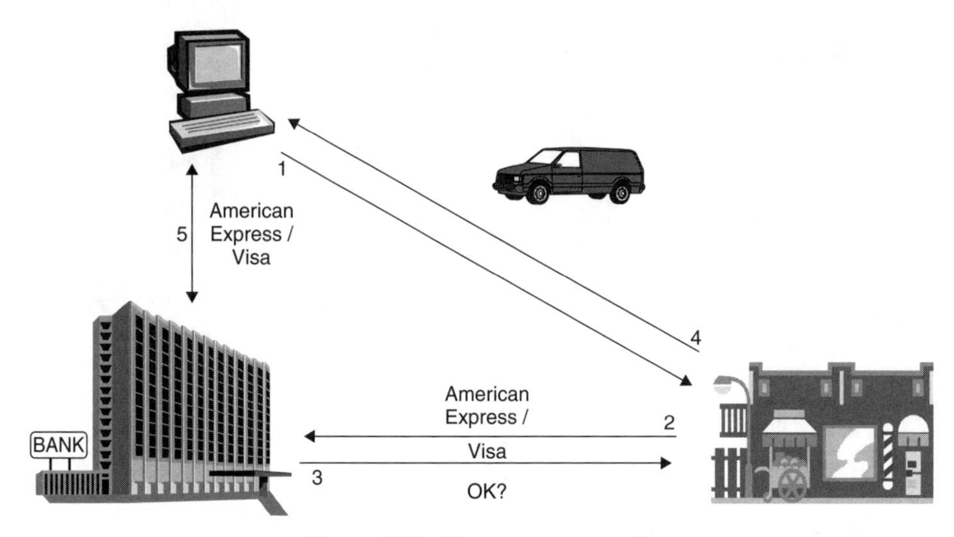

Figure 8.22 E-commerce example

of application support servers, such as bank servers, education servers, advertising servers, game servers.

Figure 8.23 shows a sequence for buying tickets for the train.

Messaging Enablers

The messaging class is not very demanding regarding the transfer since there are no time requirements. Instead, there is a demand for enablers in the network and in the terminal devices. The messages are stored and forwarded. For MMS, the size of the messages is up to 100 kbytes.

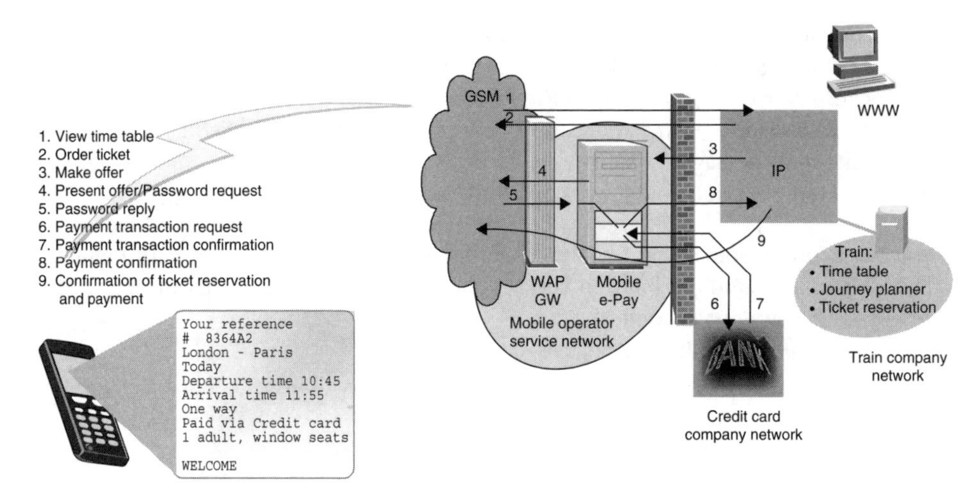

Figure 8.23 Application support (called mobile e-pay) for buying train tickets

Goal – The multimedia message service (MMS) provides an intelligent environment for the very promising multimedia mobile messaging service. MMS is primarily designed for UMTS and GPRS. Access to MMS services should, however, eventually be independent of access point. Multimedia messages should then be accessible through 3G and 2G mobile networks, fixed networks, the Internet etc. To facilitate interoperability and universal messaging access, MMS will comply with the virtual home environment (VHE) (Chapter 15).

Implementation – The multimedia message service (MMS) introduces new messaging platforms to mobile networks in order to enable MMS. These platforms are the MMS relay, MMS server, MMS user databases and new WAP gateways.

MMS requires not only new network infrastructures but also new MMS compliant terminals. MMS is not compatible with old terminals, which means, that before it can be widely used, MMS terminals must reach a certain penetration, and that will take at least a couple of years.

MMS is like SMS, a non-real time service. A relay platform routes multimedia messages to MMS servers.

The multimedia messaging architecture (see Figure 8.24) has a number of key elements that are part of a multimedia message service environment (MMSE). The interworking key elements defined by the 3GPP are:

- MMS relay
- MMS server (or servers)
- MMS store (or stores)
- MMS user agent
- MMS user databases

The MMS relay is responsible for the transfer of messages between different messaging systems. MMS relay is therefore responsible for transcoding multimedia message format, interacting and interworking with other platforms, and enabling access to various servers residing in different networks and the like.

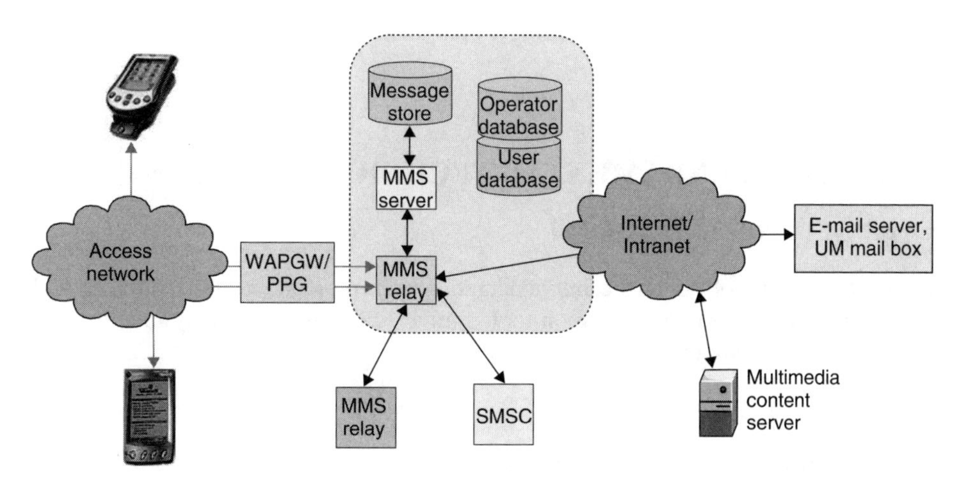

Figure 8.24 Example of structure of MMS. MMS is basically peer to peer, but has substantial network support. MMS cooperates with SMS and provides access to MM stores and servers

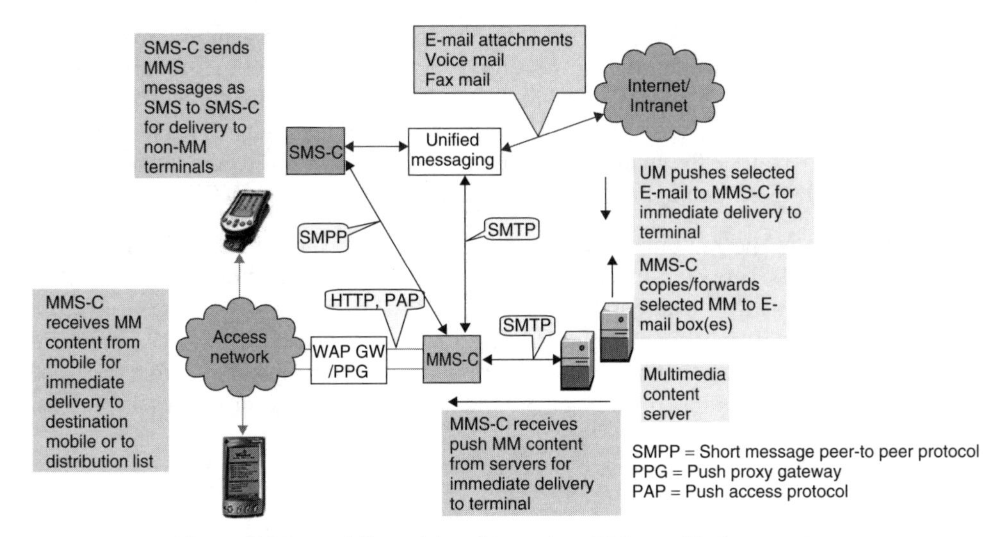

Figure 8.25 Mobile multimedia service. UM = unified messaging

The MMS server is responsible for the storage and handling of incoming and outgoing messages. Several MMS servers can be included within an MMSE, e.g. MMS server, e-mail server, SMS centre (SMSC) and fax servers.

The MMS user databases may consist of lots of different data including user profile database, subscription database and home location register (HLR) information for mobility management.

The MMS user agent is an application layer function that provides users with the ability to view, compose and handle multimedia messages (MMs), e.g. sending, receiving, deleting.

This architecture allows multimedia access to all types of information with a range of servers providing access to new and legacy services. This allows operators to consolidate access to multiple applications from a single architecture, e.g. SMS-C, unified messaging, e-mail, see Figure 8.25.

8.8 SPECIFIC FUNDAMENTAL TECHNICAL PLANS

8.8.1 PROJECT RULE HANDBOOK

For a lot of staff involved in the implementation of a project FP information is highly valuable. In fact it could be a common tool, similar to a baton in a relay race.

Such a handbook might consider:

- Business aspects
- Existing network and associated development tracks
- Target network
- Main traffic cases

- Fundamental technical plans with QoS, security, numbering, functional distribution and configuration, interconnect with roaming, SLA (service level agreement), interconnect charges, management and other FPs.

A few examples of content are given below:

- Numbering – IPv4, IPv6, PDP context with network address and access point name.
- Routing and switching – Chosen routing protocols; point-to-multipoint routing.
- QoS – Diffserv, intserv, choice of radio access bearer, policing with SLA, AAL1, AAL2, AAL5; PDP context with QoS.
- Security – Encryption; tunnelling (IPsec or L2TP); packet filtering against intrusion and denial of service attacks.
- Signalling – SIP; H.323; SDP (session description protocol, see Section 8.8.2 and 14.4.2); bearer control signalling; control layer signalling.
- Synchronization – Synchronization of packets, header marks a beginning.
- Transmission quality – Design of transmission network.
- Charging plan – See Chapter 5.
- Interworking – Address translation. See chapter 15.
- Mobility – Mobile IP; mobility management.

8.8.2 SIGNALLING PLAN

The signalling plan is to a considerable extent inherited from telephony networks. From IETF/Internet the SIP/SDP protocols are a very important contribution. Signalling is mainly used for setting up and tearing down conversational calls. Voice services develop into multimedia services. Chapter 14 contains more about signalling.

The core of the plan lies within the control layer, but some protocols are used in the connectivity layer. The reason is that the control plane controls the edge nodes, only, according to Figure 8.26. Further, the service layer might be used for some advanced control. A related protocol is called CAP, CAMEL Application Part. See Figure 8.6. This gives us strictly speaking three control planes according to Figure 8.27.

ATM Connectivity

There are three main signalling protocols used in the ATM connectivity layer (see Figure 8.28). They are the user network interface (UNI), private network-to-network interface (PNNI) and B-ISDN inter-carrier interface (B-ICI). Some more protocols are mentioned in Chapter 12. The main ones have roughly the following function:

- **UNI –** Allows subscribers access to an ATM network, UNI subscriber recognition through ATM addressing, and recognition of the QoS contract and characteristics of the data to be sent across the connection.
- **PNNI –** Provides the signalling and routing protocols required for managing and controlling the ATM network. PNNI allows for the establishment and support of on-demand, switched connections, and the mechanisms that enable every node in the

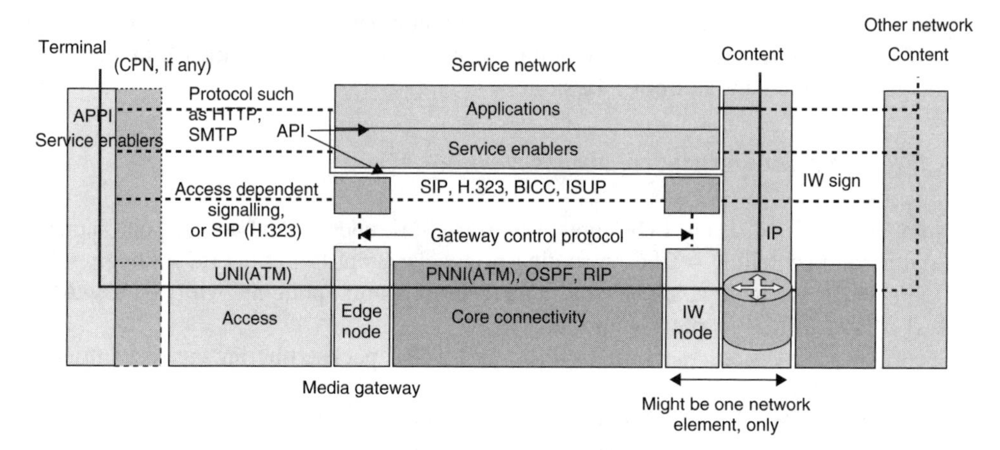

Figure 8.26 Signalling and routing systems, overview

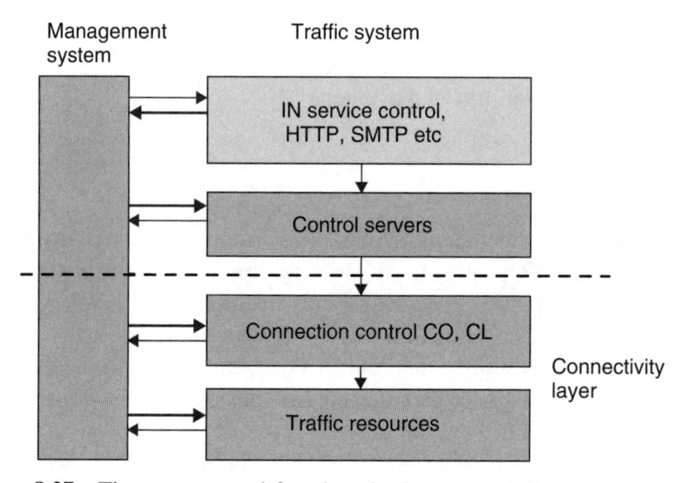

Figure 8.27 There are control functions in the connectivity and control layer

network to maintain up-to-date information about any changes in every other node in the particular network. Note similarities with OSPF.

- **B-ICI –** The signalling and routing protocol for managing on-demand, switched connections between two ATM networks. B-ICI can also be used within an ATM network to improve control of traffic routing.

Other protocols can be found in ATM as well as IP networks:

- **BICC –** The bearer independent call control protocol is used at the control layer. It is based on ISUP and is used for signalling between telephony servers. See more in Chapter 14.
- The **H.248** protocol is used by the telephony server to control the media gateways, as described in Chapter 3.

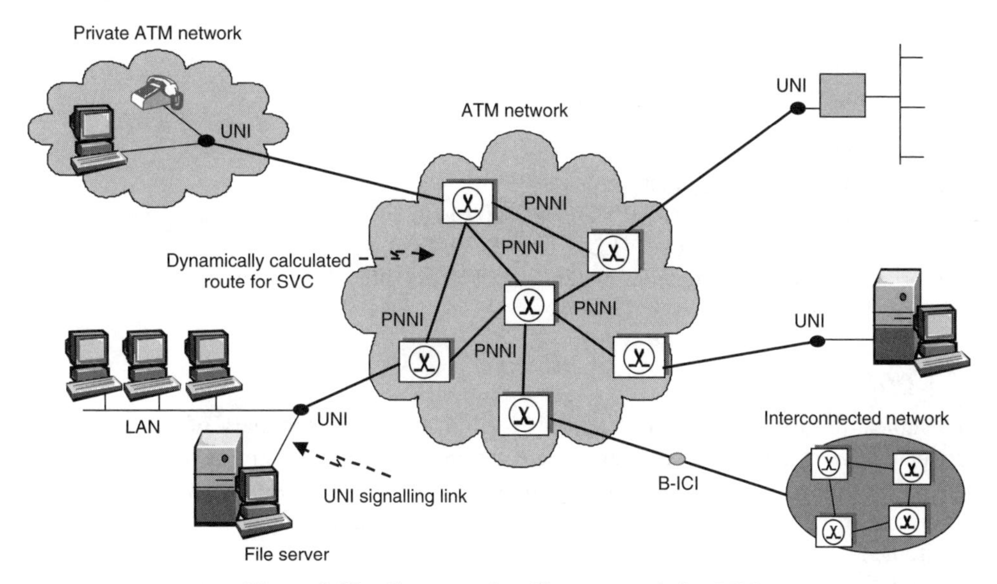

Figure 8.28 Common signalling protocols in ATM

IP-Oriented Control Layer Protocols

Some short facts about H.323:

- A collection of standards: H.225 for session administration; H.245 for multimedia control; T.120 for sharing of data (whiteboard); H.235 for security; H.450 for additional services
- Complex standard: there are many vendor selectable options which makes interoperability hard to reach.

SIP (Session Initiation Protocol)

Some short facts about SIP:

- A simple, flexible, and generic, signalling protocol.
- HTTP-style, text-based.
- SIP handles call signalling, see below.
- Handles all types of conversations.
- Scales better than H.323.

SDP (Session Description Protocol)

First, you need to tell the other party what kind of media you want to receive: audio, video, or both.

The second thing is how you want the media to be coded so that you can understand what is being sent.

The third thing you need to inform the other party about is the address and port you want the media to be delivered to.

SDP communicates several session descriptors:

- Session name and purpose
- Media type (e.g. audio, video etc.)
- Media transport protocol (RTP/UDP/IP, H.320 etc.)
- Media format (e.g. for video, H.261 or MPEG, etc.)
- Information on how to receive media (e.g. address, port, format etc.)
- Period of time the session is active

8.8.3 QOS PLAN

The subject of QoS is treated in Chapter 7. In this chapter a summary of the connection between services and QoS is given. There are in particular two groups of services that depend heavily on the QoS plan: *streaming class* and *conversational class.*

Let us start with a simple comparison between streaming and download techniques and requirements.

Adaptations are necessary when using packet networks for 'ear or eye' sensitive services. When information is streamed you look at it with a short delay. When downloading you don't look or listen during the process.

Figure 8.29 shows streaming and downloading over IP. The differences between the transfer methods are obvious: three dedicated adaptations are or can be used for streaming:

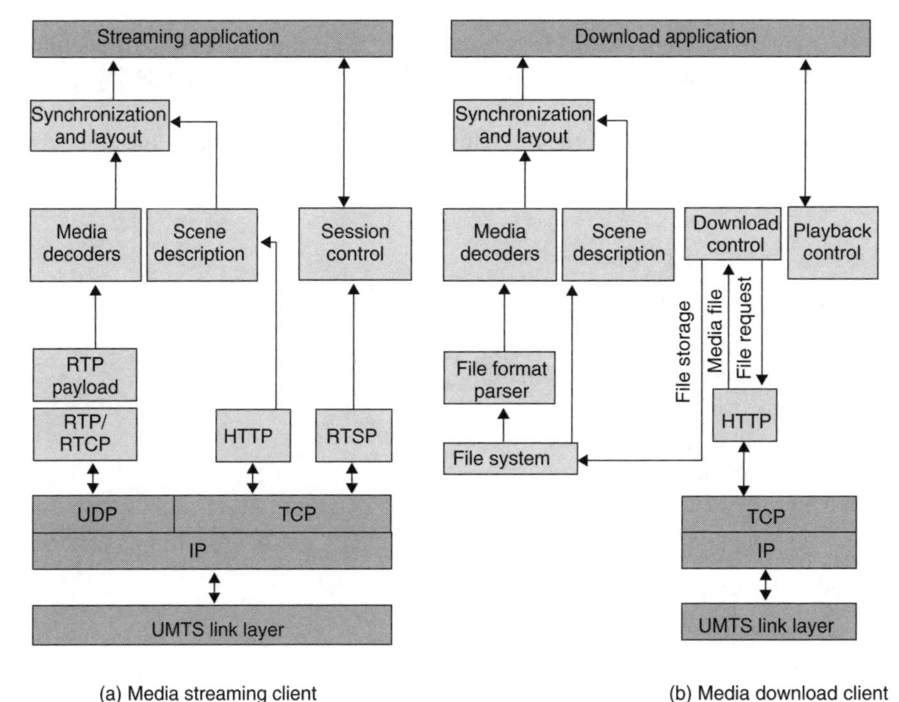

Figure 8.29 A streaming and a background application

real transfer protocol (RTP), real-time control protocol (RTCP) and real-time streaming protocol (RTSP).

Figure 7.14 gives also the relationships between adaptation protocols in IP and the underlying IP stack.

Conversational Class

Example of service implementation: Voice (VoIP, VoATM) – Voice has always been the cornerstone in telecommunications. The class is real time, conversational, which is the most demanding class.

The traditional network solution has been circuit-switched networks, initially fixed and then mobile. Today voice over IP and voice over ATM are installed in parallel with circuit-switched networks. For voice over packet short packets are used. See Figures 3.32 and 8.30.

VoIP introduction – Strictly speaking, voice over IP or IP telephony is 'just another application' using a standard data network, which uses the TCP/IP suite of protocols (Figure 8.31). It is one of many other applications that use the basic application protocols TCP, UDP and IP. However, it is an application that is very demanding in relation to the original design goals of the TCP/IP technology.

Both the telecom side and the datacom side have contributed with standards. Telecom-originated standards are H.323 and associated protocols. Datacom-originated standards are SIP and associated protocols.

The voice over IP application will therefore suffer from the inherent shortcomings of packet networks, notably lack of QoS guarantees, and has to compensate these shortcomings within the service/adaptation layer.

The basic IP telephony network (H.323) – VoIP has so far been most successful in the enterprise sector, as a tool for a common or partly common network for LAN data and voice traffic. IP-PBX and IP-Centrex solutions are becoming increasingly common. See also Figure 8.13.

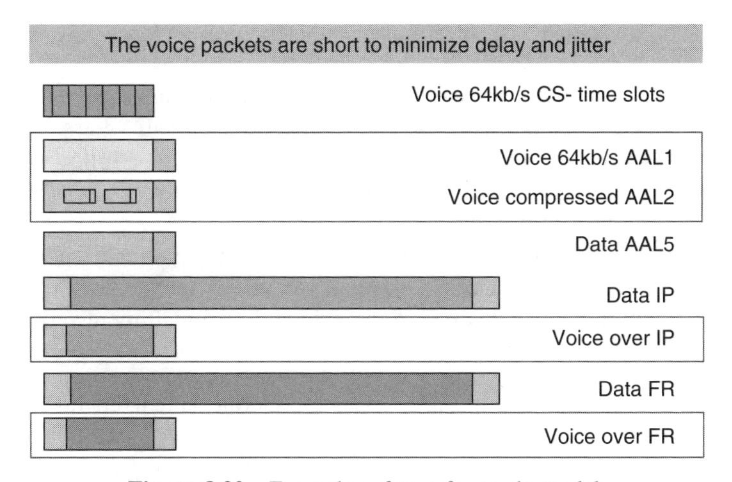

Figure 8.30 Examples of transfer mode models

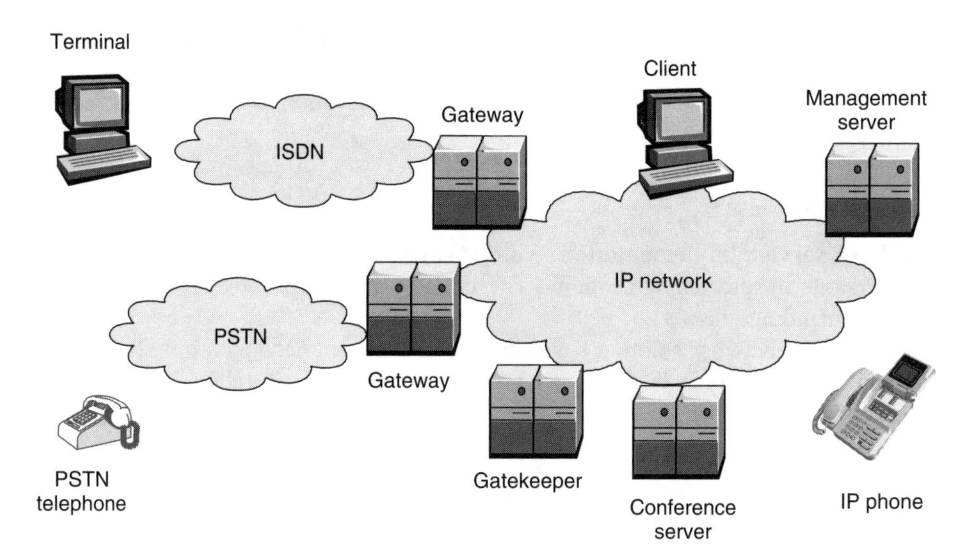

Figure 8.31 An example of a set of VoIP components

The opposite is to use IP in the central network part, taking advantage of the non-distant depending charging in IP networks. This is the voice trunking VoIP solution mentioned earlier in this chapter and represented by Figure 8.31. The IP network serves here as a carrier network for packets containing signalling and traffic data. Dedicated IP phones or IP software clients can be attached directly to the IP network.

Originally, the gateway hosted all the functions required to set up and execute the IP telephony call. However, later a clear division between media translation (coding/decoding) and all the signalling and 'administrative' functions, was seen as beneficial and thus the original VoIP gateways began to be divided into different system components.

The gatekeeper was created to be the host for all the 'administrative' tasks, thus allowing the gateway to concentrate on its main task, coding and decoding between different types of networks. The gatekeeper also offers user authentication and collection of call detail records that are used for billing.

Connected clients communicate with their corresponding gatekeeper, register to it as users and make all address requests to the gatekeeper. Users are also compelled to authenticate to be able to use certain services at the gatekeeper. Users must follow the gatekeeper directions on routing since the gatekeeper controls the available resources (bandwidth and connections).

The gatekeeper is a signalling component only – no traffic streams pass through the gatekeeper. The gatekeeper is thus a control layer server in the H.323 standard. The gatekeeper has signalling channels open to the clients (but signalling also occurs between clients directly). See Figure 8.32.

VoATM – VoATM systems (Figure 8.33) exist for mobile and fixed operators, but the approach is different.

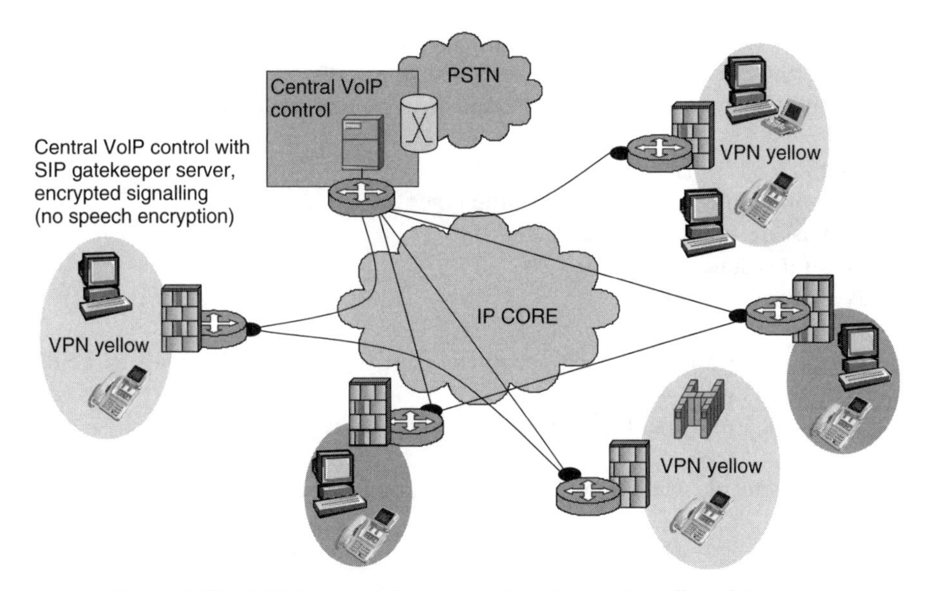

Figure 8.32 I-VPN and VoIP are added to the service offer of the actor

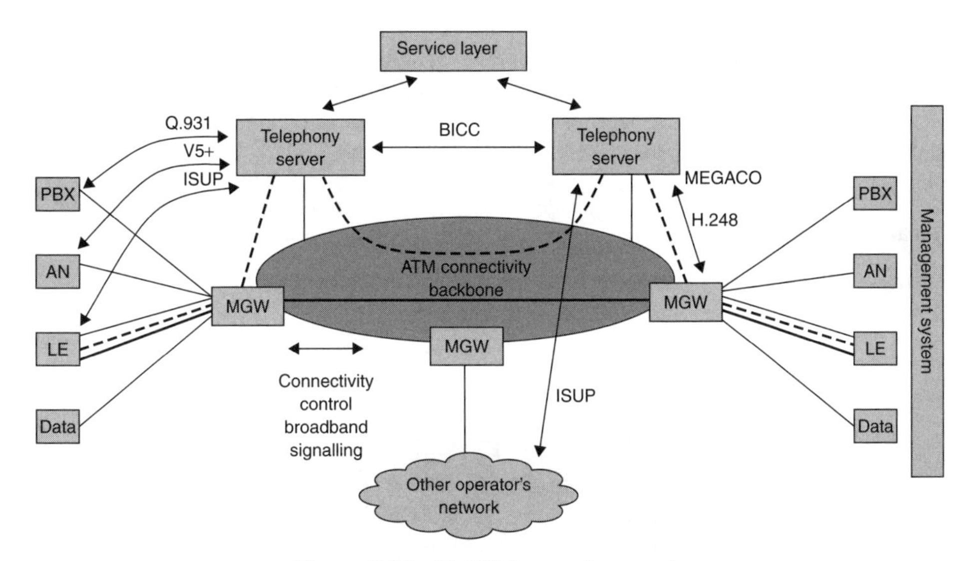

Figure 8.33 VoATM network example

Thus ATM (and IP) is used in new 3 G radio access networks for all types of services: voice, data and video. Thanks to the ongoing 3 G investments in the radio access networks the packetization in mobile networks may start in the access part. The 'mini-cell' standard AAL2 is favoured because the cell fill time for ordinary 53 byte cells becomes long. The reason is the low output bit rate from modern coders (below 10 kbit/s, for example for AMR coding).

In fixed networks the modernization normally starts in the core network. The solutions allow an existing circuit-switched network – usually a network owned by a large telecom operator – to evolve smoothly. Typical characteristics that favour modernization are:

- High traffic load in backbone network
- Targeting one infrastructure for all traffic types
- Needs to assure full service transparency
- Long-term investment
- Highly meshed network

The circuit mode traffic in fixed networks is usually PCM coded with 64 kbit/s and primarily AAL1 is used.

Two new standard signalling protocols are used (see Figure 8.27):

- The BICC protocol, which is based on ISUP and is used for signalling between telephony servers.
- The H.248 protocol, which is used by the telephony server to control the media gateways.

The same protocol can also be used in core networks connecting mobile systems.

The migration from a circuit mode core network to an ATM core network is exemplified in Appendix 3. For large ATM networks a hierarchical network structure can be adopted. The structure is similar to a modern transit layer design in circuit-mode networks, but the capabilities are of another order of magnitude. See Figure 8.34.

Interactive Class

Interactive traffic can be mixed with other classes, such as streaming.

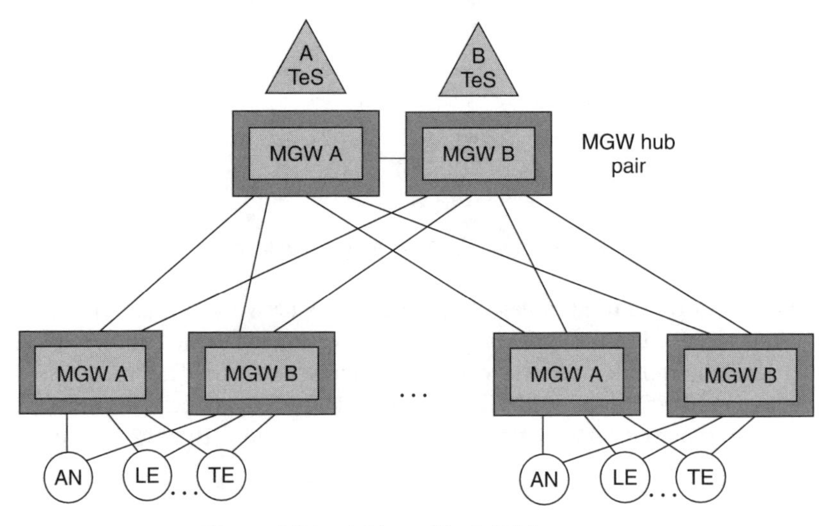

Figure 8.34 A hierarchical ATM structure

According to Chapter 5 the most critical needs are round-trip delay and accuracy. Short round-trip delay is also requested by conversational services. Interactive games are a typical application, and also ordinary WWW services. The round-trip delay issue might be somewhat problematic for TCP traffic in mobile networks. See Chapter 7.

8.9 CONVERGENCE BETWEEN FUNDAMENTAL TECHNICAL PLANS

8.9.1 NUMBERING/ADDRESSING PLAN

Regarding terminology the term 'addressing' is more common within the datacom industry.

This section on numbering/addressing is more basic than task oriented. A goal is to illustrate the consequences of the present convergence between the IP world and the telecom world (telecom in a narrow sense).

The convergence is very clear regarding numbering: two numbering plans are used, the E.164 for PSTN/ISDN/PLMN/ATM and IP numbering for IP networks (IPv4 or IPv6). Numerous cases of interworking will occur. Within IETF and ITU-T work is proceeding on mapping E.164 numbers to URLs that can be used for SIP, HTTP and SMTP.

The numbering area has grown a lot and it offers several challenges for the planner. One is the scarcity of IPv4 addresses, which is counteracted by the use of network address translation (NAT) using addresses which are valid in a local area, without reducing the overall IPv4 address space. NAT translates between authorized addresses on the Internet and private addresses, provides a type of proxy/firewall by hiding internal IP addresses, and enables a company to use more internal IP addresses. Since the addresses are used internally only (inside the proxy/firewall) there is no possibility of conflict with IP addresses used by other companies and organizations. NAT is an official IETF standard, specified in RFC 1631. NAT has no doubt prolonged the life of IPv4.

Tunnelling will be frequently used. Tunnelling can be used for private numbering schemes that are sent through publicly numbered tunnels. Another application type is tunnelling IPv6 numbers through an IPv4 core and a third type is tunnelling for protection causes.

For interconnect reasons there is a need for numbers that include operator identity.

There is always a connection between the signalling plan and the numbering plan, for example between SIP and the numbering systems. SIP uses also e-mail-like addresses and telephony addresses, which correspond to IP numbers:

- John.doe@company.se
- 0123456789@company.se
- john.doe@176.24.106.26

The connection between routing and numbering can be illustrated by the large corporate voice networks of the VPN (virtual private network) type, often covering many countries. The first goal for the call set up is to find the SCP (service control point) housing the operator VPN software package. This requires agreements with transit operators, which must route the calls based on operator identity. The next phase is to find the receiving (B-) subscriber by means of a called party number.

Numbering portability requires a flexible numbering register (FNR) for E.164 numbering. The FNR enables number portability by providing subscribers with the ability to change network operator whilst retaining the original directory number. The FNR application reroutes messages to the correct node (e.g. HLR) or network (e.g. PDC or IMT-2000) where the subscription data is kept.

8.9.2 IP NUMBERING

In a public IP network there are a lot of addresses to cope with, including addresses to application servers. 'Address management' is a huge task. DNS translates only from text address to numerical IP address, which is not sufficient. You also need a directory service such as LDAP, similar to properties/characteristics in computer applications (Outlook is an example). If informed through log-on processes, LDAP can also, together with the SIP protocol, act as a translator between a universal identity (John@company.com) and a geographical terminal location with a fixed IP address to route to or an address in the mobile network or an DHCP originated address (similar to call forwarding or the number portability feature).

The new IPv6 standard (Figure 8.35) uses 128 bits for addressing compared to the IPv4 standard with 32 bits.

APN Address

APN stands for access point name and consists of network ID + operator ID. There are two types of network ID:

- Regional APN network ID: ends with an area identifier.
- National APN network ID: ends with no area identifier.

The operator ID denotes the type of network, e.g. GPRS.

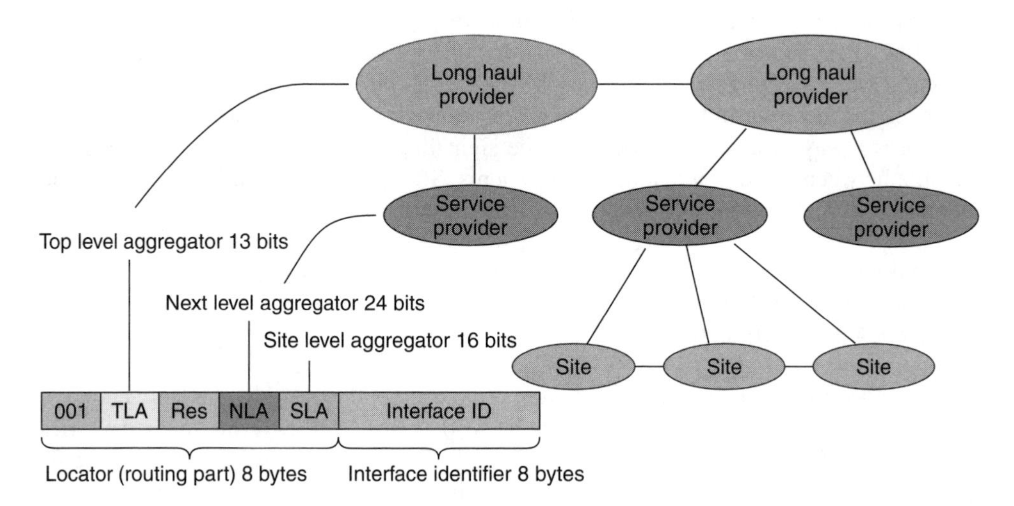

Figure 8.35 IPv6 number structure. Observe the hierarchical organization of the number with some similarity with E.164 numbers

8.9.3 SERVERS WITH NUMBERING AND ADDRESS SUPPORT

Among the servers involved in numbering are DNS, LDAP, RADIUS (optional), NAT and DHCP.

Domain Name Server (DNS)

DNS handles mapping between host names, which we find convenient, and Internet addresses, which computers deal with. It also provides several additional functions such as reverse mapping and secure querying. DNS is used by virtually all internetworking software, including e-mail, remote terminal programs like telnet, and file transfer programs like ftp (file transfer protocol). See Figure 8.36.

Lightweight Directory Access Protocol (LDAP)

LDAP is presented in more detail in Chapter 6. E-mail programs receive LDAP support to search e-mail addresses.

- We use directories of one sort or another every time we use the Internet or our own intranets.
- LDAP is an extensible, vendor-independent, network protocol standard – it supports hardware and software.
- An LDAP-based directory supports any type of data.

Remote Authentication Dial-In User Service (RADIUS)

RADIUS is an authentication and accounting system used by many Internet service providers. When you dial in to the ISP you must enter your username and password. This information is passed to a RADIUS server, which checks that the information is correct, and then authorizes access to the ISP system.

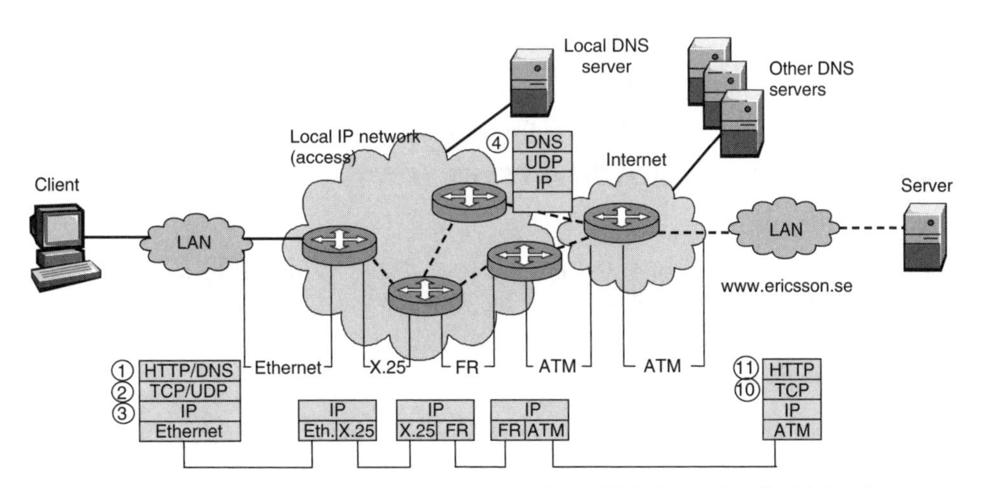

Figure 8.36 A communication with a DNS server. The DNS is hierarchical with local servers at the bottom and root servers (e.g. com) at the top

The use of RADIUS for address allocation is optional. An example is found in Figure 8.41 at the end of this chapter.

A RADIUS server could be configured to always *allocate the same IP address to a specific subscriber*. Static IP address allocation enables the operator to offer subscribers the ability to use IP addresses that have been provided by the subscriber, which can assist with their security issues. Static IP address allocation means that subscribers can provide their own IP addresses.

Dynamic Host Configuration Protocol (DHCP)

The main function here is to administer the allocation of addresses. This may be based on authentication of the entity requesting the address. This function may also apply some policy to determine the type of address allocated (private or globally routable) and the period for which the address is allocated. There are several protocol standards for IP address allocation, e.g. DHCP. DHCP enables individual computers on an IP network to extract their configurations from a server (the 'DHCP server') or servers. The overall purpose of this is to reduce the work necessary to administer a large IP network. The most significant piece of information distributed in this manner is the IP address.

Mobile IP

Mobile IP refers to the access method that is based on the Internet standard specified in RFC 2002. Here the user can use either a static or dynamic IP address belonging to its home network. The mobility options permit users to maintain their IP address no matter whether moving throughout their own or other networks.

A goal is to have the same IP address in the mobile network and WLAN network ('all-IP approach').

Simple IP

Simple IP refers to the access method in which the user is assigned a dynamic IP address from a service access provider. The user can keep the IP address within a certain network dependent area. This method permits a restricted mobility.

8.9.4 CONVERGENCE LEADS TO INTERWORKING

Numbering Interworking

There will be interworking of E.164 to/from IPv4, E.164 to/from IPv6 and finally IPv6 to/from IPv4. Figures 8.37–8.39 illustrate the situation.

8.9.5 UMTS NUMBERING

When the WCDMA/UMTS terminal is used for access to IP networks, it must be integrated in the IP numbering. Addresses can be dynamically assigned (fetched from an external server or a pool of own addresses) or statically assigned (fetched from the HLR). Address allocation is the responsibility of GGSN (gateway GPRS support node). It is

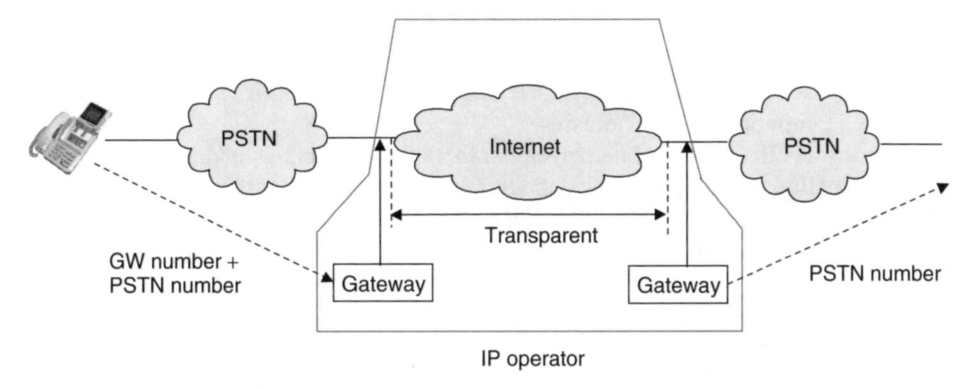

Figure 8.37 The PSTN number is sent through the IP network

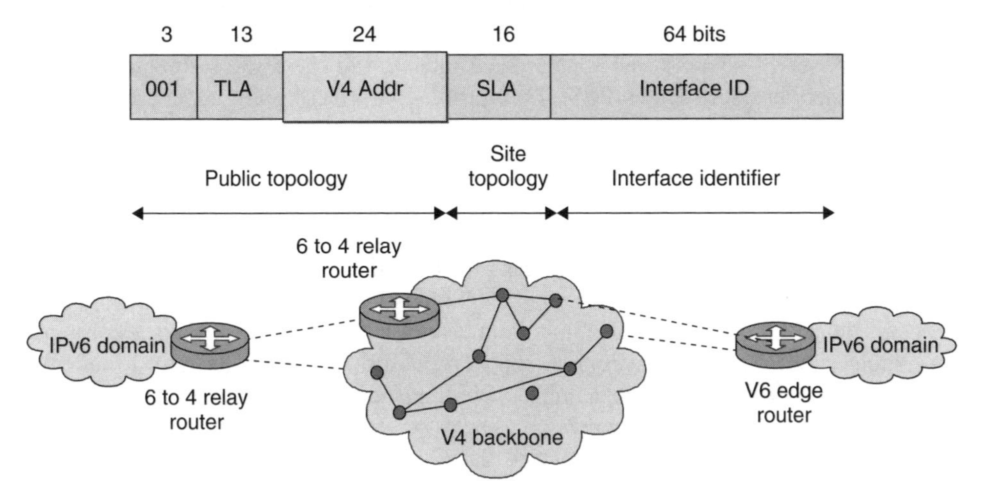

Figure 8.38 Tunnelling IPv4 through IPv6

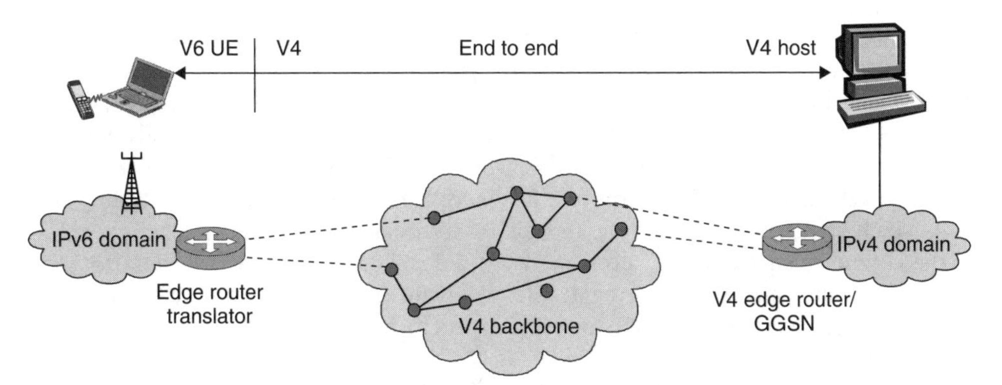

Figure 8.39 IPv4 and IPv6 domain interconnect requires translation. Tunnelling cannot solve the problem in the case of end-to-end incompatibility

done at PDP context activation. The GGSN can either allocate the IP addresses itself, or interact with a RADIUS server for end-user authentication and retrieval of IP addresses. For this purpose, the GGSN includes a RADIUS client. Dynamic address allocation is done from a common pool of addresses.

Seen from the IP domain (Internet or intranet) the terminal is addressed with an IP address. In reality, on the air interface the terminal is addressed with a temporary address based on the international mobile subscriber identity (IMSI).

An address translation function in GGSN handles the IP address to IMSI address translation for the IP packets received from the external network addressed to a terminal. Routing within WCDMA/UMTS is based on IP numbers. Send packets already have an IP address for routing.

The SGSN (serving GPRS support node) server functionality for session management is responsible for establishment, maintenance and release of end-user PDP contexts. This includes interworking with the GGSN for address allocation if dynamic IP addresses are used.

The SGSN server supports the standardized interface to the HLR for management of end-user subscriber data such as IMSI, QoS profile and access point names.

8.10 TRAFFIC CASES

The traffic cases presented here are simplified. The text below describes an early UMTS traffic case version. It is not the traffic case in itself that is important, instead the case exemplifies the management of the fundamental plans in a call, such as numbering, security, QoS, signalling and routing. An important task is to compose the packets accordingly, also adding appropriate low layer protocols. The description uses the term termination establishment. A termination is a number of user plane protocols necessary to establish in a node, typically in a media gateway.

The packet design is based on service description parameters. Service description parameters are for example bearer type (e.g. ATM/AAL2), codec (e.g. AMR) and B-number in telephony-oriented networks. If the coding is changed along the path a transcoder is linked into the path. In packet-oriented networks the necessary packet call set-up information is called packet data protocol (PDP) context. It includes for example numbering and QoS information. IP addresses for packet traffic to/from attached devices are often provided by a RADIUS server connected to the GGSN node in UMTS. A RADIUS server also has AAA functions. Other security functions are creation of tunnels, which is one of the main protection methods in WCDMA/UMTS. Tunnelling protocols are added to the packet stack. See the central part in Figure 8.15 starting with IP and ending with PPP, which are extra protocols. The tunnels form virtual private networks.

The resources in the radio access network are called radio access bearers (RABs). An early version has the following RABs: for real-time services an ATM AAL2 path is available (conversational RAB for AMR speech) and also $n \times 64$ kbit/s channels (n starting with 1).

As UMTS progresses with issues like QoS, it will be possible to also provide real-time packet-based services based on IP. A common IP path for all services is seen as the ultimate solution.

The 1×64 kbit/s bearer service is called conversational RAB for 64 kbit/s multimedia.

A combination of several of these 64 kbit/s bearers is possible by using the multi-call service. The main use is for circuit-based multimedia (including streaming services) according to H.324 M. The service can be connected to IP networks (using H.323), ISDN networks (using H.320 or H.324I), and to other networks with H.324 M.

For *non-real time services* a much cheaper best effort packet mode is acceptable and therefore preferable. To start with the packet services will be 'best effort' services. Both point-to-point and point-to-multipoint services will be possible.

A packet mode path will offer a shared bearer up to 2 Mbit/s. Initially the maximum bit rate will be 384 kbit/s (interactive RAB up to 384 kbit/s).

8.10.1 ORIGINATING WCDMA/UMTS CALL TO PSTN

Figure 8.40 shows an example of a real-time service call set-up in UMTS

1. The user equipment requests a call set-up and sends the call requirements (bearer type, codec, B-number).
2. Based on the B-number the MSC/VLR server selects MGW.
3. The MSC server sends CALL PROCEEDING including the selected codec back to RNC.
4. The MSC server creates a termination T1 in MGW (properties: AAL2, AMR, Context1). Termination here means that the MGW creates links between this call and its own seized resources. All connectivity layer resources assigned to this call are linked together. If another MGW is interconnected the linked resources there will be treated independently.
5. The MSC server orders the RAB assignment.
6. RNC establishes the bearer to the MGW.
7. Notification from RNC to MSC that RAB is established (assignment complete).

Add new termination T2 (properties: PCM, AAL2, Context1). MGW links in a transcoder, because of different coding (AMR/PCM) in T1 and T2. MSC server sends BICC IAM (initial address message) with B-number and MGW1 to TSC server.

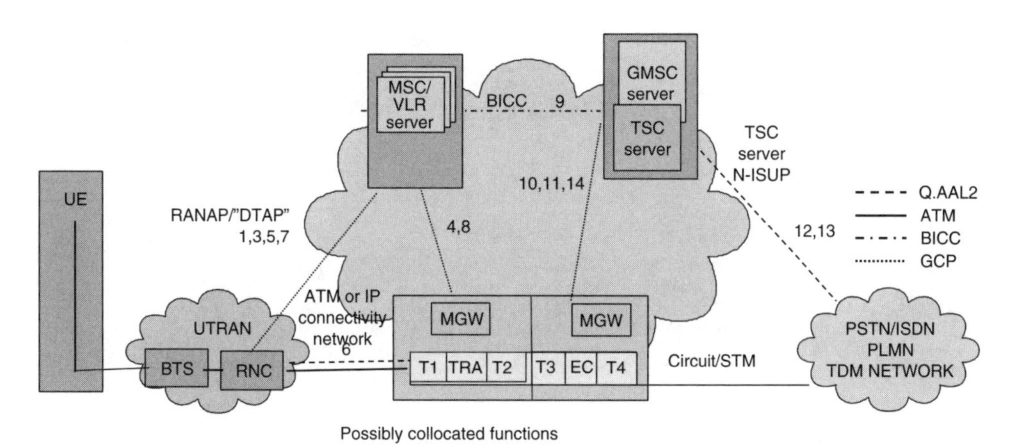

Figure 8.40 Example of a real-time service call set-up in UMTS

8. The TSC server creates a new termination T3 with properties PCM, AAL2, and new context and connects to termination T2.
9. The TSC server creates termination T4.
10. The IAM is sent to the external network.
11. The ACM (address complete message) is received from the external network.
12. The TSC server links in the echo canceller (EC).

8.10.2 PACKET WCDMA/UMTS CALLS

In this example, the control signalling between the SGSN and RNC is routed via the media gateway.

Assuming that the WCDMA/UMTS user equipment is attached, any packet call set-up is treated as PDP context. In the example the GGSN will not be physically split so the server and the MGW part of it will co-reside.

PDP Context Activation

See Figure 4.41. PDP context: type of network, network address, access point name, QoS, radio priority, etc.

1. An Activate PDP context request is sent from the UE to the SGSN server.
2. A Create PDP context request is sent from the SGSN server to the GGSN via the connectivity layer (routers).
3. The GGSN server requests an IP address from the RADIUS server.
4. The GGSN server requests (internal communication) the GGSN MGW to create a GTP tunnel end point.
5. The GGSN server sends a Create PDP context response to the SGSN server.
6. The SGSN server requests WRAN to set-up a RAB.
7. Acknowledgement on RAB request from WRAN.

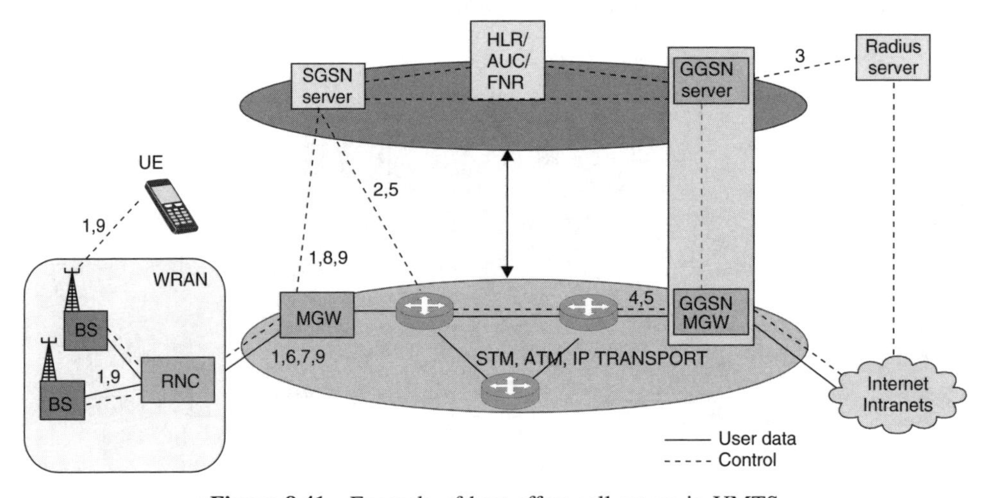

Figure 8.41 Example of best-effort call set up in UMTS

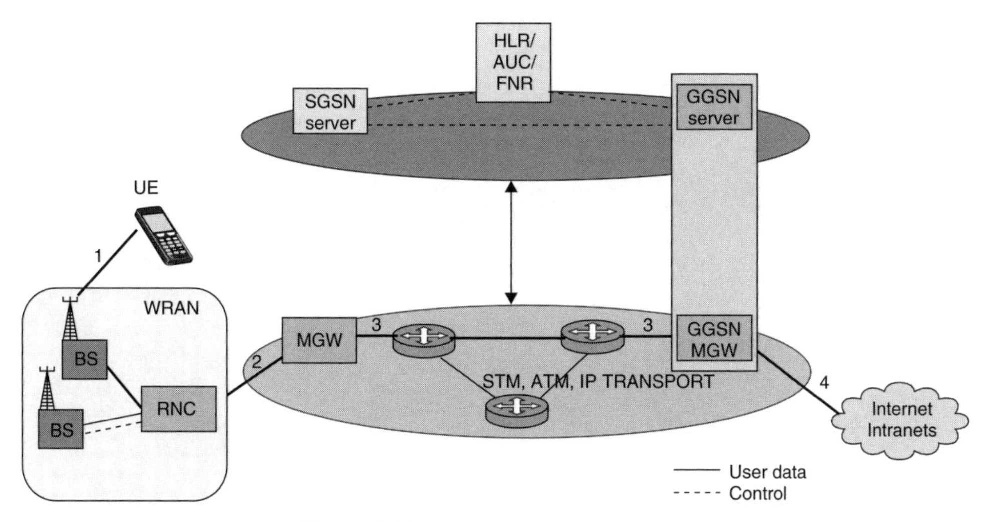

Figure 8.42 Packet data transfer phase

8. The SGSN server requests the MGW to create a GTP tunnel transfer point.
9. Activate PDP context response is sent back to the UE from the SGSN.

Packet Data Transfer, Uplink

See Figure 8.42.

1. A packet data unit (PDU) is sent from the UE.
2. The PDU is sent over the GTP-U tunnel between the RNC and the MGW.
3. The PDU is sent over the GTP-U tunnel between the MGW and the GGSN MGW.
   The PDU is sent over to the Internet/Intranet.

9

Service Network

'Overcoming the challenges of handling a multitude of services and business-related partners.'

9.1 OBJECTIVES

The overall objective is to understand the increasing role, the architecture, way of working and the potential of the service network and all its 'domains'. Such techno-economical understanding is crucial for successful service development and business. The problem in real life is indeed to limit the area.

What is described in this chapter is more a kind of ideal future situation than a description of what has yet been implemented in real networks.

9.2 CONNECTION TO PRECEDING CHAPTERS

An introduction to the service network area is found in Chapter 3, see for example Section 3.11. The area is also treated in Chapter 8, section 8.7. This chapter develops the subject further.

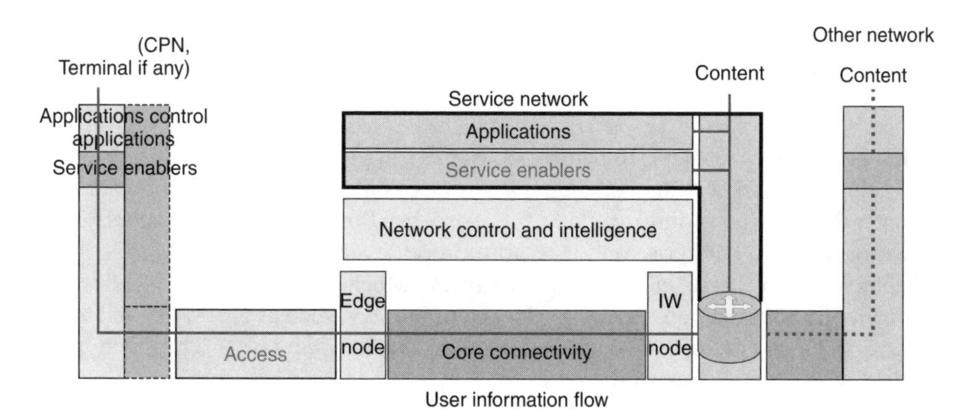

Figure 9.1 Position of the service network in the reference model

Understanding Changing Telecommunications – Building a Successful Telecom Business. Edited by A. Olsson
 ISBN: 0-470-86851-1

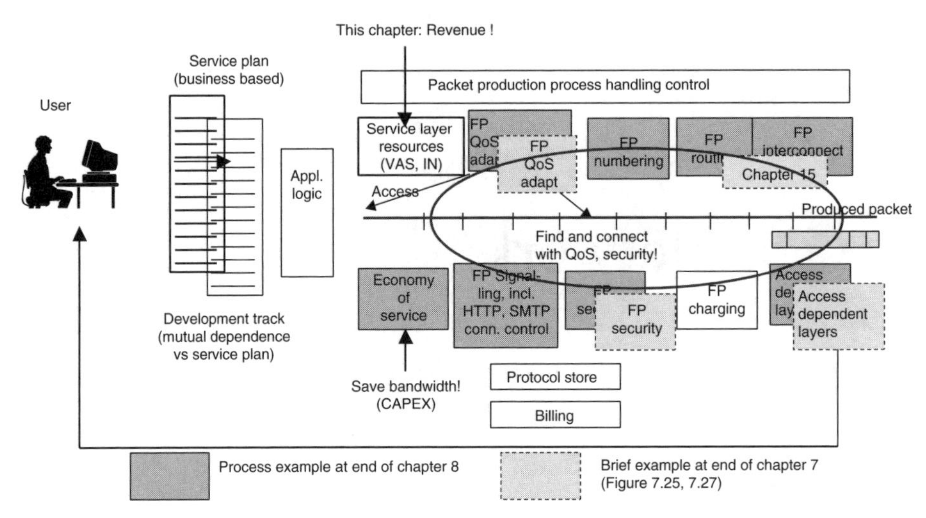

Figure 9.2 A process view that was introduced in Chapter 8 – FPs involved in traffic cases are indicated

Regarding the *technical* side, the chapter pursues the successive communication system build-up in Chapters 5, 6, 7 and 8. The parent service layer has been treated in previous chapters. Let us use the process view of Figure 9.2 as a bridge between technique in preceding chapters and this one.

Pedagogically, a process view like this one can hopefully give the reader a feeling of animation. Traffic cases are often described as sequential processes, and can be used to further support the process thinking. The traffic case in Chapter 8, section 8.10 contains many fundamental plans, indicated in Figure 9.2 with a darker shade. This figure also indicates the early position of the service layer in the flow, by means of a thick arrow.

Regarding the *business* side the following references in other chapters are relevant examples.

- Chapter 1 – Easy-to-use, affordable, making life easier (Mäkitalo), society-oriented services (Malm, Thorngren).
- Chapter 2 – No specific answer on consumers is possible. The human end user as a traffic generator is relevant. Regarding consumers in general it is recommended to start by testing the market using existing systems (see Malm, Thorngren in Chapter 1).
- Chapter 2 – xSP according to Table 2.1. Also IS/IT outsourcing, support to internal processes, mobile enterprise and enterprise processes (SCM, ERP, CRM).
- Chapter 4 – Figures 4.2 and 4.3 are pertinent. Much of Section 4.4 is relevant, in particular the introduction. Figure 4.15 and terminology, such as HE-VASP and MVNO. Figures 4.25 and 4.26 and consequences for the service layer. The main part of Section 4.6 is relevant as well.

9.3 WHAT IS A SERVICE NETWORK?

A service network is a logical IP-based network for end-user services that goes 'beyond voice'. Service networks serve public subscribers as well as enterprises and society. The

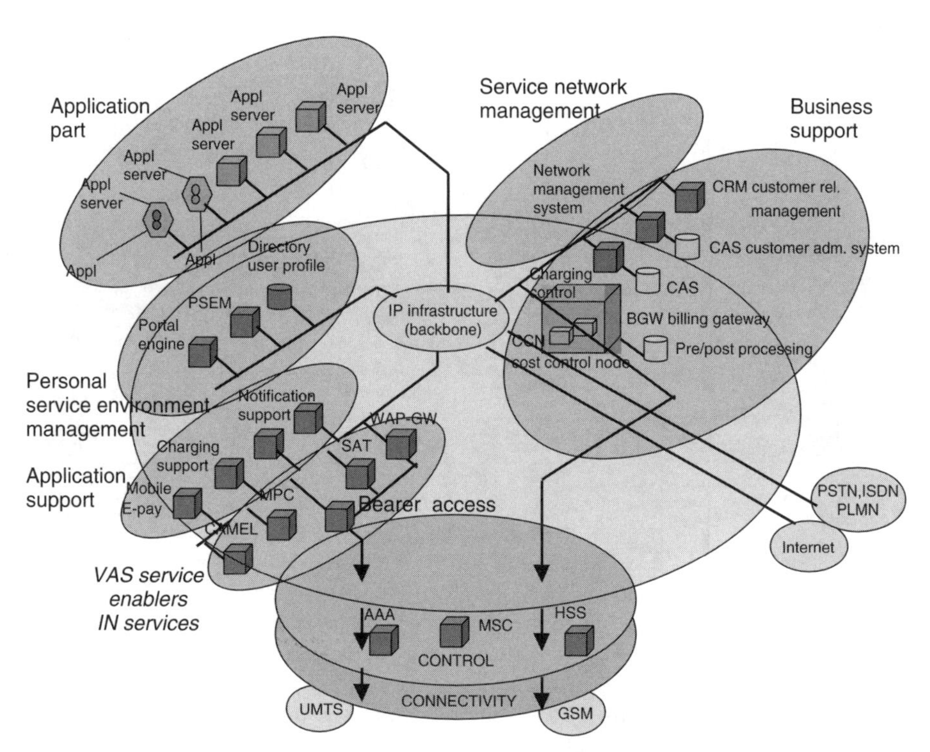

Figure 9.3 An example of a service network structure. Applications, management and service enablers are main functions of the server-rich service network. Everything is connected by means of an IP infrastructure backbone network

networks provide and support end-user services, support partner cooperation, B2B, B2C, provide service management, end-user management and core network access.

The owner of a public service network may be a mobile operator, an ISP, a portal or a virtual network operator (VNO). In the enterprise segment, the mobile VNO or systems integrator may take on important roles.

The word 'enabler' should be borne in mind, with location as a good example. Enablers are the main tools for the service network, creating value beyond voice. Semantically, also the often-used term 'middleware' has a flavour of enabler, by offering API, service handling and end-to-end protocols (e.g. XML) to the applications.

The service layer can also be regarded as a promoter of business contacts via telecom on behalf of direct contacts or contacts via other media. It participates in the contacts as an enhancing middleman.

Many presentations do not show the traffic flow between the server in the service network and the client. Such flow has been included in the reference model, and also a hint regarding the interaction between service network and terminals. In fact, there are strong dependencies between the terminals and the service network, creating needs of combined application development.

Figure 9.3 gives an example of the server-rich structure of the service network and also the traffic flow referred to above.

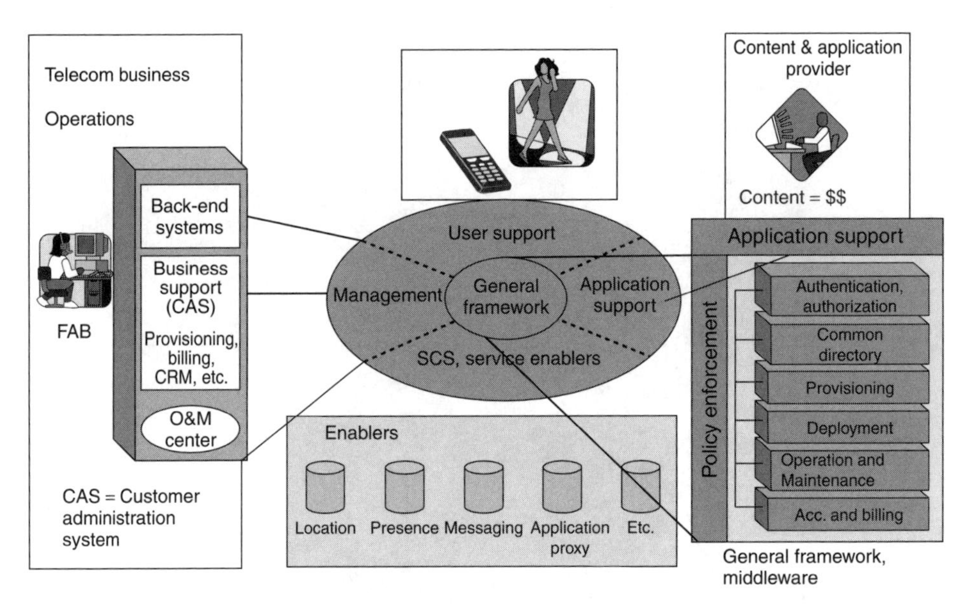

Figure 9.4 The spider role of the service network based on an OMA structure

Whereas Chapter 8 primarily is a technique integrator in this book, Chapter 9 is an *overall* integrator. Chapters 2 and 4 become especially important now because of the spider role of the service layer and its main component, the service network.

As mentioned before, the service network is a product of the converging telecom, datacom and media industries. The first stage was the functions of the ISP, as the telecom intermediary between the user and the content source. The business models are not yet stable.

This meeting point for users and the different actors in the value chain is a melting pot of technologies, from general computing trends to telecom-oriented technique. Let us look at all the standard bodies that are involved in service layer matters: here we find 3GPP, W3C, IETF, Parlay, ETSI and OMA, to take a number of important ones. One of the hottest technical areas is *software as a service*, such as web services.

This is a promising area, indeed. However, it needs further standardization and implies some security and QoS risks when software from different sources need to cooperate.

Company business technique is found in the idea of the Open Market Alliance (OMA), which offers enabling technology for delivering relevant, end-to-end high-performance services (see Figure 9.4). OMA was formed on 12 June 2002, but it is already a very important player in the area.

9.4 SERVICE NETWORK DOMAIN AND PRINCIPLES

The service network can be structured into a number of 'domains' that cooperate and fill the original space in the interfaces between the converging parties. A domain is justified provided it gives smooth support to the business. Let us take some examples of needs:

- The user needs support to securely sign on to the network, to upgrade the terminal with new applications, and to communicate with sufficient QoS. A personal portal is also appreciated.
- The application provider and value-added service provider do not know the network, and need assistance to utilize all the possible enablers when creating new applications (there are other media industry oriented roles as well).
- The actor running the service network needs support from an IP network to get all necessary servers to cooperate properly.

All domains in this chapter appear in Figures 9.3 or 9.4.

- The service network owner domain has two shapes: the technical solution domain and the management function domain. The technical solution ideally provides a common service framework machinery for services over all access systems. The framework supports virtually any type of service, that is telecom, data, Internet, multimedia. Sometimes the framework is referred to as a *middleware framework* or the *service network platform.*
- The user domain.
- The domain of cooperating roles, such as content provider, value-added service provider, application provider etc.
- The domain of the core (and access) network.
- The domain of the IP network that connects the elements of the service network.

Basic ideas and tasks of the service network are:

- To be a kind of spider that connects *application providers*, representing society needs and integration of telecommunications, with the *users* with their personal needs, with the *service providers* with their needs to be able to efficiently administer the exploding service area, and finally with the *network* for transport to the user.
- In the spider role, to ensure that everything works together based on interoperability between communication segments and providers.
- Employment of common utilization of resources for cost-efficiency reasons.
- Implementation of the OMA concept and/or Parlay3GPP Open System Architecture and Virtual Home Environment (OSA/VHE) towards application providers.
- Provision of access independence and personalization of the subscribed-to services to the user, catering for end-user integrity, privacy and security.

Scalability and robustness, flexibility, and a decreased cost of ownership are other improvement goals when comparing service solutions in vertical networks versus horizontal networks.

The service network is also affected by what is called the processing dimensioning in this book. OMA has established a workgroup for mobile web services. The group creates a specification suite to aid developers in applying web service technology and standard development tools to discover, access and leverage service enablers within the OMA framework. This results in much richer application services. The specification suite provides developers with bearer-independent service components such as location, content downloading, device management and presence.

Other new workgroups include requirements, architecture and interoperability. The requirements group identifies and specifies technical requirements for the development of mobile services. The architecture group defines the overall OMA system architecture including evolution, integration and maintenance. The interoperability group ensures end-to-end interoperability. See http://www.openmobilealliance.org/docs.

The sections on XML for business documents and web services and the section on MExE (mobile execution environment = WAP and Java) in this chapter are also part of this dynamic subject area. Regarding XML see also Chapter 3, Section 3.11.2

A number of support services exist in the service network to aid the proper operation of the IP network that connects all the servers. Service networks are initially deployed together with mobile systems.

9.5 TERMINOLOGY

For the purposes of this book, the definitions and abbreviations given below apply:

- **Home environment:** actor responsible for overall provision of services to users.
- **HE-VASP:** home environment value-added service provider. This is a VASP that has an agreement with the home environment to provide services.
- **Personal service environment:** contains personalized information defining how subscribed services are provided and presented to the user.
- **Service capabilities:** bearers defined by parameters, and/or mechanisms needed to realize services. These are within networks and under network control.
- **Service capability feature:** functionality offered by service capabilities that are accessible via the standardized OSA interface.
- **Service capability server:** functional entity providing OSA interfaces to an application.
- **Value-added service provider:** provides services other than basic telecommunications service for which additional charges may be incurred.
- **Virtual home environment (VHE):** concept for personal service environment portability across network boundaries and between terminals.

9.6 THE ARCHITECTURE OF SERVICE NETWORKS

9.6.1 GENERAL

What is described here is the 3GPP/Parlay architecture. The OMA concept is similar as seen in Figure 9.4.

The heart of the technical solution is a *framework* or *platform* with many supporting functions for service handling, security, QoS, SLA and other tasks. The basic architecture is called VHE/OSA. VHE/OSA supports three of the domains mentioned above: the network owner that could be called the home environment operator; cooperating roles such as content and application provider, VASP and HE-VASP (and other partnerships); and finally the core network access. VASP and HE-VASP are defined above.

9.6.2 IMPLEMENTATION OF VHE/OSA

Open service architecture (OSA) is a concept defined by 3GPP and Parlay. It specifies services available to applications on application servers over standardized programmable

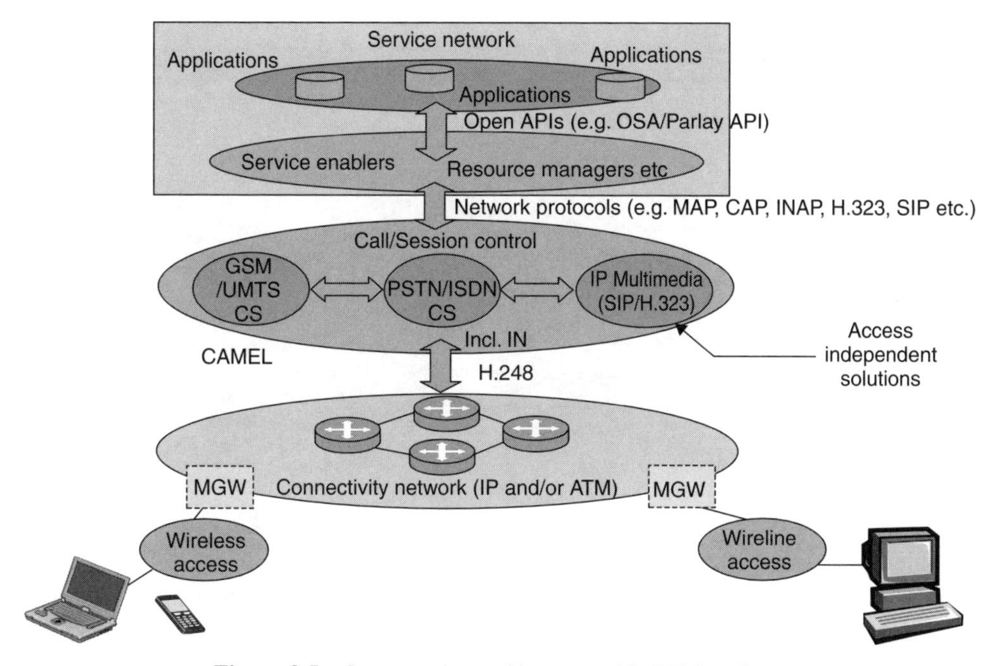

Figure 9.5 Open service architecture with API interfaces

interfaces (application programming interface, API) See Figure 9.5 and Chapter 3. The services provided by the service network respond to the needs of the actor:

- New marketing and sales channels.
- Efficient tools and development environments.

Another related important concept is virtual home environment (VHE), offering service roaming. The VHE is defined as a concept for personalized service portability across network boundaries and between terminals. The idea with VHE is that subscribers are consistently presented with the same personalized features, user interface and services, regardless of their access network or terminal (within the capabilities of the terminal). Bundling sets of services forms personalized service portfolios for subscribers belonging to a home environment. VHE is further treated in Chapter 15.

The standardized OSA interface needs to be secure, independent of vendor-specific solutions and independent of programming languages, operating systems etc. used in the service capabilities. Furthermore, the OSA interface is independent of the location within the home environment where service capabilities are implemented and independent of supported server capabilities in the network.

9.6.3 THE SERVICE CAPABILITY SERVICES (SCS)

A kind of bridge between applications and network resources is required. Services and applications negotiate the resources with the control layer over standardized APIs. Resources in

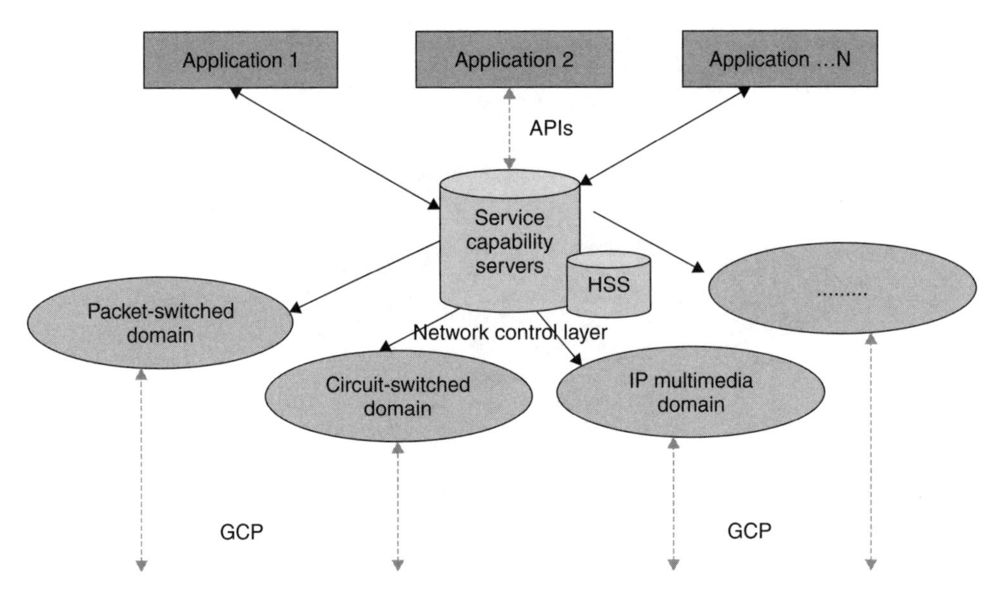

Figure 9.6 Bridge between applications and bearer resources

the packet, circuit or multimedia domain can (ideally) be negotiated. The bridge is illustrated in Figure 9.6. API is explained in Chapter 3. See more below.

The bridge consists of software enablers connected to a network. The original name was service capability services (SCS), which reside in service capability servers. Today it is perhaps common to speak about the more open-ended and wider expression service enablers rather than SCS. SCS constitute the basic set of network-related APIs offered to applications.

Network functionality as offered to applications is thus defined as a set of service capability features (SCFs) in the OSA interface. These SCFs reside in service capability servers. The servers provide access to the network capabilities on which the application developers can rely when designing new applications (or enhancements/variants of existing ones). The different features of the different servers can be combined as appropriate.

SCS hide network-specific protocols and offer connectivity to circuit-switched networks, packet-switched networks and multimedia network resources. Typically TCP/IP connectivity is offered. SCS are initially gateways to mobile bearers, only.

An example of a service capability feature is 'user location'. 'User location' provides an end-user service with information concerning the location, e.g. in terms of universal latitude and longitude coordinates.

9.6.4 APPLICATION SUPPORT FUNCTIONS – FRAMEWORK FUNCTIONS – SECURITY

Application Support Services (ASuS)

Application support services are similar to SCS with the exception that they interface with service network resources or with external resources/systems other than core networks, e.g. a billing system or billing gateway.

Application support servers provide support services for applications (end-user services) in the service network. They provide functionality that ensures the service capability features offered by the service capability servers are for example accessible, secure and manageable.

Application support services consist of (but are not limited to) the following functionality:

- Charging support (= OMA)
- Security: authentication, authorization (= OMA)
- Notification
- Directory access support (= OMA)
- Registration
- Discovery

Charging support services receive charging information from applications and create standard charging streams that could be forwarded to billing systems.

The overall security is reduced when opening up between applications and core networks. Therefore some kind of firewall is necessary to protect the core networks from attacks from unauthorized users of the service network.

All applications must therefore first become authenticated. Only authorized applications are permitted at the border. Before an application can use the network functionality made available through the service capability servers, authentication between the application and the framework is needed. After authentication, the discovery service feature enables the application to find out which network service capability features are provided by the service capability servers.

The registration support service enables a service capability feature (e.g. user location) to register its existence.

The discovery support service enables applications (end-user services) to identify all the service capability features that have been registered in a service capability server.

9.6.5 APPLICATION SERVICES

Application services (end-user services), for example VPN, conferencing, and location-based applications, are implemented in one or more application servers. The end-user service is provided by executing the service logic on the application server. Service logic and data for end-user services reside on the application server.

Application control servers provide the control for the application and application content servers provide the content for the application (e.g. text, video clips etc.)

9.6.6 SUMMARY

The overall architecture so far can be presented as a stack. See Figure 9.7.

9.6.7 THE WAP/MExE/SAT LAYER

The physical enablers are sometimes considered not to be SCS. Therefore they are treated as a separate layer here. The name service enabler is still correct. Two important enablers

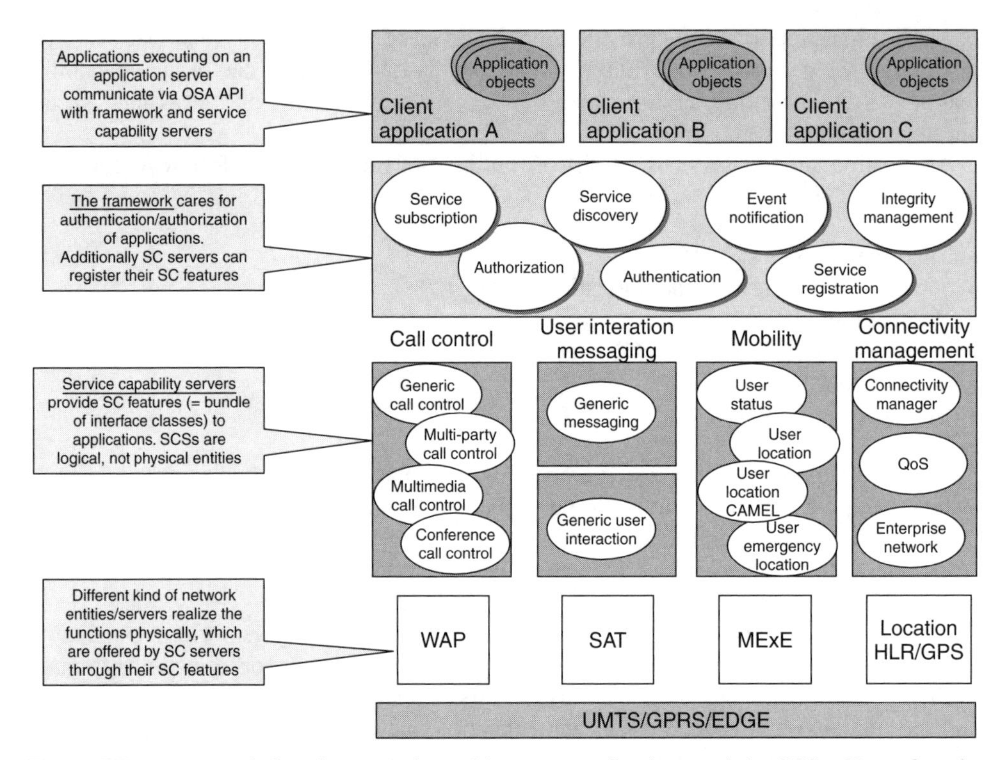

Figure 9.7 Framework functions are situated between applications and the SCSs. Very often the 'WAP layer' is included in SCS and service enablers

will be treated here: SAT and MExE. WAP is briefly treated in Chapter 5. OMA uses the name service enabler as well.

SIM Application Toolkit (SAT)

SAT offers the flexibility to update the SIM to alter the services and download new services over the air (OTA). For example, network operators can remotely provision the user's wireless terminal by sending codes embedded in short messages from the server. Within the SAT specification, SMS is a key mechanism for personalizing the SIM in each user's phone.

A SAT browser resident on the SIM card can execute different Internet applications. SAT-enabled terminals are able to access the same services as WAP terminals.

SAT comprises two components:

- SAT in the network.
- SAT browser in the terminal.

SAT connects to Internet applications via HTTP. SAT makes it possible to download applications to the terminal and perform high security operations.

Web applications are downloaded to the SIM card using standard protocols. Once the SIM card receives a web application from a content provider, the web application is executed, resulting in a number of SIM toolkit commands sent to the mobile phone, allowing interaction with the user.

Mobile Execution Environment (MExE)

MExE is a standardized execution environment in a mobile device. The aim of MExE is to provide a comprehensive and standardized environment on mobile terminals, for execution of operator or service provider specific applications. MExE is designed as a full application execution environment and builds a Java Virtual Machine into the client mobile terminal.

MExE enables a standardized open architecture for the MS (mobile station) including content and application download mechanisms and an environment to execute those applications. ETSI has determined that two technologies should be accommodated within the MExE framework:

- Wireless application protocol (WAP), which brings Internet content and advanced services to digital mobile phones and other mobile terminals.
- Java, which brings platform independence and built-in network awareness.

On the network side, MExE comprises the WAP gateway.

9.7 THE NEEDS OF THE USER DOMAIN

In this book the user is queen/king, so these needs are crucial. Important needs are:

- Consumers: ease of use, single sign-on functionality, personalization, availability, security and service range.
- Business: Cost-efficient internal processes, mobility, personalization, IS and IT outsourcing.

9.7.1 EASY-TO-USE-SERVICES

The services must be easy to use, for example according to the 0,1,2,3 solution with single sign on to content in a personalized portal:

0. No manuals needed for configuration. Device Management and configuration are performed by means of an OTA server.
1. One-button access to mobile Internet (AAA, single sign-on, MSISDN forwarding, handsets with a mobile Internet button and support for OTA or equivalent).
2. Two seconds to access personalized portal.
3. Three clicks to desired content on personalized portal.

9.7.2 SINGLE SIGN ON

It is tedious to log in to many different services. Single sign on enables a single secure log-on.

Single sign on has evolved as a cost-saving solution to minimize support calls, and at the same time simplifies the administrative process of authentication and authorization.

For a large enterprise, as the number of passwords each user is required to maintain increases, so do the support calls. With each of these calls having an associated operational cost of $32 (according to the Gartner Group), and the increasing number of applications in use, businesses cannot afford the productivity lost through continuous password resets.

9.7.3 MANAGEMENT OF PERSONAL SERVICES

A service network offers control of network resources to the applications. Such a control implies also a possibility to receive personalized services, provided the user preferences are known and paid for.

From an end-user point of view, the service network owner becomes a single point of contact for service provisioning and matters related to a user's personal service environment.

The main responsibility of personal service environment management is to handle end-user service provisioning and service management. It is an end-user's contact point for managing his/her personal service environment. As such, it is not coupled to a specific access type or underlying core network, and is reachable via several accesses and core networks.

9.7.4 AVAILABILITY OF SERVICES

This is a prioritized area for OMA. 'No matter what device I have, no matter what service I want, no matter what carrier or network I'm using, I can communicate and exchange information'. The OMA solution is interoperability across technologies, products, services and value chains.

9.7.5 ENTERPRISE B2B COMMUNICATION AND WEB SERVICES

According to common business language, CBL (a framework for standardized business documents such as B2B) business documents are sent as XML (eXtensible Mark-up Language) based messages as a standard. The XML documents are structured in the same way as traditional electronic data interchange (EDI) documents, in order to facilitate the migration from EDI to CBL based systems.

XML web services are a tool for service interchange (service here stands for a functional software module). XML web services could become, for the service layer, what IP is for the transport layer. This will have impact on terminals as well.

9.8 THE NEEDS OF THE SERVICE NETWORK OWNER

9.8.1 TELECOM BUSINESS DOMAIN

The main needs are:

- Support for business development and planning – business layer in the TMN (telecommunications management network) pyramid for the service network.

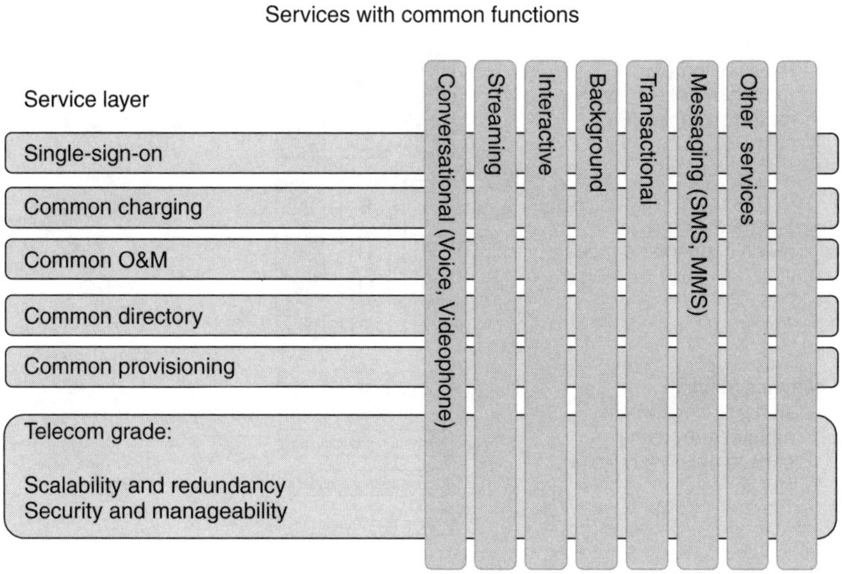

Figure 9.8 Cost efficiency in the service network. A third dimension is provided by for example location-based services, which can be combined with the vertical services

- Service extensions to stay competitive – new enablers.
- Reduced OPEX, CAPEX.
- A future-safe framework and service infrastructure.
- FAB support

9.8.2 COMMON USE OF RESOURCES

A common use of the resources and processes provided by the service network gives a better ROCE. See Figures 9.8 and 9.9. The common resource goal is most applicable for public service networks. Enterprise service networks tend to be more tailored, and even more vertical.

The common resource goal is most applicable for public service networks. Enterprise service networks tend to be more tailored, and even more vertical.

Management

The business 'pyramid' is under transformation (see Figure 9.10). When services beyond voice take a larger part of the overall service market, service and business management will start migrating to the service layer. Development, launching, billing and maintenance of new services will be increasingly located to the service management layer.

Service network management can be divided into application and network management and service management for service provisioning and service data management.

Service provisioning is defined as the process of handling the data in the network that is related to subscriptions, and service data management as operations on service data.

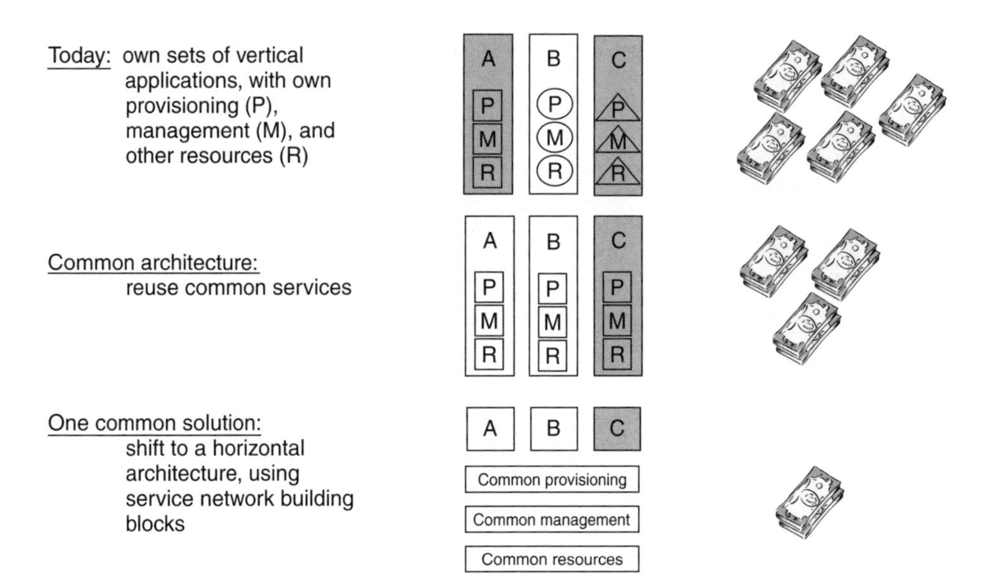

Figure 9.9 Common utilization of resources

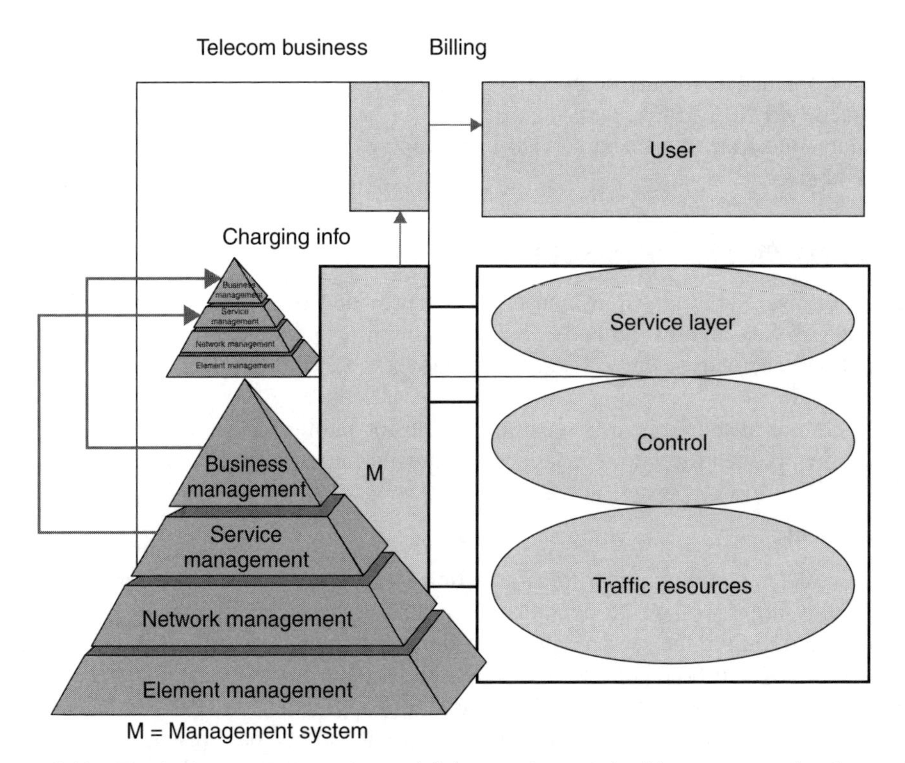

Figure 9.10 The basic management 'pyramid' is complemented with a new one for the service layer, including the service network, and service and business process management start migrating upwards

9.9 SERVICE NETWORK IMPLEMENTATION

9.9.1 GENERAL

Summing up the network structure the application servers and other service network components are interconnected over an IP network, which allows for a distributed deployment of the service capability servers and the application servers. The interconnection may be in an Intranet or between Intranets (e.g. tunnelled over the Internet).

Based on the definitions of a 3G service network, a service network solution must be able to efficiently handle four interfaces. First, there is one interface to the mobile networks. The enablers called resource managers/service capability servers are directly connected to the core network and offer that interface.

Secondly, there is one interface to the end users, where they can administrate their services. This is done through the portal solution, which is one of the building blocks in personal service environment management (PSEM).

Thirdly, there is one interface to the applications, which is also the interface for the application developers. This is taken care of by the applications/services interface, which is also hosting a operator specific developer zone solution.

And the fourth interface is to operation and management (the operators' interface). This is done through the service network operation system.

The service network embraces network internal transport via an IP network. There is also a logical interface to other networks. See Figure 9.11.

In the larger perspective the service network is located on top of the control network, but the application could also be remote and connected to the service platform by means of the Internet. Also, the connection from the service network to the connectivity network can be shown in a similar way as in Figure 9.11. See the bearer 'pipes' in Figure 9.12, which shows a distant application, supported by home service network resource management and application support.

For an overview see Figure 9.13.

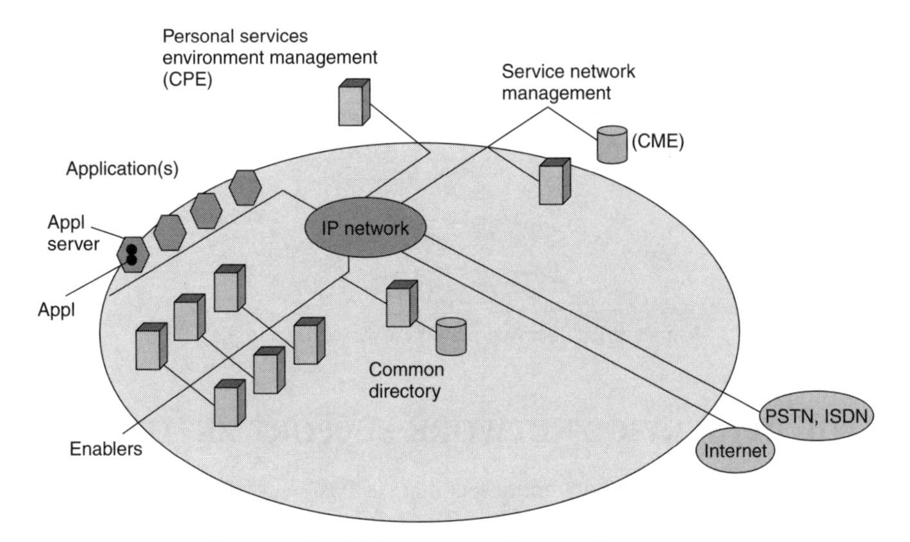

Figure 9.11 Major functions in a service network

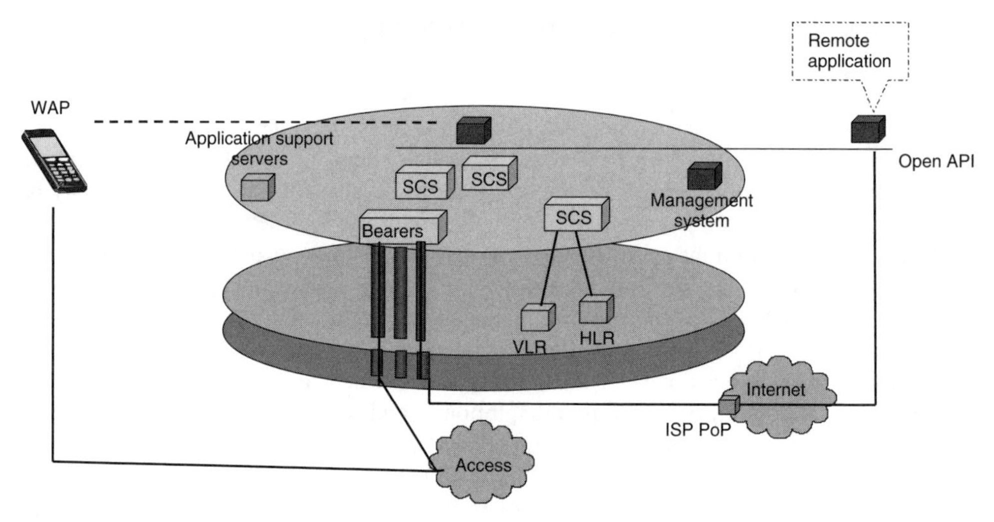

Figure 9.12 The service network cooperates with, and supports, remote applications

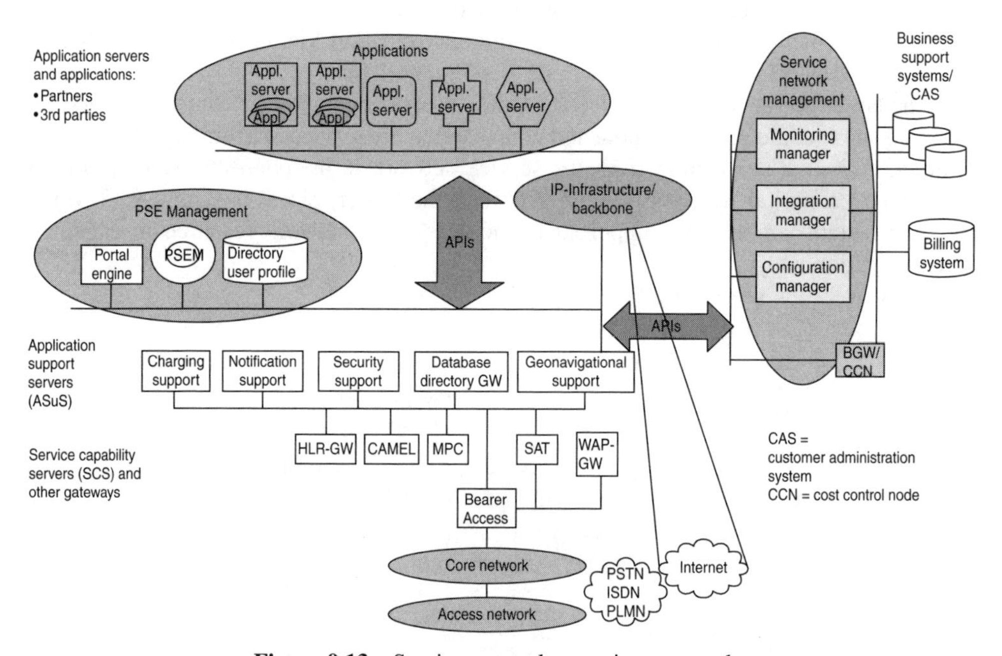

Figure 9.13 Service network overview, example

9.10 THE (IP) SERVICE NETWORK SUPPORT ENTITIES

The (IP) service network support entities contain a basic set of infrastructure services that can be utilized by the service network. The services are performed by infrastructure servers, such as:

- RADIUS – a RADIUS server performs user authentication, and optionally accounting and IP address assignment. It can be located in the packet core network, in the service network and in corporate networks.
- Firewall – protects a service network from unauthorized users. They often work in combination with security gateways utilized for VPN tunnels.
- Proxies – hides internal IP addresses from users and applications external to the service network, thus performing a very important security function, and caches web contents to minimize response times for common web/WAP pages.
- Security gateway – a device that handles IPsec tunnelling and security, often used for building IP VPNs.
- Network address translation (NAT) – performs address translation, a feature that enables the use of private IP addresses within the UMTS domain.
- Dynamic host configuration protocol (DHCP) – provides configuration parameters to Internet hosts.
- Domain name system (DNS) – one of the most fundamental Internet services. IP addresses are not easy to remember by humans, and therefore logical names are used instead. These logical names are translated into IP addresses by a DNS.
- Certificate servers – A Certificate Authority (CA) issues certificates. In order to generate keys and issue certificates, a certificate server (certificate management system) is needed.
- Network access server (NAS) – a tunnelling server for L2TP and IPSec. It is primarily used for accessing a service network over the Internet and from dial-up accesses.
- User authentication and IP address assignment – in addition to the GSM/UMTS authentication procedures (USIM – core network), a service network authentication of the user should be considered. This authentication guarantees that the current possessor of the user equipment is allowed to access a service network.
- Policy and multicast servers.

9.11 EXAMPLES OF SERVICE IMPLEMENTATION

9.11.1 FEDERATED APPLICATIONS

Federated applications are built upon cooperating local and distributed services across the Internet. See Figure 9.14.

Let us exemplify, using listening to music via the mobile phone:

- Connect via the WAP gateway to the wireless Internet portal and select your favourite music site.
- Click on the tune you want to hear and the tune is streamed to the phone.
- Whilst listening, the mobile advertiser sends a message saying that you have the opportunity to purchase the CD containing this song at a discount price if you respond within 10 minutes.
- Click on the 'buy' link and connect to the CD store's page. You buy the CD via the mobile e-pay secure m-commerce platform.
- Finally you request to have the payment sent to your mobile phone bill. The operator sends the payment transaction data to its partner bank/credit card agency, which clears the payment to the CD store.

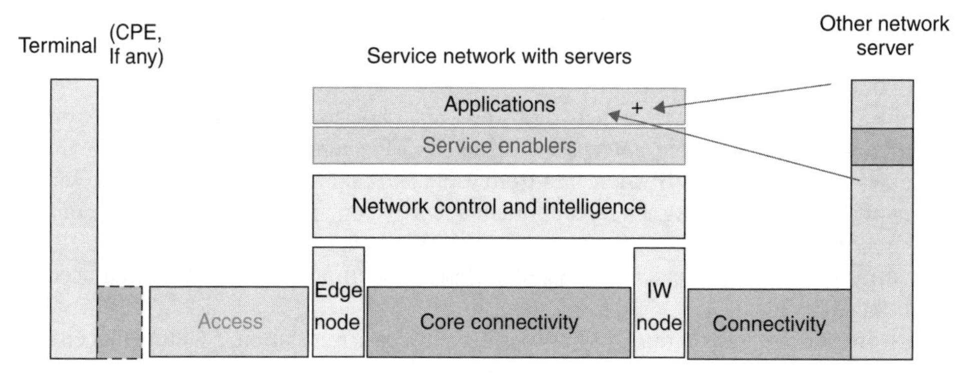

Figure 9.14 Federated applications

The user should not be aware of the products and processes involved in this complex series of transactions. But achieving this is not easy. This demonstrates the power of the service network concept where all applications share common resources and interwork smoothly with each other.

9.11.2 MORNING NEWS APPLICATION

As another example, we take a typical push service, a subscription-based news delivery using WAP technology (see Figure 9.15).

The news is customized for the user by using a number of areas of interest as a sorting parameter. The service is delivered cheaper if the user has accepted to receive local

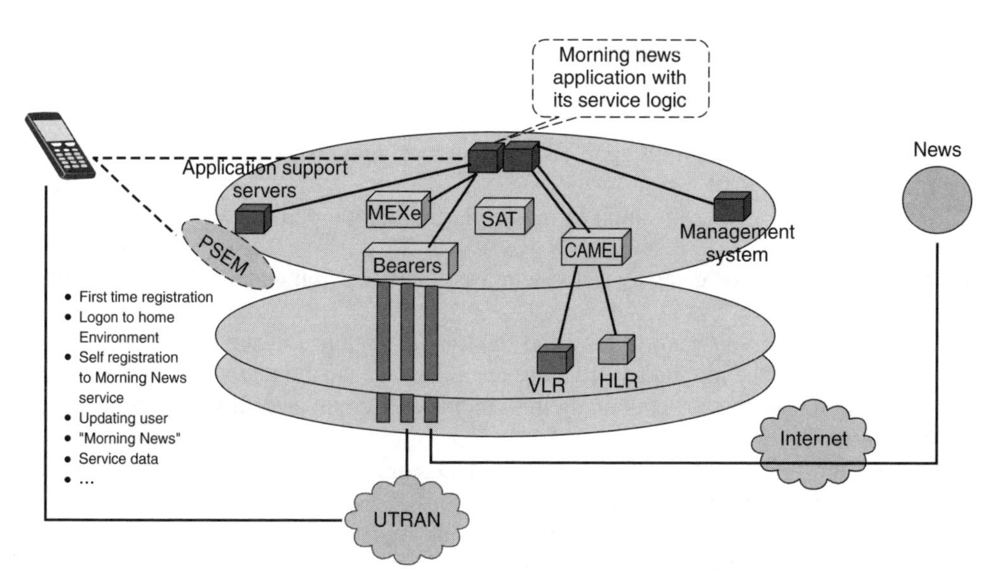

Figure 9.15 The morning news application 1: personalized newspaper

adverts (‘Check out the extra discounts at Arlanda tax free’). The news is delivered in the morning independent of which time zone the user happens to be in.

The outline of the service is that a message is received showing the headlines of telegrams. By clicking on the terminal the full text with pictures and graphic appears (if possible on the terminal).

What are the network requirements for such a service? Of course there is a need for setting up and maintaining a type of subscription. Both the selected type of news and what has been received on the terminal are input to a proper charging structure.

News has to be adapted to the terminal, size and width of screen, picture or graphic capabilities.

To be able to deliver news in the morning or at lunch or whatever decided time, there is a need for the system to change delivery times depending on user location.

The morning news application has the following simplified traffic and data flow. The application is divided in two major parts. One part is the front-end for collecting news from different sources on the Internet or home-made. It can be based on agreements between a service provider and the network provider. The other one is the back-end for the actual delivery to the end user.

Front End

The application server front end collects information and updates a news database, a database with classified (economic sport etc.) news. It runs almost independent of the back end.

Back End

First the application is set up as a service for the user using the service management function. A personal profile is set up, which can be changed by the user, creating changes in the application support server.

The user data sets a trigger, e.g. time of day, and the application keeps track of the user in which time zone (here e.g. roaming operator’s home country or visitor location register (VLR) address) by having a dialogue with the user’s home location register (HLR).

When the trigger is released, it is morning time where the user is located. A request is sent to the VLR where the user is updated, to find out if he is active. If he is active a WAP session is set up using the MExE service capability server setting up the appropriate bearer, as defined in user data (terminal dependent). The session continues with the user clicking for news (see Figure 9.16).

The session is recorded for charging purposes. Charging data is transferred to the charging function in the management systems.

The addition of local news is added by another application; the box beside the ‘Morning News’ application box.

The *positioning* application (see Figure 9.17) will have the following simplified traffic and data flow.

Here positioning is a part of a more general information service. First we set up WAP session using a bearer and then through a WAP SCS. Then we ask for a map with information about the nearest automatic teller machine (ATM). That triggers the application

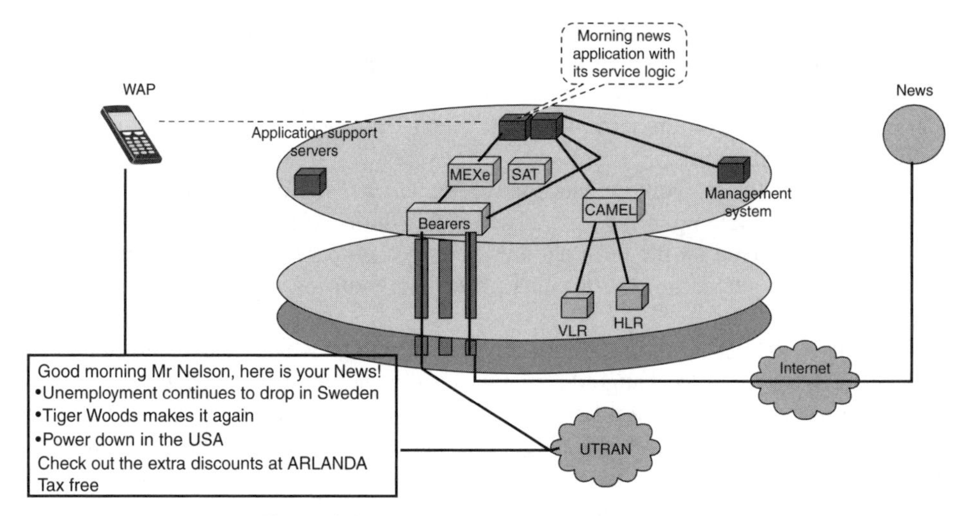

Figure 9.16 The morning news application 2

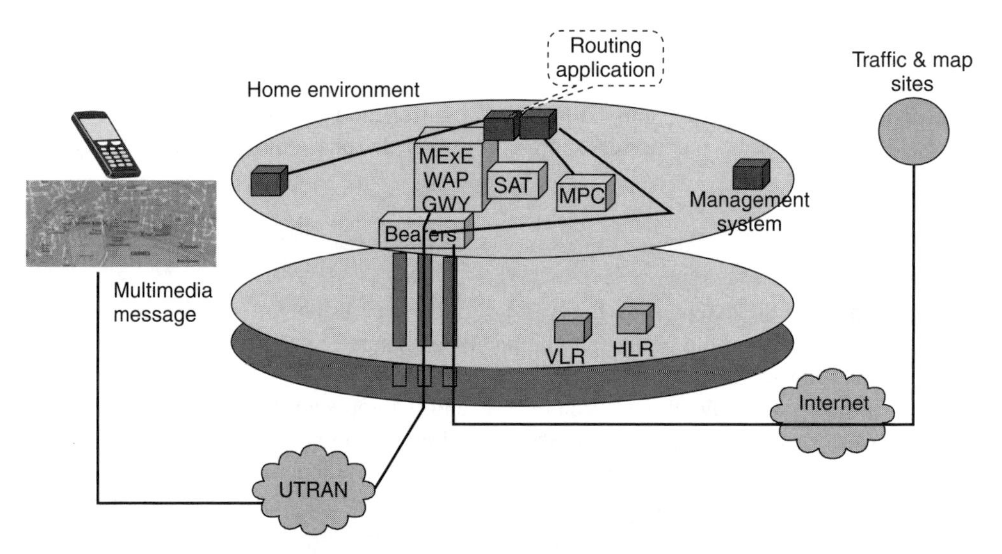

Figure 9.17 The positioning application

server to look up our position. The user's position is information that resides in the core network, therefore a request is sent to the VLR.

The application receives a response, retrieving a map if the terminal has that capability. Otherwise data is assembled in another format that is suitable for the terminal. Then the application adds the specific data, in this case my location and the location of the ATM. In this step the map data can be collected from a common database for all applications, in an application support server.

Finally the information is transferred to the user over the WAP session.

10

Terminals

'Soon we will need only five computers for the entire country.'

Thomas Watson, Sr, 1947

'The future terminal will be able to sense the presence of a user and calculate his/her current situation.'

Wireless World Research Forum

10.1 WHAT IS A TERMINAL?

- The chapter includes mobile terminals, fixed terminals and customer premises networks.
- Terminals are the man–machine interface for the subscriber. See Figure 10.2 (M2M is different, see below).
- Modern terminals are/will be part of the service layer. See Figure 10.3. As such they have a terminal–terminal or client–server interface and service enabler functionality. Service layer functions such as browsing are in principle not network dependent, but adaptations may be necessary (e.g. WAP).

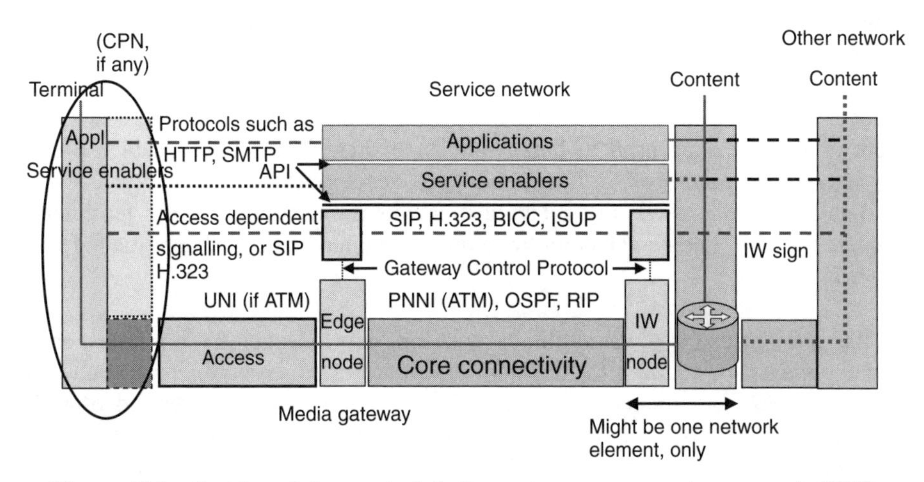

Figure 10.1 Position of the terminal device and customer premises network (CPN)

Understanding Changing Telecommunications – Building a Successful Telecom Business. Edited by A. Olsson
 ISBN: 0-470-86851-1

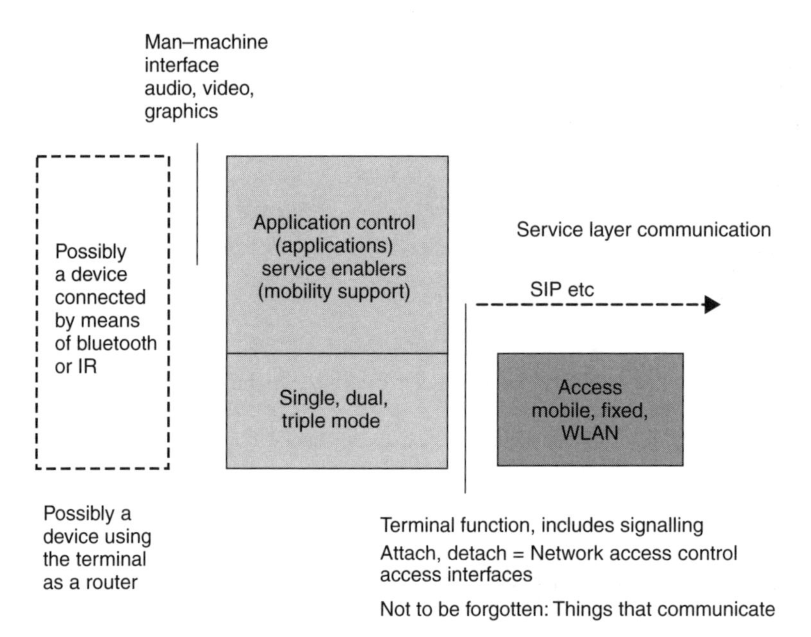

Figure 10.2 Main terminal functions and interfaces

- Terminals connect at the lower layers to a public or private network by means of a terminal–network interface.
- There is a vast range of terminals from simple telephones to complex enterprise networks. The service explosion makes service and application downloads a crucial issue to achieve flexibility.
- There is a trend from simple terminals (such as fixed telephones) and an 'intelligent' network, towards 'intelligent' terminals/terminal clusters and a variety of networks with varying capabilities depending on the traffic types. Examples of intelligent terminals are computers and the new mobile terminals, both of which use the whole TCP/IP protocol stack. Terminals with cameras or video cameras are already here.
- There is an increasing need to build home networks. The service enabler Bluetooth may play an important role in facilitating such development.
- Terminals can act as network relaying elements connecting to other terminals, for example when connecting laptops to mobile terminals or in what is called ad hoc networks.

Another strong trend is towards 'things' (washing machines, coffee machines, engines) that communicate. In addition to 'classical' customer premises equipment, such as telephones, faxes, PCs, etc., many microelectronic controlled devices (stereo/video products, home theatre, white goods, light control, burglar alarms, person identification/location systems, etc.) are used in a household. Most of these systems are currently stand-alone. Integrating these devices into an integrated communication environment using existing or easy-to-install networks (e.g. power lines, wireless systems) will allow major enhancements to existing services. It will also allow the creation of new applications and services and thus provide a significant push to all players in the field.

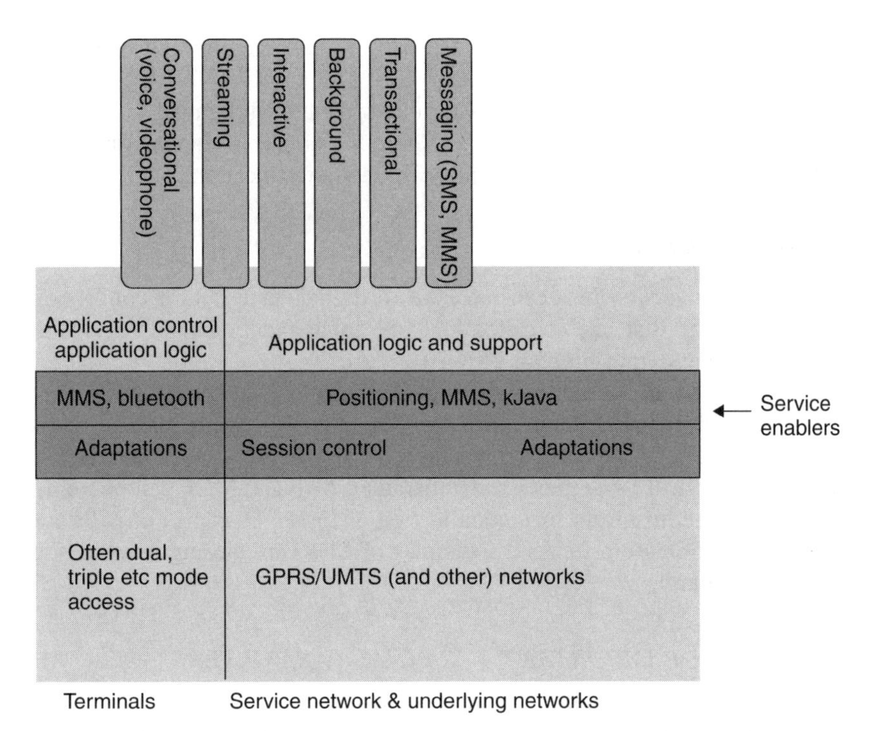

Figure 10.3 Modern terminals are an important integrated part of the overall communication 'machine'

10.1.1 TERMINAL CAPABILITIES

The capabilities depend on application logics, the service enablers in the service layer, but also on extra facilities, which exist without network support, such as cameras. When such facilities are integrated with communication, it gives added value compared with an ordinary camera.

Downloading applications to thin clients with Java (or corresponding) gives flexibility but requires thorough authentication of servers.

Memory capacity needs are almost being solved, but used to be a limitation for mobile terminals.

10.1.2 HUMAN INPUT AND OUTPUT FUNCTIONS

Human input function is used to interpret input from human end users. It is tailored to the needs of human beings. Examples of human input are interpretation of keys pressed on a phone and fingerprints on a fingerprint reader. The most important future human–system interface are probably speech interfaces that can replace many complicated button commands. Speech recognition, synthesis and control can be used in combination, and may also be used in conjunction with other types of interfaces.

Mobile phones and communicators already have speech control. Speech recognition and synthesis have still to be enhanced, e.g. to allow a user independent broad vocabulary.

But some problems may never be solved. How to differentiate between 'eight' and 'eat', spoken by people with strong dialects?

Human output function is used to provide output to human beings. These outputs are tailored to the needs and tastes of human end users. Examples of human output are audio output on a phone, graphics display of an area, video of a location.

10.1.3 M2M FUNCTIONS

M2M requires a conversion between machine to digital value and a conversion in the reverse direction. The first one is called machine input function (MIF) and the second one is called machine output function (MOF).

Machine input function is used to read input from machines. MIF and MOF are tailored to the machines and their behaviour. MIF converts physical properties like temperature, an optical scene, voltage into understandable digital values. MOF converts the digital information into physical properties like temperature. MIF and MOF can be implemented in hardware, software, firmware or a combination of these. Examples of MIF are smoke detector sensors and pressure sensors. Examples of MOF are closing a valve, toggling a switch, raising an alarm.

10.1.4 SUMMING UP IMPORTANT TERMINAL PARAMETERS

- What hardware (desktop, laptop, mobile, handheld)?
- What operating system (Palm, Microsoft XP 2000, Linux, Symbian, Windows CE.)? An operating system may contain many communication and security functions. Encryption may be used much more.
- What program language, for example Java?
- What access method(s) to the network?
- Access to other devices? Bluetooth, Infrared? Number of bands in the device?
- Automatic device configuration and single sign-on?
- Ring signals?
- What application(s)? Internal resources will be controlled via API interfaces from the applications.
- What service enablers?
- What man–machine interface (or machine–machine)? Display size? Voice control?
- What security properties and risks? Chapter 6, section 6.3.1 offers an overview.

10.2 BUSINESS ASPECTS

The capabilities, design and usefulness of terminals/devices are increasingly important for the telecom business. Comparisons between new terminals are frequently published. Lack of terminals seems to be one of the most hampering factors in the telecom development. Terminals are indeed very complex devices, and it is easy to underestimate the necessary development efforts.

A fairly new trend is that the big operators want to sell the devices with the name of the operator on them, although someone else manufactures the main part. Examples are Sprint in the USA, NTT DoCoMo in Japan and Vodafone in the UK. The operators usually subsidize the devices when sold with a subscription.

Some important success factors for terminals are:

- Flexibility in upgrading, with configuration over the air
- Power consumption – stand-by time
- QoS
- Security
- Ease of use – single sign on.
- Design, 'sexy devices'
- Personal assistant function/feeling

10.3 HISTORY

In order to get a perspective of the fast-changing pace, let us take a look at what was written a few years ago.

Figure 10.4 shows a multifunctional device presented by the French TV channel Canal+ at the IBC'98 expo. It is particularly intriguing that the device also is a TV set.

In 1998 Telia described the development of the mobile service demands from 1990 to 1998 and then to 2003. One single terminal can handle all communication needs at reasonable tariffs. See Figure 10.5.

10.4 TERMINALS FOR MOBILE NETWORKS

The following subsections describe the current range of categories. In the coming years, the market will be even more segmented, with products tailor-made for specific applications, but there will also be a convergence in terminal functionality as well as multifunction terminals. Let us start with some existing terminal types. See Figure 10.6.

10.4.1 FEATURE PHONE

In this category there are products that combine good voice capability with text-based messaging and data. Micro-service support by WAP is used for non-voice applications

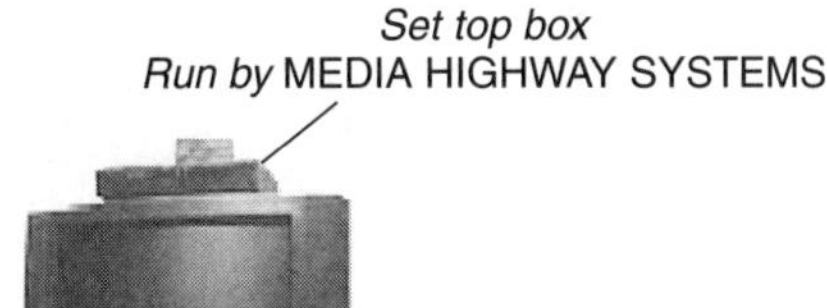

Figure 10.4 The human being has in reality been somewhat conservative regarding new use of the TV set compared with this forecast

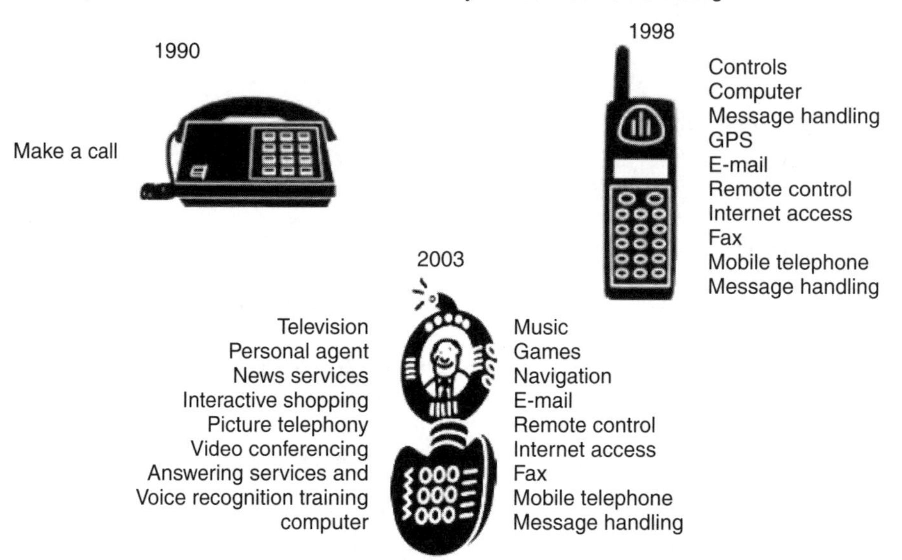

Figure 10.5 Camera function was not foreseen in 1998, otherwise it was a very good forecast

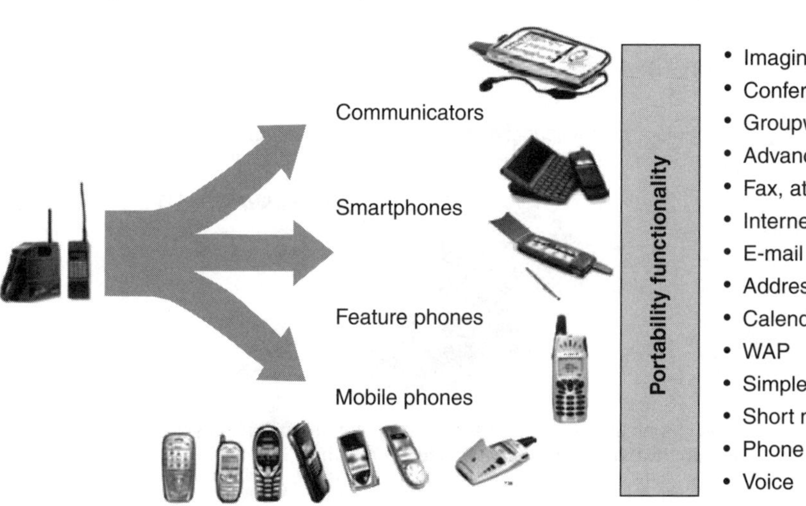

Figure 10.6 Segmentation of mobile terminals

together with built-in modems. This category is equipped with a slightly larger display to run messaging applications and images. WAP enables good usability with a limited display. For external interfacing it primarily utilizes Bluetooth. Limited personal information manager.

10.4.2 SMARTPHONE

The term *smartphone* has been coined to describe devices, which, while being primarily a mobile phone, incorporate elements of functions found in paper-based personal organizer systems or in modern electronic personal digital assistants (PDA). Typically, a smartphone contains a calendar, an address book, e-mail and messaging functions, and a browser for the wireless application protocol, together with a range of ancillary functions. It works simply as a mobile telephone. A smartphone is roughly the same size as a standard business mobile phone, and similar to a standard phone, it allows users to operate its basic functions in a one-handed fashion. Extra functionality is accessed via the large touch screen, which differentiates it from standard phones.

Connectivity with peripheral devices is accomplished primarily through Bluetooth. One or more camera devices are built in. Even with the added message-centric functionality, voice communication is still the primary function.

10.4.3 COMMUNICATOR

At the functional high-end of the market there is a family of messaging-centric wireless devices that combine communications of voice, text and multimedia, and are open to third-party applications. The product concept is distinct from PDAs and other hand-held computing products through the superior and integrated communication capabilities. The communicator differs from mobile phones and smartphones by a further extended user interface suited for messaging and information access and exchange. Distinct 'attributes' are user interface/ease of use and technology applied to real user needs.

10.4.4 VOICE TERMINAL

The voice terminal is focused on excellent voice quality in combination with enhanced voice services, i.e. voice activated dialing etc. Good voice quality in combination with convenient size and weight makes this product very easy to use; which is an important competitive factor in this category. Voice terminals are available in all different price segments, since factors other than features give these products a competitive edge. Such factors are industrial design, and possibilities for the end user to individualize the product. Some limited types of messaging are also of interest for these customers, e.g. SMS, especially to a younger group since they have limited possibilities for investing in a feature phone but may still want the 'chat' type of communication they have on the Internet. WAP is fully supported and Bluetooth is considered to further enhance the voice capability as well as improving the messaging services.

10.5 PDA DEVELOPMENT

A typical range of functions in a personal digital assistant (PDA) may be mobile phone, music, calendar, GPS positioning, SMS-EMS-MMS messaging, WWW, camera, voice recognition, translation, digital wallet, and possibly TV and/or printer.

10.6 TERMINAL CONVERGENCE

Convergence in terminal capabilities is prevalent for most services. For mobile devices we have the grouping into wireless Internet and mobile media. Starting with Internet WWW services we may find such capability in 'everything' from PDAs to desktop machines to laptops to WAP phones to phone browsers to web television units to specialized browsers for users with disabilities.

We see that the 'web terminals' can also be used as media terminals, such as streaming video services in the mobile terminal.

A new word is 'wearables'. The idea behind wearables is to design gadgets out of simple electronic components in place of 'computers'. Thus, wearable computers can be seen as propagations of the human senses.

Terminal convergence can also be viewed for transactional services. See Figure 10.7. Some trends point at terminal interworking and roaming between different technologies such as UMTS and WLAN. Roaming GSM-UMTS is already established. Different deployment alternatives for the two radio interfaces may appear on the marketplace: separated terminals for different radio interfaces and multi-mode terminals, whereby multi-mode terminals integrate both cellular radio interface and WLAN into one physical device. The user subscriber information, which is stored on the USIM, needs to be accessible from both radio interfaces.

10.7 THE CHANGING ROLE OF TERMINATING DEVICES

In the good old days the conversational mode of service (such as voice calls) dominated. This is a peer-to-peer connectivity service and the terminals were simple, especially in the fixed network. The communicating humans added the rest of the required intelligence or 'protocols' to the terminating points.

In the modern 'information society' terminals also often appear as a client in a client–server relationship, for example when downloading from the Internet. The terminals

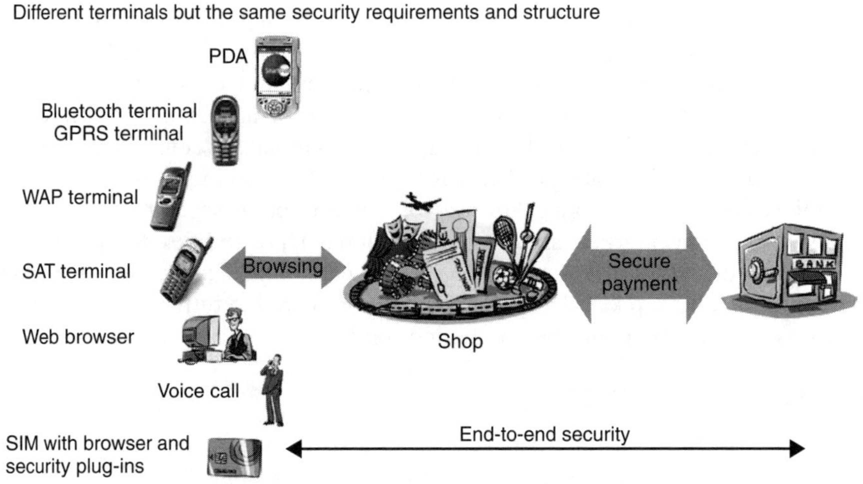

Figure 10.7 Convergence in terminals with transaction service capabilities

communicate with application servers using application protocols. The communication may or may not include interaction. Other relationships are called peer-to-peer and distributive. Napster is an example of a peer-to-peer service (but with network support) and web services are an example of distributed services. The terminal can be equipped with WAP, Bluetooth, cameras, different applications and operating systems.

There is in addition a trend towards ad hoc networks, and because of a lack of worldwide standards the conclusion is clear:

- From being peripheral the terminals become more and more a central point.
- It seems increasingly tricky to define the capability range of a specific device. Also at the transmission interface there is a need for flexibility.

To solve the problem a first step has already been taken with Java, .NET and other program languages.

Reconfigurable, downloadable protocol stacks would be a future key enabler for flexible network architecture, allowing an easy adaptation to the application's demands, including radio access (software defined radio). Thus, it ensures a future-proof network architecture, which can keep pace with the application innovation process by changing the mobile station's protocol stacks remotely. The research in this area includes software architectures and an investigation of the hardware impacts.

10.8 WHAT IS A CUSTOMER PREMISES NETWORK?

A customer premises network (CPN) is a privately owned network. It is generally managed and controlled by the customer. In some cases, it can be managed and controlled by an external party. It provides services for end users belonging to the CPN domain.

CPN can range from simple user equipment (terminals) to a fully deployed and managed network offering end-user applications and services.

CPN may contain a number of network technologies like WLAN, Ethernet, Bluetooth etc. to provide connectivity services. The defining criterion of the CPN is the ownership of the network.

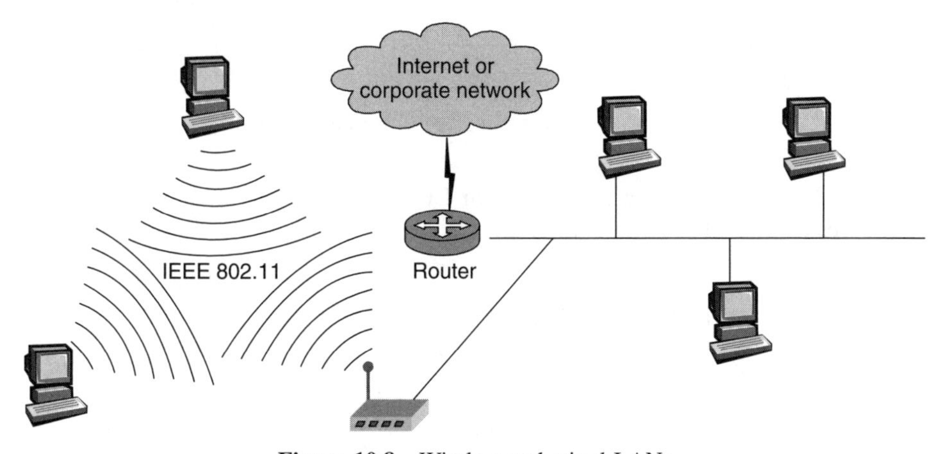

Figure 10.8 Wireless and wired LANs

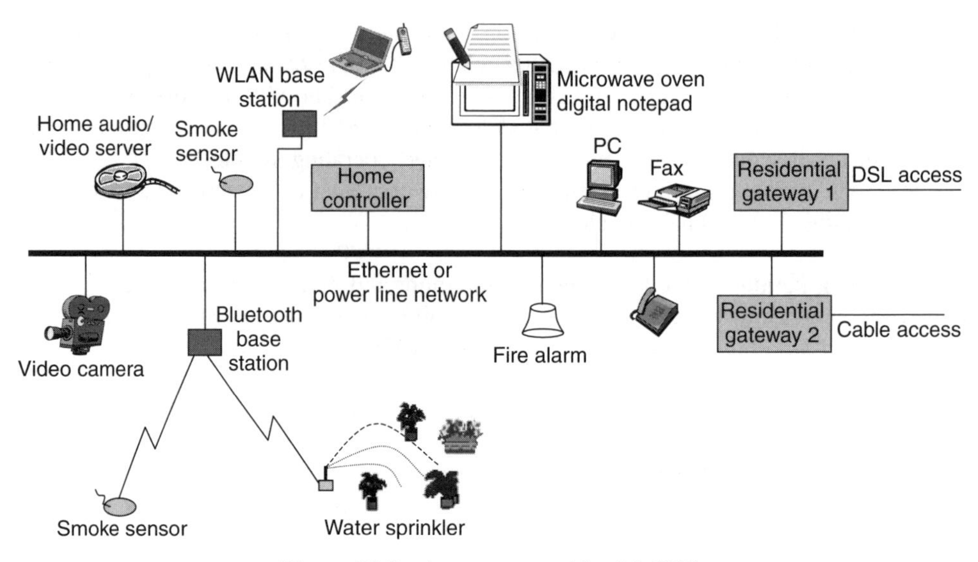

Figure 10.9 A customer residential CPN

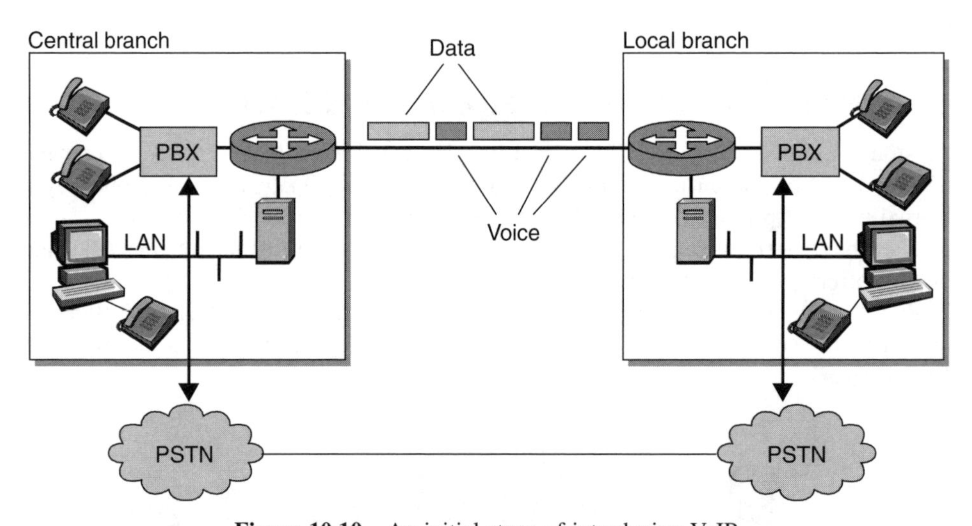

Figure 10.10 An initial stage of introducing VoIP

Wired and wireless LANs are examples of CPNs (see Figure 10.8). An example of a customer residential CPN is shown is Figure 10.9.

10.8.1 VoIP AT CUSTOMER PREMISES

VoIP has entered the scene, at least for intra-company voice traffic. A gradual and safe transition to VoIP is expected to be common. Figure 10.10 shows an early step with VoIP primarily for company internal voice and traditional circuit mode voice for public calls. Figure 10.11 shows some VoIP termination types and applications.

IP telephony terminals

- Development of IP telephony terminals
  - 1st generation PC-PC
  - 2nd generation PSTN-PC
  - 3rd generation PSTN-PSTN
  - 4th generation dedicated IP telephones
  - 5th generation wireless IP telephony
- Software and hardware phones
- Functionality included
  - Coding/decoding
  - Addressing and call set-up functions
  - Multimedia functions
  - Network interface

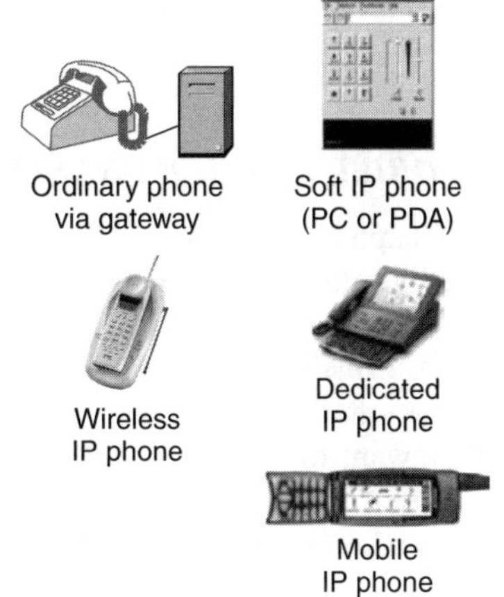

Figure 10.11 Some basic facts about IP telephones

10.8.2 AD HOC NETWORKS

Ad hoc networks means units that:

- occasionally are close to other units
- establish a connection between them

Ad hoc networks can only function when the units are within the mutual area of coverage of the cooperating units.

There are no specific access points, and the network functions without traditional routers, switches, address plans and firewalls. An example is piconet with Bluetooth.

Coverage ranges from 10 metres to several hundred metres depending on the environment.

10.9 SOME ENABLERS

10.9.1 (MOBILE) EXECUTION ENVIRONMENTS

- Java, WAP, .NET.
- Java is an industry-standard object-oriented language for development or deployment of applications, applets and APIs.
- Java is created for thin clients. The mobile terminal is an example of a very thin client.

10.9.2 SECURITY ENABLERS

- SIM-USIM. See Chapter 6, Section 6.14.2.
- AAA client
- HTTP/TLS/WTLS. See Chapter 6, Sections 6.7.4 and 6.7.5.
- Cryptography. See Chapter 6, Section 6.10.
- Virus via OTA is a danger.

10.9.3 *QoS ENABLERS*

- RTP, RTCP, RTSP, RSVP. See Chapter 7, Sections 7.5.1 and 7.5.4.
- CC/PP. See Chapter 7, Section 7.6.1.

10.9.4 *MOBILE SERVICE ENABLERS*

See also Chapter 8 on service enablers. The main features of a presence service could be described by a number of questions:

- Who is on line?
- Where are they?
- Do they want to be contacted?
- If so, how? Voice, text, video, instant message? (Menu is presented on the screen if action is taken to make contact)

Device Management

For OTA, over-the-air configuration, the settings are sent over air to the terminal device from the operator. The device interprets the data and performs a configuration accordingly.

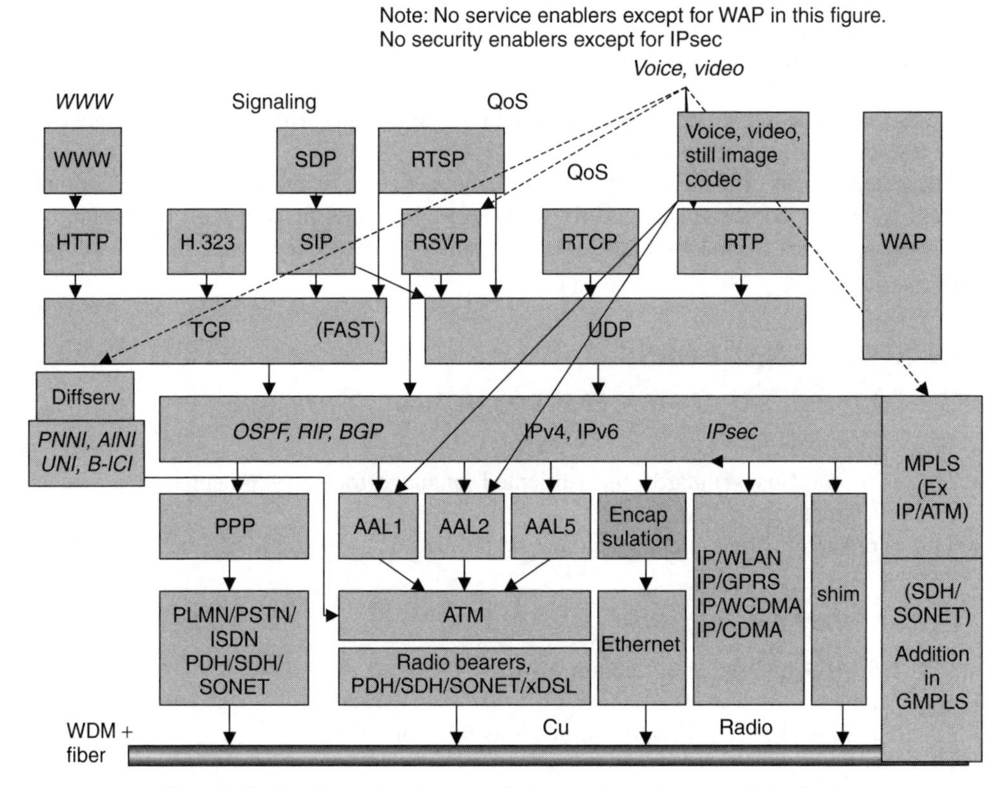

Figure 10.12 Example of protocols in previous chapters of the book

10.9.5 PLUG AND PLAY TECHNOLOGIES

Plug and play technologies enable ad hoc networking of different devices. They are based on existing technologies like TCP/IP, HTML, Java VM or XML.

Using such architecture, users will be able to plug printers, storage devices, speakers and any kind of device directly into a network and every other computer device and user on the network will know that the new device has been added and is available. Each pluggable device will define itself immediately to a network device registry. The operating system will know about all accessible devices through some network registry.

10.10 TERMINAL FUNCTIONALITY – EXAMPLE

With the new central position of the terminal in mind, there is a need for a huge number of protocols in modern terminals. By means of Figures 10.12, 10.13 and 10.14 it is possible to group the terminal functions in a way that corresponds to what has been applied in previous chapters of the book.

Some functions in Figure 10.12 are core oriented according to Chapter 12 (ATM, SDH, SONET, MPLS, GMPLS) but an application of such standards in a CPN or a terminal could not be excluded. Security and service enabler functions are included in Figures 10.13 and 10.14.

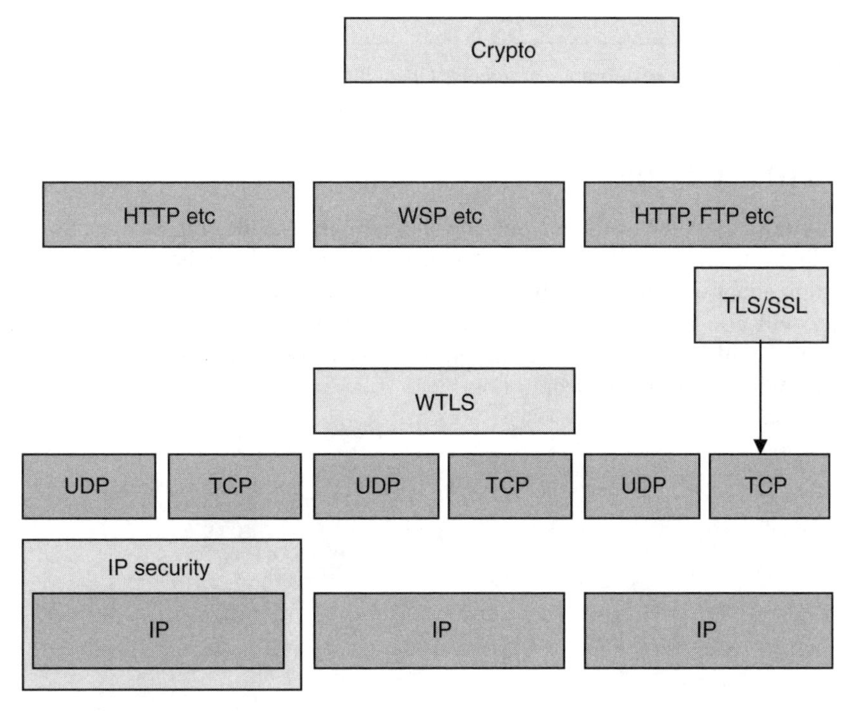

Figure 10.13 Security protocols and TCP/IP protocols in a mobile terminal, example

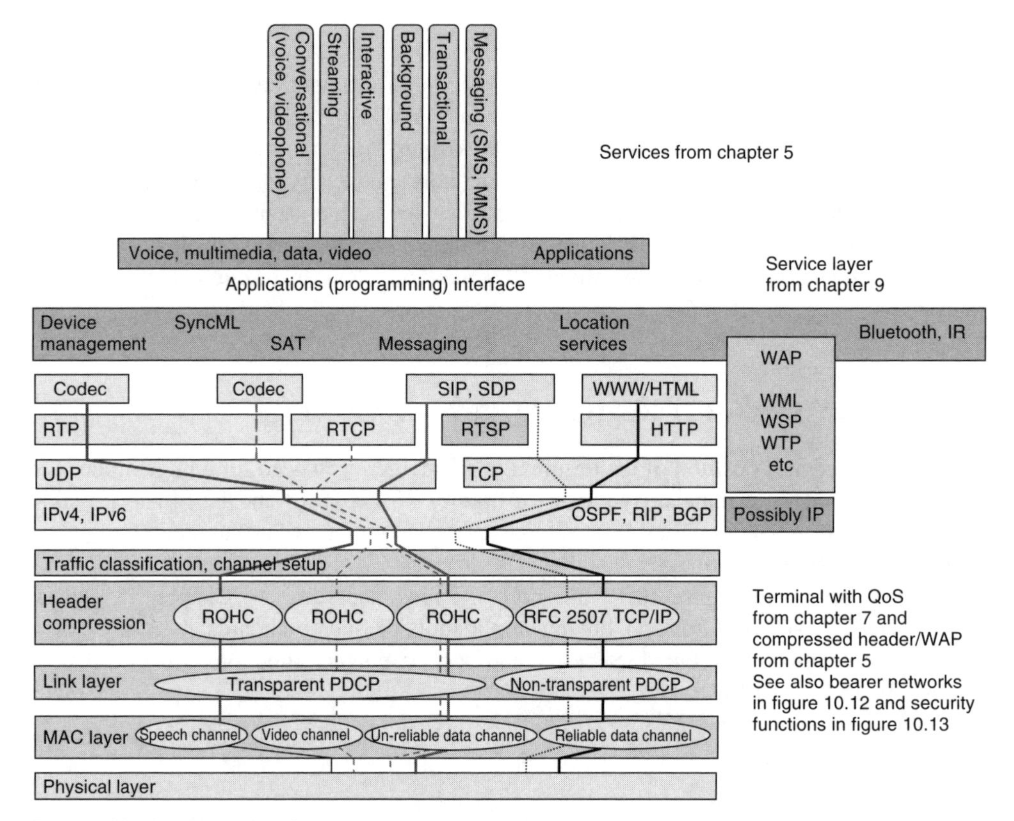

Figure 10.14 Example of terminal protocols. Operating system and execution environment are not incuded

10.11 THE FUTURE

'In the future computers will not only be increasingly mobile, but information will be accessible from any mobile position. Devices will recognize who we are and obtain information about us. There will be a mixture of interoperable fixed and mobile devices. A device will also be more aware of its user and more aware of its own environment. The terminal will be able to sense the presence of a user and calculate his/her current situation' (Wireless World Research Forum).

11

Edge Nodes

11.1 INTRODUCTION

11.1.1 WHY A SEPARATE NODE-ORIENTED CHAPTER?

Edge nodes/media gateways have developed into a key component in the new horizontal network in many types of networks embracing an impressing functionality capable of almost any type of conversion. They also operate at the payload level, with recoding capabilities. The media gateway constitutes a resource bank that is controlled by the media gateway control, for networks with such architecture. The typical position in the network is shown in Figure 11.1. Understanding media gateways and other similar edge nodes

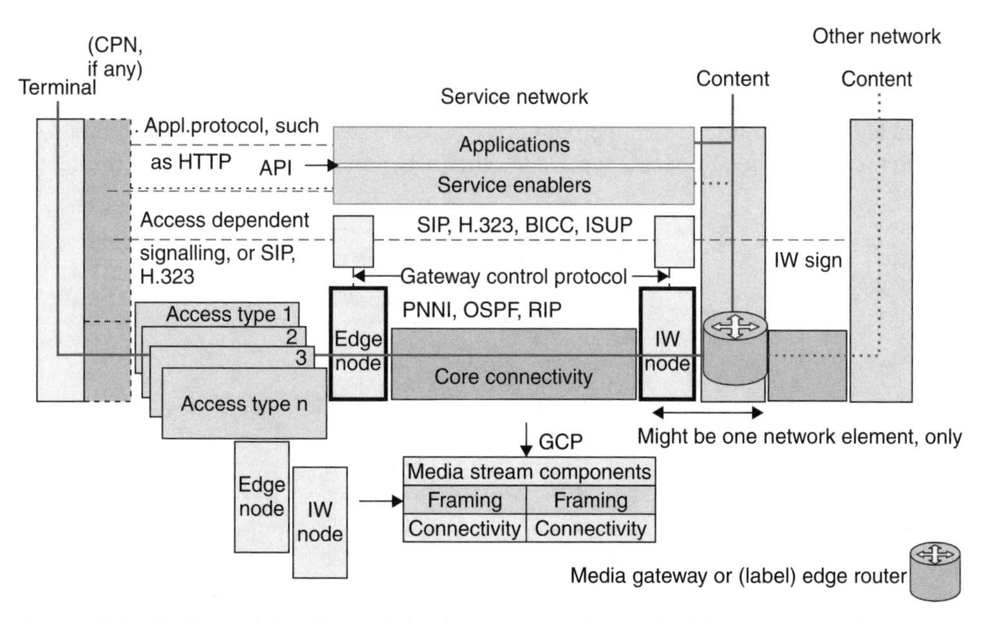

Figure 11.1 Position of the edge node/media gateway and the principle for a conversion stack in the lower part

Understanding Changing Telecommunications – Building a Successful Telecom Business. Edited by A. Olsson
 ISBN: 0-470-86851-1

is necessary in order to understand the new network. The subject of edge nodes/media gateways was introduced in Chapter 3.

This chapter will be complemented by a media gateway dimensioning to be found in Appendix 4.

11.1.2 TERMINOLOGY

The term media gateway (MGW) has nothing to do with for example transmission media. It stands for a conversion of the media streams of voice, data, video or multimedia information. However, this is not necessarily what the MGW does. A true conversion of for example voice would imply a recoding, let us say from AMR to PCM (pulse code modulation) code. It is true that a media gateway contains coders, but the professional network designer tries to avoid recoding for QoS reasons (time) and for security reasons (the packet must be unpacked).

So, the information or media stream can often be retained intact, whereas the transfer mode might change. Transfer mode can be seen as the way of packing information. An example is repacking PCM voice from TDM into ATM AAL1 cells. A more correct term is media framing. See the lower part of Figure 11.1. For a repacking of circuit mode voice to packetized voice the term circuit emulation is often used. An 'all-access' emulation to ATM is shown in Figure 11.2.

Another common term is softswitch. This is a generic term for any open application program interface (API) software used to bridge a public-switched telephone network (PSTN) and a voice over Internet path (VoIP) by separating the call control functions of a phone call from the media gateway. The separation is usually performed by means of the H.248 protocol, presented in Chapter 3. See Figure 11.3.

Concluding this introduction, the 'all-access network architecture' with a common core interfacing a multitude of access types and other networks, is the basic reason for the use

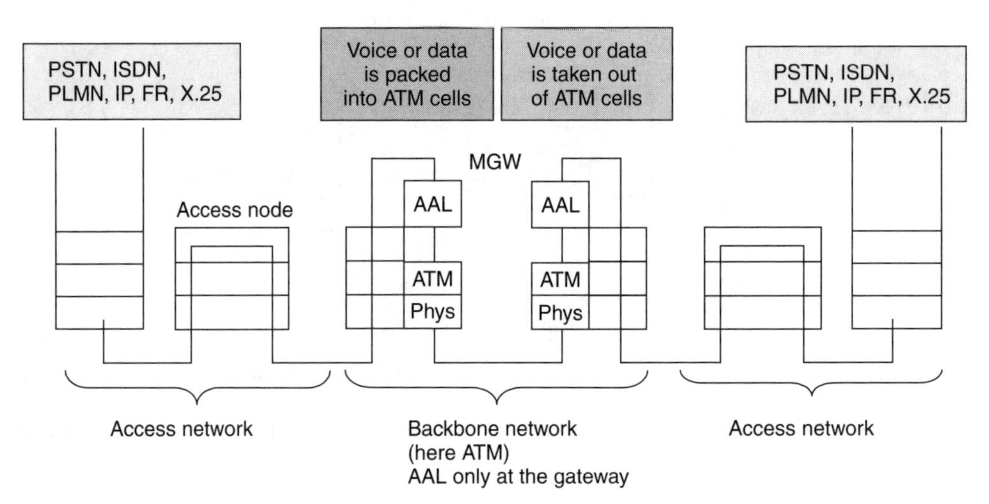

Figure 11.2 Illustration of a possible network structure: various access types, a gateway and a uniform ATM backbone. Layer 3 below AAL in the gateway is reserved for possible layer 3 signalling and addressing

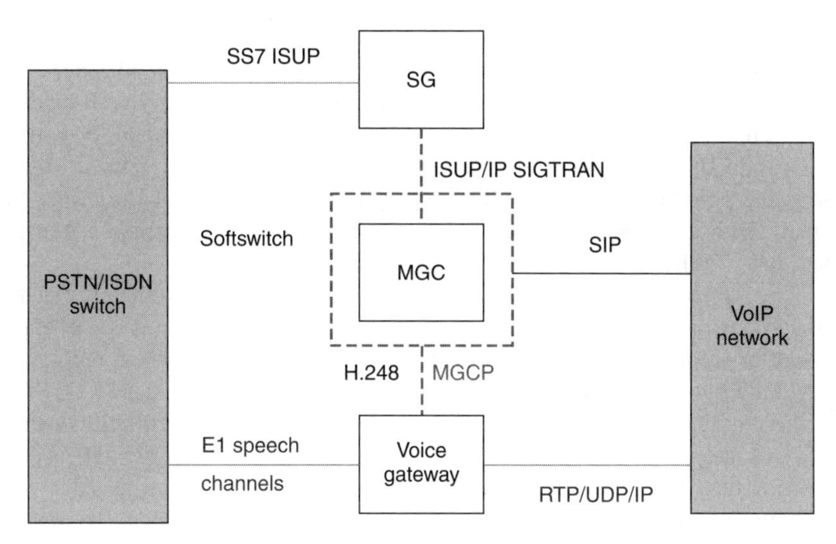

Figure 11.3 A softswitch configuration

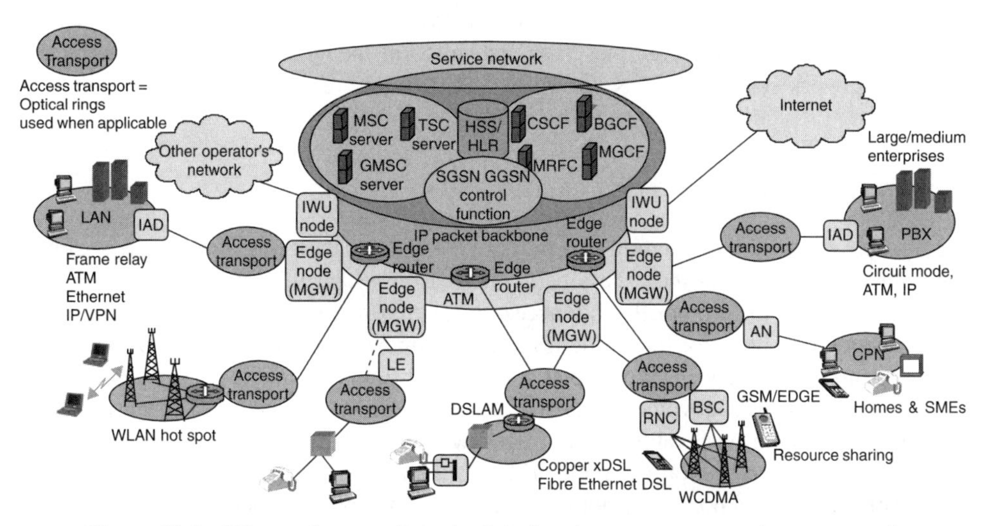

Figure 11.4 Edge nodes constitute the interface between access and core network

of edge nodes. See Figure 11.4. When MGW and edge router are collocated the router is often called a site router.

11.2 ACCESS AND BACKBONE NETWORKS

11.2.1 ACCESS CONNECTIVITY

During all processes of change, the expensive access network has been a conservative power. Let us mention the digitization, which by definition is not yet complete for PSTN in the access part. The convergence of the existing networks towards a packet network with broadband capability must be profitable with strict control of investments. The access network investments are by far the most critical in order to reach that goal, since, for traditional operators, the access networks represent more than 50 % of the overall investments. Fortunately all the fixed transmission media, whether originally designed for narrowband or higher bandwidths, have proven surprisingly feasible in the new telecom world. An example is the possibility to install VDSL (very high bit rate digital subscriber line) on copper pairs with a bit rate of about 50 Mbit/s. This requires of course a transmission technique that is tailored to the transmission medium.

For mobile systems the investments can be based on upgrades utilizing existing frequency bands, with GPRS and EDGE as examples. This alternative is the cheapest but might not offer sufficient bandwidth for new services. The UMTS system employs a new frequency band but also makes better use of the existing band around 1800 MHz.

Thus, the conservative access impact is not always true. The ongoing investments in mobile 2.5 G and in particular 3 G access systems, may in fact turn the access side more modern than the backbone. And other new access types such as Bluetooth and WLAN strengthen such a development. Existing media will successively be upgraded to a broader bandwidth.

In any case, the number of access alternatives will increase rather than converge to a few standard alternatives. Mobile or fixed wireless, Cu pair, fibre and cable-TV will all remain as access media alternatives. The transmission media will for a long time carry different network services (PSTN, GSM, IP, ATM, frame relay, GPRS, UMTS).

11.2.2 CORE CONNECTIVITY

For the network core the situation is different. The bit transport cost in this part is almost negligible compared with the access part. The reason is that this traffic is highly aggregated, compared with the mostly thin accesses. Therefore we can afford to build a fairly homogeneous backbone for the traffic transfer, opting for a standard type of packet transfer mode through the core, which meets present and future demands. This transfer mode must be able to carry both voice and data (and video) separately and also carry multimedia communication with acceptable quality. The two main alternatives are ATM and IP. A closely related further option is called MPLS. See the Chapter 12.

The converging backbone does not imply that existing core network types will disappear overnight. Therefore interworking edge nodes are also necessary, for connection to legacy networks, such as PSTN/ISDN.

Most of the required network conversions will take place in the edge nodes/media gateways. Seen from the access side the MGW could be called an access switch. Another name is multi-service gateways, where ‘service’ means network service (ATM, FR, PSTN) and not end-user service. MGW belongs to the access network with part of its functionality and to the backbone network with other functions.

In IP networks with matching access and core networks, there is a limited need for MGW functionality. However, the IP network architecture often contains an edge node with QoS and aggregating functions at the border between access and backbone. For an IP-based core network two variants of the MGW exist, the edge-MGW towards the access network and the interworking-MGW towards other networks.

11.2.3 LOCATION OF THE MEDIA GATEWAY

The MGW is by definition located at the border between access and core connectivity, or at the border of other networks. To settle the access–core border point there is a need for a geographical partitioning into access areas. For traditional operators the location of the local exchange or mobile switching centre is acceptable as a partitioning point between access and core.

If node locations and coverage areas of access switches are not given from the beginning, a first step is to study the forecast of traffic sources of various kinds in a selected area. Integrated access nodes or radio base stations are located close to the gravitational centre of their suitable coverage area. Then the access switch/MGW is positioned in a suitable available building. Provided most of the subscriber traffic comes from access nodes and not from direct access lines the exact position of the access node is not very critical. However, it must be located close to a high capacity optical ring. The capacity of the access switch must be dimensioned to support all the access nodes in addition to own possible direct access lines.

11.3 MGW INTERFACES

11.3.1 ACCESS

On the access side there is an abundance of interfaces. Using reference architecture makes the situation much simpler. See Figure 11.5.

11.3.2 BACKBONE

Although the backbone converges towards IP/ATM over DWDM, it can still take many different shapes. See Figure 11.6.

11.3.3 INTERFACE TO THE MEDIA GATEWAY CONTROL H.248

This interface is described in the key enabler part in chapter 3, section 3.14. See also Figure 11.7.

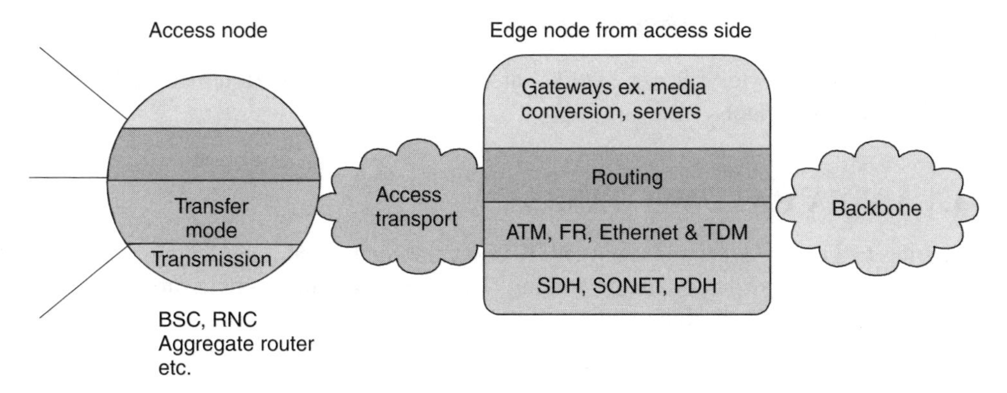

Figure 11.5 The access side of the edge node

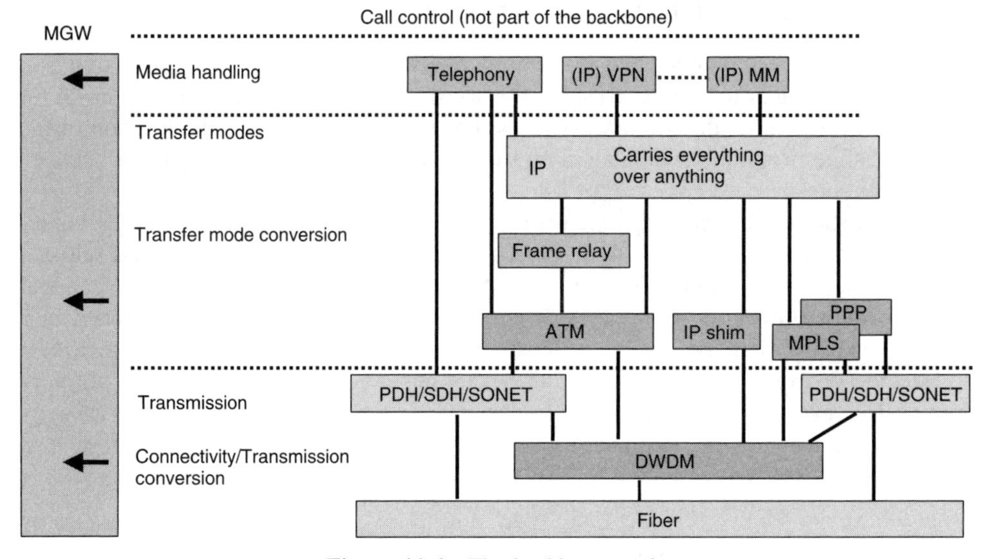

Figure 11.6 The backbone stack

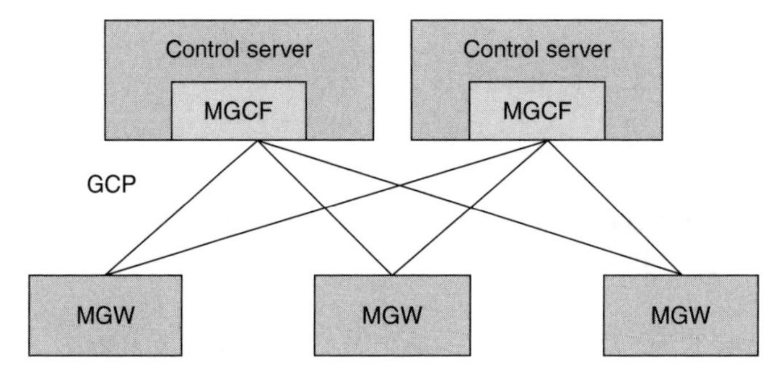

Figure 11.7 Decoupled media gateway and media gateway control – initially the server and the gateway were integrated in the same unit

11.3.4 INTERFACES TO OTHER NETWORKS

Common interfaces to other networks are the ones towards Internet and PSTN/ISDN/PLMN. See Figure 11.8.

11.4 MEDIA GATEWAY TASKS

Referring to Figure 11.1 some main MGW tasks are media handling, framing and connectivity, including switching and/or routing. The MGW can for example function as one or any combination of the following:

- ATM switch
- Packet data handler

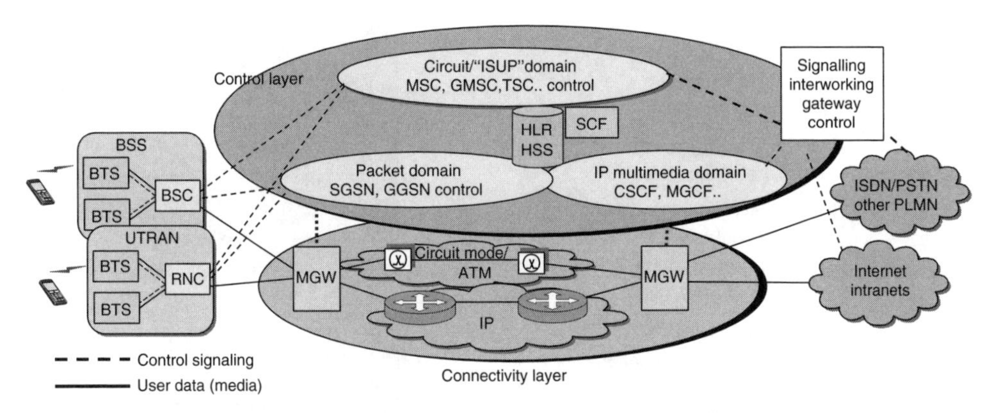

Figure 11.8 A summary of the interfaces to the media gateways

- Real-time IP router
- Media stream handler

Among framing components are ATM, AAL1, AAL2, AAL5, IP, UDP, RTP, L2TP (layer 2 tunnelling protocol), MPLS-LER (multi-protocol label switching-label edge router), GTP (gateway tunnelling protocol) and TDM. It should be stressed that it is important to distinguish between transfer mode and transmission system. A transmission system such as SDH can carry all types of transfer mode.

In GPRS and UMTS there are two types of MGW on the access side: the MSC MGW and the SGSN MGW. Figures 11.9 and 11.10 give examples of their tasks.

Obviously, the MSC MGW has many media stream functions.

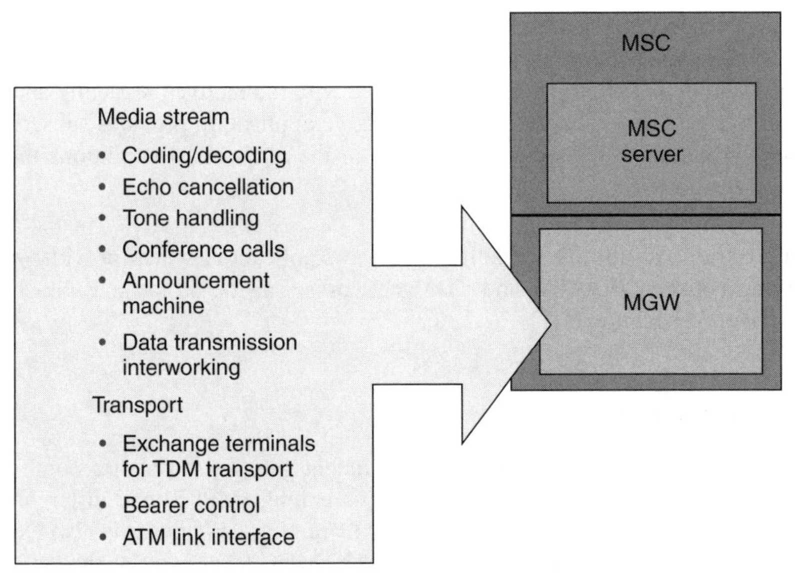

Figure 11.9 Example of MSC MGW functions

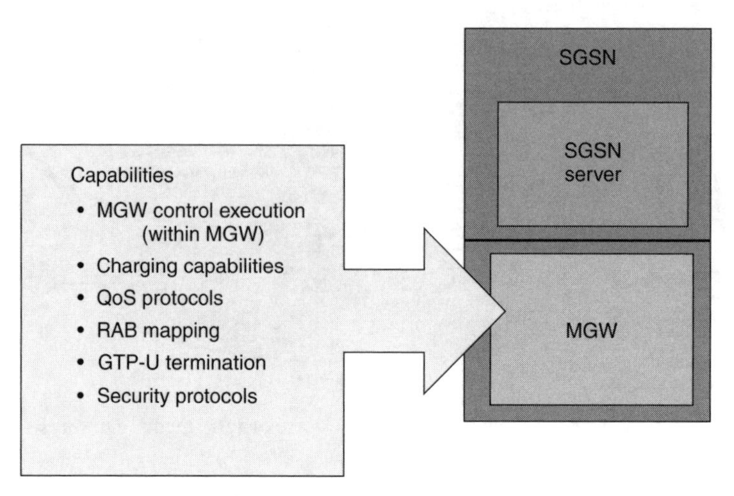

Figure 11.10 Example of SGSN MGW functions

Example of media stream components are:

- Voice coder. For UMTS the adaptive multi-rate (AMR) voice codec is the default coding/decoding algorithm.
- Echo canceller. Echo cancellers attenuate echo generated from the conversion between four-wire and two-wire transmission in the PSTN, and reduce mobile crosstalk (see Book 2 Part b, Chapter 4.3).
- Circuit-switched data. (See Book 2, Chapters 1.3.3 and 4.3.2).
- Multiparty call. Multiparty call supports the conversation between more than two parties.
- Tone sender/receiver. The tone sender/receiver provides tones to be sent to and received from end users. DTMF tones can be sent to the far end of the connection as requested by a mobile station. DTMF tones can also be used with interactive messaging applications.
- Interactive messaging. The interactive messaging application provides subscribers with information messages on special conditions in the network or conditions that pertain to the service in use.

To reduce the footprint, the signalling gateway application, which provides signalling interworking between IP, ATM and TDM networks, can be co-located with the media gateway node. See Figure 11.11.

11.4.1 QoS FUNCTIONS

Quality of service (QoS) is supported by the media gateway through a combination of ATM traffic management, multi-protocol label switching (MPLS) and differentiated services (Diffserv) for IP. The selection of the AAL framing component is also QoS related as well as the protocols RTP, RTCP and RTSP. The MGC level controls the QoS handling in the MGW.

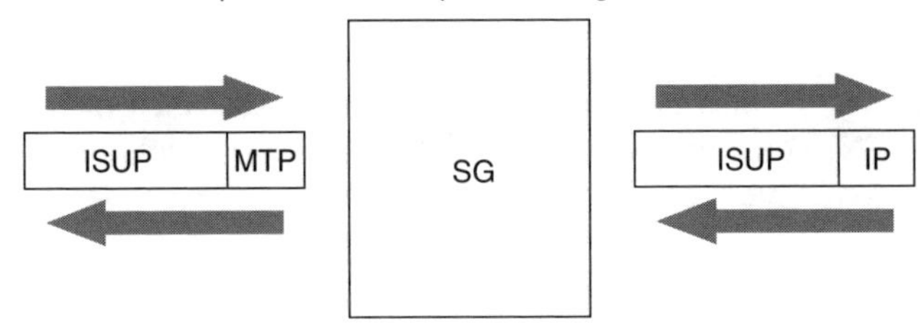

Figure 11.11 Changing the transport protocol of ISUP in a signalling gateway

11.4.2 SECURITY FUNCTIONS

The MGW security resources embrace various types of tunnelling, such as L2TP and GTP-U, IPsec and encryption. The MGC level controls the security measures taken in the MGW.

11.4.3 MANAGEMENT, BILLING

CDR (call detail record) generation is one of the functions.

11.4.4 ACCESS SERVERS

In general, present-day fixed access servers can be separated into two categories:

- narrowband access servers (NAS); and
- broadband access servers (BAS).

Narrowband access servers, are used for Internet dial-up applications and for providing Internet access to businesses via E1, primary rate access (PRA) or fractional E1 access. As dial-up traffic increases, local exchanges are subject to longer holding times (typically 20 instead of 3 minutes per call). When used for Internet traffic, the capacity of the circuit mode transport part of the network is not being used in a cost-effective way – a 64 kbit/s channel is used per modem, whereas on average only a fraction of this capacity is utilized because of a bursty traffic pattern.

Many Internet service providers have also expressed interest in outsourcing access server capabilities (Internet offload) to a telecom operator, and the narrowband access server migrates towards the local exchange or even towards remote subscriber switches (See for example Figure 13.5).

Broadband access servers, which are part of a DSL solution, are used to consolidate permanent virtual circuits (PVC) and to provide frame relay, ATM or IP connectivity over the backbone. They are also used for selecting ISPs and corporate networks. Also the broadband access server can be situated close to the local exchange. Moreover, since many functions of the broadband and narrowband access servers are similar, they can be provided in a common product.

11.4.5 VOICE GATEWAYS

The main purpose of a voice gateway is to convert packet voice (and similar services) for use in the PSTN. Voice gateways can reside within the access node (serving a local exchange via the V5.2 interface) or at the edge node. Some examples of voice gateways are the:

- VoATM gateway using ATM adaptation layer 1 (AAL1);
- VoDSL gateway using ATM adaptation layer 2 (AAL2); and
- VoIP gateway using H.323 (an ITU recommendation for multimedia applications) or the session initiation protocol SIP.

11.4.6 ROUTERS

The backbone and access networks will require the edge node to provide some routing capabilities:

- routers close to the subscribers will take care of leased-line and IP-based virtual private network (IP-VPN) traffic;

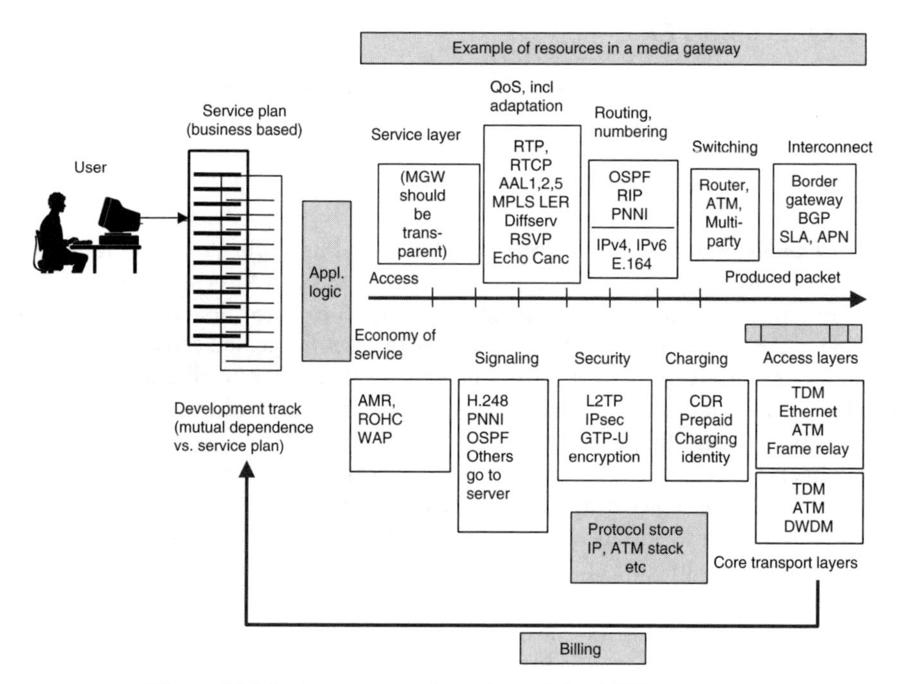

Figure 11.12 A more complete view of the MGW resources

- label edge routers (LER) and label switch routers (LSR) will handle connectivity toward a multi-protocol label-switching (MPLS) backbone network.
- egde routers will collect for example xDSL and WLAN traffic. Edge routers can contain LER/MPLS functions. They can be collocated or integrated with MGW.
- IP multicast routers will provide end users with streaming video or other multimedia applications.

11.5 SUMMARY

Finally let us use the fundamental plans to sum up the main tasks of the MGW. The approach is the one presented in Chapter 8. The scope of resources is impressive according to Figure 11.12. In a way it is easier to say what is not included, such as service enablers, application logic.

12

Packet Backbone

12.1 OBJECTIVES

The packet backbone is the main solution regarding core connectivity. Figure 12.1 shows its central network position. What are the technical and business requirements on the packet backbone? What protocol stack is available to fulfil the requirements, especially the IP and ATM enablers? What about control and management?

The main requirements on the backbone are low cost, high capacity, high quality and reliability. The fulfilment of these needs is feasible:

- Low cost: low cost per circuit because of low fibre cost and very high capacity in a fibre.
- High capacity: with new techniques, such as dense wavelength division multiplex (DWDM), 10 million users can communicate simultaneously over a fibre optic cable with 10 Mbit/s (!) bandwidth per user.
- High quality: normally fibre based with very few bit errors. However, the combination of real-time services and packet switching is still a challenge, requiring a set of QoS enablers and fairly sophisticated traffic engineering especially for IP transfer of traffic with multiple QoS requirements.

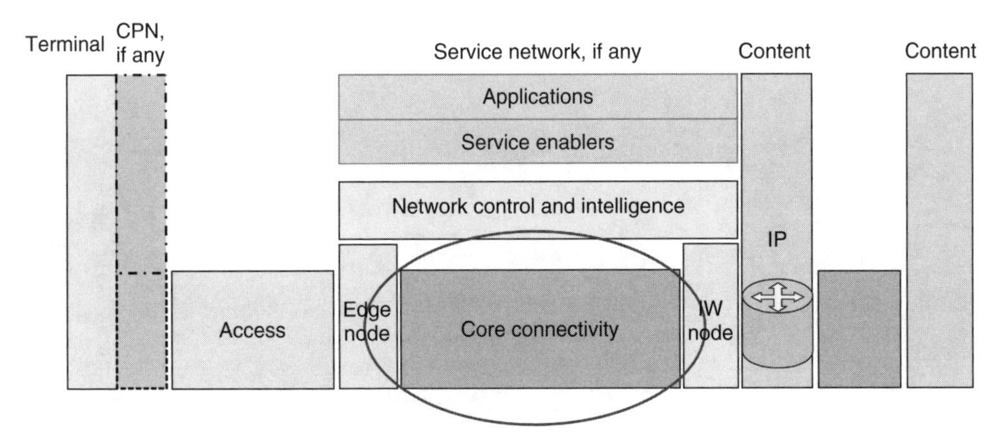

Figure 12.1 Position of the packet backbone

Understanding Changing Telecommunications – Building a Successful Telecom Business. Edited by A. Olsson
 ISBN: 0-470-86851-1

- Reliability: redundant capacity in ring/meshed networks, protection logic and cross-connect equipment for fast rearrangement in case of cable breaks etc.

See also the CD IP Internetworking, especially Chapters 11 and 12.

The main objective of this chapter is therefore to gain further knowledge about providing a bit-transporting backbone that fully responds to the demands of the service plan as cheaply as possible, regarding CAPEX and OPEX (CAPEX and OPEX are explained in Chapter 4). This is illustrated in Figure 12.2.

12.2 SERVICE PLAN VERSUS PACKET BACKBONE

See also Section 5.13 in Chapter 5, and many parts of Chapter 8.

As mentioned above, the crucial services are real-time services, traditionally voice but also in the future video telephony and similar multimedia real-time services. As usual QoS and security are in focus.

The heritage from the previous century is a number of vertical networks, which often carry the same media traffic, and therefore need interconnection. See Figure 12.3.

In this chapter the design of a universal backbone is in focus. Thus, instead of designing many vertical networks, including their respective backbone, let us look at a common multi-service backbone serving the 'all-access architecture'. Figure 12.4 also shows its role as a bridge to and between other network types. Thus, let us disregard the fact that many access network operators will tend to have tailored backbones as well.

A common backbone can be designed in numerous ways, with ATM as a common example of a transfer mode today. In principle, it should be possible to connect all types of accesses and other backbones to this network. A balanced traffic engineering and dimensioning with smooth scalability gives the best ROCE.

A multi-service packet backbone network must be capable of handling different types of traffic with widely differing characteristics, for example:

- real-time traffic for person-to-person communication
- Internet traffic

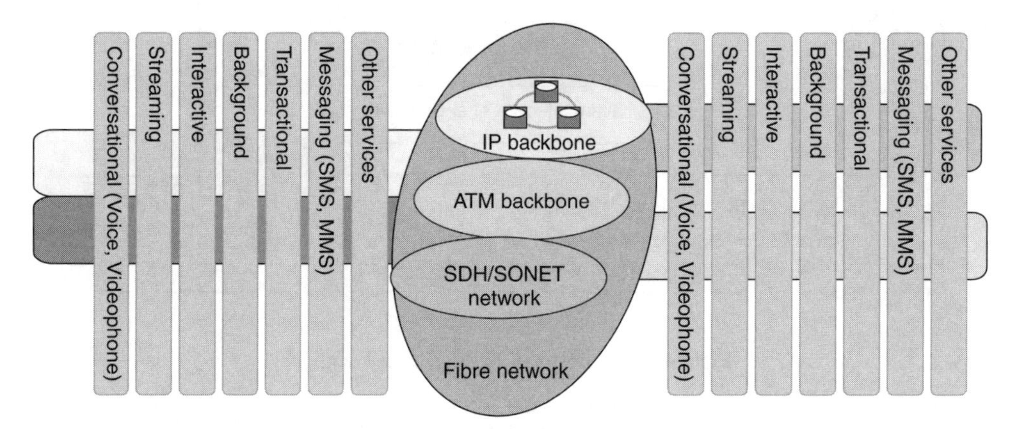

Figure 12.2 The wanted service classes and a common packet backbone structure today

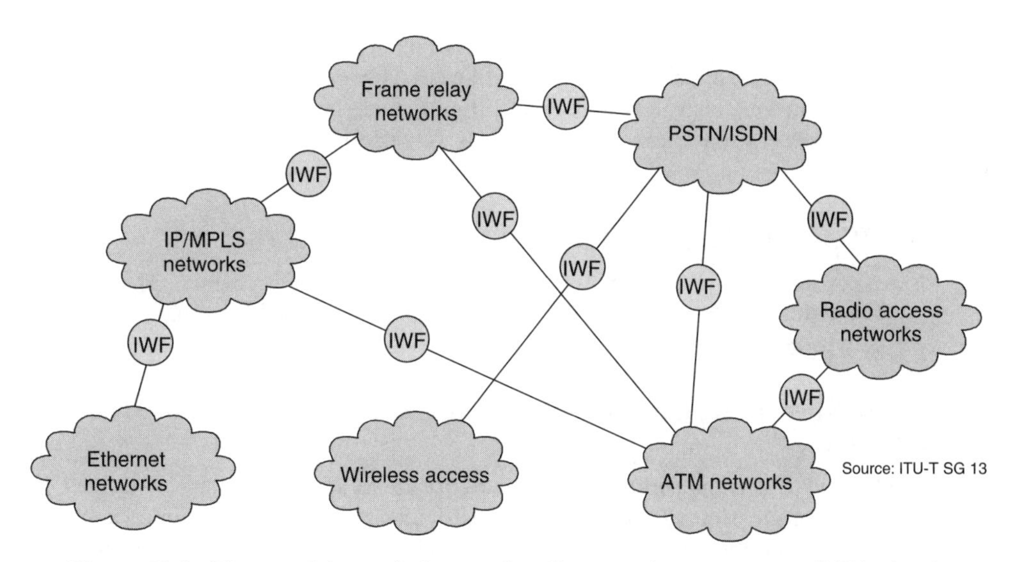

Figure 12.3 The era of the vertical networks offer many interconnect possibilities/needs

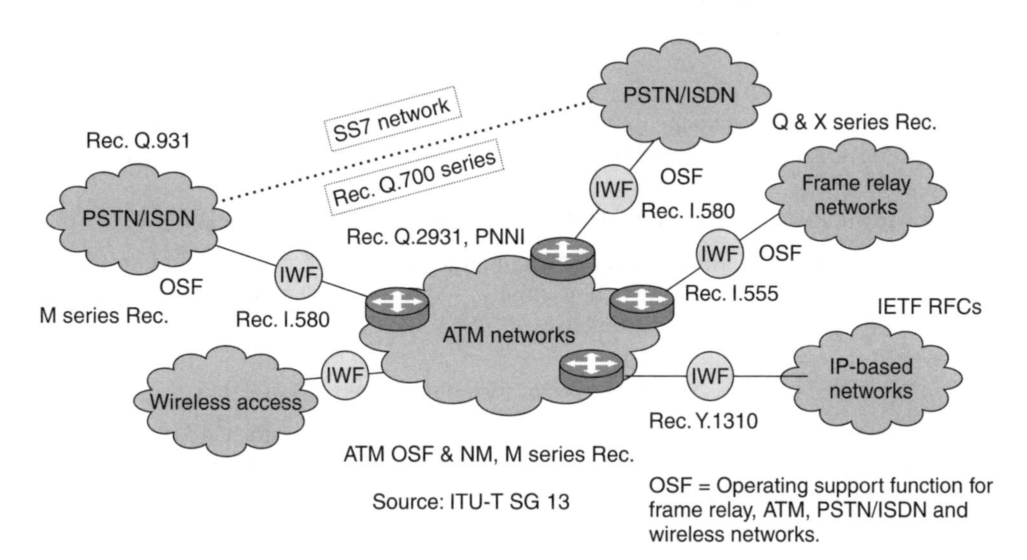

Figure 12.4 A common packet backbone simplifies the interworking needs

- enterprise communications
- video and voice streaming

Multi-service packet backbone networks need advanced QoS and high network resilience. Single faults should not impact the quality.

Terminology: multi-service does not mean multimedia! In this case service equals a network service, such as PSTN or frame relay.

The backbone does not really operate at a service media level, rather on a 'pipe level'. The pipe characteristics, priority etc. are set by the edge nodes and the control layer. If the backbone is operated exclusively with best-effort pipes there is no need for a control layer.

The mediation between backbone and access is performed in the edge nodes. See Chapter 11.

12.3 CAPACITY DEVELOPMENT

For the packet backbone there are two important dimensions to consider: The layered dimension and the distributed dimension. In order to understand the situation, a brief history may be useful.

Early Phase After the Digitization of the Transmission. Switched Circuit Mode in Fixed TDM

It is a well-known fact that ordinary voice calls for quite some time have been sent in time slots in a transmission system utilizing a particular transmission medium, such as a window in an optical fibre. Normally, the network designer did not care very much about the layering of this traffic within the medium. The transmission system was inflexible with manual cross-connection facilities in digital distribution frames in the nodes. Some traffic was through-connected and other traffic was dropped to the switching function of the node.

Much attention was/is required regarding resilience in the case of faults and the hierarchical design of the backbone network at a local, transit and international level.

The traffic was/is bundled by the switches into routes between the switching points.

Packet Traffic in Leased Time Slots. Switched Packet Mode in Fixed TDM

The packet traffic (X.25, frame relay, ATM) often used/uses a certain leased transmission bandwidth between the packet switching points. A pure vertical solution uses dedicated transmission systems and cables as well.

The SONET/SDH Phase. Managed TDM. Managed Cross-Connection

The introduction of SONET/SDH removed the inflexibility of the manual cross-connection by introducing a new managed network level with 'cross-connect switches' operated by commands. The concept includes programmed automatic protection switching in case of faults.

Stratification within the Packet Domain

Both the frame relay and ATM transfer modes gained a strong foothold in the backbone. Now one could see the beginning of stratification, where frame relay traffic is carried in ATM cells, that are carried in SONET/SDH paths with capacities from the basic PCM level (T1 or E1) and upwards.

The IP Layer. IP over ATM over SONET/SDH. IP over FR over SONET/SDH

The IP protocol 'carries everything over anything'. 'Anything' could for example be frame relay or ATM. The stratification increases when IP packets are distributed to ATM cells utilizing the AAL5 adaptation layer. ATM provides QoS to IP or other traffic, such as IP over AAL5 and voice over AAL1 or AAL2 adaptation layers. IP over ATM becomes a common stratification.

Optical Point-to-Point Lines Develop into Optical Networks. One Window gives a 'Lambda' Capacity. Optical Cross-connections for Windows and Fibres

The optical fibre capacity is split into wavelengths. The technique is called wavelength division multiplex (WDM). A recent development manages 160 wavelengths in one fibre. Figure 12.5 shows even greater capacity. Since such capacity is not often required, cheaper technologies can be chosen especially in metropolitan areas. An example is CWDM (coarse WDM) with 16 wavelengths. Figure 12.6 shows the development towards optical networking.

The forecasts around the millennium shift indicated a heavy increase in overall backbone traffic. For fixed broadband and corporate networks there were forecasts with more than 100 % yearly increases during the first years of the new millennium. A new managed

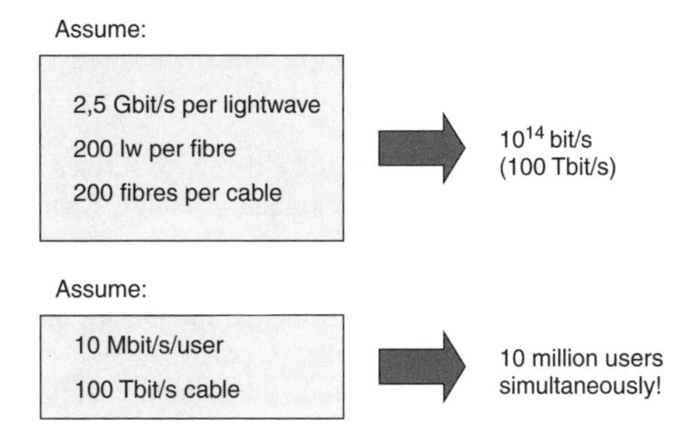

Figure 12.5 Potential capacity of one fibre cable

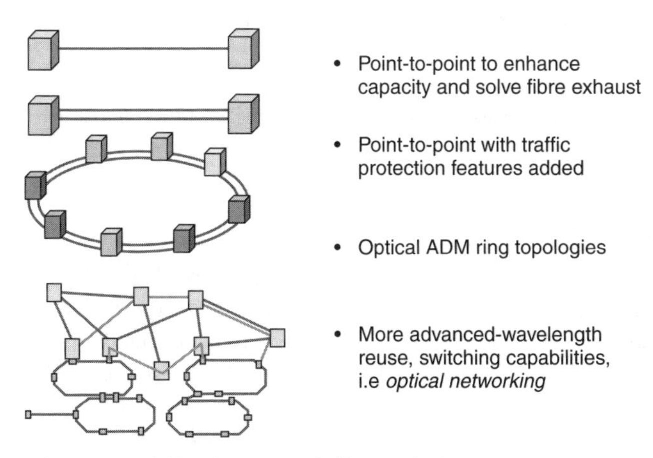

Figure 12.6 Development within the area of fibre technique – a new managed network level appears, with optical switches/cross-connects

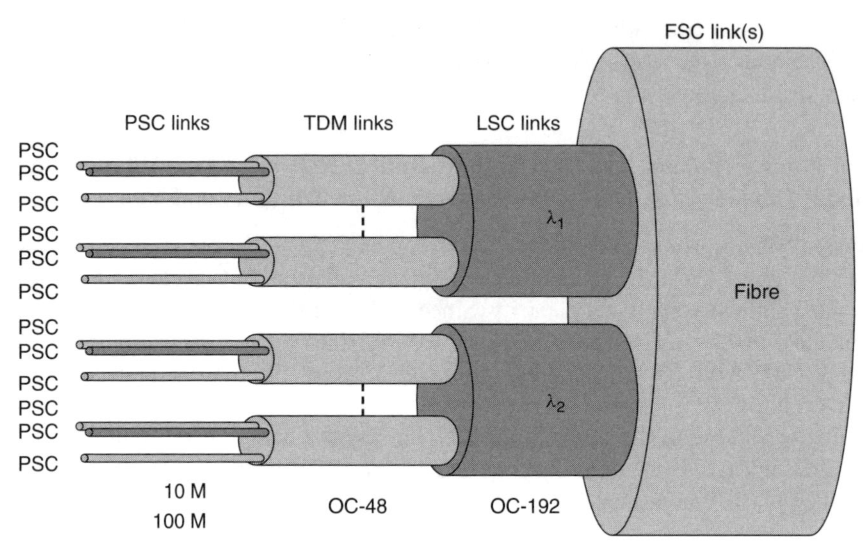

Figure 12.7 The available capabilities: fibre switching capability FSC, lambda light-wave (wavelength) switching capability LSC, time division multiplexing TDM and packet switching capability PSC

network level appeared with a tremendous capacity, thanks to further development of optical networking of the wavelength division technique and higher speeds. The result was a significant and fast extension of the backbone networks, bringing lots of over-capacity when the forecasts failed. The overall capabilities are illustrated by Figure 12.7. OC-192 (OC = optical carrier) has a capacity of 10 GBit/s and the OC-48 of 2.5 GBit/s. OC-192 is an established capacity in backbone networks.

Certainly, capacity is not the problem, rather the handling of the resources, that is control and management.

12.4 CONTROL FUNCTIONS IN THE PACKET BACKBONE

12.4.1 OVERVIEW

Figure 12.8 shows that the control functions are distributed to many layers, including the connectivity layer.

12.4.2 CONTROL SPLIT

Figure 12.9 shows an example of a border area functionality split between backbone connectivity control and call control. The exemplified standard (BICC) is described in Chapter 14. In the connectivity part the traffic resources are handled by means of bearer signalling. There are also QoS and security resources in the backbone. ISN is an interface-serving node, that is to say a gateway/edge node. This border area automatically brings the question of connectionless versus connection-oriented backbone network design.

For pure data IP connectionless networks are fine. However, IP will apparently conquer a much broader range of services. If considering all the services according to Figure 12.2,

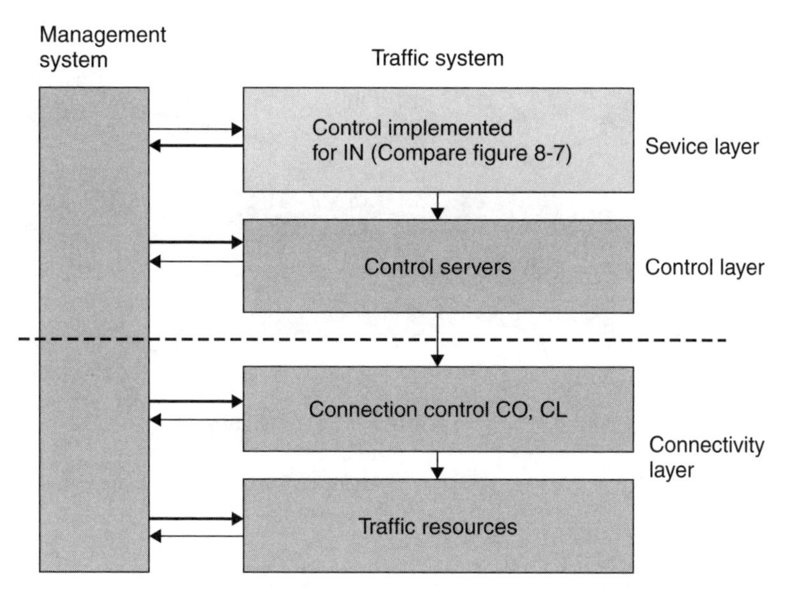

Figure 12.8 The principle split between call control and connectivity layer functions

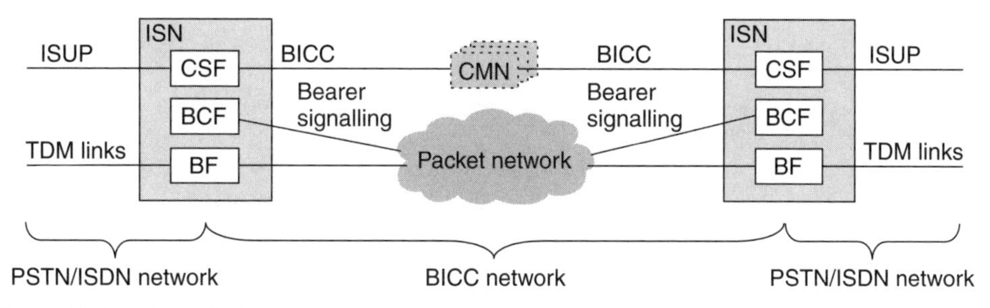

Figure 12.9 The split in control between control and connectivity layers, as illustrated by means of BICC

then connection-oriented design gives the best QoS and traffic engineering tools. This means that there is a need for connection-oriented IP network design, One of the most interesting standards is called multi-protocol label switching (MPLS). The reason for the name is that MPLS label switching can be achieved using any L2 protocol such as ATM, frame relay, Ethernet, and point-to-point links. An introduction to MPLS is presented in Chapter 7. The connection-oriented feature with paths is illustrated in Figure 12.10.

The backbone is now defined as the lowest elliptical layer in Figure 12.11 with some additions indicated in the text below the figure: QoS, routing, numbering and security solutions in the backbone, and bearer signalling for path set-up.

Regarding QoS, users will actually perceive the performance of the backbone network, especially in long-distance connections, where most of the transport distance utilizes the backbone.

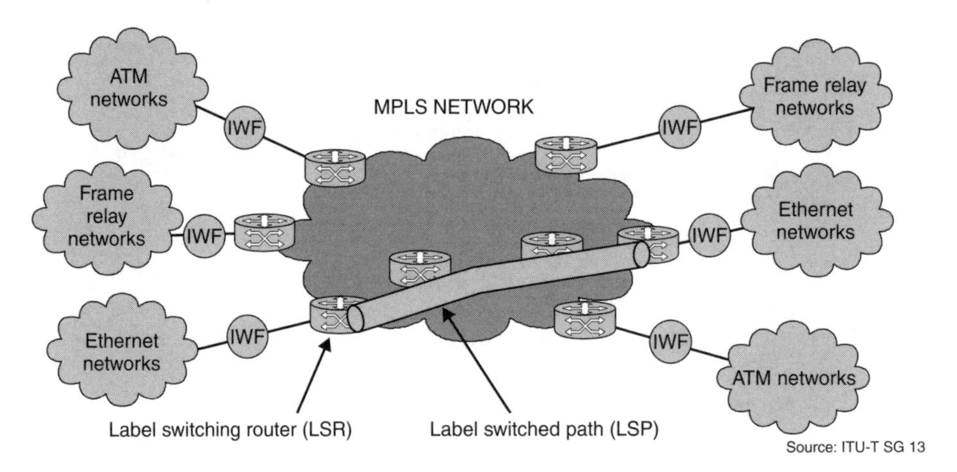

Figure 12.10 The MPLS principle to create paths in the backbone network

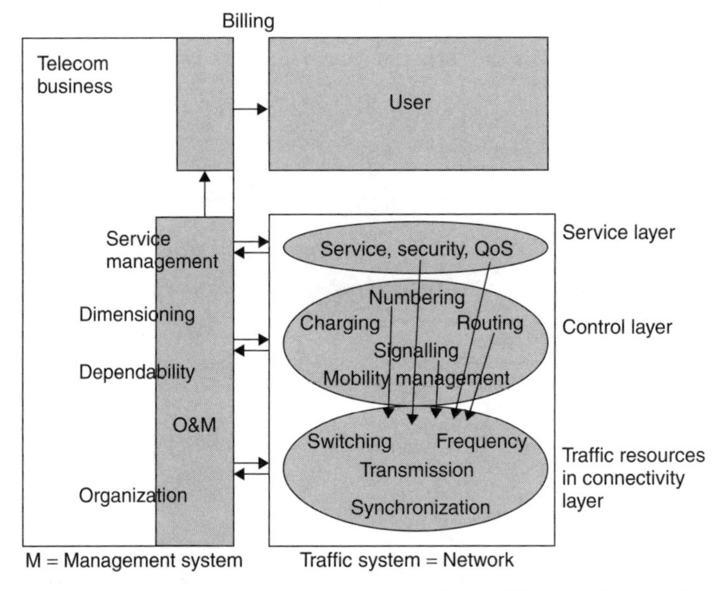

Figure 12.11 The FP is refined with connectivity control signalling, routing, numbering, QoS and security functions in the connectivity layer

12.4.3 DESIGN SPLIT

The functional split corresponds to a design split, where a core network design is normally split into:

- Control layer design, including MGW from the connectivity layer
- ATM and IP network design and dimensioning
- SDH/SONET/PDH network design

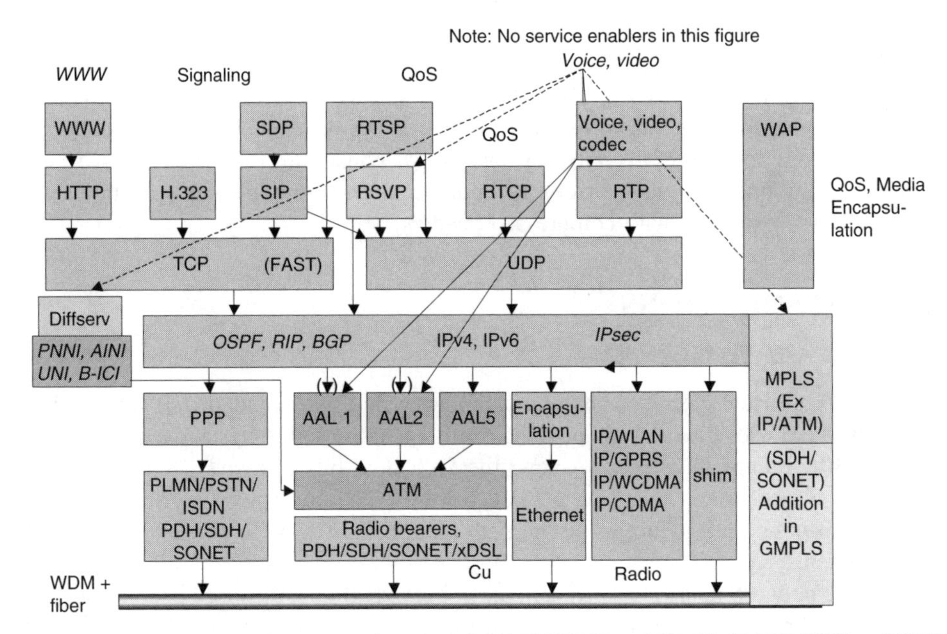

Figure 12.12 Examples of stack combinations: IP/ATM/SDH or SONET, IP/PPP/SDH or SONET (called PoS, packet over SONET/SDH), IP/MPLS/WDM and IP/Integrated Shim layer/WDM. A main task of the shim layer is framing

With a new standard GMPLS a more integrated planning is necessary. See next section.

12.4.4 LAYERING OF THE PACKET BACKBONE

See Figure 12.12. A similar figure has been used before, but protocols to the left and right have been added.

With all possibilities and layers, facilities management becomes somewhat cumbersome. An overall approach for multi-layer traffic engineering is called GMPLS, generalized MPLS, from IETF. GMPLS is briefly described in section 12.11.

A combination of stack reduction and a more general approach to resource handling, like GMPLS, would have impact on CAPEX and OPEX. This will be covered below.

12.5 THE DISTRIBUTED DIMENSION

Most of the figures show a limited backbone network. In reality it is global. A single operator can operate networks with hierarchical switches/routers. The similarities with the large circuit-switched telephone networks are clear.

See for example the router hierarchy in Figures 12.15 and 15.4.

12.6 TRAFFIC

Of the total inter-node traffic the largest volume is originated by fixed residential and SOHO broadband xDSL and cable modem accesses, and corporate accesses over

(S)HDSL, Ethernet over fibre, frame relay, ATM also over fibre, or possibly radio, such as LMDS. See Chapter 13.

Fixed voice traffic decreases its share in comparison with fixed broadband. Traffic from mobile networks constitutes a lower share than one would expect, since the data traffic is still very low and therefore not very bandwidth consuming. Provided it is designed for that, the (mobile or multi-service) backbone transports mobile voice traffic as ATM AAL2 cells or IP packets with compressed headers.

12.7 ATM SOLUTIONS

12.7.1 ATM AND IP OVER ATM

ATM is a VC (virtual circuit) switching technology that was standardized starting in the late 1980s. ATM uses fixed-length payloads with a length of 48 bytes and a 5-byte header, yielding 53-byte long ATM cells. Among the 40 header bits of a cell, 28 are reserved to identify the virtual circuit to which the cell belongs. The corresponding fields are called VCI/VPI (virtual circuit identifier/virtual path identifier). The VCI/VPI fields are updated at each switch.

The prime network niche for ATM is in the packet backbone. It is used in data backbones and voice backbones. In the access network there is stiff competition at layer 2 from the cheaper Ethernet.

ATM responds well to the requirements of a packet backbone. The equally long packets make it easier to work at high speed. The QoS capabilities were included from the start of system design.

For IP over ATM traffic we need a definition of the encapsulation of IP packets in ATM cells, i.e., how to put IP data inside ATM cells. See Figure 12.13. Encapsulation is performed by an *adaptation layer* (AAL). Most IP packets are too large to fit in a 53-byte ATM cell. Therefore, IP packets must be cut into smaller pieces in a process called segmentation before they can be encapsulated and put in ATM cells. The last router on the path of IP packets must reassemble the fragments to reconstitute the original IP packets. Segmentation and reassembly (SAR) is a complex and time-consuming process.

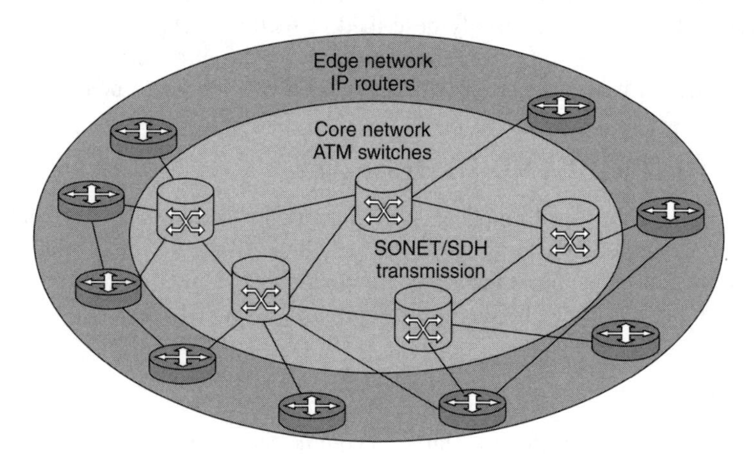

Figure 12.13 IP over ATM

12.7.2 THE ATM NETWORK AND ITS INTERFACE PROTOCOLS

See also Chapter 8, especially Figure 8.28.

In a common design with a bearer signalling system called PNNI (see below), the ATM nodes are grouped into peer groups (PGs). Each peer group consists of a group of ATM nodes that serve a particular geographical area. In order to keep the network, signalling and routing simple, the maximum number of nodes in a peer group is typically set to 10. This number includes existing and planned nodes for the geographical areas. Each PG is identified with a peer group identifier (PGI). Within a peer group a peer group leader (PGL) is selected. A peer group hierarchy is shown in Figure 12.14.

The ATM Forum defines a number of interface protocols, most of them previously introduced in Chapter 8, Section 8.7.2:

- private user–network interface (UNI)
- private network-to-network interface (PNNI)
- public UNI
- public NNI
- B-ISDN inter-carrier interface (B-ICI)
- ATM inter-network interface (AINI)

See also Figure 8.28

UNI signalling is used between a device and an ATM switch. It is based on Q.2931 ITU-T specification and is associated with layer 3 in the OSI model. Routing is provided with the aid of a logical address.

PNNI signalling is used between ATM switches. PNNI is an ATM Forum standard protocol applicable to private ATM switching networks that may also be used in public

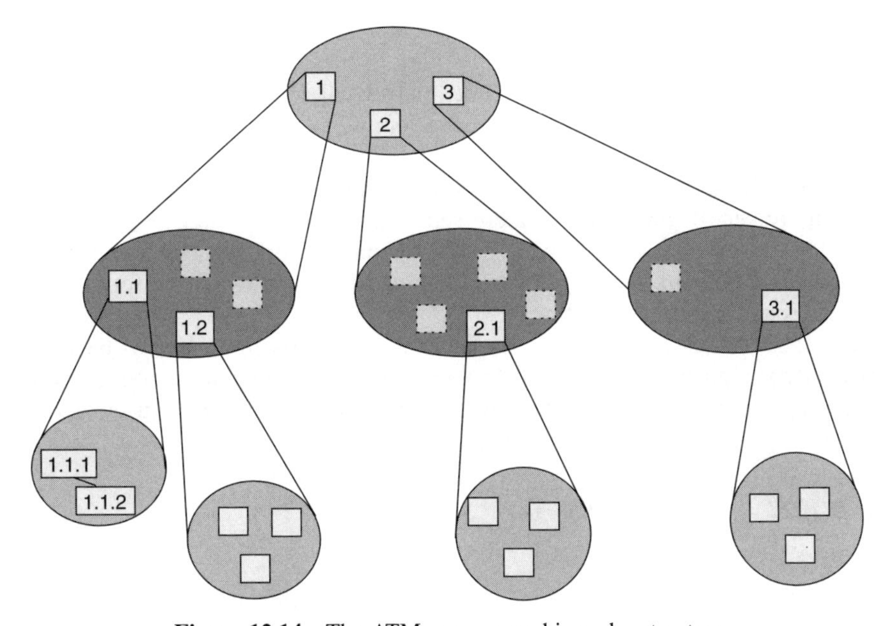

Figure 12.14 The ATM peer group hierarchy structure

networks. The PNNI signalling system consists of two protocol categories: routing and signalling. This protocol is further explained below.

B-ICI is the interface between different ATM providers.

The AINI protocol is based on PNNI signalling and is used between ATM networks.

A permanent virtual circuit (PVC) is a logical connection between ATM endpoints. Once the connection is set-up it remains up, until it is manually torn down. PVC is suitable for connections that are frequently used.

A switched virtual circuit SVC is a temporary logical connection established on demand. In each ATM node the information will be analysed in order to find the next point to be connected to. When the connection is not needed any more it is disconnected.

PNNI Routing Protocol

The routing part of the PNNI protocol is responsible for distributing topology information between switches and clusters of switches.

The PNNI routing protocol computes the routing of switched connections in the connectivity network based on the QoS requirements specified in the call set-up request. Depending on the size of the network there can be several levels in the routing hierarchy, as seen in Figure 12.14.

PNNI Signalling Protocol

The signalling part of PNNI handles the signalling between ATM switches in the network. It establishes point-to-point or point-to-multipoint connections across the ATM network.

The PNNI signalling protocol is used to establish and release connections in the connectivity network. A signalling control channel (SCC) is used to transport the PNNI messages between ATM switching nodes in the connectivity plane.

12.8 IP ROUTING

Routing is probably the most important function of IP. It is the act of moving information across the Internet from router to router using routing tables. A route is a path from the sending device to the receiving device.

The routing decision is based on the destination address of the packet and the routing table on the router. If the packet is addressed to a system on a network that is directly connected to the router, it will place the packet on that network segment. If it is addressed to a device on a network that is not directly connected, then it will be forwarded to an appropriate router for further routing.

IP routing protocols are dynamic. Dynamic routing calls for routes to be calculated automatically at regular intervals by software in routing devices. This contrasts with static routing, where routers are established by the network administrator and do not change until the network administrator changes them.

12.8.1 ROUTING PROTOCOLS

Overview

A number of routing protocols are in use today, e.g. interior gateway routing protocol, (IGP), exterior gateway routing protocol, (EGP), routing information protocol (RIP), open shortest path first (OSPF) and border gateway routing protocol (BGP). Once the destination address is resolved the routers determine the best way to reach that address through a network of nodes.

Routing algorithms are used to exchange information about the network topology. When *distance vector routing algorithms* are used each router builds and maintains its own routing table (a vector) that gives the best-known distance to each destination in the network and how to get there. These tables are updated by periodically exchanging information with their neighbours.

In *link state routing algorithms* each router learn the link state in the whole network. This is achieved by flooding the link state changes via the entire network. Flooding means that the information is copied from node to node in a tree structure. Each router builds its own forwarding table from its own perspectives.

Routing Information Protocol (RIP)

The RIP is a relatively old, but widely used interior gateway protocol (IGP) created for use in small, homogeneous networks. This is a classical distance-vector routing protocol. RIP allows hosts and gateways to exchange information for computing routes through an IP-based network. An entry in the routing table can represent a host, a network or a subnet. RIP uses broadcast user datagram protocol (UDP) data packets to exchange routing information.

Open Shortest Path First (OSPF)

The development of the OSPF routing protocol began in 1987. In order to be able to build large OSPF networks, an OSPF routing domain is split into regions called areas. Routers within a common area all maintain identical copies of the network map, or topology database. OSPF is considered to be an interior gateway routing protocol (IGP), meaning a routing protocol normally implemented on a network under the control of a single administrative authority.

The OSPF routing protocol adjusts to network changes more quickly than RIP and is more robust. In RIP the routers have to wait until the route is unavailable before action is taken, while in OSPF the routers maintain a database with the actual topology of the area. Each router updates the rest of the network with information on the direct connections it has to its neighbours. However, poor OSPF deployment causes worse network performance than the RIP protocol.

OSPF can route real-time traffic one-way and other type traffic a different way, performs load balancing by splitting the load over several multiple lines and supports hierarchical systems when needed.

The OSPF represents the actual network with a graph consisting of several areas. Each area is considered as an autonomous system (AS), and every AS has its own backbone called area 0. All areas are connected to a special area called a backbone area. Areas are connected to the backbone using an interface on one of the routers in an area. See Figure 12.15.

Border Gateway Protocol (BGP)

Between the AS the BGP is used. This is the mandatory exterior gateway protocol used between ISPs. The BGP is a path vector routing protocol. With path vector routing, the

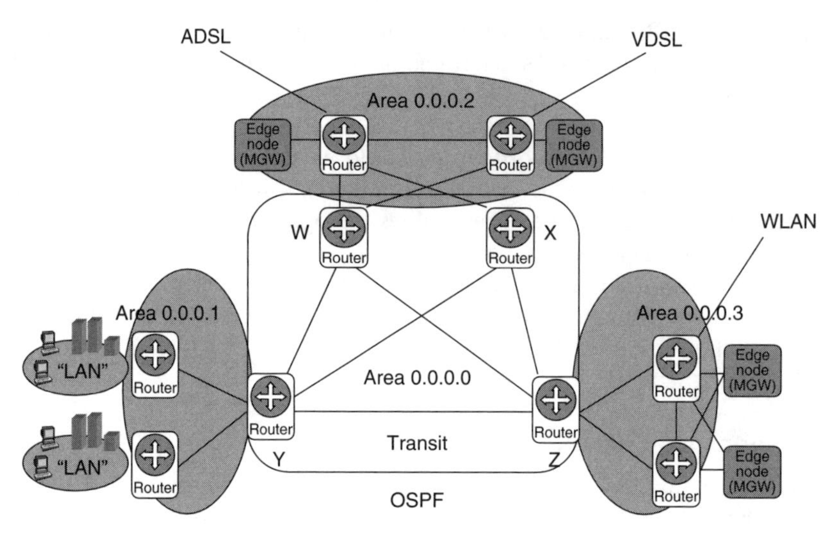

Figure 12.15 OSPF areas in hierarchical IP router networks

routers simply add the AS number before sending route information. This enables the receivers to choose the shortest path. The path vector routing carries information about all the autonomous systems it passes on the way to the destination. Therefore the router can choose not only the shortest path, but also it can choose paths depending on the policies between two ISPs. BGP neighbours are called peers. The connection between the peers is provided by TCP.

12.9 IP QoS

12.9.1 RSVP

RSVP was briefly described in Chapter 7. RSVP is not used very much in the backbone, since it does not scale very well. A special signalling protocol, RSVP-TE, is used in MPLS. TE stands for traffic engineering.

12.9.2 DIFFERENTIATED SERVICES

A brief introduction to differentiated services is supplied in Chapter 7. Essentially, differentiated services (Diffserv) specify the transfer priority of a packet as it passes through each network device on its journey through the network.

When using Diffserv a number of mappings are necessary. First, the actual types of traffic stream, such as VoIP and IP-VPN, are mapped to a number of traffic flows, called behaviour aggregates (BA). Traffic belonging to a certain BA is treated in the same way. These service classes are then related to a code point value. This value is called the Diffserv code point (DSCP) and it is carried in the IP headers. It is often referred to just as code point.

Finally, the DSCP values are mapped to the alternative priorities for transfer across the Diffserv domain. These priorities are called 'per-hop behaviour' (PHB). PHB is thus a name for a certain way to treat a packet stream. The treatment contains essentially loss, delay and bandwidth characteristics.

The routers offer a number of PHBs to the traffic, and this selection is called a PHB group. A PHB can also be called a forwarding class. Thus, a forwarding class has a certain DSCP value and a certain PHB.

The byte used for carrying the DSCP is the "type of service" field in IP version 4 and the 'traffic class' field in version 6. See Figure 12.16. Only six of the eight bits are used, which gives a maximum of 64 classes.

Every network has a traffic limit for each forwarding class. The need for traffic policing at the entrance to a Diffserv domain is obvious, as well as agreements between the traffic sources and the Diffserv domain operator on setting of priorities. See Figure 12.17.

All nodes will sort incoming packets according to their service class and treat them differently depending on their needs. The sorting processes in the traffic plane are called packet classification and traffic conditioning.

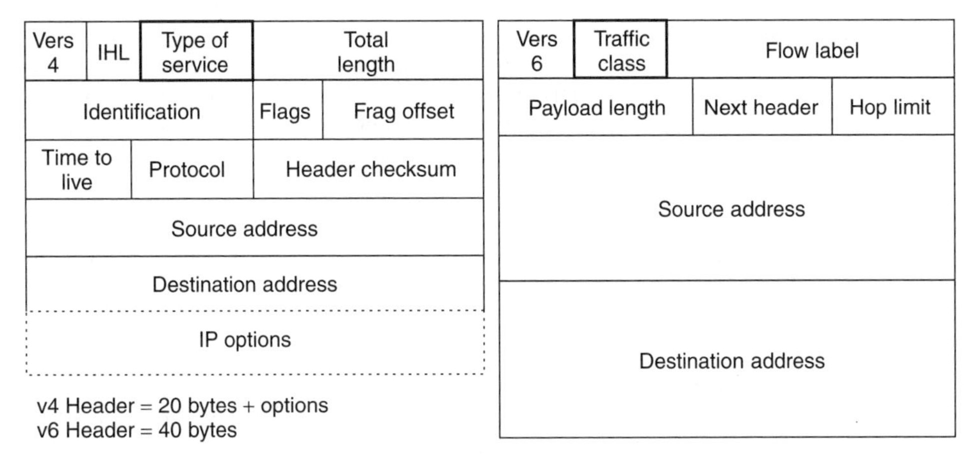

Figure 12.16 Code point locations in IPv4 and IPv6 headers

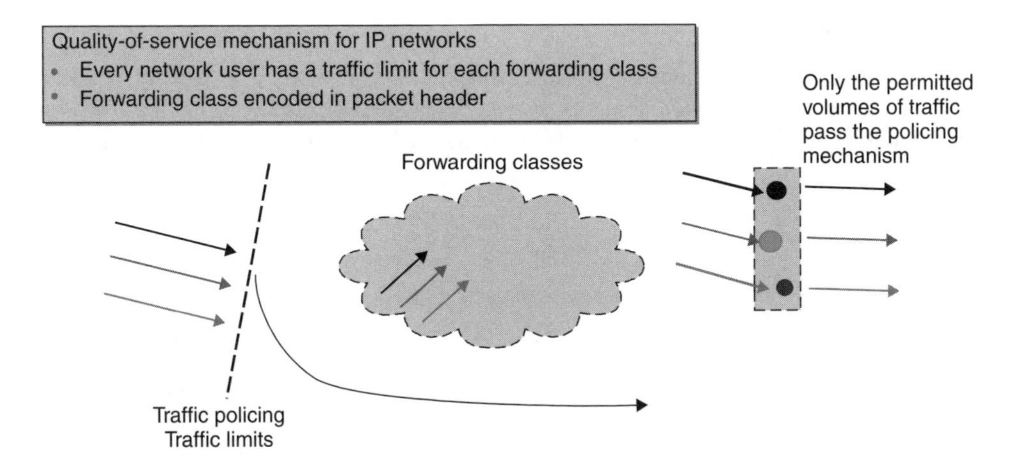

Figure 12.17 Diffserv principles

Packet classification is the process used to determine from the application what QoS level should be applied. Packets from multiple applications/sources could belong to the same BA.

Traffic conditioning is used to reshape a traffic stream to fit the traffic restrictions.

Default PHB should result in best-effort service. Some PHB types are:

- **ELL:** emulated leased line
- **EVLL:** emulated virtual leased line
- **QAMS:** quantitative assured media playback service
- **EF:** expedited forwarding

EF is a defined PHB, like assured forwarding. EF PHB offers low loss, low delay and low latency. EF could be used for VoIP, possibly also for IP-VPN, for example based on business aspects such as SAP traffic.

Otherwise assured forwarding (AF) could be used for IP-VPN and other traffic types with a need for certain priority. AF PHB offers a kind of PHB internal priorities, expressed as AF 1, AF 2, etc. From the point of view of a customer AF offers the possibility to tell what traffic could be dropped first, if the network is congested.

Diffserv does not give any guarantees. It is like travelling by train without knowing when you will arrive. The standard is a compromise between reservation (you get a seat, but this could also be cancelled) and prioritization (you travel faster than packets with a lower priority). Diffserv is considered a scalable technology that can offer differentiated services for various needs in large networks. It is therefore a viable alternative in the packet backbone.

Diffserv does not use any signalling.

Diffserv is an IETF standard. It only addresses layer 3 and makes no assumptions regarding the underlying transport, as does MPLS.

Network Design

One or more Diffserv domains can form a Diffserv region with some type of common traffic-handling rules. Between domains, service level agreements (SLAs) and traffic condition agreements (TCAs) are used. Each corporate or ISP network is a DiffServ domain.

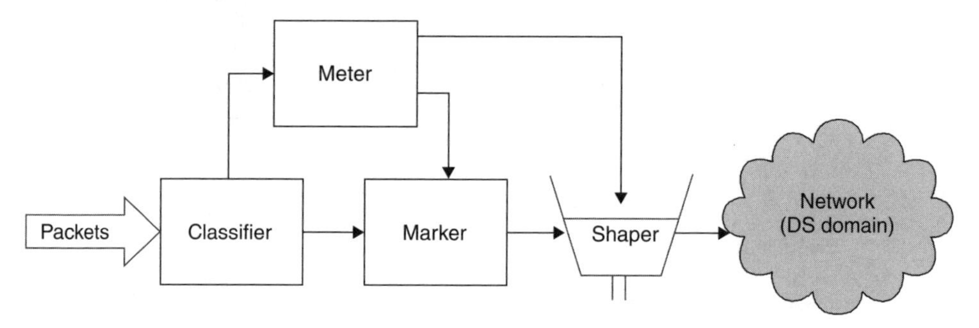

Figure 12.18 DiffServ traffic conditioner block (TCB)

For true QoS, the entire IP path that a packet travels must be DiffServ enabled. An example of service policy might be that EF gets 10 %, Gold 40 %, Silver 30 %, Bronze 10 %, and best-effort traffic (default class/PHB) the remaining 10 % of the bandwidth. Gold, Silver and Bronze could be mapped to AF classes AF1, AF2 and AF3 for example.

Figure 12.18 illustrates the typical traffic conditioner at the edge of a DS domain. A DS internal node enforces the appropriate PHB by employing policing or shaping techniques, and sometimes remarking of profile packets, depending on the policy or the SLA.

12.10 MULTI PROTOCOL LABEL SWITCHING (MPLS)

An introduction to MPLS is included in Chapter 7. MPLS is used at speeds up to 10 Gbit/s.

In classical IP over ATM the IP packets need to be put together again (from cells) in each IP node, which is cumbersome. The need for a new switching technology with IP packets to replace IP over ATM was identified in 1996 and standardization of MPLS started in 1997. MPLS runs over many existing network hardware like Ethernet, frame relay and ATM and supports the forwarding of IP packets over virtual circuits. MPLS is implemented in IP routers. Essentially, MPLS improves IP scalability and quality of service by creating virtual label-switched paths (LSP) across a network of label switching routers (LSR).

In MPLS, each packet carries a virtual circuit identifier, called a *label*, as a field of a *shim header* inserted between the IP header and the link layer header of a packet. A single packet can carry more than one shim header. An MPLS header is 32 bits long and labels take only 20 bits out of these 32 bits, therefore leaving room for further information inside the header. The 'label' is a number, which identifies a path across the network. The packets are routed more quickly based on a label, than when having to look up the destination address in the IP packet.

MPLS handles labels just like all other virtual circuit identifiers are handled in other virtual circuit switching technologies. Consider an IP packet sent through the network in Figure 12.19. The packet is forwarded through an MPLS network or *domain*. When the packet arrives at the first MPLS router, also called the *ingress label edge router* (ingress LER) of the MPLS domain, the source and destination IP addresses of the packet are analysed and the packet is classified in a forwarding equivalence class (FEC).

All packets within the same FEC use the same virtual circuit, called the *label switched path* or *LSP*. There are LSPs for both individual and aggregated traffic streams. Separate LSPs are used to serve flows with different QoS requirements.

Suppose a virtual circuit has already been established for the FEC of a packet. Then the ingress LER inserts or *pushes* an MPLS header (L1) on the packet. Subsequent routers of the MPLS domain update the MPLS header by *swapping* the label L1 against L2, L2 against L3 etc. Finally, the last router of the LSP, called *egress LER*, removes or *pops* the MPLS header so that the packet can be handled by subsequent MPLS-unaware IP routers or hosts.

The main benefit of using MPLS in an IP core is optimization of resources through dynamic bandwidth allocation and traffic engineering. Other benefits include scalable IP over ATM support, separation of routing and forwarding and support for VPNs.

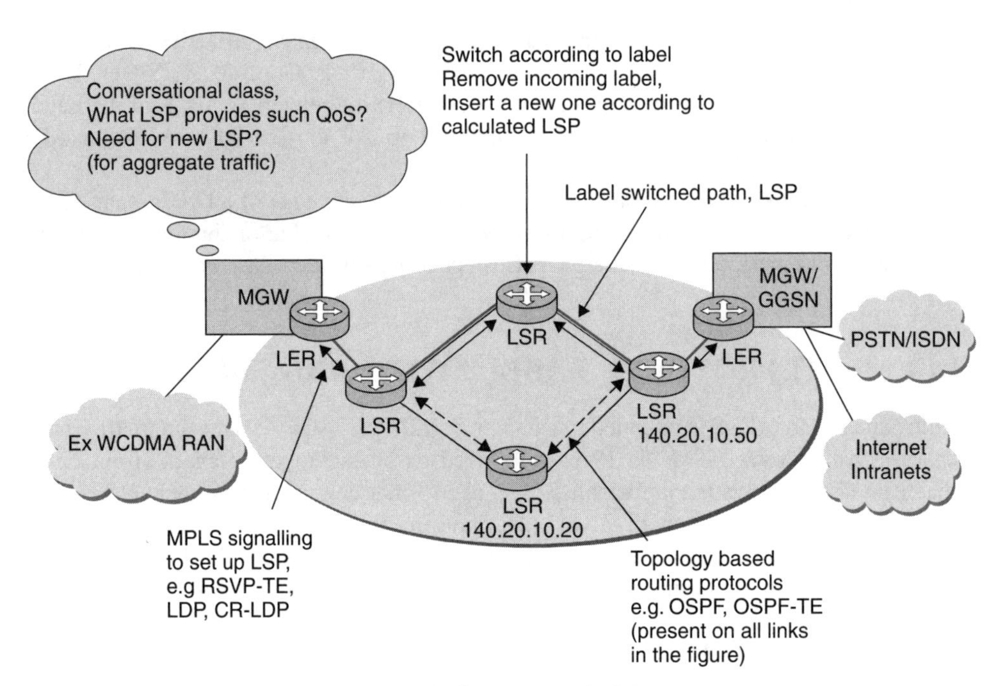

Figure 12.19 MPLS principles

12.10.1 SIGNALLING AND ROUTING

LSPs are set up between MGWs based on topology information provided by routing protocols, such as OSPF or OSPF-TE (TE = traffic engineering) for GMPLS. Each LSP carries traffic from only one service class. The LSPs are created with a specific set of characteristic parameters (carried in MPLS signalling) according to the type of traffic (voice, signalling, etc.) it is intended to carry. LSPs may be set up either statically or dynamically according to the actual bandwidth needed between the media gateways.

For dynamic set-up there is a need for signalling. In GMPLS the signalling is provided by the protocols RSVP-TE and/or CR-LDP (CR = constraint-based routing).

Not all types of traffic carried through the network require the use of LSP paths. For real-time applications and signalling the use of LSPs enables proper service provision, while for non real-time traffic e.g. best effort, the use of LSPs may be omitted to enable more flexibility and reduce the network management costs.

MPLS trunks are used to apply adequate traffic engineering, security and resource management on the core network.

12.11 MULTI-LAYER CONTROL

12.11.1 GENERALIZED MPLS, GMPLS

As the requirements for bandwidth increase there is a need to manage and control large optical networks. MPLS-based control intelligence is being adopted for use within optical networks to shorten provisioning times.

Ordinary MPLS does not embrace the optical layer. In fact, of the capabilities in Figure 12.7 only the packet-switching capability is included. However, the development pushes a need for a control and management plane that extends from IP at layer 3 right down to the optical transport level at layer 1. This includes fibre-switching capability, lambda light-wave (wavelength) switching capability, time-division multiplexing, and packet-switching capability.

The developing standard generalized multi-protocol label switching (GMPLS) aims to meet this need to coordinate capabilities at different layers.

GMPLS' primary enhancement to MPLS is its capability to establish connections at layer 1.

GMPLS can be deployed in two ways: overlay model or peer model. In an overlay model, also called a UNI, the router is a client to the optical domain and interacts only with the directly adjacent optical node. In the overlay model, the actual physical light path is decided by the optical network and not by the router.

In the peer model, the IP/MPLS layer operates as a full peer of the optical transmission layer. Specifically, the IP routers are able to determine the entire path of the connection, including through the optical devices.

The following resources may be included:

- **Switching resources –** packet switches FR, ATM, IP, time switches/SONET/SDH, wavelength switches, OXC, optical cross-connect, PXC, photonic cross-connect.
- **Transmission resources –** transmission resources are expressed in LSP hierarchies. Large LSPs carry smaller ones. Aggregation is performed upwards in the network in discrete units, 1 to 10 Mbit/s, 100 Mbit/s.
- **Business aspects –** GMPLS is expected to help operators dynamically provision bandwidth and capacity, improve network restoration capabilities and reduce operating expenses. Services such as optical VPNs may become profitable.

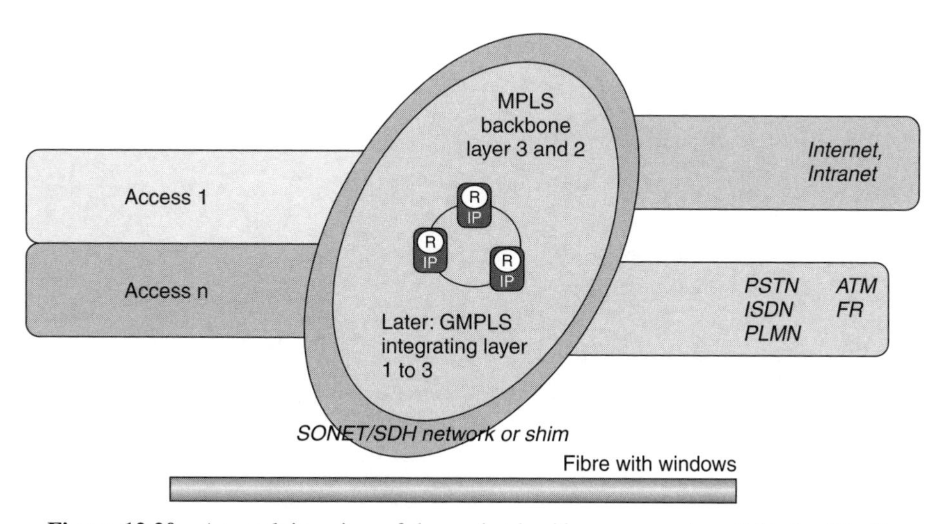

Figure 12.20 An evolving view of the packet backbone, managing all kinds of traffic

Demand for GMPLS should grow as IP traffic and services increase. A challenge for operators is to coordinate optical transport and IP administrative domains.

For management a link management protocol (LMP) has been developed. There will be thousands of LSP to manage.

Other supporting standards are:

- OSPF-TE (TE = traffic engineering) – auto-discovery of network topology
- RSVP-TE for establishment of LSP
- CR-LDP – constraint-based routing

Figure 12.20 illustrates a possible MPLS/GMPLS based packet backbone design as an element of an all-service and all-access network.

13

Access Network

13.1 OBJECTIVES

The main objective of this chapter is to create an improved understanding of access networks at an overview level. Such understanding includes the fragmentation in access alternatives, some common architectural features, and ways to improve ROCE in the expensive access segment. Figure 13.1 shows its position.

This chapter is closer to standardized systems than other chapters in the book. Deeper descriptions of particular access systems can easily be found in the rich literature that is available on the market.

13.2 INTRODUCTION

In order to retain a balance between different network segments in this book, there is no point in going too deeply into individual system solutions in such a fragmented area as access has become. The number of parameters includes for example the range of services to be offered, transmission media, availability of frequency bands, the layers above including the signalling system, multiplexing and aggregation technique, subscriber

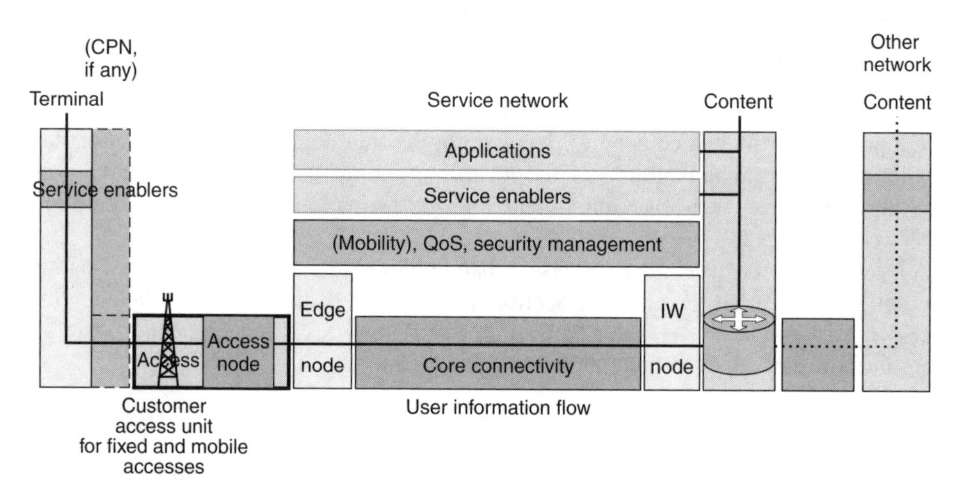

Figure 13.1 Position of the chapter, and the main access internal nodes

Understanding Changing Telecommunications – Building a Successful Telecom Business. Edited by A. Olsson
 ISBN: 0-470-86851-1

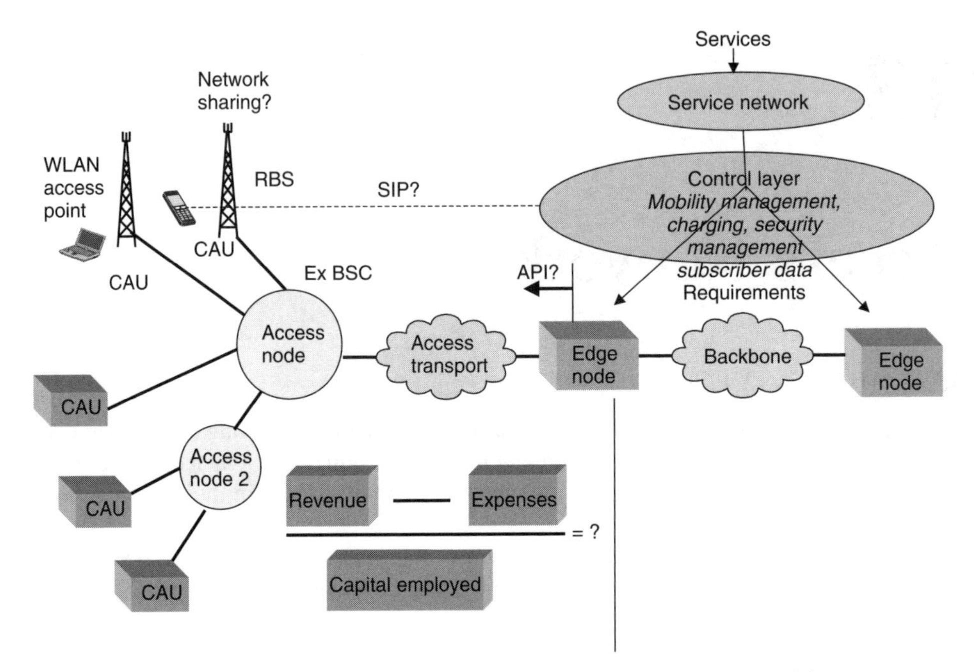

Figure 13.2 A broader view of access architecture, environment and issues

velocity, variations in business conditions such as ARPU, adaptation to geographical topology and subscriber density, distance to the edge node, cost development and new enabling technologies. The result is thousands of different possible access network designs. Therefore stress is put on the more generic aspects of access, complemented with overview information on some specific systems. In the long run access independence is more or less necessary, and the development is ongoing. A fairly generic access architecture with some hot issues is shown in Figure 13.2.

The access network is continuously evolving and adapting to meet emerging needs. Some drivers of this evolution are business aspects (expressed as ROCE etc.), service development and the related need for bandwidth, the Internet protocol (IP), the regulatory development such as unbundling of the local loop, competition; emerging technologies; QoS and reliability; security requirements and the formation of a multi-service and all-access core network.

A brief part on the access network as the bandwidth bottleneck is included in Chapter 3, Section 3.13.2. See also Chapter 4, Section 4.7.3, which deals with cost efficiency in the access network. Finally, Chapter 8, Section 8.3, Figure 8.12 and associated text gives more introductory input to the area of this chapter.

13.3 WHAT IS AN ACCESS NETWORK?

In the reference model used, the access network extends between the subscriber premises and the point that corresponds to a mobile switching centre (MSC), local exchange (office), edge node or media gateway (MGW).

In this book, the term access network (AN) is used to represent only the basic access connectivity service over various infrastructure technologies.

What is called an ad hoc network is briefly covered at the end of this chapter. An ad hoc network means units that occasionally are close to other units and establish a connection and relationship between them. An example would be to utilize an (idle) near-by TV screen as a display for a mobile service, by attaching it to a mobile device. Bluetooth is a main enabler for ad hoc networks.

Most of what happens in the telecom world has a business-oriented explanation, often service-oriented. The access area is no exception. Let us therefore take a brief look at access as a serving network segment to the driving services. Figure 13.3 shows four main steps from voice over voice optimized systems to multimedia over IP systems.

The trend is a development from 'all services over circuit mode' to 'all services over IP', although there is a pronounced uncertainty, for example for QoS reasons. At the same time the narrowband bandwidth is successively complemented by broadband accesses to fixed and mobile subscribers. The development towards IP is shown in Figure 13.4.

1. Voice over voice optimized system (fixed PSTN developing towards cellular PLMN)
2. Modem, dialled up over voice optimized system
3. Broadband data/ Internet 'Always on' etc. over data optimized system
4. Voice and data over compromise system (cells)
5. Voice and data over IP with QoS

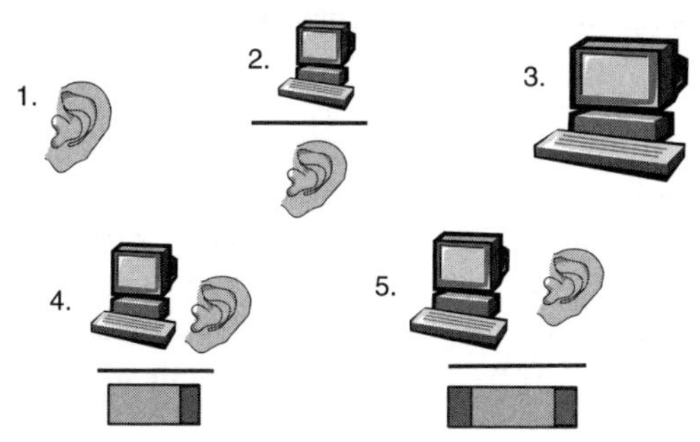

Figure 13.3 Access network service drivers

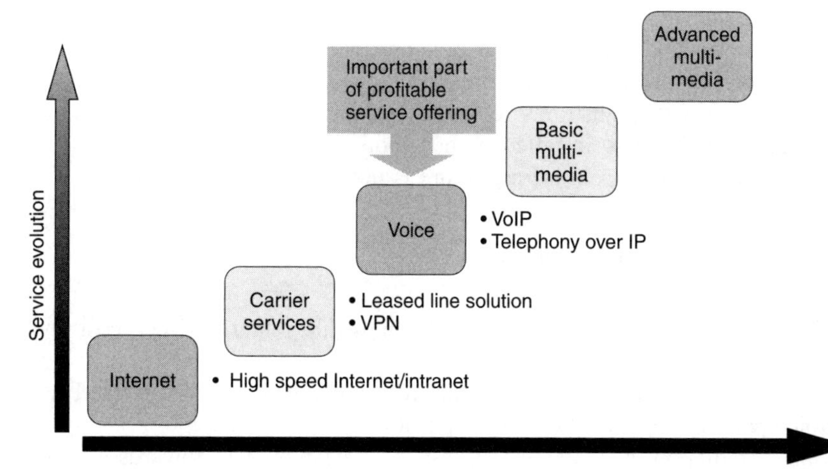

Figure 13.4 Two main applications of IP today are Internet access and IP-VPN, with VoIP and MMoIP (multimedia over IP) in the pipeline or being implemented

As seen from previous chapters, VoIP (as a replacement for PSTN/GSM etc.) requires new signalling in the network (SIP or H.323) and the multimedia may require support from the IMS in for example UMTS (see Chapter 14).

13.3.1 HISTORY AND FUTURE

The access area includes many old assets, such as copper cables. And still they play a role; just look at the xDSL systems. 'Copper is buried, but not dead'. So let us look quickly at the history. See also Chapter 3. The phases below are overlapping in time.

- Phase 1: PSTN over copper.
- Phase 2: Data over PSTN or leased lines using modem.
- Phase 3: Introduction of dedicated switched data networks: X.21, X.25, and later frame relay and ATM. Mainly used for business (intra-company) traffic.
- Phase 4: A connectionless datacom technique, IP, appeared. Universities and defence were among the users.
- Phase 5: ISDN was the first attempt to integrate voice and data. At that time (mid-1980s) no killer data application was available to fill the capacity.
- Phase 6: Mobile voice was introduced early, but the real worldwide break-through came during the 1990s.
- Phase 7: ATM was developed during the 1980s as next integrating system. The initial deployment concerned data. During the 1990s the technology found a fixed-voice application in backbone networks and was used in some mobile systems from the terminal. Utilization: data over ATM and voice over ATM rather than integration.
- Phase 8: The killer application, WWW, changes the telecom world during the 1990s. E-mail supports. Suddenly all accesses need to find a path to the WWW servers and acquire an e-mail address. See Figure 13.5.
- Phase 9: Now the killer applications are in place and dedicated broadband data accesses are developed, such as 3G mobile systems and WLAN. On the wired side cable modems and xDSL systems find a large number of subscribers and Ethernet over fibre solutions offer even more bandwidth and promises.
- Phase 10: Nation-wide transitions to broadband radio access coverage require a lot of investments. WLAN is installed in hot spots. The business case for such broad extensions of the radio access infrastructure is not clear, causing some delay.

Among the reasons for uncertainty are:

- There is no linearity between revenue and bandwidth, rather between cost and bandwidth.
- Instead of increasing, the ARPU has been fairly stable during recent years. The income from voice per user declines in many countries.
- The issue of radiation is frequently discussed. A development towards smaller abundant (low cost) base stations would solve the problem. This is indicated in Chapter 1, Section 1.3.

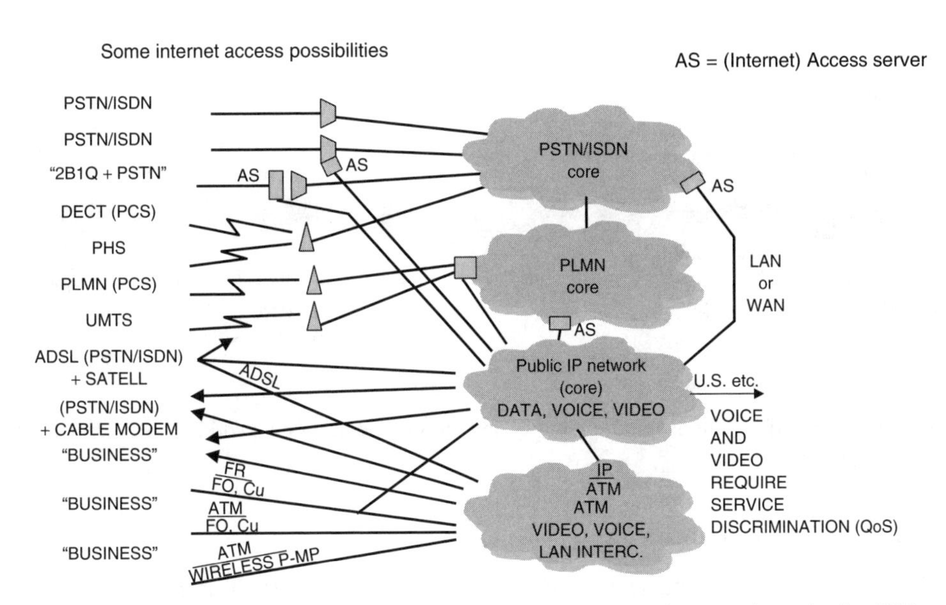

Figure 13.5 Internet access over all accesses quickly became the a top issue in the 1990s

Some support for an improved business case has been covered in previous chapters:

- Chapter 4: Re-using existing investments. Convergence fixed-mobile access. Network sharing.
- Chapter 5: 'Service continuity' over different systems. Stepwise extensions.

The fact that voice and data have quite different network requirements has been reflected by dedicated access system developments. However, since access is the most expensive part of the network, the interest in integrated access solutions has been substantial. Integration attempts include finding common transmission as well as common transfer mode. Circuit-switched data technologies (X.21) and several packet switched voice trials testify to the ambitions. The success of the circuit and packet solutions as final integrated solutions was initially limited and a cell-switched mode (ATM) was tested as a compromise. However, it was the IP packet technology that became part of all computers, thus becoming the strongest access standard at the network layer, with Ethernet as the strongest link layer technology.

With voice becoming a minor part of the total traffic, and packet technology becoming the dominating transfer mode the question is if VoIP is the final solution for voice access? For an enterprise there are obvious savings in cabling and overall operation costs. The access systems for mobile and fixed telephony can quite easily carry multiple mode solutions, such as voice in circuit mode over ADSL combined with data in packet mode over IP (or ATM) over ADSL. A corresponding situation prevails for different releases in UMTS. The conclusion is that it is a matter of finding an economically comfortable conversion pace towards an assumed 'all-IP access'.

Also QoS and security are important access issues. See Figure 13.6. The development is supported by the operating systems in the computers that take on more and more network-oriented tasks, also in QoS and security.

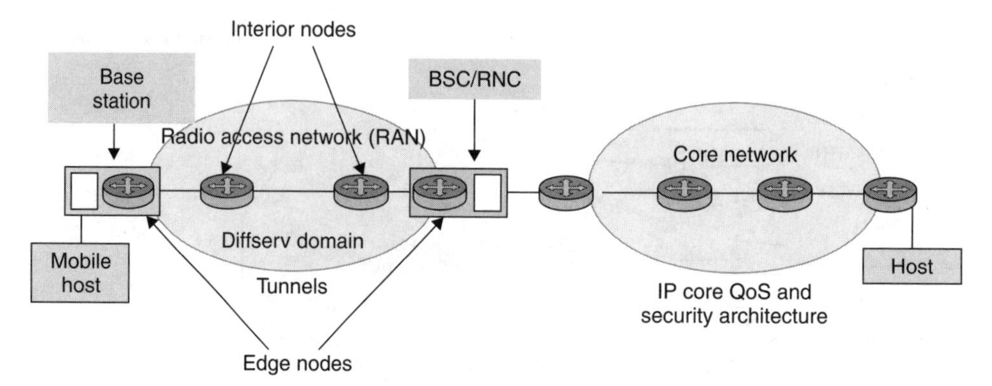

Figure 13.6 All IP solutions require security and QoS in the access network, such as tunnels and Diffserv respectively

13.3.2 THE SUBSCRIBERS

The subscribers can of course be segmented in many ways, such as residential users, mobile users, appliances, small and medium size enterprises, corporate regional, national or global enterprises with considerable intra-company traffic, and so on. See Chapter 2. Lately the expression 'hot spot' has become common, since specific solutions target this segment. Examples of hot spots are found in Figure 13.7.

A somewhat related segmentation differentiates between rural, suburban and urban subscribers. A metropolitan area is a common description of large cities. Regarding Figures 13.7 and 13.8, 3G/EDGE and WLAN etc are just examples of a number of possible solutions. See Figure 13.8.

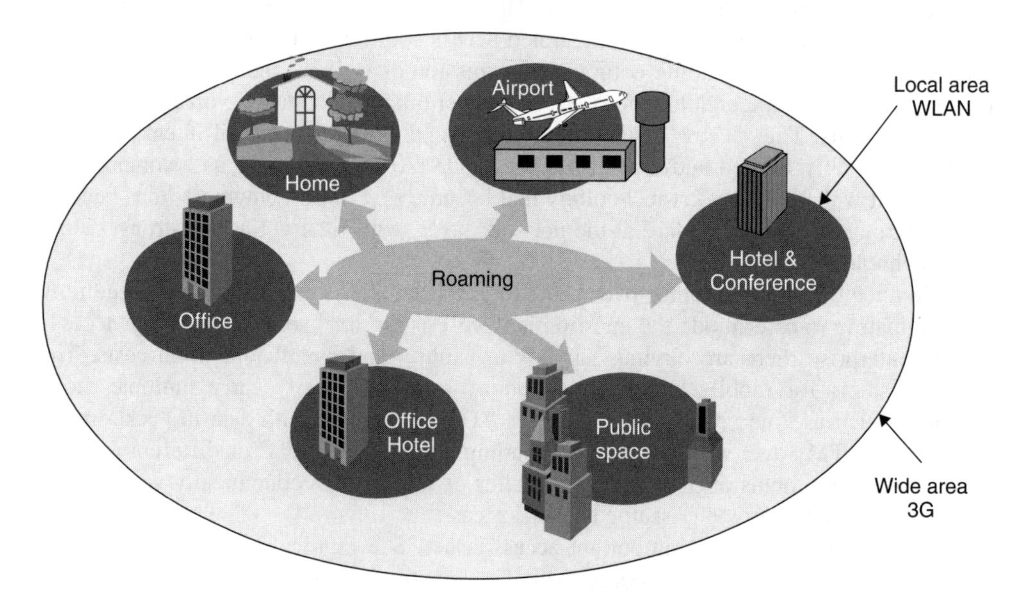

Figure 13.7 Example of hot spots

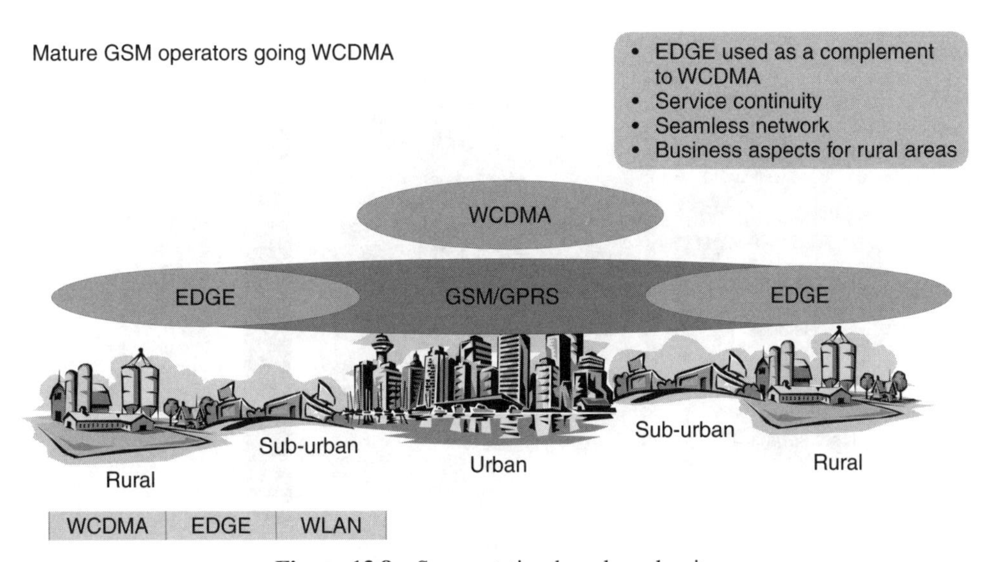

Figure 13.8 Segmentation based on density

There are many other parameters used when describing the subscriber segments:

- ARPU.
- Existing investments in telecom cables, cable-TV cables, radio sites and equipment and power line systems.
- The number of established operators in the area. This makes investments in new cables extremely risky, since you don't know exactly where your new subscribers will be located. This favours radio solutions, as an intruder. Among wireless solutions PLMN 2G, 2.5G and 3G, LMDS (local multipoint distribution system) and WLAN (hot spots) are wireless alternatives to cover subscribers which have not subscribed to wired systems.
- Length of possible waiting lists.

13.4 ACCESS SYSTEM FRAGMENTATION

After proceeding with business cases based on appropriate parameters including state-of-the-art access solutions, the operators selected a large number of access solutions. This has led to a fragmentation in subscriber access networks. See Figure 13.9.

Especially the three top alternatives show a sub-fragmentation, sometimes denoted with 'x'. xDSL includes ADSL, HDSL, SDSL, SHDSL, RADSL and VDSL to take some variants. WLAN systems exist in a number of 802.11 standards. Mobile systems grow with new standards, while the older ones are still in operation. An example is the analogue Nordic NMT system 450 MHz with an excellent coverage. The evolution looks like Figure 13.10.

The fragmentation will be treated at an overview level later in this chapter.

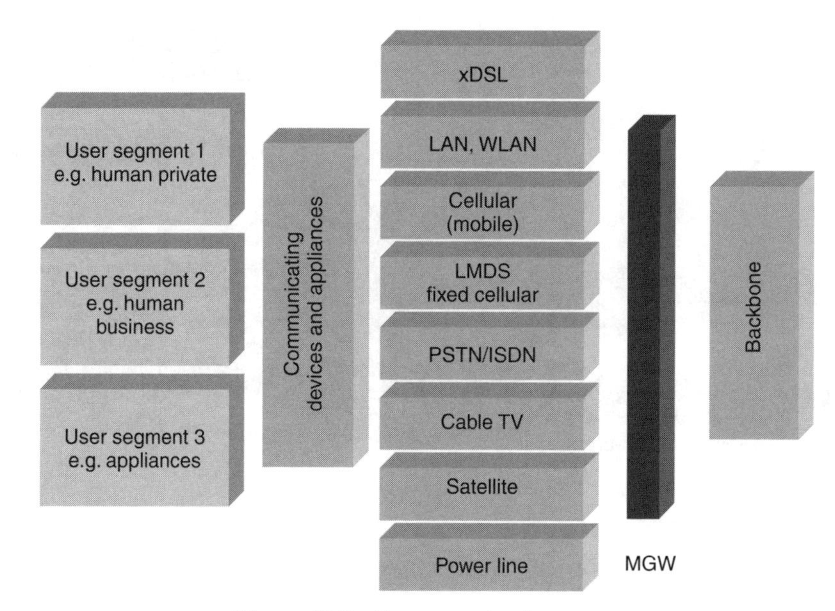

Figure 13.9 Fragmentation in access

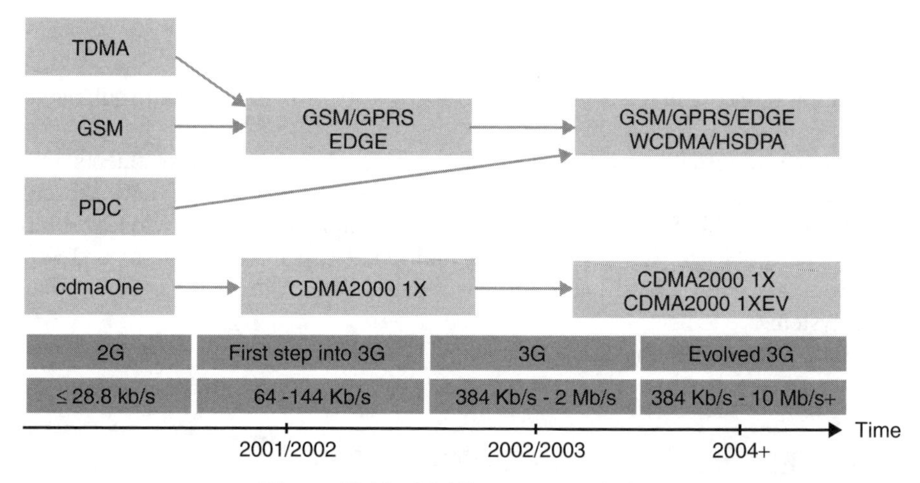

Figure 13.10 Mobile system evolution

13.5 UNIFICATION

At the same time there is an opposite converging trend in the core network. This requires a kind of cross-connect between access systems and the core, with three interfaces to the access network:

- A signalling interface, developing towards SIP, or possibly H.323.
- A circuit mode user traffic interface.
- A packet mode user traffic interface.

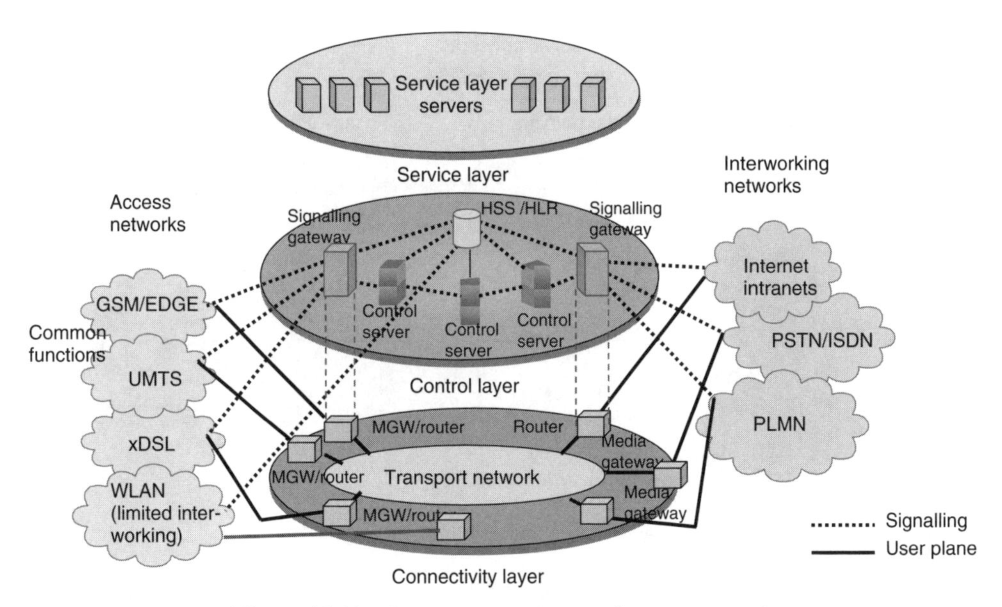

Figure 13.11 Access connections to the core network

What is described could be logically illustrated by means of Figure 13.11, which, however, does not separate the two user traffic cases. In the long run a unification to an all-IP interface is foreseen, such as 3GPP, release 5 and 6. Some mobile systems are briefly described in Section 13.11.

Access independency is a goal. It also becomes a unification factor, since a common access-core type interface is required. The signalling is key to access independence. As mentioned, a future SIP solution is expected.

Instead of only describing the access fragmentation (which would require a book in itself if done thoroughly) this chapter also looks at some common access features. Some unification trends will be covered under the heading 'Fundamental plans'. The more commonality, the better are the possibilities to utilize common resources in the expensive access part, with ROCE impact.

13.6 THE DISTRIBUTED DIMENSION

13.6.1 STRUCTURE OF THE ACCESS NETWORK

See Figures 13.2 and 13.12. A customer access unit (CAU) can for example be located at the business or residential premises or at a pole or mast. The CAU serves as a termination point for the public network. The unit can provide access to a number of services (voice, data, video). The unit can be composed of several sub-units, such as network termination, a set-top box and devices connected to a local area network. The CAU can also refer to a device shared by several end users, for example, a hub located in the basement of a multiple tenant building, a mobile radio base station or a WLAN access point. The mobile access points are widely segmented in a range from indoor picocell to 2G type coverage.

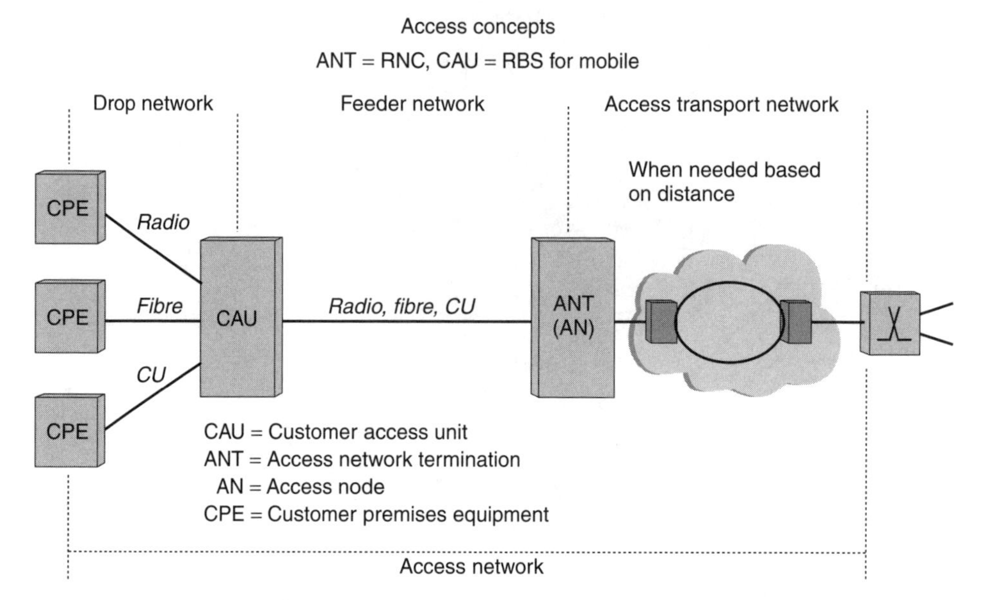

Figure 13.12 A viable terminology in the access network

A multi-service access node 2 (Figure 13.2) can be part of the access infrastructure, providing an interface for various types of CAUs. It multiplexes or concentrates traffic and can convert protocols between different types of traffic, for example between circuit mode and cell mode (ATM).

The access node provides connection of customer access units, and connection to the access transport part of the network and to the edge node. It can be used by more than one operator. Internally the access node will be able to multiplex and/or switch circuit, packet (including Ethernet and IP) and cell mode. Conversion functions, such as circuit emulation, are normally not located here, but in media gateways/edge nodes. In the long run the media gateway and router function will probably migrate to the access node.

Towards the edge node, the access node is often connected to a star point-to-point network structure, mainly over fibre with copper as an alternative. For reliability reasons, a connection to an optical access transport ring via an add/drop multiplexer is preferable but somewhat more expensive.

The edge node, which connects the access network to backbone networks, provides a variety of functions, including multiplexing, concentration, switching and narrowband and broadband access servers and gateways. The edge node also contains media gateway functions. See Chapter 11.

The access management is supported by a set of element managers, network and sub-network managers of the access network. This layer also provides interfaces to higher order management functions.

For the various access network parts the following terms may be used:

- A customer drop between customer and customer access unit.
- A feeder part between access node and customer access unit. In mobile access a *lower radio access network* is sometimes used.

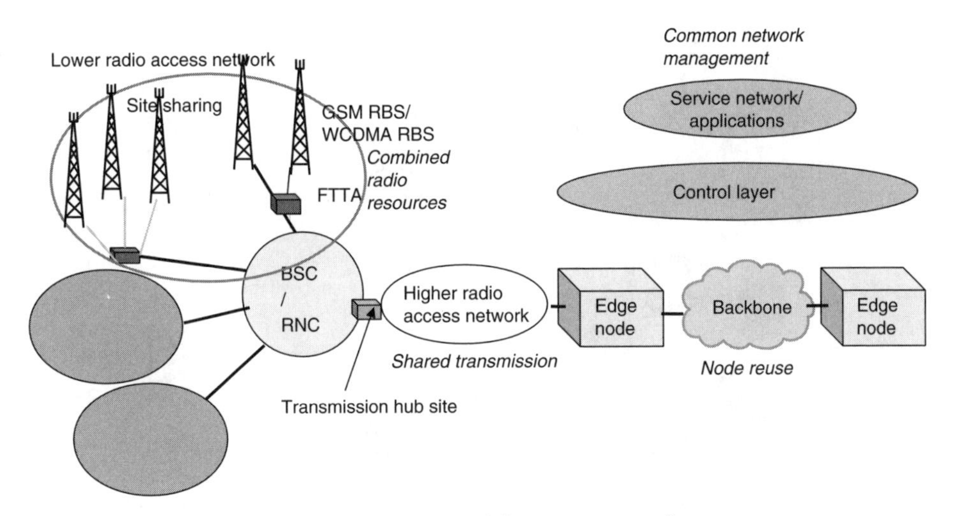

Figure 13.13 A common mobile access network structure

- An access transport network part (if necessary) between an access node and the edge node In mobile access a *higher radio access network* is sometimes used.

A mobile access network can look like Figure 13.13.

The higher radio access network is usually fibre based. The lower radio access network is often based on point-to-point microwave links or point-to-multipoint microwave of the LMDS type. These (fixed) radio technologies are briefly described at the end of this chapter.

The general access network design is based on economical criteria, such as ROCE. All possible synergies between different accesses should be used. Concentration and multiplexing are other key features in saving money. For mobile systems the initial concentration is achieved by means of the attach–detach procedures ('concentration in the air'), as opposed to for example an ATM access multiplexer in the fixed network, differentiating between idle cells and traffic carrying cells from the subscriber side. See Figure 13.14.

The efficiency at transmission level can be measured in bits per bandwidth unit.

A summary and example of common structures is given in Figure 13.15.

13.7 THE LAYERED DIMENSION

A specific transfer mode can use different transmission systems. This is a main reason for the richness in implementation of fixed and mobile access systems. One of the access vendors has calculated the number of access variants in its own portfolio at 7000.

In general, public traffic needs a layer 3, which is intended for public addressing. Historically we have had E.163 for PSTN, E.164 for PSTN/ISDN, X.121 for X.25 and now IPv4 and IPv6 for the Internet. ATM and the upcoming Ethernet are both on layer 2. ATM can, however, also use E.164 addressing.

For corporate private network traffic there is no need for layer 3 addressing, since layer 2 offers a sufficient addressing for limited networks, at least in frame relay, ATM

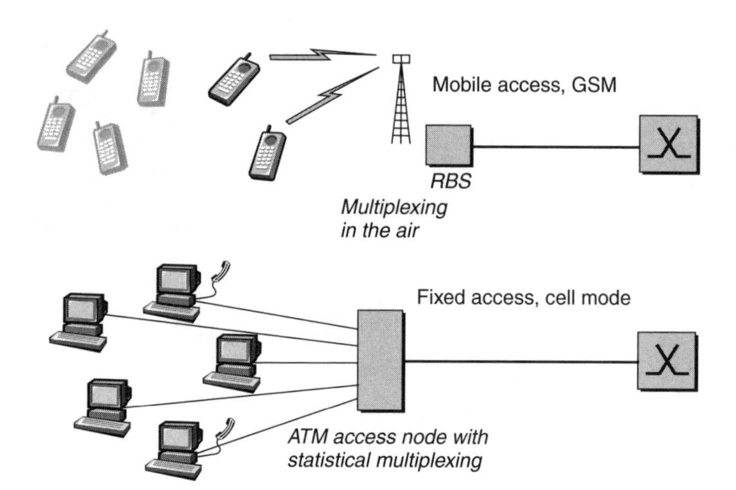

Figure 13.14 Different methods but the same goal: traffic concentration

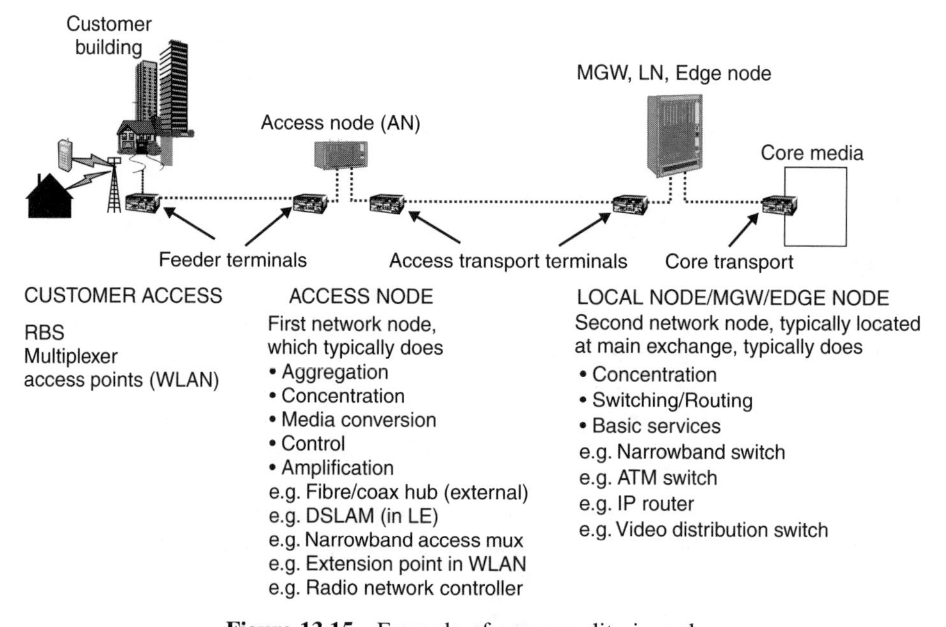

Figure 13.15 Example of commonality in nodes

and possibly also in Ethernet. Observe that the network is seen as limited to the public domain, where public numbering rules. More specific (tunnelled) private addresses or IP NAT addresses are used to find the end-user terminal within the private domain.

In the case of public addressing using private networks there is a two-stage procedure. First the call is often routed in the private network to a point as close as possible to the public address. Where the connection leaves the private network and enters the public network, this is known as 'break-out'. This point is called the break-out point. The new IMS control part is equipped with a break-out call function.

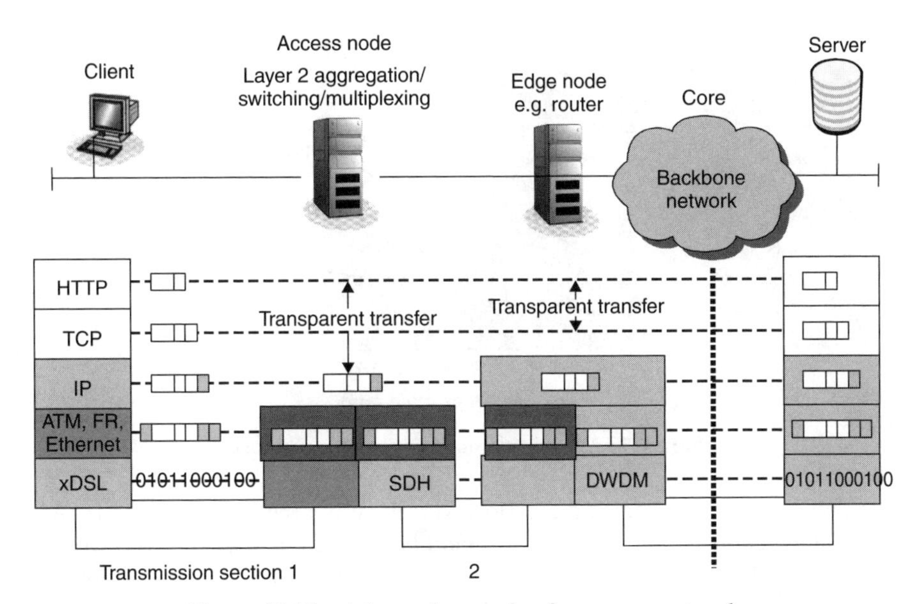

Figure 13.16 A layered example of an access network

There is always a need for layer 2 functions, such as framing (numbering and error control) in all types of packet-mode networks. The error control needs are large at the access part, since the quality of the line is lower there (radio, copper) than in the core network (fibre). A lower quality means more bit errors. X.25 layer 2 (LAPD) is an example of a strong layer 2 protocol, permitting transport over low quality lines. Figure 13.16 provides an example of an end-to-end layered view with access focus.

Framing can be done by layer 2 in for example frame relay, ATM, Ethernet, PPP/SLIP.

Layer 2 might change many times over a long connection, mainly depending on the type of traffic carried and its requirements on error rate, QoS differentiation needs and quality of the transmission line.

Layer 1 can be PSTN base band, ISDN base band, xDSL above the frequency band of PSTN or PSTN/ISDN, SHDSL, PDH base band, STM-1, STM-X, SONET to take a few. Layer 1 normally changes many times over a long connection, mainly depending on transmission media and capacity need.

Below layer 1 is the transmission medium, possibly split into wavelengths using dense wavelength division multiplex (DWDM). The physical layer and the link layer terminate at the edge node (almost by definition). The edge node/media gateway takes care of conversions to core technology in layers 1 and 2 and possibly even higher.

The network layer should preferably be end to end, since it carries the necessary number to the called subscriber. In addition it often carries a security protocol (IPsecurity), for which end-to-end continuity is preferred.

13.8 FUNDAMENTAL PLANS IN ACCESS NETWORKS

The access part is designed according to plans that primarily reside in the connectivity layer, with licenses for frequencies and rules for synchronization methods, transmission

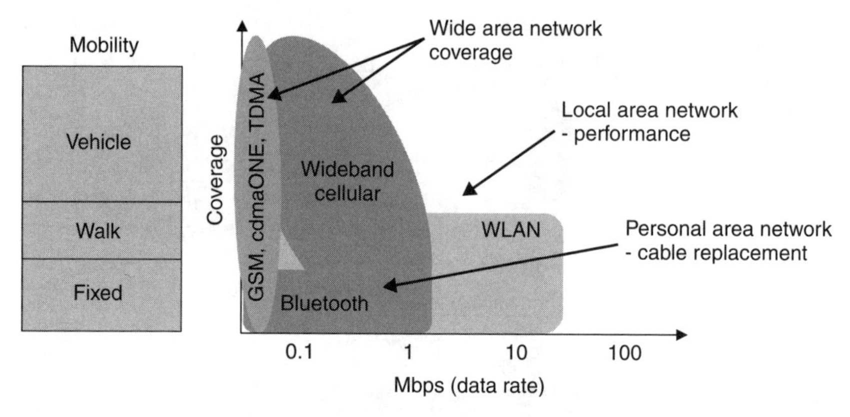

Figure 13.17 Positioning of wireless techniques

systems and multiplexing/aggregation. However, there is a need for support from many other fundamental plans. For example, it should be recalled that security and QoS in fact are involved at all layers. Further, access nodes might need numbering and simple routing functionality. Certainly, a good management system is required.

13.9 MOBILITY

The large added value of mobile systems is of course the mobility. This parameter varies widely between different system solutions. See Figure 13.17. The coverage also corresponds to a degree of mobility.

13.10 ACCESS TECHNOLOGIES IN MOBILE NETWORKS

There are different technology standards available for access to a mobile system: FDMA, TDMA and CDMA. The main characteristics are illustrated in Figure 13.18.

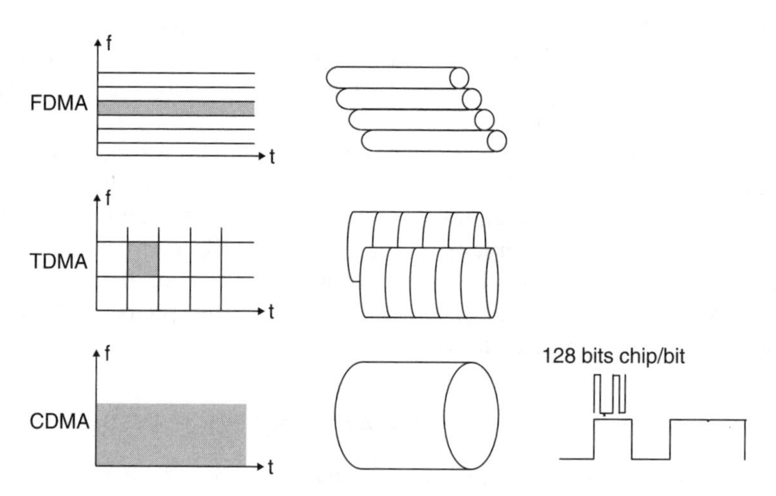

Figure 13.18 Access methods

13.10.1 FREQUENCY DIVISION MULTIPLE ACCESS (FDMA)

FDMA is common in the first generation mobile communication systems, the analogue systems.

One physical channel is allocated per subscriber and is used during the entire duration of the call. The channel is unavailable for other subscribers during that time. The physical channel is released at the end of the call and is then available for the next subscriber.

13.10.2 TIME DIVISION MULTIPLE ACCESS (TDMA)

TDMA is the most common technology in the second generation of mobile communication systems, the digital systems. The available spectrum in TDMA is divided into physical channels of equal bandwidth. See Figure 13.19.

The subscriber is allocated a time slot. The originally continuous subscriber information (speech/data/signalling) is divided into blocks. In GSM a voice block is 20 ms long and sent in bursts when the assigned time slot is available. TDMA requires strict timing of the burst transmission in order to avoid overlapping of adjacent time slots.

In cellular systems with large cells the varying time delay for terminals with different distances to the base station must be compensated. A precise synchronization between the mobile station and the radio base station is required.

13.10.3 CODE DIVISION MULTIPLE ACCESS (CDMA)

The basic concept of CDMA is to simultaneously handle several users' mobile stations without dividing the radio carrier into time slots. This is achieved by using codes to distinguish between different signals to enable many users to share the same radio channel.

The information for each user is spread with a unique code, called chip. Spreading of the information means that the information is 'multiplied' with the codes. CDMA is also known as a spread spectrum technology.

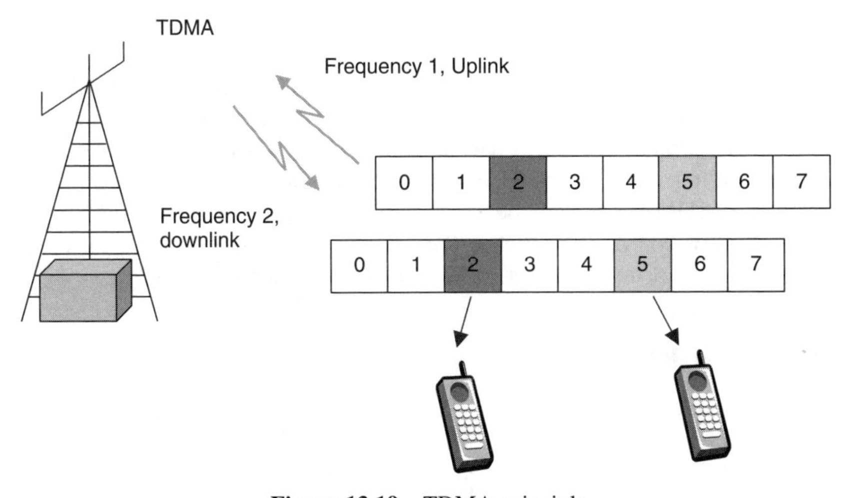

Figure 13.19 TDMA principle

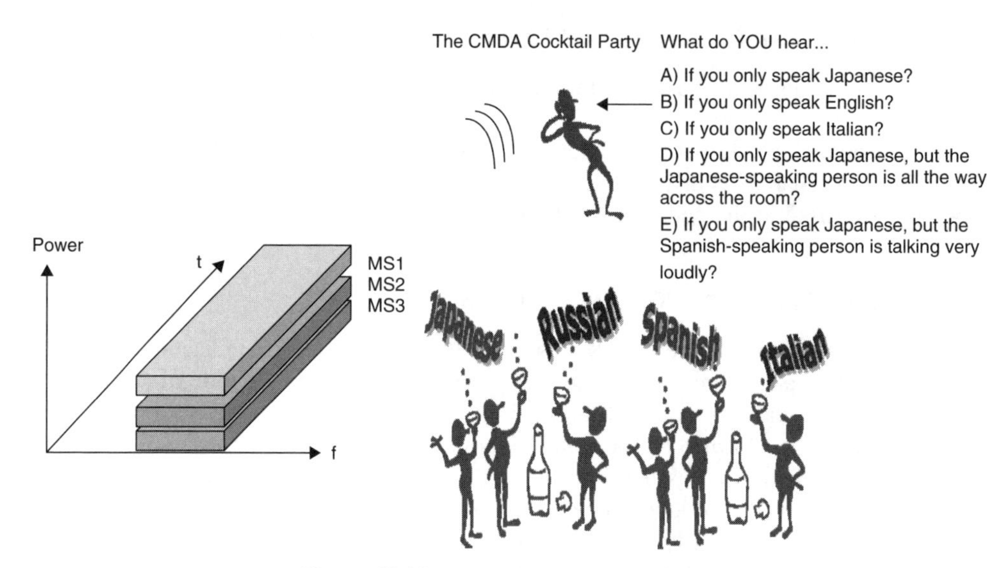

Figure 13.20 The CDMA cocktail party

The information for several mobile stations is transmitted downlink at the same time. Functions in each mobile station can then be used to analyse the information and to decode only the information that is intended for that particular subscriber. Security must also be ensured. Since a mobile station does not have the decoding key for other mobile stations, it will not be able to decode information to other terminals.

When the number of users of the same carrier increases, the more difficult it becomes for a mobile station to decode its own information. The decoding is enhanced by a wide bandwidth when using CDMA solutions. This leads to the term WCDMA (wide-band CDMA).

Another way to understand the concept of CDMA is to use the analogy of an airport international arrivals lounge where everyone is talking in different languages at the same time. If a French-speaking person arrives one may hear a French conversation across the room. See Figure 13.20.

Similarly if an English person arrives, one may hear an English conversation. For all these conversations to be intelligible by listeners, each speaker's volume must not exceed that of the others, which 'explains' why power control is vital to WCDMA networks, since each transmitter is an interference source to all others.

The type of WCDMA that will be used in most cases, at least in the beginning, is called frequency division duplex (FDD). This means that UMTS will use different bands for uplink and downlink just like GSM. The duplex distance is 190 MHz.

13.11 SYSTEM EVOLUTION

In 1999, the International Telecommunication Union (ITU) approved an industry standard for third-generation wireless networks. This standard, called International Mobile Telecommunications-2000 (IMT-2000), is composed of three standards – commonly referred to as WCDMA, CDMA2000 and TD-SCDMA – based on CDMA technology.

In general, mobile systems develop from voice carriers to IP-based multimedia carriers in a number of phases, called releases by 3GPP (for WCDMA/UMTS type systems) and 3GPP2 (for CDMA type systems).

The evolution towards increasing bandwidth for GSM and cdmaOne type systems was shown in Figure 13.10.

13.11.1 GSM

The second generation of cellular systems is based on the transmission of digitally coded information. The radio transmission itself is analogue – there are no square-type radio waves. GSM (global system for mobile communication) systems were launched on the market in 1992. In 1993/94 commercial PDC (personal digital communication) systems were implemented in Japan and D-AMPS (digital AMPS), later called TDMA systems, were implemented in the USA.

In second-generation systems speech is still dominating, but fax, short message service (SMS) and data transmission are also available. Several supplementary services comparable to the fixed network can be used, and fraud prevention and ciphering of user data are possible.

The GSM network is divided into two systems, the switching system and the base station system.

The switching system is responsible for performing call processing and subscriber-related functions. The base station system performs all the radio-related functions.

A cell is the basic unit of a cellular system and is defined as the area of radio coverage given by one base station antenna system. See Figure 13.21.

As GSM has grown worldwide, it has expanded to operate at three frequency bands: 900, 1800 and 1900 MHz.

Various frequencies are used by mobile networks in the first and second generation. See Figure 13.22.

In GSM each radio frequency can be used by maximum of eight subscribers. Each subscriber is allocated a time slot or channel and uses this channel for speech (13 kbit/s) or data transmission (9.6 kbit/s).

13.11.2 GPRS

The GPRS system provides a basic solution for IP communication between mobile stations and Internet service hosts or a corporate LAN.

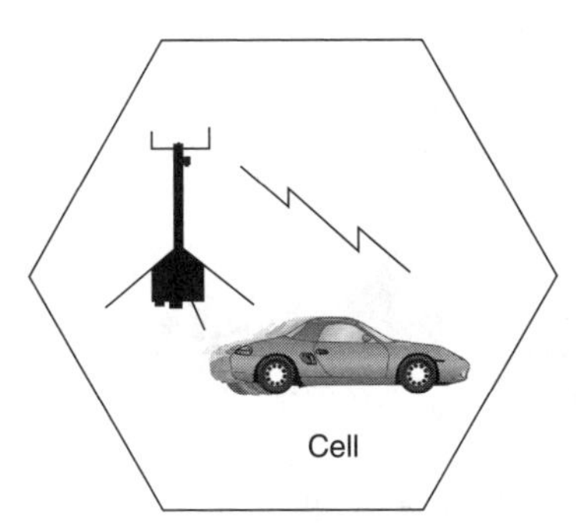

Figure 13.21 The cell is covered by one antenna system

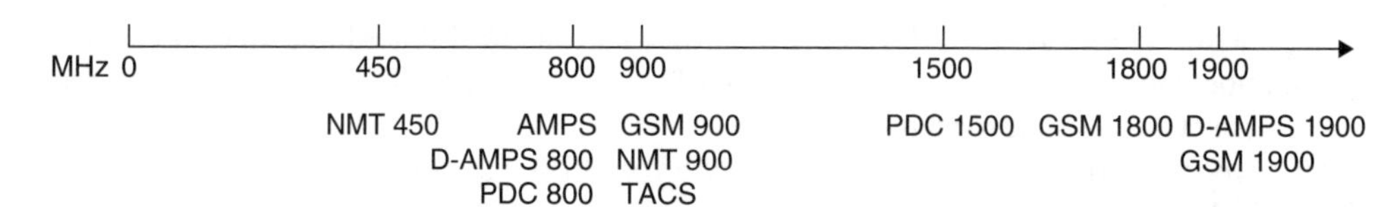

Figure 13.22 Frequency allocation. D-AMPS was the digital evolution of the US AMPS analogue mobile telephony system, now known as TDMA, and also known as IS-136

While the GSM system uses circuit mode air interface for telephony and data, the GPRS system uses preferably packet mode air interface for data. A time slot is then used as a packet data channel. A packet data channel consists of a multi-frame pattern that runs on time slots assigned to GPRS. The allocated packet data channels can be shared by every GPRS user in the cell. A GPRS network can be seen as an extension of a GSM system.

The main new nodes in the core network are the serving GPRS support node (SGSN) and the gateway GPRS support node (GGSN). Both are edge nodes as seen in Figure 13.23. Their initial task is to support data traffic. When the control network has evolved to UMTS and has been provided with IP multimedia support (see Chapter 14), all types of traffic will be supported. More information is found in Chapter 11.

Theoretically, data speeds could increase to 115 kbit/s. The real figure depends on how many time slots are available, and how much traffic is occupying the network.

Since packet mode is used for data transfer over the radio interface it is natural to proceed with packet mode through the core network part. In GPRS, therefore, data traffic is separated from voice in the base station controller and the data traffic continues in packet mode through a new IP packet-switched core network. From the packet-switched core network we can reach the Internet and other data networks via GGSN.

With packet mode several users can share the same physical channels and make use of statistical multiplexing. We can stay online and pay only for the packets we send instead of paying for the time being connected. Figure 13.24 gives an overview of architecture and terminology in GPRS.

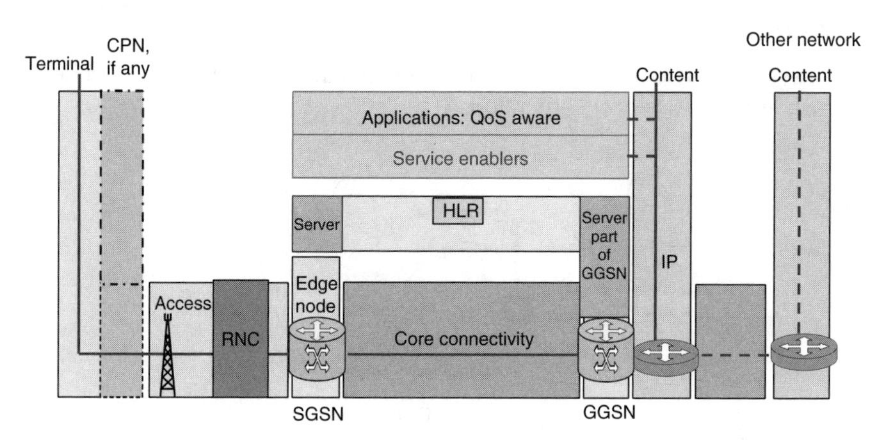

Figure 13.23 The new nodes SGSN and GGSN initially take care of data traffic, later on multimedia traffic is also included

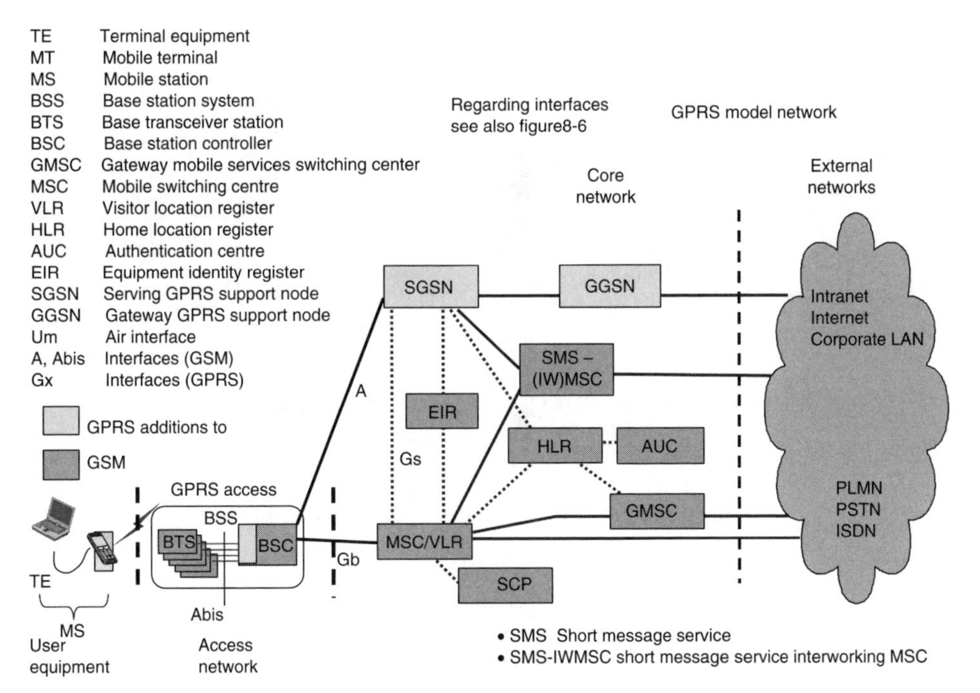

Figure 13.24 An overview of the GPRS system with the new parts

13.11.3 EDGE

EDGE is considered a 3 G type system. It is an extension of GPRS, and the main additions are related to the base station/base station controller part. While UMTS will utilize a new spectrum, EDGE will only be deployed in the existing frequency bands, such as 900 MHz, 1800 MHz and 1900 Mhz. EDGE is increasing the speed in the air interface through a new modulation technology with eight phases (eight-phase shift keying) Each phase represents a sequence of three bits. See Figure 13.25. Up to 384 kbit/s can be achieved when all eight time slots are used. GMSK modulation (Gaussian minimum shift keying) for GPRS offers maximum 115 kbit/s.

EDGE uses the same TDMA frame structure, logical channel and 200 kHz carrier bandwidth as today's GSM networks.

The base station controller needs a software upgrade from GPRS functionality, and the BTS needs EDGE-capable hardware. All other parts of the network are identical to GPRS. Also the radio planning can be maintained. No new sites are needed.

Implementing GPRS caused an impact on operators, such as:

- packet-switched data in a cellular network
- new node types
- new business models
- marketing and selling new types of services
- pricing and charging new types of services

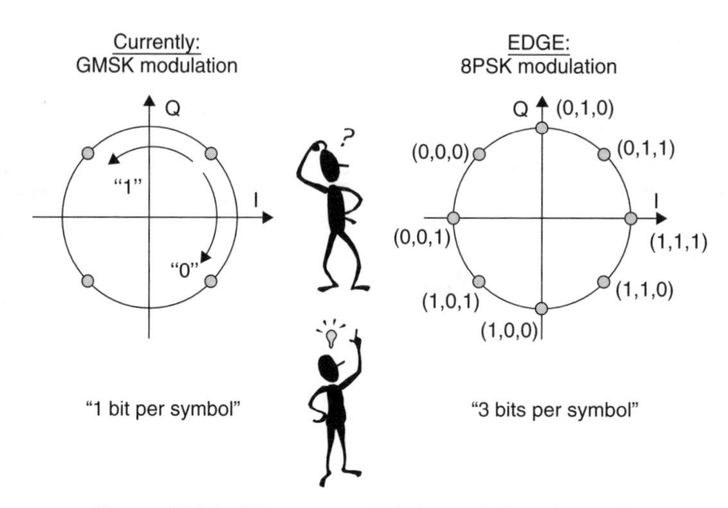

Figure 13.25 The new modulation technique in EDGE

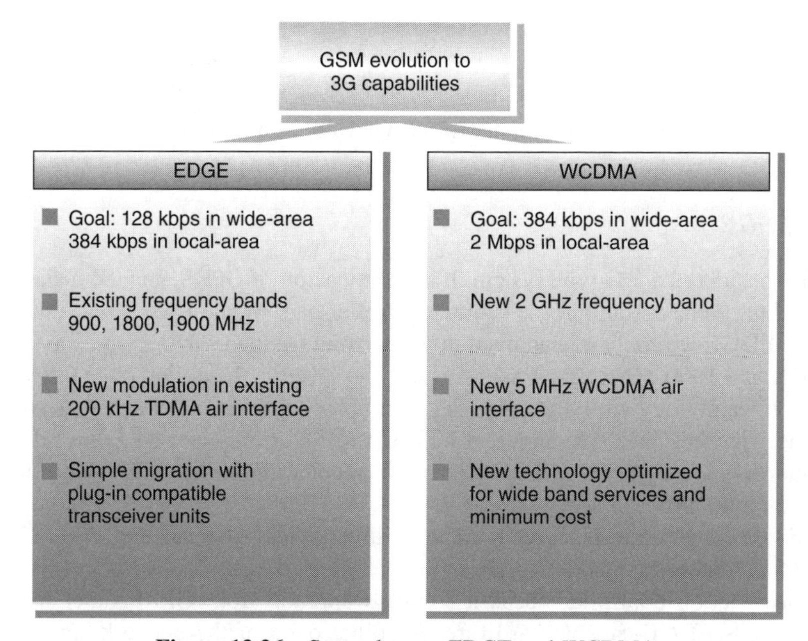

Figure 13.26 Some data on EDGE and WCDMA

Introducing EDGE is a capacity upgrade of the radio interface reusing what is already implemented for GPRS.

Therefore, the additional complexity of GPRS-EDGE cannot be compared to the significant evolution from GSM to GPRS.

Figure 13.26 compares EDGE and the next evolution step WCDMA, creating a bridge to next subject.

13.11.4 WCDMA/UMTS

Two basic approaches in this book are the all-access network architecture and ROCE as a profitability tool. This implies a modern 3G/NGN oriented core network and a multitude of new and existing access types. It means also that the UMTS core characteristics are part of many chapters in the book. Chapters 3, 5, 7, 8, 9, 11, 12 and 14 are such examples. The development track GSM–GPRS–UMTS is briefly described in Chapter 8, section 8.2. The UMTS access does not have a corresponding strong position on the fragmented access side, but a strong expansion is expected. The business case is carefully studied and developed with ingredients such as network sharing between operators, sharing of resources between access technologies and intersystem (for example GSM-UMTS) roaming. The WCDMA access for UMTS is schematically shown in Figure 13.27.

13.11.5 WCDMA RADIO ACCESS NETWORK (UTRAN)

The WCDMA radio access network, known as UTRAN (universal terrestrial radio access network) will provide access speeds at up to 2 Mbit/s in the local area and up to 384 kbit/s wide area access with full mobility. These increased data rates compared with GSM/GPRS require a wide radio frequency band, which is why WCDMA with 5 MHz carrier has been selected, compared with 200 kHz carrier for narrowband GSM. The new frequency allocation is shown in Figure 13.28.

To begin with, voice will probably continue to be transferred in the GSM networks of the 3G operators, but voice will also be efficiently carried over WCDMA.

UTRAN consists mainly of radio network controller (RNC) nodes and radio base station nodes. The RNC manages radio access bearers for user-data transport, manages and optimizes the resources of UTRAN and controls mobility. Other parts are the radio access network operation support (RANOS) and tools for radio access management (TRAM). The RNC is connected to the core network via the Iu interface. Internally within radio

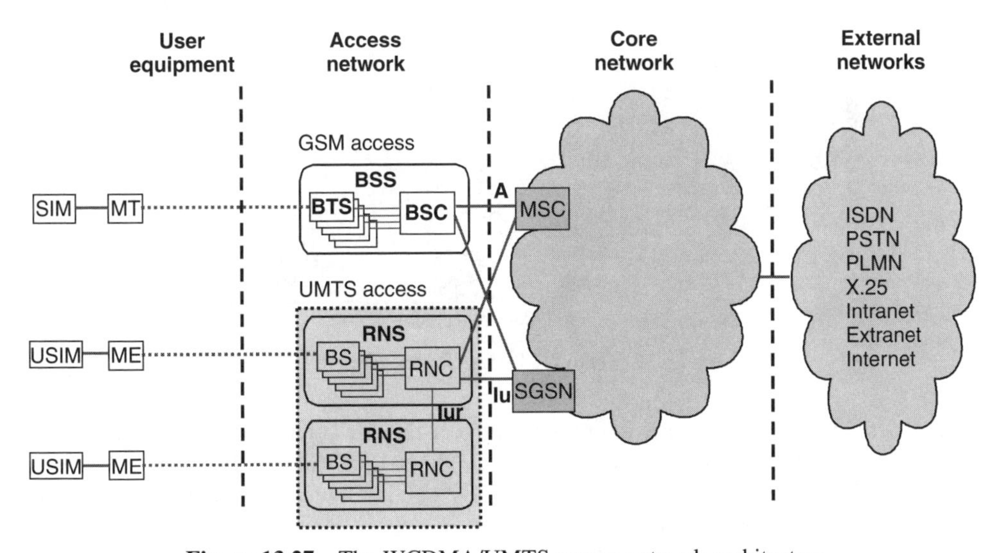

Figure 13.27 The WCDMA/UMTS access network architecture

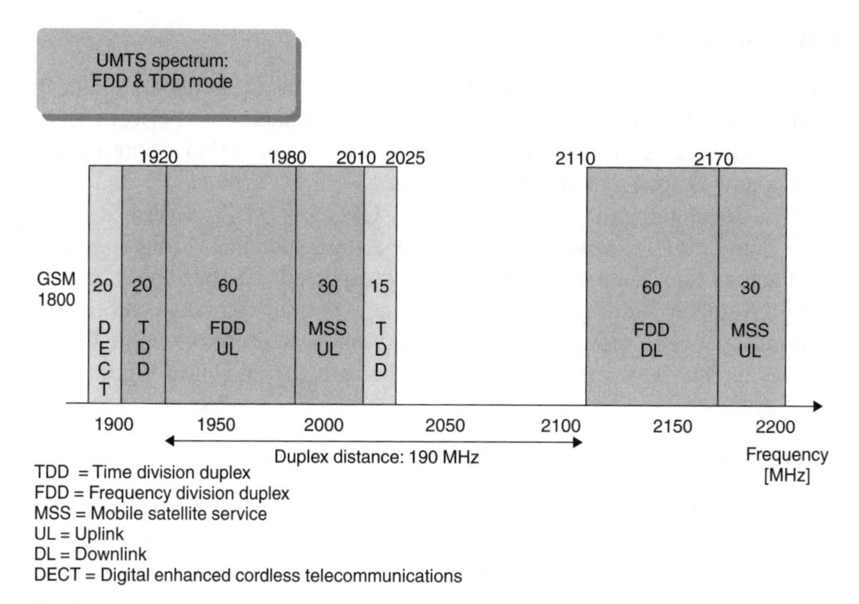

Figure 13.28 WCDMA/UMTS access network (UTRAN) utilizes a new frequency band above 2.1 GHz

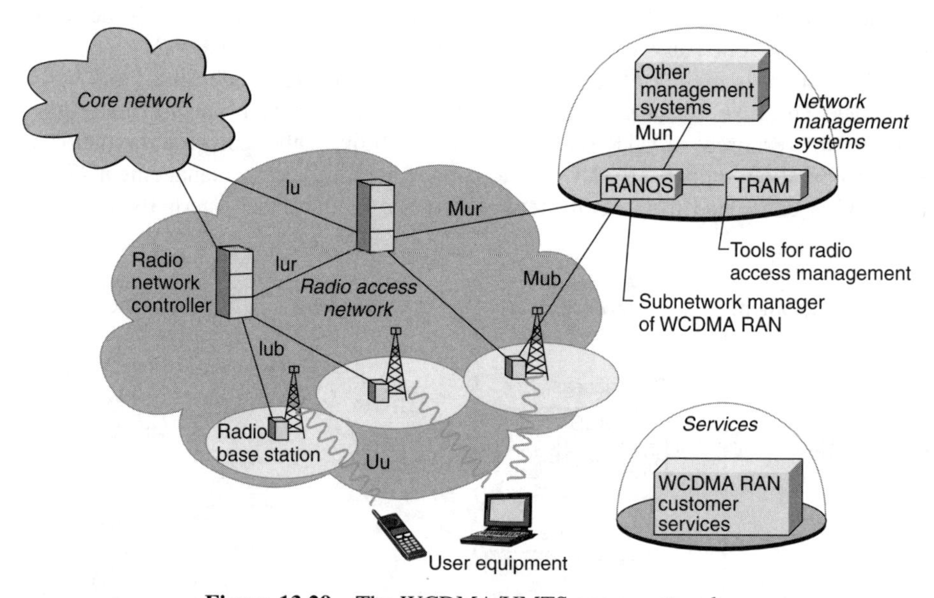

Figure 13.29 The WCDMA/UMTS access network

access network, the RNCs are interconnected via the Iur interface and the radio base stations to the RNC via the Iub interface. See Figures 13.27 and 13.29.

The RNCs are the central intelligence of UTRAN and will have even more functionality than GSM base station controllers.

The radio base station provides the actual radio resources.

The RANOS software package takes care of day-to-day operation and maintenance. The position is shown is Figure 13.29.

13.11.6 CDMA

Within the CDMA2000 standard, several phases have been defined to support the ITU requirements for third-generation services. Figure 13.30 shows the evolution of the CDMA 2000 standard, CDMA2000 1X, which is based on the IS-2000 standard. It:

- is backward compatible with the original cdmaOne
- offers up to twice the voice capacity of cdmaOne systems
- supports always-on packet-data sessions
- provides data rates of up to 144 kbit/s – with peak over-the-air data rates (including overhead) of 163.2 kbit/s (IS-2000 Rev. 0) and 307.2 kbit/s (IS-2000 Rev. A).

Beyond CDMA2000 1X, the Third Generation Partnership Project 2 (3GPP2) specifies two (1xEV) standards: 1xEV-DO (data only) and 1xEV-DV (data and voice).

CDMA2000 1xEV-DO comprises a separate data carrier that provides best-effort packet-data service with a peak over-the-air data rate of 2.4 Mbit/s. CDMA2000 1xEV-DV provides integrated voice and data with real-time data services and a peak over-the-air data rate of 3.1 Mbit/s.

The main radio network parts are:

- a base station controller
- radio base stations
- a radio network management system

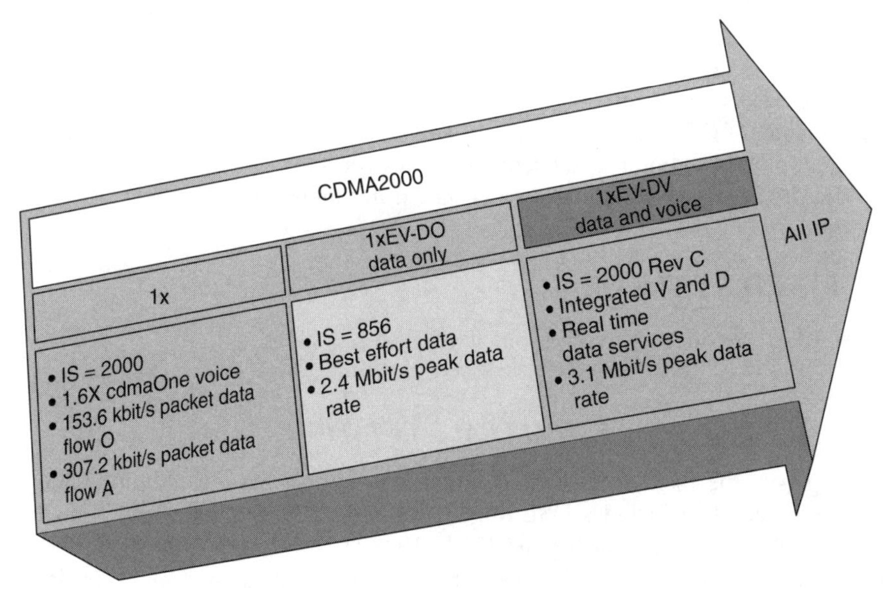

Figure 13.30 Evolution within the CDMA systems

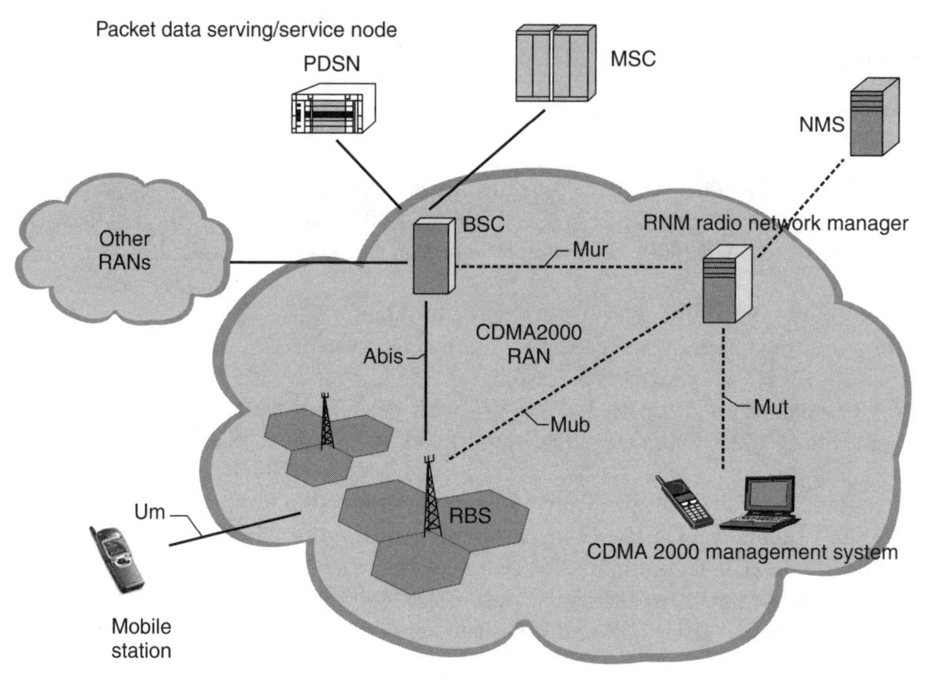

Figure 13.31 A CDMA access network structure

Typical base station controller functions are: control of the radio base stations, management of radio network resources and provision of user mobility.

Base station controllers may also perform voice compression (vocoding), process handovers, manage power control to ensure efficient use of network capacity, control timing and synchronization within the radio access network, and provide interfaces to the radio base stations, the radio network manager and packet-data service/serving nodes (PDSN).

The radio base stations provide the radio resources.

A key feature of CDMA2000 1X packet-data service is that the packet-data connection between the mobile station and the network is always on.

The CDMA access network structure is shown in Figure 13.31.

13.12 FIXED SYSTEMS

Figure 13.32 provides an overview of fixed access systems.

13.12.1 COPPER-BASED SYSTEM FRAGMENTATION

The drop part of the fixed accesses is dominated by copper-based solutions. Many solutions have a name that includes DSL (digital subscriber line). Very efficient modulation schemes enable bandwidths far beyond ISDN 2B + D.

There are also fixed radio solutions (LMDS, fixed cellular systems), coaxial cable systems, often called cable modem systems or cable TV (CATV) systems, and finally fibre

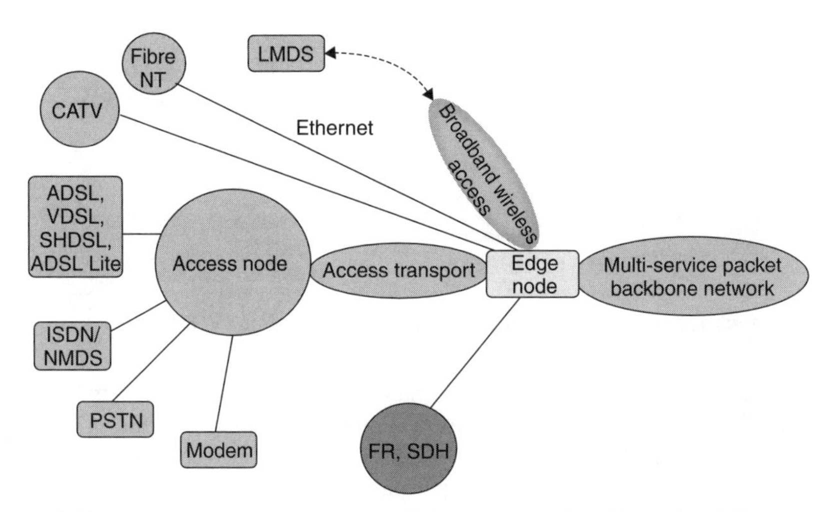

Figure 13.32 Fixed systems overview (NMDS: narrowband multi-service delivery system)

systems to enterprises and to basements, curbs and other points very close to residential subscribers.

A possibly upcoming solution is a power line.

13.12.2 ADSL + G.LITE

Asymmetric digital subscriber line (ADSL) provides broadband access on the existing PSTN pair cable to the home. At the same time you may continue using your ordinary telephone. Separation is achieved by means of a 'splitter'. The data is sent on unused frequencies above 3400 Hz. This interface is called 'asymmetric' since the downstream flow is much bigger than the upstream flow. Often you download more than you upload, so this is a preferred division for most users. For peer-to-peer communication and small office/home office activities with a balanced traffic in-out a symmetric DSL solution may be better.

The actual bandwidth of ADSL depends on technique but also on operator policy. In addition to the original standard there is a standard called G.Lite with lower bit rate but cheaper, since it is a splitterless design. G.lite is also known as universal ADSL, and referred to as G.992.2 by the ITU.

Downstream transmission rates from the carrier to the subscriber premises range between 0.5 Mbit/s and 8.0 Mbit/s depending on type, whereas upstream transmission rates are normally up to 1 Mbit/s. The possible speed is proportional to the distance from the exchange or from a remote access unit. (Low speed) ADSL can be used over distances up to 5.5 km. In Europe ~90 % of all subscriber lines are shorter than 3 km.

13.12.3 SHDSL

Single pair high bit rate digital subscriber line (SHDSL) delivers high-speed digital information for data and voice services over copper lines. The transmission is symmetric (downstream speed and upstream speed are equal), providing a rate of up to 2.3 Mbit/s.

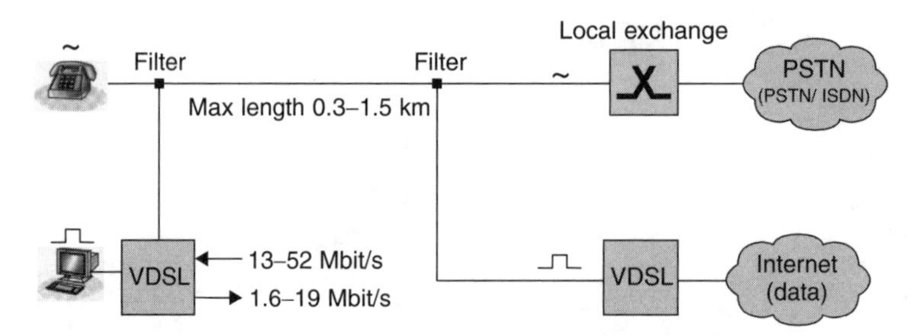

Figure 13.33 Very high bit rate digital subscriber line (VDSL). Example

SHDSL is also known under its standard name G.SHDSL or ITU-T standard G.991.2.

SHDSL delivers greater bandwidth over longer distances than current 2B1Q-based SDSL and traditional HDSL systems and enables operators to provide symmetric, high-speed data transmission over existing copper pairs.

The standard is a multi-rate DSL technology that allows data rates from 192 kbit/s to 2312 kbit/s and can transport T1 (1544 kbit/s, E1 (2048 kbit/s), ISDN, ATM and IP signals.

Symmetric solutions fit small and medium enterprises with roughly the same traffic volume in and out.

13.12.4 VDSL

An even faster technique is the very high bit rate digital subscriber line (VDSL). See Figure 13.33. Here the standard defines the full bit rate to 26 Mbit/s in each direction for symmetric traffic, or different speeds downstream and upstream for asymmetric traffic.

13.13 FIBRE-BASED SYSTEMS

There was a lot of talk of fibre to the home (FTTH) during the 1980s. However, fibre was fairly expensive and the pressure from the subscribers was modest. Figure 13.34 shows the dramatic cost reduction during the last few years. Since then many competing solutions have been developed, as seen from Figure 13.34.

An increasingly popular fibre-based solution uses Ethernet in layer 2. It can also offer home access.

X in FTTx may be basement, curb, home, office, radio base station and so on. What is called fibre Ethernet systems benefit as well.

13.14 ETHERNET

- 95 % of all IP traffic utilizes Ethernet.
- All business offices use Ethernet.
- Virtually all home PCs have Ethernet.
- Public Ethernet is being built all over the world.
- Well-proven, mature and robust technology.

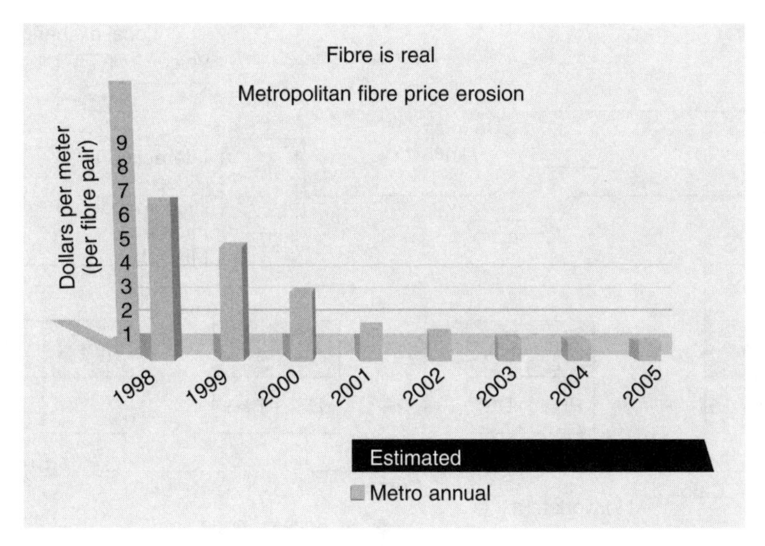

Figure 13.34 The fibre cost development makes FTTx systems competitive

Table 13.1 Cost comparison for equipment only

	Equipment ($/Mbit/s)	Bandwidth management and provisioning ($)	Annual maintenance upgrades ($)	Bandwidth on demand
IP/ATM/SONET	8–40	5000	750–3750	Hard
IP/SONET	6–35	5000	750–3750	Hard
IP/Ethernet	1–3	1000	150–450	Easy
GigE advantage	8:1–13:1	5:1	5:1–8:1	Easy

Some data:

Ethernet	10 Mbit/s	100 m
Fast Ethernet	100 Mbit/s	0–10 km (fibre dependent)
GbE	1000 Mbit/s	0–70 km (fibre dependent)
10 GbE	10 000 Mbit/s	0–40 km (fibre dependent)

In the long term Ethernet is expected to take a dominating role in access networks. The equipment price for Ethernet is favourable in comparison to other alternative IP carriers. See the indicative cost comparison in Table 13.1.

13.15 COMBINED ADSL OVER COPPER AND ETHERNET OVER FIBRE SOLUTION

By combining ADSL over copper at the drop part and fibre Ethernet between access node and edge node the ADSL part stays short with a potential for a high speed up to 8 Mbit/s. See Figure 13.35.

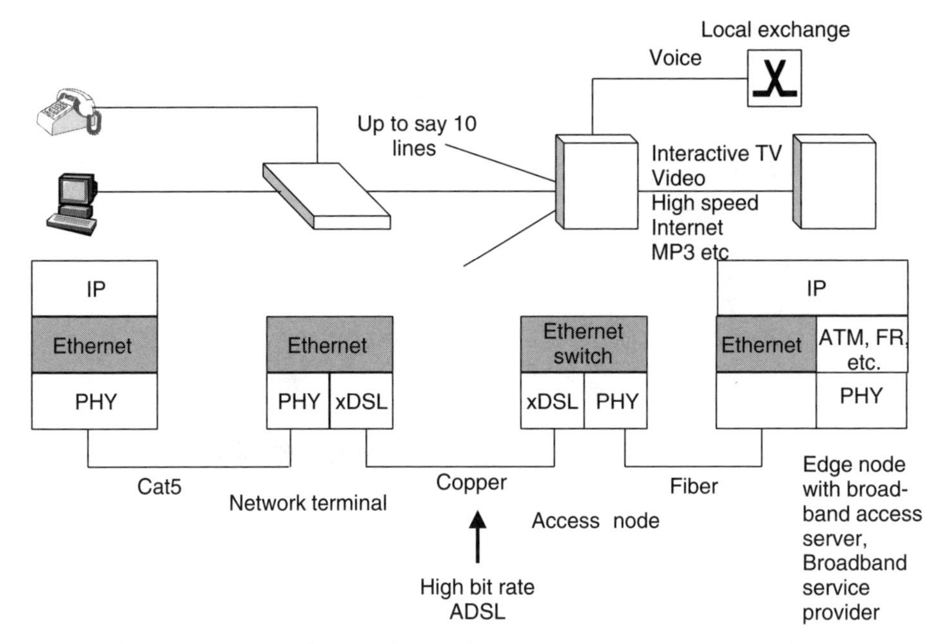

Figure 13.35 Combined ADSL – Ethernet solution

13.16 CABLE MODEM

A cable modem (or cable TV) solution extends from the modem to the head-end. The head-end is a central distribution point for a CATV system. Video signals are received here from satellites and other sources, frequency is converted to the appropriate channels, combined with locally originated signals, and rebroadcast onto the hybrid fibre-coaxial (HFC) plant. The head-end is where the cable modem termination system (CMTS) is normally located. A CMTS is a system of devices located in the cable head-end that allows CATV operators to offer high-speed Internet

Older CATV systems were provisioned using only coaxial cable. Modern systems use fibre transport from the head-end to an optical node located in the subscriber neighbourhood to reduce system noise. Coaxial cable runs from the node to the subscriber. The fibre plant is generally a star configuration with all optical node fibres terminating at a head-end. The coaxial cable part of the system is generally a trunk-and-branch configuration. The solution is called HFC (cable network).

The upstream frequency used to transmit data from the cable modem to the CMTS is normally in the 5–42 MHz range for US systems and 5–65 MHz for European systems.

The downstream frequency used for transmitting data from the CMTS to the cable modem is normally in the 42/65–850 MHz range depending on the actual cable plant capabilities.

Data over cable service interface specification (DOCSIS) is the dominating cable modem standard. It defines technical specifications for both cable modem and CMTS.

13.17 WLAN

The increasingly popular WLAN or wireless LAN is used as a common LAN in office environment but also as a home LAN and as a public access network to the Internet at hot spots, such as airports and conference areas. The WLAN network consists of base stations, called access points and wireless clients. The clients can be 'anything', such as PDAs and portable or stationary PCs. Soon most new portable computers are expected to be equipped with WLAN chips.

IP telephony and WLAN is a new combination.

The access points are bridging wireless and wired communication. They can be connected as an extension to an existing wired network, but they can also be connected to other base stations by means of radio in totally wireless networks. Some possible accesses are illustrated in Figure 13.36.

The WLAN development is shown in Figure 13.37. The first 802.11 standard came out in 1997. It specifies wireless communication with the speeds of 1 and 2 Mbit/s in the 2.4 GHz band. The next standard was called 802.11b. It specifies radio communication with speeds of 5.5 and 11 Mbit/s in the 2.4 GHz band. 802.11b is backward compatible with 802.11.

802.11g is an extension to 802.11b, the basis of the majority of WLANs in existence today. 802.11g broadens 802.11b's data rates to 54 Mbit/s within the 2.4 GHz band using OFDM (orthogonal frequency division multiplexing) technology. Because of backward compatibility, an 802.11b radio card will interface directly with an 802.11g access point (and vice versa) at 11 Mbit/s or lower depending on the range. Upgrading newer 802.11b access points to be 802.11g compliant is possible via relatively easy firmware upgrades.

A complete upgrade in an area would create a need to move the access points closer together and include additional ones to accommodate higher data rates.

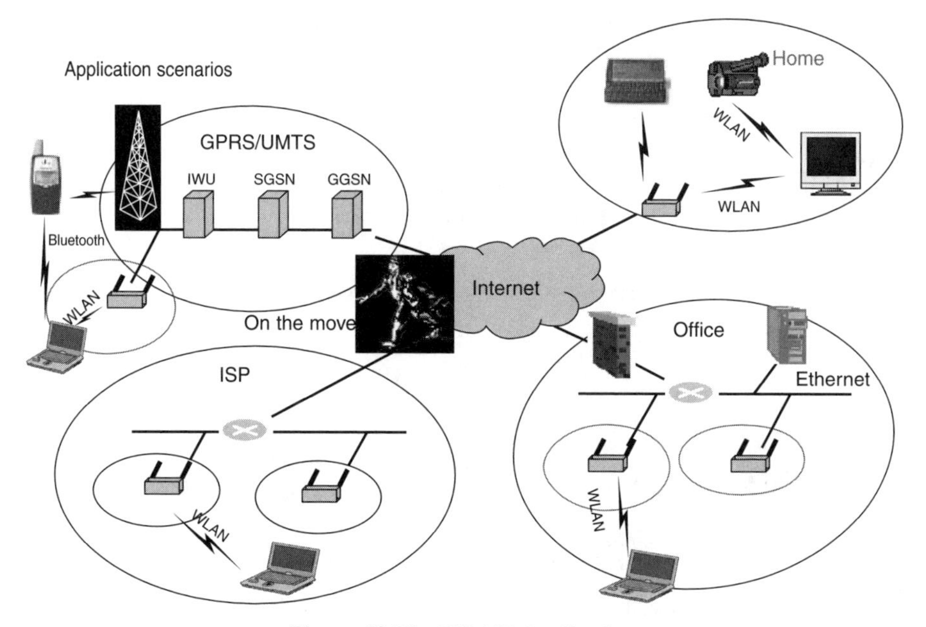

Figure 13.36 WLAN Applications

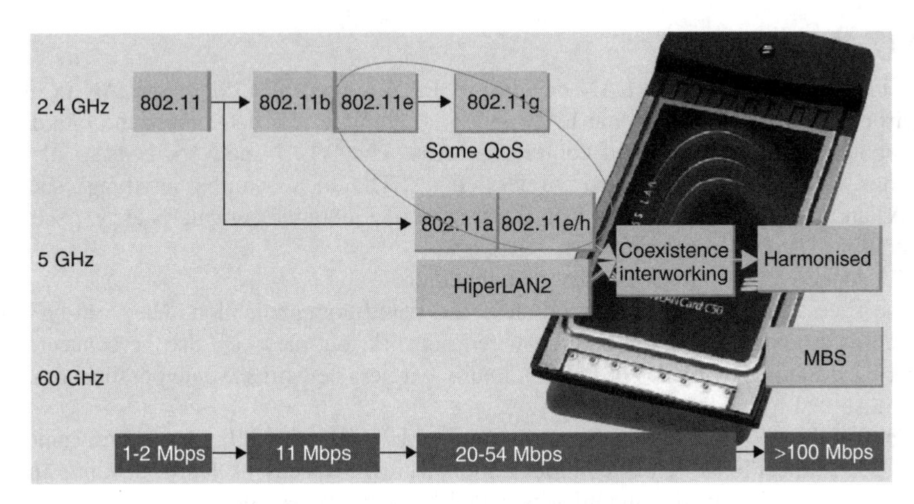

Figure 13.37 WLAN development

A big issue with 802.11g, which also applies to 802.11b, is considerable radio frequency interference from other 2.4 GHz devices, such as the newer cordless phones.

802.11a is a standard that operates in the 5 GHz frequency band with 12 separate non-overlapping channels. As a result, you can have up to 12 access points set to different channels in the same area without them interfering with each other. In addition, radio frequency interference is much less likely because of the less-crowded 5 GHz band.

Similar to 802.11g, 802.11a delivers up to 54 Mbit/s, with extensions to even higher data rates possible by combining channels. Due to higher frequency, however, the range (around 25 m) is somewhat less than lower frequency systems (i.e., 802.11b and 802.11g). This increases the cost of the overall system because it requires a greater number of access points, but the shorter range enables a much greater capacity in smaller areas via a higher degree of channel reuse.

802.11a is not directly compatible with 802.11b or 802.11g networks. In other words, a user equipped with an 802.11b or 802.11g radio card will not be able to interface directly to an 802.11a access point. The real operational bit rates are much lower than the ideal ones. The gross digit speeds are summarized in Figure 13.38.

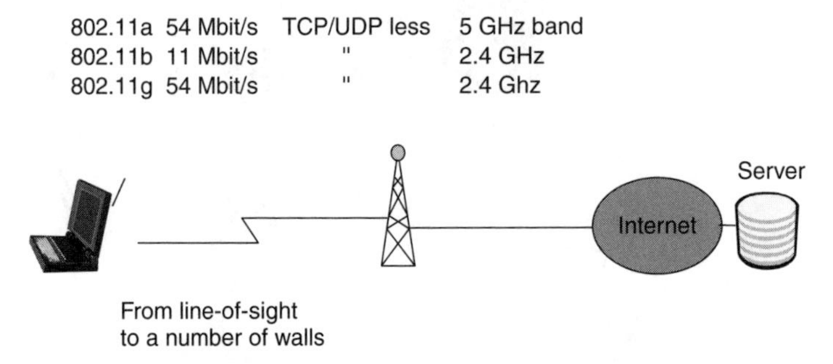

Figure 13.38 Gross digit speeds – the net speed with TCP/IP traffic is considerably less than the gross digit speed

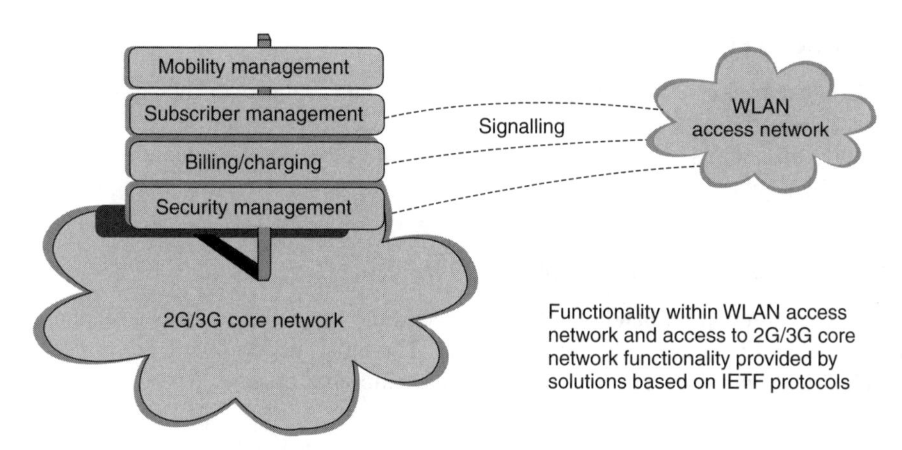

Figure 13.39 Cooperation between WLAN and 2 G/3G mobile systems

WLAN as an Extension to Mobile Systems (ex GPRS and UMTS)

For a mobile operator, WLAN is a complementary access technology, especially for city area packet data service offerings. The solution gives broadband mobile public access to the Internet and to corporate intranets.

Combined cellular and WLAN could be interesting to business professionals. They already use mobile phones and they usually bring their laptop computers along when they travel. Trouble-free access to the Internet, intranet, e-mail and corporate applications is becoming an everyday necessity.

Combined cellular and WLAN allows GPRS and UMTS customers to take advantage of WLANs at indoor environments (hot spots) such as airports and hotels conference areas, and roam between cellular and WLAN networks without interruption. Figure 13.39 gives an idea what UMTS functions and FPs could support WLAN access. Authentication (part of security) and VPN functionality mechanisms are some specific functions that are offered to WLAN.

New user segments are expected to pick up WLAN, such as young people using WLAN for interactive gaming and entertainment

UMTS operators can provide two access forms to mobile packet data services: one via the wide area UMTS network (WCDMA-based) and the other one via the high capacity WLAN access network. The packet data service can be seamlessly provided between these two access forms. Network management for WLAN access can be integrated into the existing network management routines.

13.18 SATELLITE TECHNOLOGIES

13.18.1 SMALL DISH SATELLITE INTERNET

Small dish satellite Internet services (such as Direct PC) provide high-quality compressed digital services delivered directly to the home. These services currently promote throughput of up to 400 kbit/s.

13.18.2 GEOSTATIONARY ORBIT SATELLITE

The GEO system uses a satellite positioned at a specific point over the equator to give it the largest footprint. The technology has a proven reliability. Video service including TV is currently the main source of revenue. Internet and other multimedia are also expected to utilize GEO.

13.18.3 LOW EARTH ORBIT BANDWIDTH

These solutions require a band of satellites around the globe. The idea is to capitalize the advantages of LEO such as absence of latency. The most well-known LEO project so far was called Iridium with satellite telephony as a main application.

13.19 HIGH SPEED FIXED RADIO

13.19.1 POINT-TO-POINT

Typical capacities range from E1/T1 to STM-1/OC-3 using radio frequency bands ranging from 7 GHz to 38 GHz. Point-to-point radio links can be deployed in ring, star and tree topologies, and offer built-in traffic routing functionality. P2P radio has a typical application area in the lower radio access network, sometimes in competition with P2MP radio below.

13.19.2 POINT-TO-MULTIPOINT

The local multipoint distribution system (LMDS) is a ‘last-mile’ fixed wireless technology that offers line-of-sight coverage over distances up to 3–5 km. It can deliver data and telephony services to some 80 000 customers from a single node. LMDS services include delivery of voice, data, Internet and video services in the 25-GHz and higher spectrum (depending on national licensing). It can also feed radio base stations in mobile systems.

LMDS is weather dependent. The range is in the area of around 5 km.

It is physically a point-to-multipoint network, and the available bandwidth depends on the number of subscribers connected at the same time. The bandwidth can be up to some tens of Mbit/s downstream. See Figure 13.40.

13.19.3 BLUETOOTH

Bluetooth is a cable replacement technology for data and voice communication in the unlicensed 2.4 GHz frequency band. It is based on a low-cost short-range radio link, built into a microchip, and facilitating protected ad hoc connections for stationary and mobile communication environments. Some possible application areas are shown in Figure 13.41.

Bluetooth technology allows for the replacement of the many proprietary cables that connect one device to another with one universal short-range radio link. For instance, Bluetooth radio technology could replace the cumbersome cable used today to connect a laptop to a cellular telephone. Printers, PDAs, desktops, fax machines, keyboards, joysticks and virtually any other digital device can be part of the Bluetooth system. Bluetooth radio technology also provides a universal bridge to existing data networks, a peripheral

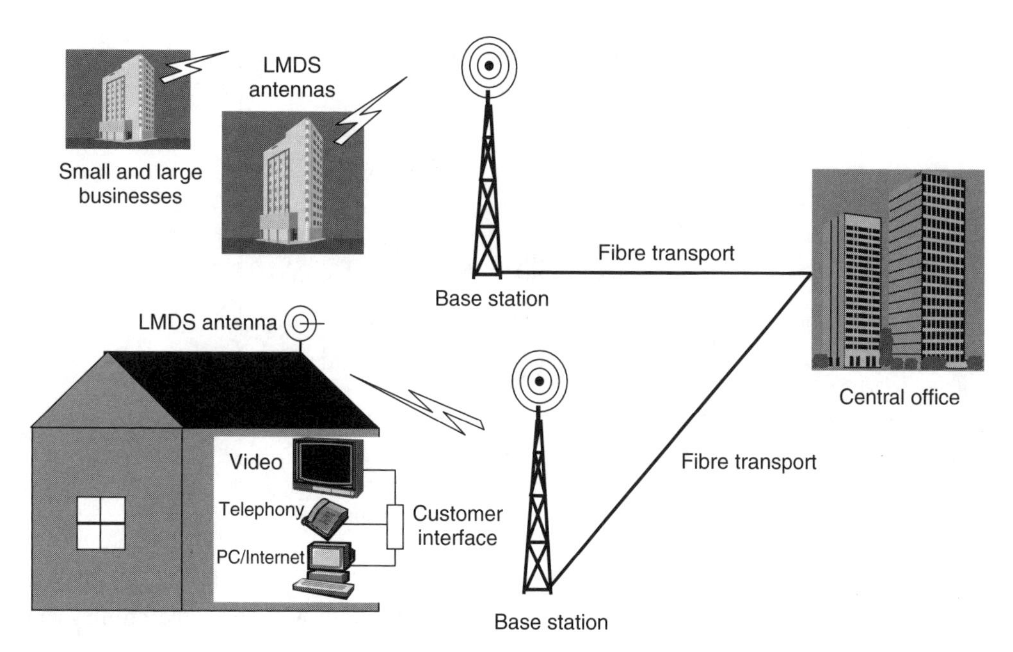

Figure 13.40 LMDS network structure and services. LMDS can also feed radio base stations

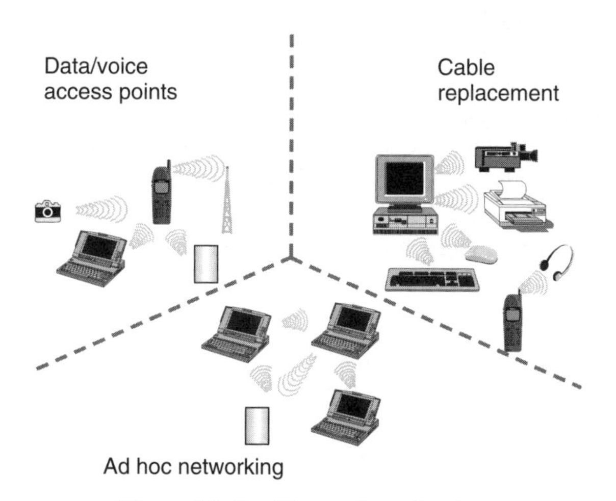

Figure 13.41 Bluetooth applications

interface, and a mechanism to form small private ad hoc groupings of connected devices away from fixed network infrastructures.

Designed to operate in a noisy radio frequency environment, the Bluetooth radio uses a fast acknowledgement and frequency hopping scheme to make the link robust. Bluetooth radio modules avoid interference from other signals by hopping to a new frequency after transmitting or receiving a packet. Compared with other systems operating in the same frequency band, the Bluetooth radio typically hops faster and uses shorter packets. This makes the Bluetooth radio more robust than other systems.

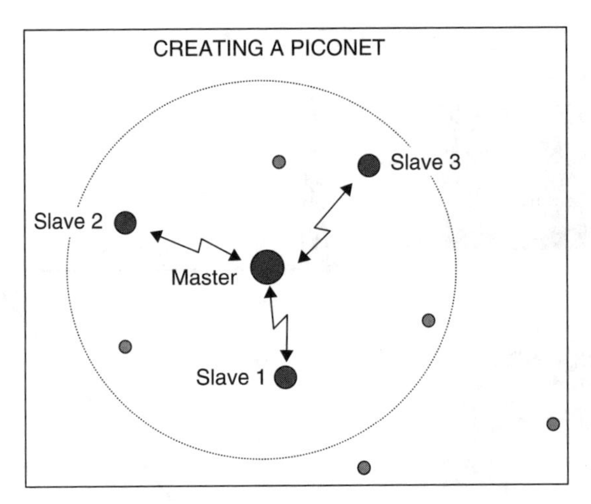

Figure 13.42 Example of a piconet

Interoperability with other Bluetooth-capable devices is supported by means of profiles, indicating what functionality a device supports. Devices with the same profiles can communicate.

Bluetooth can also be used for the creation of ad hoc networks. Let us take piconet as an example. See Figure 13.42.

A Piconet is a collection of devices connected via Bluetooth technology in an ad hoc fashion. A piconet starts with two connected devices, such as a portable PC and cellular phone, and may grow to eight connected devices. All Bluetooth devices are peer units and have identical implementations. However, when establishing a piconet, one unit will act as a master and the other(s) as slave(s) for the duration of the piconet connection.

Several piconets can be established and linked together ad hoc, where each piconet is identified by a different frequency hopping sequence. All users participating on the same piconet are synchronized to this hopping sequence. The topology can best be described as a multiple piconet structure.

14

Control Network

14.1 INTRODUCTION

In Chapters 11, 12 and 13 we reviewed traffic resources. We now need to understand how to control and use them in order to create a successful business. The services develop towards multimedia handling. The control layer responds with the intelligence of an IP multimedia subsystem and deployment of more advanced control network signaling, mainly based on the SIP protocol.

An introduction to the subject of control was included in Chapter 3 (See section 3.2.1).

The control network cooperates with the service layer, subscriber equipment and with other networks, in order to create adequate end-to-end connections in the connectivity layer. The cooperation requires advanced inter-node signalling communication. To get an adequate connection one may have to define aspects such as addressing, routing, QoS, security and billing. The choice of parameters is related to what particular service is chosen. In mobile networks, such as GPRS, this FP type information is conveyed in messages called PDP context. Similar information for the IP connectivity core is carried in BICC IP Bearer Control Protocol (IPBCP) (See also sections 14.3, 14.5 and 15.8.).

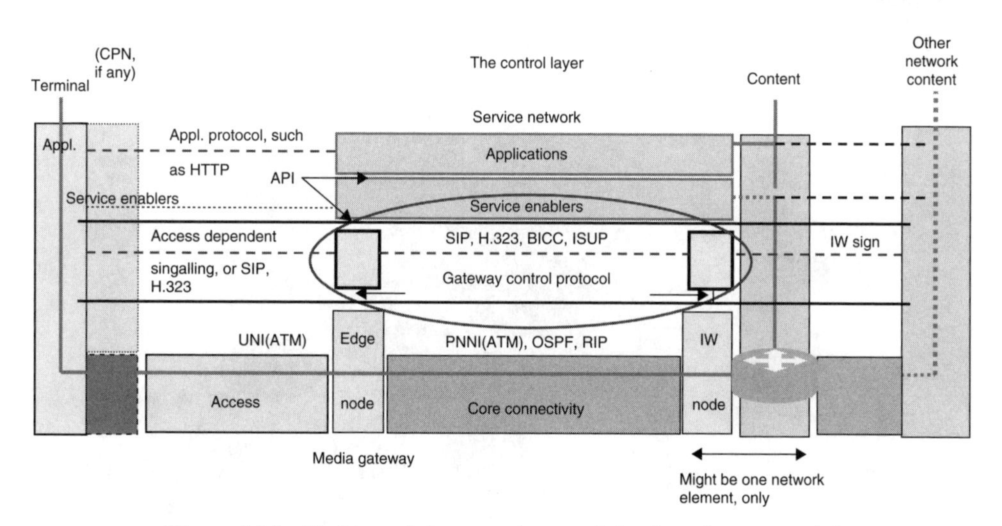

Figure 14.1 Position of the control network in the reference model

Understanding Changing Telecommunications – Building a Successful Telecom Business. Edited by A. Olsson
 ISBN: 0-470-86851-1

Regarding intelligence, the traditional IN SCP nodes are part of the control layer, assisting the control servers. In addition, IN based control functions are- or will- be implemented in the service layer, See for example Figure 8.7 in Chapter 8.

Much traffic can manage without the control layer. Connectionless best effort IP traffic is a good example, leased lines is another. As indicated in Chapter 12, there is an internal control function in the connectivity layer as well, serving data oriented traffic in IP, ATM and other networks.

Both ordinary voice and data communication have managed without a service layer, but this will change with an increasing number of service offerings beyond voice. The development will increase the business role of the service layer. The control layer and the connectivity layer will serve the service layer more and more by means of APIs. For interconnected networks see Chapter 15.

The control layer is related to MGW/edge node control (See Figure 14.1). There isn't normally a 1:1 relation between edge nodes/MGW:s and media gateway control nodes/control functions, unless the nodes are built in one physical unit. Figure 11.7 gives a better perception. However, GGSN and releases of SGSN and other nodes are typically implemented in one physical unit. In homogeneous environments there is no need for a gateway, see for example Figure 14.5.

Figure 14.2 shows a number of control parts or domains in the control layer, the common home subscriber server (HSS) function and a signalling gateway server function. The signalling traffic gateway is, strictly speaking, part of the connectivity layer.

The description of control in this chapter covers control of the following domains of networks:

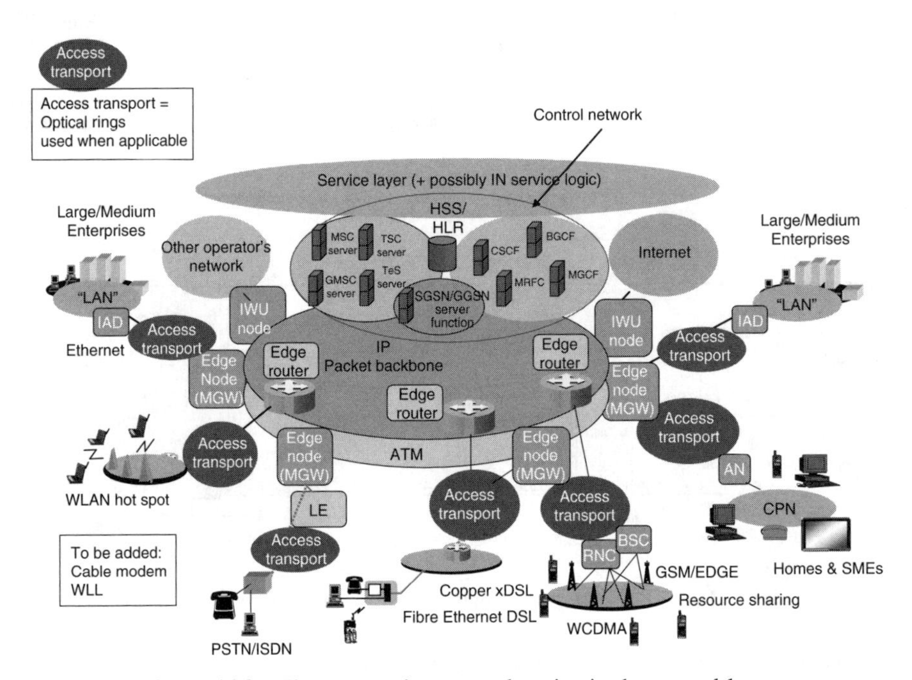

Figure 14.2 Three control parts or domains in the control layer

- Mobile and fixed voice networks ('circuit-switched domain'). ISUP/BICC signalling. Data uses the connectivity layer
- Mobile data networks ('packet-switched domain') as introduced in GPRS
- Mobile and fixed IMS multimedia networks. Primarily SIP signalling. Carries media individually and as multimedia
- Signalling over IP solutions. Gateways SIP-ISUP. Sigtran. 'Softswitch' with SIP/H.323 signalling. See Figure 14.3, central part. Voice trunking with H.323 gatekeepers according to Chapter 8, Section 8.8.3 Figure 8 will not be further treated (See also Chapter 11, Figure 11.8).

14.2 THE ENVIRONMENT OF THE CONTROL NETWORK

The introduction of the horizontal network is performed in an environment dominated by legacy networks with numerous signalling systems, especially on the access side (See Figure 14.3). A mixture of legacy and new (SIP, H.323) signalling systems will be common. This will require signalling gateways. A typical signalling gateway connects ISUP and SIP. ISUP, V5 and Q.931 will be briefly treated when describing the circuit domain. The signalling traffic should be carried in protected paths in the connectivity network. The interfaces upwards and downwards are new. ISUP, V5 and Q931 will be briefly treated when describing the circuit domain. The signalling traffic should be carried in protected paths in the connectivity network. Compare the No 7 network where the user parts are carried by the MTP carrier. The interfaces upwards and downwards are new. Media gateway control protocol (MGCP), also known as H.248 and Megaco, is the call control protocol between the MGW and the MGC (See Chapter 3). For the interface to the service network the Parlay and JAIN standards treated in chapter 3, are available. The SIP technology is also used here, for example in the interface between call control and application. The work of OMA must be considered.

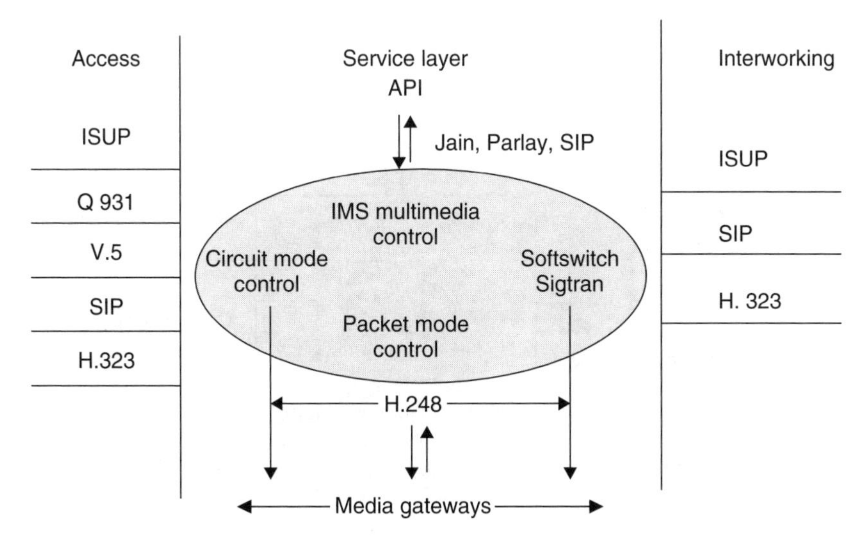

Figure 14.3 A typical communication environment of the control network

14.3 FUNDAMENTAL PLANS IN THE CONTROL NETWORK

Let us start by repeating a paragraph from chapter 8, Section 8.5:

> 'It is also confirmed when looking at the need for information at the start of a packet communication after attaching, let us say, a mobile terminal in GPRS to a packet network. The information referred to is called PDP (packet data protocol) context. It might be a logical association between a mobile station and a public data network running across a GPRS network. The context may define aspects such as routing, QoS, security, billing, type of network, network address, access point name and radio priority. Largely, we are again back to the fundamental plans.'

(See also the centre of Figure 14.4). A control layer for multimedia traffic requires a number of central functions, as opposed to the original Internet, often considered be a stupid network. A key future component is the call state/session control function/server, CSCF. The CSCF has similarities with the traffic control subsystem in ordinary telephone exchanges. A corresponding parallel can be found for other main traffic-handling functions. Let us look at a few FP oriented tasks performed by the CSCF:

- Numbering plan: Address Handling (AH). The Address Handling (AH) function handles analysis, translation, modification if required and address portability
- Routing plan: Routing of incoming calls
- Routing and interconnection plan: Cooperation with BGCF (border gateway control function) for calls to the PSTN to determine the MGCF (media gateway control function) to use
- Transmission/QoS plan: Analysis of need for a transcoder for mobile to mobile calls
- Switching plan: Controlling the MRF in order to support need for multi-party calls

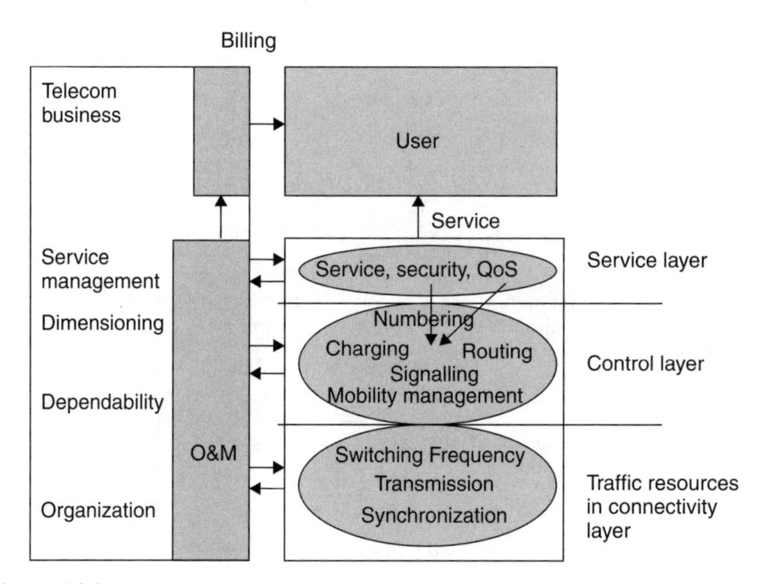

Figure 14.4 Security and QoS are also part of the control network and its tasks

- Charging: (and security) plan: Call monitoring and logging for billing, verification, auditing, intercept or other purpose
- Security plan: Is the requested outgoing communication allowed given the current subscription?
- QoS: Authorize the bearer resources for the appropriate QoS level. Determination of allowed QoS classes is done by means of subscriber profile information

At the end of the chapter the distribution of functionality to the numerous individual nodes will be treated at an overview level.

Let us have a very brief look at some specific FP features or references concerning the control layer.

14.3.1 SIGNALLING

Signaling is a focal point in this chapter. See Section 14.4.

14.3.2 MOBILITY MANAGEMENT: ROAMING

Roaming between GSM and UMTS is one of the important tasks, where the control layer is involved.

The inter-system roaming solution allows the subscriber to roam from the UMTS network to the GSM network and vice versa. The MSC Server supports mobility management in order to enable attachment/detachment and roaming within the UMTS network, between UMTS networks, and between UMTS and GSM networks. See more in Chapter 15.

14.3.3 MOBILITY MANAGEMENT: HANDOVER

The MSC server supports inter-MSC and intra-MSC handover from UMTS to GSM and vice versa. The handover function is vital for the offered QoS.

14.3.4 WLAN ISP (WISP): ROAMING

Islands of public WLAN coverage have already established roaming between their respective networks, based on unilateral agreements.

There are also companies establishing themselves as intermediary roaming players. That can be a choice for many WISP operators who are very local or who want to establish global roaming quickly. Supposing that visited and home WISPs do not have any roaming relations; the roaming broker can make direct agreements with each involved WISP.

14.3.5 MOBILE OPERATOR: WLAN ROAMING

For a mobile operator, the well-established GSM (and GPRS) roaming concept can be extended to also cover WLAN roaming traffic.

WLAN roaming traffic costs can be integrated with other GSM roaming traffic charges and billed on regular invoices by the home operator.

14.3.6 CHARGING

The servers provide flexible charging mechanisms. The existing charging capabilities allow the operator to charge a subscriber for network access and network usage. The system collects data, creates call detail records (CDR) and delivers them to external media, such as billing systems or a mediation system that can process the collected data for different purposes. Charging is treated in more depth in Chapter 5.

14.3.7 SECURITY AND QoS

CSCF provides predefined PDP context parameters via SIP signaling to user equipment.

Just as with dial-up roaming, proxy RADIUS is a common method for remote authentication for WISP roaming – at least in the short to medium term (See also Chapters 6 and 7).

14.3.8 ROUTING AND NUMBERING/ADDRESSING

See Chapter 8.

14.3.9 IN SERVICES

In general IN services were successful during the 1990s. However, many services have lost popularity with some exceptions, for example:

- cashless calling with prepaid, payment by card or billing afterwards. Prepaid is a ‘killer application’
- access screening
- number portability
- freephone, which is a must in the USA
- premium rate information services with fixed charging.
- televoting
- universal access number

A number of other IN services have been placed in ‘maintenance mode’. The complicated IN-VPN service, with company internal telephony traffic at reduced rate as the main application, has often been replaced by special agreements with carriers.

At least the successful IN services are migrating to IP based networks. The No 7 signalling system is a key enabler for IN, therefore VoIP systems must support and interoperate with SS7 in order to take advantage of what is already there. Interactions between IN and IP-based networks are subject to international standardization. See Section 14.9.

14.4 A SIMPLE TARGET CONTROL NETWORK SIGNALLING

A target network example can look simple, if the network is homogeneous. See Figure 14.5.

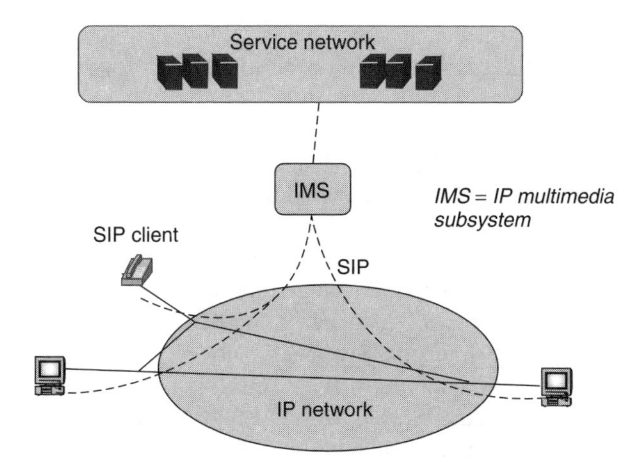

Figure 14.5 The future is simpler than the present convergence phase

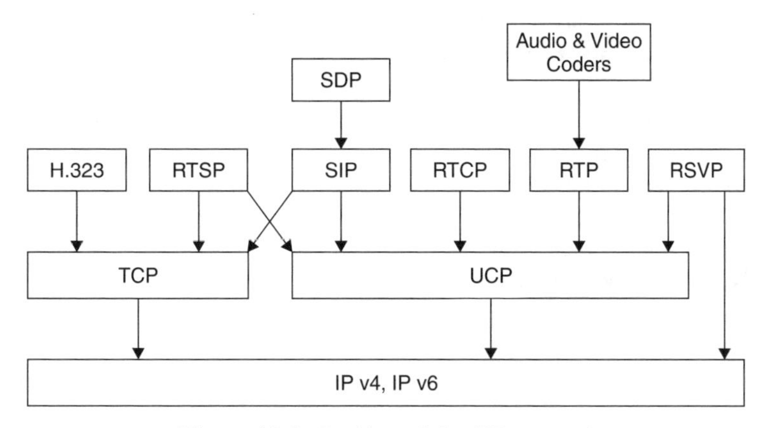

Figure 14.6 Position of the SIP protocol

The session initiation protocol (SIP) is a key standard in the IP telephony architecture, backed up by for example IETF and 3GPP. See Figure 14.6. H.323 is an alternative. In this book SIP gets most focus.

Both SIP and H.323 are signalling and control standards for Internet telephony. H.323 is an ITU recommendation, and SIP is developed by IETF. Both protocols provide a similar set of services, but SIP is often considered simpler with less logical components.

H.323 is an ITU standard. In H.323 architecture, each client belongs to a zone, and there is a gatekeeper in each zone. All the clients of a zone are registered to its gatekeeper. The gatekeeper provides address translation (allowing aliases to be used), admission control and bandwidth control.

14.4.1 SESSION INITIATION PROTOCOL (SIP)

The IETF Multiparty Multimedia Session Control (MMUSIC) working group has specified an IP telephony architecture based on SIP. The two basic protocols are the session description protocol (SDP), which is specified in RFC 2327, for channel and terminal capability handling, and the session initiation protocol (SIP) for registration, admission, status and Q.931 messages.

SIP uses a client server model similar to the HTTP. It is used in conjunction with other protocols such as SDP, RTP and RSVP. SIP can set up all types of sessions, e.g. voice, video, text or combinations, and manage any of these types of sessions, regardless of the media type.

SIP is a direct competitor to H.323. An IP telephone call is considered a kind of multimedia session in which voice is exchanged between the parties. The voice path is independent of whether SIP or H.323 is used for signalling.

SIP is independent of the transport layer and only requires an unreliable datagram service, as it provides its own reliability mechanism. It is typically transported over UDP or TCP, but could also use frame relay, ATM AAL5 and X.25.

SIP supports personal mobility and negotiation of the capabilities of the end users. It also supports the fundamental security services: authentication, access control, confidentiality and integrity.

SIP is a client-server protocol. Clients issue requests and servers answer with responses. A limited number of request types, also referred to as methods are defined for SIP:

- INVITE is used to ask for the presence of a certain party in a multimedia session.
- The ACK method is sent to acknowledge a new connection.
- OPTIONS are used to get information about the capabilities of a server.
- The REGISTER method informs a server about the current location of a user.
- A client sends a BYE method to leave a session.
- The CANCEL method terminates parallel searches.

A SIP dialogue can look like Figure 14.7.

When a server receives a request, it sends back a response. A code number identifies each type of response. There are six main types of responses. They are listed in

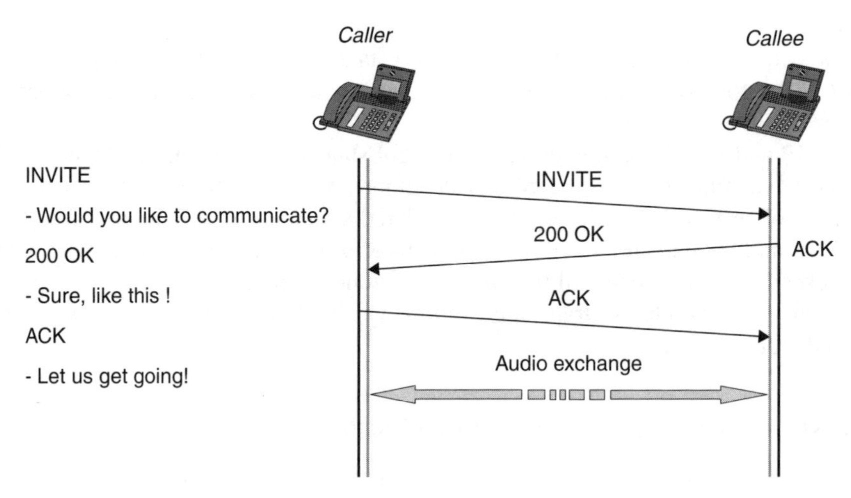

Figure 14.7 Example of the SIP protocol dialogue

1xx	Informational
2xx	Successful
3xx	Redirection
4xx	Request failure
5xx	Server failure
6xx	Global failure

Figure 14.8 SIP responses

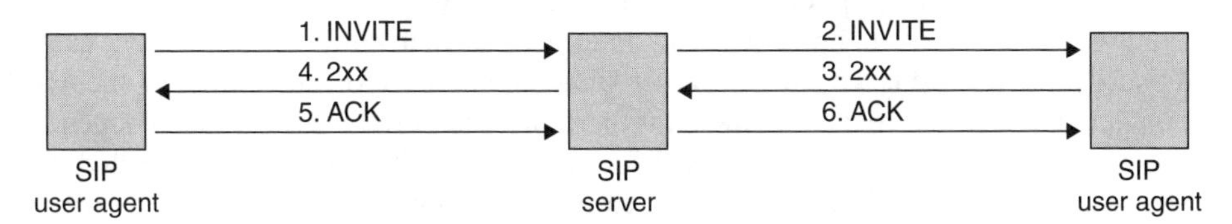

Figure 14.9 Proxy mode SIP operation

Figure 14.8. The server keeps the client informed of the status of the call by means of these responses.

There are two modes of operation in SIP: when servers are used with a proxy server or with a redirect server. The proxy server returns responses on behalf of the user. The redirect server informs the client of the current location of the user. Then, the client can reach the user directly. The proxy server takes care of the location of the user. Thus, the process is transparent to the client. There is another mode of operation when no servers are used. The user agent can send directly requests to the other user agent. Even when a SIP server is used in the first exchange of messages between the parties, the subsequent exchanges may be addressed directly to the user agent, without traversing any server. The normal exchange of SIP messages is illustrated in Figure 14.9. Translation between telephone number and SIP address is performed in a gateway at the PSTN-IP border.

SIP uses other protocols for specific tasks, for example DNS for user-domain location and SDP for media selection. 3GPP is dedicated to using SIP for call control.

14.4.2 SESSION DESCRIPTION PROTOCOL (SDP)

Session descriptions have a list format containing information about the session (see RFC 2327). Message headers are in plain text and look similar to e-mail headers. SDP is designed to convey session information to recipients and media information that pertain to the session. It allows more than one media stream to be associated with a session. SDP inform the receivers of its messages the existence of a session. It also conveys sufficient information to enable joining and participating in the session. See also the signalling plan in Chapter 8.

14.5 CIRCUIT MODE DOMAIN

The circuit domain embraces mobile and fixed networks mainly intended for voice. The typical circuit mode feature is the functions of the servers, whereas the traffic path normally migrates to become packet based. Data traffic can manage with the connectivity layer. See Figure 14.10.

Let us start with the mobile servers, which are called MSC and TSC/GMSC (TSC = transit switching centre).

14.5.1 MSC SERVER

The MSC server remotely controls the circuit-switched part of the MGW and initiates set-up/release via the gateway for the user plane. The control is supported by the gateway control protocol (GCP, H 248). When needed, the MSC server connects devices to the payload path (e.g. codecs, tone senders). Figure 14.11 gives an overview.

The MSC server also supports the radio access network application part (RANAP) for control signalling over the RNC-MSC interface, so that radio access bearers can be established and released. The MSC server also supports MAP signalling to the HLR and equipment identity register (EIR). Furthermore, the ISDN and bearer-independent call control (BICC) protocols can be used for control signalling between the call-control servers in the circuit-switched domain and external integrated services digital networks.

14.5.2 TSC/GMSC SERVER

The circuit mode PLMN traffic often has PSTN/ISDN as the originating or terminating point. The TSC/GMSC server selects the path across the core network and suitable MGWs with suitable gateway facilities.

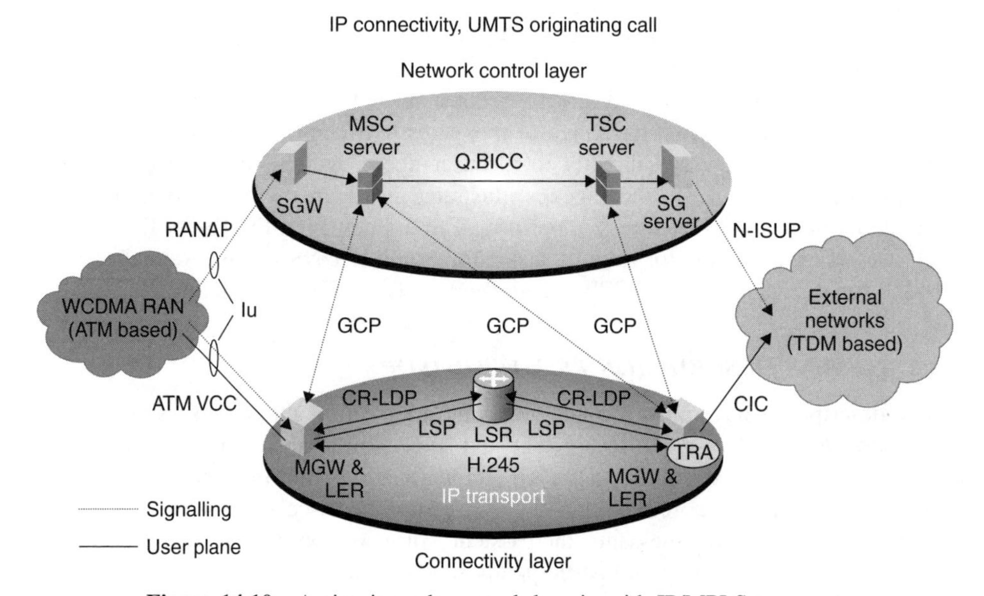

Figure 14.10 A circuit mode control domain with IP/MPLS transport

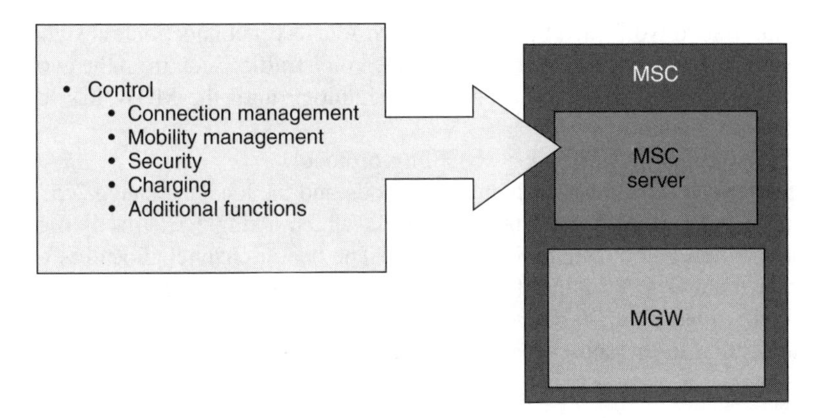

Figure 14.11 MSC server functions, example

The GMSC and TSC server functionality is often housed in the MSC server.

14.5.3 TELEPHONY SERVER IN FIXED NETWORKS

The telephony server (TeS) functional component handles circuit-oriented calls, for example call signalling, call routing and termination, maintenance of relevant call state, event handling and the invocation of network services. TeS might control ATM as well as IP connectivity networks. See Figure 14.12.

14.5.4 SIGNALLING INTERFACES

For PSTN/ISDN subscribers a subscriber loop signalling system between a user terminal and the edge node is common. See Figure 14.3.

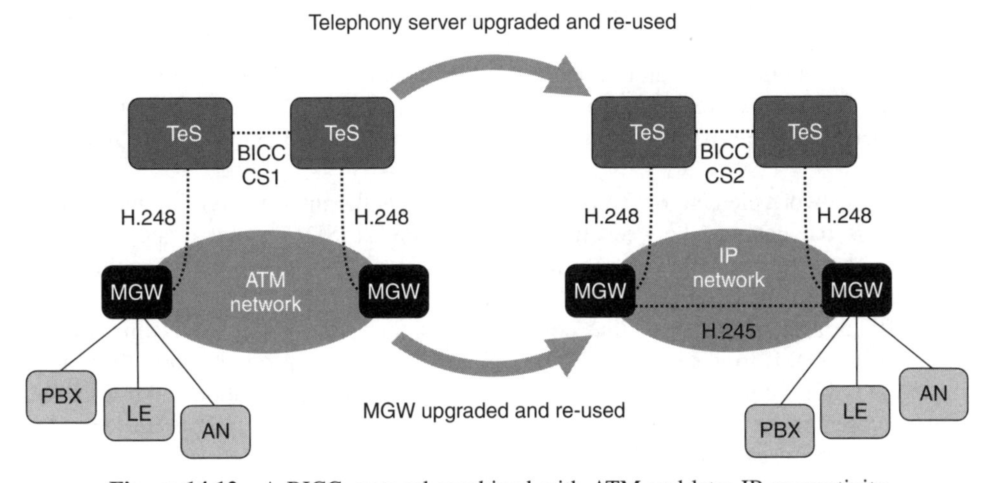

Figure 14.12 A BICC protocol combined with ATM and later IP connectivity

Signalling system No 7 (SS7) is a well-known four-level connectionless signalling system, primarily used for control of inter-exchange voice traffic. Seen from the evolving core network SS7 will often appear as an access technology until the MGW has 'conquered' the present local nodes.

Q.931 is the ISDN network layer signalling protocol.

The V5 interface is used between an access node and the local exchange. V5.1 interface is based on a static multiplexer type principle in which a single E1 link is used to carry the bearer channels and the signalling channels. The bearer channel allocation to the V5.1 subscriber is fixed. The V5.2 interface is based on a dynamic concentrator type principle, typically consisting of 1 to 16 E1 links carrying bearer and signalling channels. The bearer channel allocation to the subscriber is dynamic on a per call basis.

14.5.5 BICC AND BICC IP BEARER CONTROL PROTOCOL IPBCP

BICC is a signalling protocol based on ISUP. The basic issue leading to the development of BICC was that network operators had experienced heavy growth of their voice traffic mainly due to the Internet. PSTN/ISDN networks required major investments to accommodate the growth, while network operators did not want to invest too much in 'old telephony' networks.

BICC can support PSTN/ISDN/PLMN services over a broadband connectivity network. It is specified by ITU-T in recommendation Q.1901 and allows any bearer to be used, such as IP, ATM and TDM.

The BICC protocol allows operators to offer the complete set of PSTN/ISDN services with operator-grade quality, including all supplementary services, over a variety of packet networks. The BICC network is scalable.

BICC supports all currently deployed services on circuit mode, ATM and IP networks, including third-generation wireless, yet does not limit the future introduction of new multimedia, multi-mode services and applications.

Figure 12.9 shows the basic BICC network architecture. The general idea of a BICC network is the separation of call signalling on the one hand, and bearer signalling and user traffic on the other hand. The BICC protocol is used for call signalling functions, such as indicating the requested type of ISDN service and call routing, based on the dialled phone number. A bearer signalling protocol is used to set up a connection over the packet network to transport the speech or data payload. The bearer signalling protocol depends on the type of packet network for example ATM AAL1, ATM AAL2, IP or TDM.

The key network element of the BICC architecture is the interface serving node (ISN). The ISN is the gateway between the traditional PSTN/ISDN network and the BICC network. Three functions have been distinguished in an ISN: the bearer function (BF), the bearer control function (BCF), and the call serving function (CSF).

The BF is a media gateway that converts the signals transported over time-division multiplexed (TDM) trunks in the PSTN/ISDN network to an appropriate form that can be transported over the packet network. Also functions like echo cancellation and transcoding may be present in the BF. The BCF controls the bearers by using the appropriate signalling protocol for the packet network and can establish and remove bearer connections over the packet network. The CSF handles the call signalling, that is, ISUP signalling to the

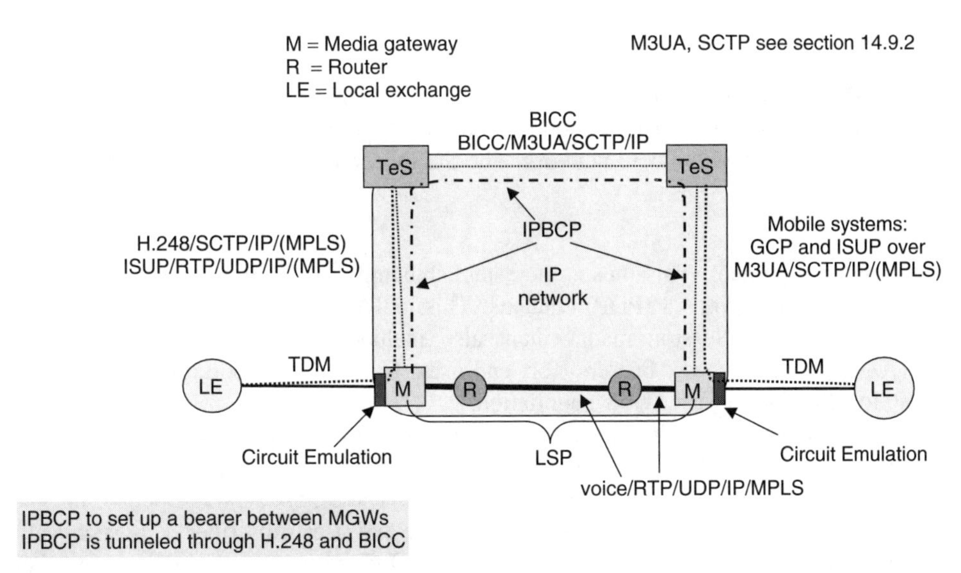

Figure 14.13 Example of signaling protocols and signalling and voice transport in an IP core as defined in Figure 14.12

PSTN/ISDN network and BICC signalling in the BICC network. The call mediation node (CMN) can be used to route the BICC messages and may be useful in a large-scale BICC network with a large number of ISNs.

Two other nodes have also been defined for the BICC architecture: the transit serving node (TSN) and the gateway serving node (GSN)

The TSN is a node that handles BICC on both sides. A TSN may be used to offer intelligent network based services. A TSN may also convert the bearer from one type of packet network to another.

The GSN is similar to the TSN since it also handles BICC on both sides. A GSN acts as a gateway between network operators in a full BICC environment.

BICC IP Bearer Control Protocol (IPBCP) (ITU Q.1970) is used for the exchange of media stream characteristics, port numbers, and IP addresses of the source and destination of a media stream to establish and allow the modification of IP bearers. The information exchanged with IPBCP is done during BICC call establishment. In addition it may be exchanged after a call has been established. IPBCP uses the Session Description Protocol (SDP) defined in RFC 2327 to encode this information. See Figure 14.13. Section 15.8 describes the similar PDP context. Signalling transport is described in section 14.9.2.

14.6 PACKET MODE DOMAIN

The packet mode domain is introduced in GPRS and it is conveyed into UMTS. In the IMS phase the SGSN and GGSN node become the main core nodes. GGSN will then partly be controlled by IMS CSCF.

14.6.1 SGSN

The SGSN server, which is located in the packet-switched domain, handles control layer functions related to packet-mode communication services at the border between the access network and the core network. The SGSN server may host the following main functions:

- Session management. This means the establishment, maintenance and release of end-user packet data protocol (PDP) contexts. This includes interworking with the GGSN for IP addresses. Session management also includes functionality for establishing WCDMA radio access bearers for end-user IP data transportation, as well as functionality for end-user QoS negotiation.
- Mobility management. This means the functionality that supports roaming within and between GSM and UMTS mobile networks.
- Integrated VLR and subscriber data management. The SGSN server supports the standardized interface to the HLR in order to manage end-user subscriber data, such as the international mobile subscriber identity (IMSI), QoS profile, access point names, and so on.
- GGSN interface-the gateway tunnelling protocol-control (GTP-C) supports control signalling between SGSN servers and the GGSN. GTP-C is transported by UDP/IP and contains functionality for SGSN-to-GGSN tunnel management and control.
- MAP and RANAP control signalling. The SGSN server supports the RANAP protocol for control signalling over the RNC-SGSN interface for establishing and releasing radio access bearers. The SGSN server also supports MAP signalling to the HLR, EIR, MSC server and SMS centre (SMS-C).
- Media gateway control function. The SGSN server supports the GCP protocol for controlling the packet-handling functionality of the media gateway.

See Figure 14.15.

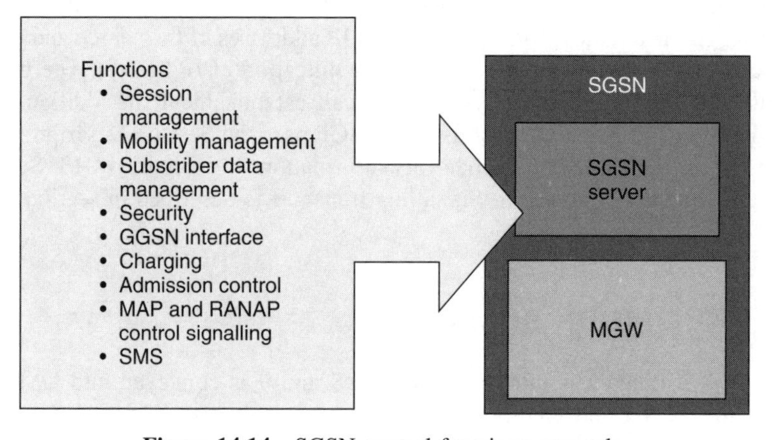

Figure 14.14 SGSN control functions, example

14.6.2 GGSN

The GGSN provides functions such as routing, packet session establishment and termination, message screening, service measurements and billing.

Tunnels are dynamically established between an SGSN and GGSN using the GPRS tunnelling protocol (GTP), and are deactivated when no longer needed.

The GGSN functions include assignment of an IP address to the user equipment, maintaining packet data protocol (PDP) context, and collection of information for billing.

GGSN controls the interface for IP bearer path to other networks, such as the public Internet, a VPN, or another wireless service provider's network. The GGSN handles routing of wireless IP multimedia subscriber packets and supports the QoS classes. It supports QoS interworking with external networks to support end-to-end QoS requests.

14.7 IMS DOMAIN = IP MULTIMEDIA SUBSYSTEM

14.7.1 MCS

The goals of the MCS (multimedia core system) are to:

- Allow telephony service providers to extend their services portfolio with more advanced, IP-based, multimedia services.
- Provide support for a number of successful multimedia services to ensure that these systems will continue to focus on multimedia, rather than telephony over IP.

The MCS provides complete solutions for multimedia access, communication and services with transparency between fixed and mobile communication services.

The architecture and specifications are based on the IM subsystem defined by the 3GPP for UMTS. SIP is used for call/session control signalling, and MEGACO architecture is used for edge nodes/media gateways. It has an open service creation environment. By using standardized APIs such as Parlay, the customer can develop its own services or bring third-party services into the network.

MCS includes the elements necessary to support IP multimedia services in UMTS as defined by the 3GPP. MCS complements, but does not replace, MSC-based mobile telephony systems that will continue to provide voice-centric services.

14.7.2 MULTIMEDIA CORE NETWORK FUNCTIONS

See Figures 14.15 and 14.16.

The IM core network functions include:

- **Call state control function (CSCF) –** Basic functions includes call routing, SIP session set up, interface to media gateway, HSS and MRF for call connection and control.
- **Home subscriber server (HSS) –** HSS contains authorization and authentication information for a user, terminal-based information, and information about the SGSN/VLR that a user is currently being served by.
- Other functions such as border gateway control function (BGCF), media resource function (MRF) and media gateway control function (MGCF).

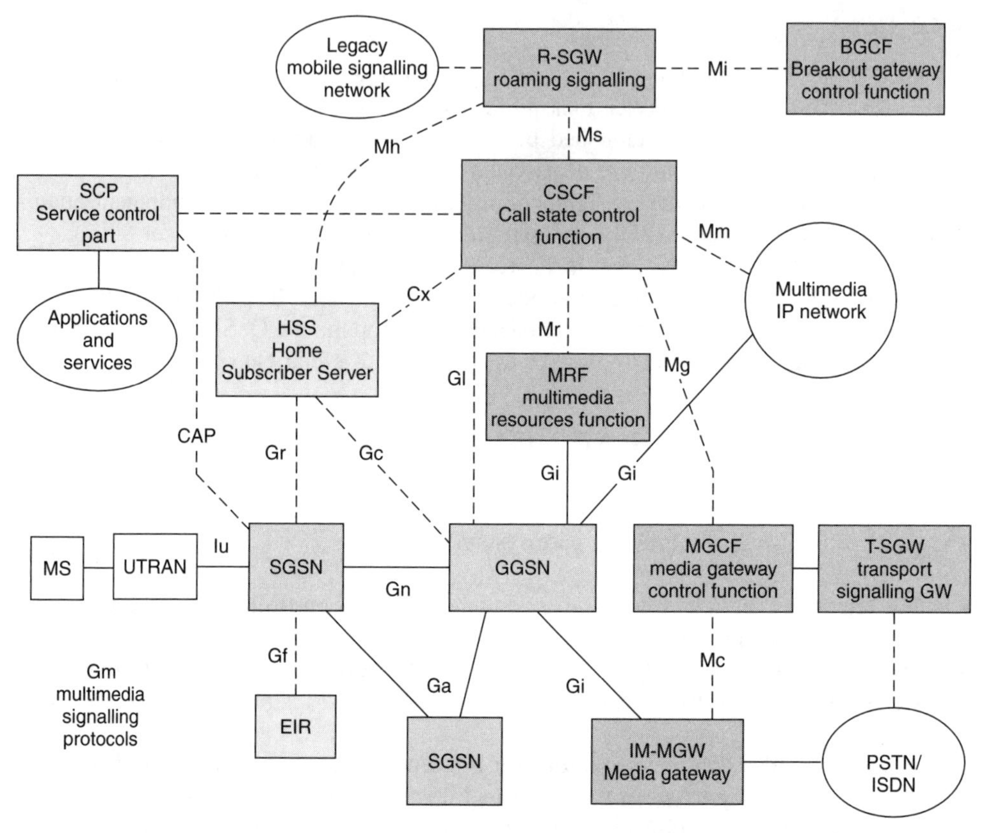

Figure 14.15 IMS functions and other capabilities

BGCF

The BGCF is a signalling entity for call/session control and acts as a router (at the SIP level) to another SIP entity. When multiple MGCFs exist within the network operator's network, BGCF acts as a 'SIP in, SIP out' entity trying to find the best MGCF for interworking with the circuit-switched network for a call from a UE to a PSTN address.

CSCF

A CSCF in a network may perform one or more of the following roles:

- **Interrogating CSCF (I-CSCF)** – the I-CSCF function interrogates an external location-service function to determine which CSCF is serving at a given time and acts as a SIP proxy firewall to fulfil network security and privacy functions.
- **Serving CSCF (S-CSCF)** – the S-CSCF is the serving network element with which subscribers register in order to be reached when roaming. It temporarily stores user profile-related data, which is downloaded from the HSS as registration takes place. The S-CSCF also triggers the call- and session-related services to which the user has subscribed.

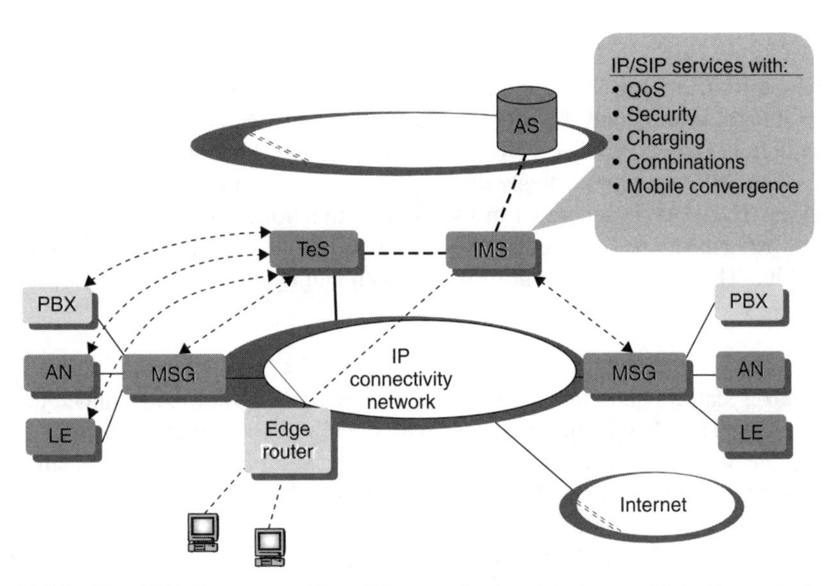

Figure 14.16 The IMS domain position. IP networks used to be stupid in the central network part, but with IMS we are heading towards a central intelligence again

- **Proxy CSCF (P-CSCF) –** the P-CSCF, which contains a very limited CSCF function, is the only Internet multimedia function placed in the network being visited; that is, when the subscriber's unit is roaming outside of its home network. The P-CSCF contains address translation functions in order to proxy the session request directed at the I-CSCF in the home network.

Figure 14.17 shows possible CSCF locations. The upper half of the diagram shows the UE served at the home network and the lower half of the diagram shows the UE roaming to a visited network.

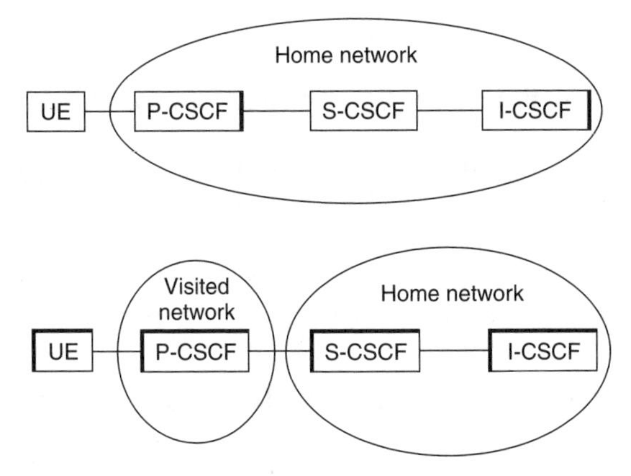

Figure 14.17 CSCF locations for mobile-terminated sessions

MRF

The MRF performs multiparty call, multimedia conferencing and transcoding functions. It is responsible for bearer control (with, for example, the GGSN and MG) in the case of multiparty/conference sessions. The MRF is also responsible for performing announcement functions. The MRF communicates with the CSCF for call control of announcements and conferences. The MRF is separated in two logical functions: MRC (media resource control) and MRG (media resource gateway). The MRG is the media-related part, co-located with the MG. The MRC is the control part, which can be located with the server nodes.

MGCF

MGC controls call-related state information that pertains to the call connection in the media gateway. The MGCF communicates with the CSCF for call information. It requests the media gateway to handle the bearer connection. See Figure 14.15. 'A call between a mobile roaming subscriber and for example, an ISDN subscriber, may involve many control networks. An example with a home control network, a visited control network, a gateway ISDN network and their common connectivity layer is presented in Chapter 15, Section 15.10.

14.8 HLR/HSS FOR ALL PREVIOUS DOMAINS

14.8.1 THE HOME LOCATION REGISTER (HLR)

The HLR is a network database for mobile telecommunications. The HLR holds all mobile specific subscriber data and contains a number of functions for managing this data, controlling services and enabling subscribers to access and receive their services when roaming within and outside their home PLMN. The HLR communicates with the SGSN, MSC and other network elements via the MAP protocol.

14.8.2 THE HOME SUBSCRIBER SERVICE (HSS)

The HSS is responsible for keeping a master list of features and services (either directly or via servers) associated with each user, and for tracking location and means of access for the subscribers. It holds user-specific information including identification, security, location and service profiles.

For reliability reasons there must be more than one HSS/HLR. The node can also support other systems such as WLAN.

The 3G HLR is part of the HSS functionality. It also needs to communicate via new IP-based interfaces. Like the HLR, the HSS contains or has access to the authentication centres/servers (e.g. AUC, AAA) and SMS nodes, as well as the FNR register.

The HSS may consist of the elements and interfaces shown in Figure 14.18. The DNS and the location server are sometimes referred to as the user mobility server (UMS).

14.9 THE DOMAIN OF (VOICE AND) SIGNALLING OVER IP

A typical feature of this domain is that digitized voice encapsulated in RTP (real-time transport protocol) packets is sent over carrier-grade IP networks using the services for

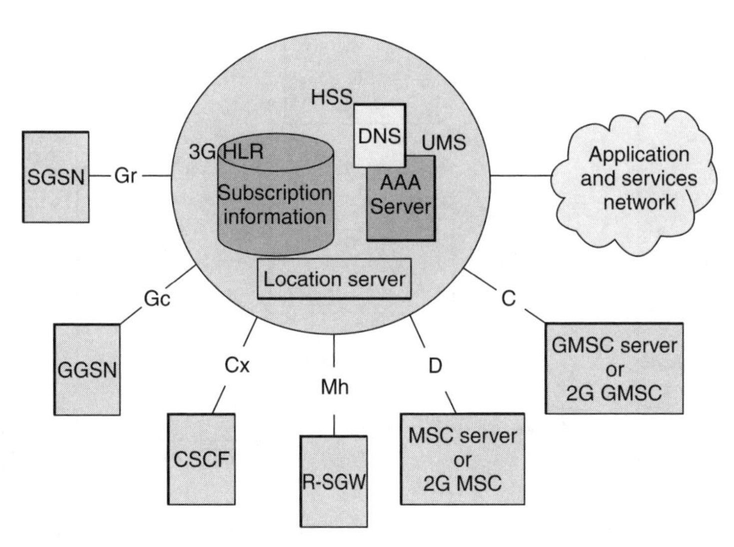

Figure 14.18 The central role of HSS as a spider component for short- and long-term subscriber data, such as location, security, numbering and mobility management

transport, provided by the layers below RTP (UDP, IP). At the receiving end of the VoIP network the packets are de-capsulated, and the original information is recreated.

Another associated feature is the use of the SIP protocol for signalling with H.323 as an alternative. SIP often interfaces ISUP signalling, where IP and PSTN/ISDN meet.

So far, large-scale public voice over the Internet has not been used very much for QoS reasons.

The VoIP architecture of the pertinent domain is called Softswitch. Softswitch is the concept of separating the network hardware/traffic resources from network software/hardware control. Softswitch is nowadays often interpreted as a specific control layer component, also called media gateway controller or call agent. See Figure 14.19.

The separation of traffic resources and control is valid for other domains as well.

14.9.1 USER PART MAPPING BETWEEN ISUP AND SIP

In order to provide compatibility between ISUP and SIP signalling systems, there is a need for a user part gateway connected to PLMN/PSTN/ISDN (sometimes denoted GSTN, general switched telephone network), and the Internet. This gateway has to handle calls, which originate in either of the two sides.

The mapping between ISUP and SIP is based on the type of the message received when a connection is being established (e.g. ACM, IAM, INVITE, BYE). For example an ACM (acknowledgement) message is equal to SIP 183.

There are different ways to arrange the MGW, the MGC and the signalling gateway (SGW). See Figure 14.20. The SGW and the MGC can be combined as shown in Figure 14.21. See the MGC–SGW combined version in Figure 14.22. This approach allows the installation of one SGW/MGC that controls several distributed MGWs.

The SGW role is to establish and join one or more IP-SS7 links and to maintain the state of the connection between the two networks. Maintaining the state implies sequence

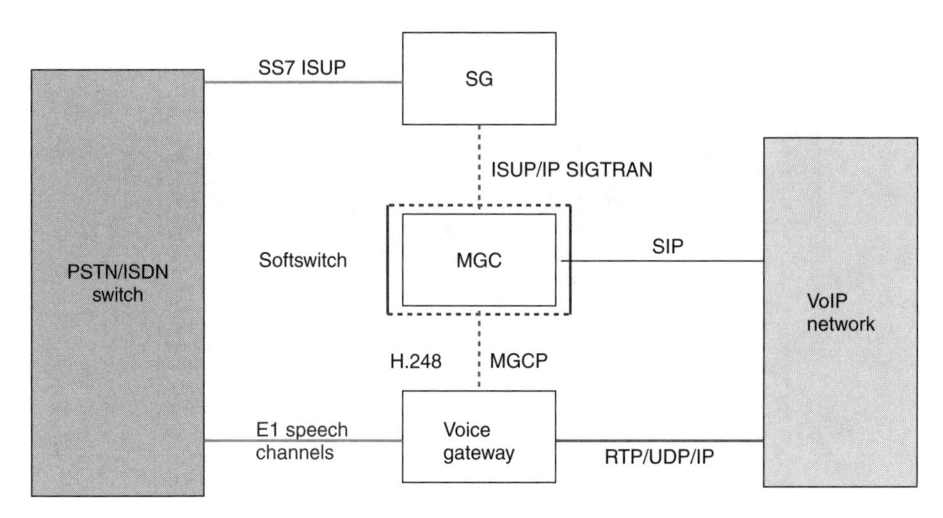

Figure 14.19 A typical interface between VoIP and PSTN/ISDN

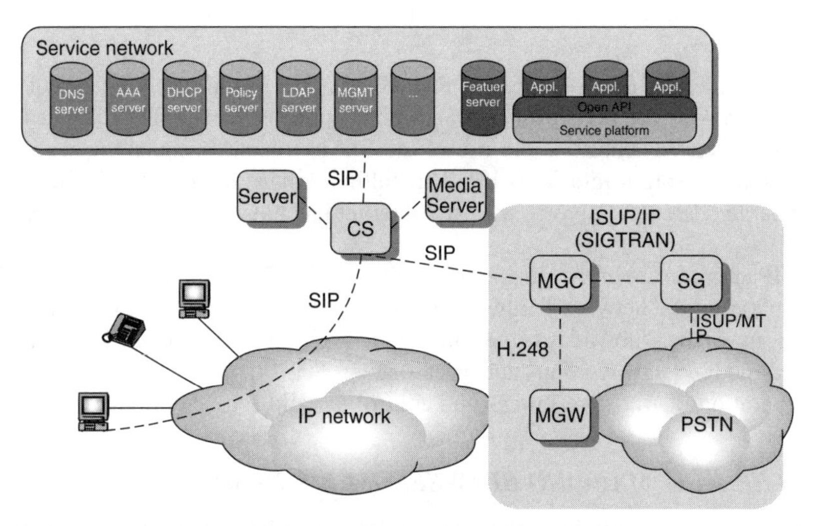

Figure 14.20 A 'softswitch' part connected to an all SIP/IP network. Support servers at the top

numbering, acknowledgements, retransmissions and notifications if anything is out of sequence. IP congestion control, the detection of session failures and security are other important functions performed by the SGW. An SGW also contains signal transfer point (STP) functionality, in order to convey SS7 messages over the message transfer part (MTP3 and MTP3b) in TDM and ATM networks as well as over the stream control transport protocol (SCTP) in IP networks. The SGW may terminate SS7 signalling or translate and relay messages over an IP network to an MGC or another SGW. Because of its critical role in integrated voice networks, SGWs are often deployed in groups of two or more to ensure high availability. The MGC uses the signalling information from the SGW (e.g. ISUP, SIP, H.323, BICC) to control one or more edge nodes.

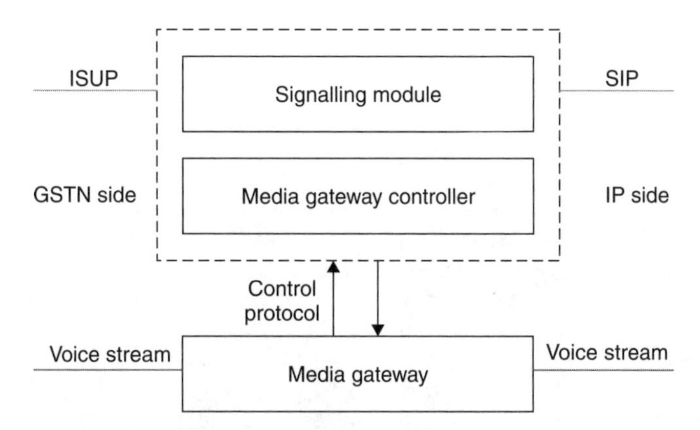

Figure 14.21 Gateway architecture

14.9.2 TRANSPORT PART MAPPING

When using IP as a transport technology there is a need to transfer SS7 signalling (BICC/ISUP, MAP, etc.) over IP. VoIP networks carry SS7-over-IP using protocols defined by the Signalling Transport (Sigtran) working group of the IETF. The SS7 over IP stack is shown in Figures 14.22 and 14.23. See also Figure 14.13.

SCTP

The stream control transmission protocol (SCTP) is designed to transport PSTN/ISDN/PLMN signalling messages over IP networks, but is capable of broader applications. SCTP is an application-level datagram transfer protocol.

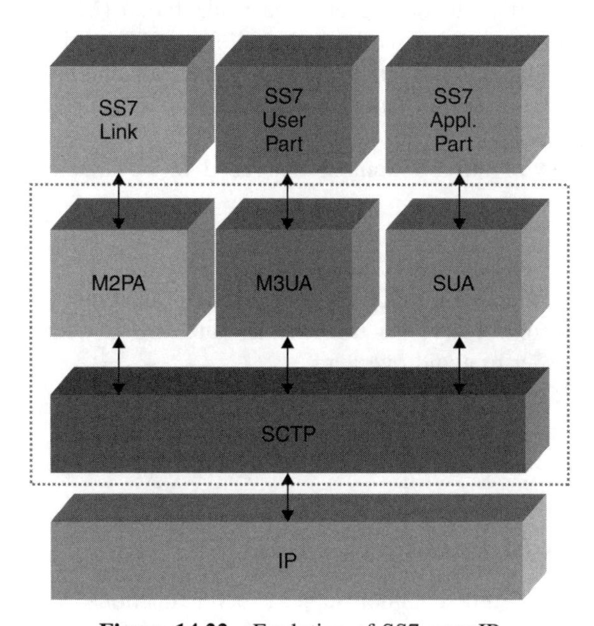

Figure 14.22 Evolution of SS7 over IP

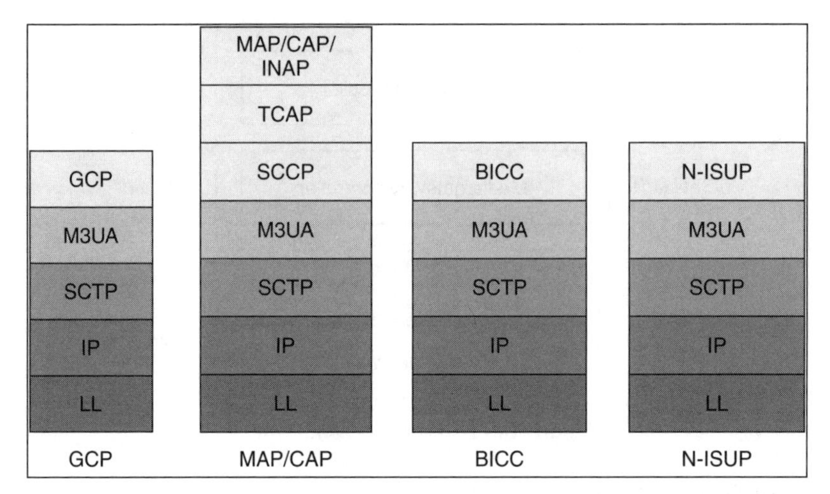

Figure 14.23 CS Protocol stacks for signalling over IP

M3UA Protocol Layer

The message transfer protocol 3-user adaptation (M3UA) protocol layer implements the SS7-like signalling network. It provides the same connectionless service to its users as MTP3. M3UA establishes connectivity between adjacent signalling nodes.

M3UA adapts the SCTP to the MTP level 3. Consequently ISUP protocol SCCP can run directly over IP without having to adopt an SS7 topology. While M2PA requires the maintenance of the topology of SS7 and SS7 links, M3UA doesn't. M3UA only requires SS7 endpoints, thus participating only at the services level.

The message transport protocol 2, peer-to-peer adaptation (M2PA) layer matches the SCTP generic transport to a primitive level that allows MTP level 3 to run over it.

Another evolving adaptation layer is the SCCP user adaptation layer (SUA), which allows SCCP and global title translation to run directly over it. TCAP can run on top of SCTP or on top of M3UA.

The protocols used in the CS domain that need to be adapted for transport over SCTP/IP are:

- MTP3 users (SCCP and its users, ISUP).
- GCP uses M3UA on top of SCTP. It may include functions such as support of a standardized load control package that enables an MSC to reduce traffic towards a certain MGW in order to avoid overload.
- BICC uses M3UA on top of SCTP.

These protocol stacks are given in Figure 14.23. (See also Figure 14.13).

14.10 COMMON SUPPORT FUNCTIONS

This section summarizes some common support server functions. Most of the servers can be seen in Figure 14.20. The servers normally belong to a fundamental plan. Therefore the treatment in the book is distributed to chapters dealing with FP security (Chapter 6) and FP in general (Chapter 8).

Gateway location. The main function of gateway location is to provide information that enables the selection of the appropriate gateway. The information provided could include route, cost, congestion and other parameters.

Directory server. The main function of a directory server is to provide a listing of information. The most common information model for directories is based on LDAP. LDAP directory services are based on a client-server model. See the Chapter 6.

Name server. The main function of the name server is to map names to addresses. This function essentially consists of two parts: a database to store the mapping, and a query resolver program that answers requests to translate the name to an address. The de facto standard for name server function is the DNS, which is briefly described in Chapter 8.

Address allocation. See DHCP in Chapter 8.

Security function. See AAA in Chapter 6.

Accounting function. This function addresses the problem of accounting. Accounting provides information about the utilization of resources by a user. Accounting can be based on providing information at specified intervals, end of session and the information may be aggregated to an agreed level. This information may be used for charging purposes, statistics or others.

Policy function. The policy function provides capabilities to state and enforce rules to manage the behaviour of heterogeneous systems (in terms of QoS, security etc.), devices, applications and network resources. The enforcement of policies on multiple managed entities is intended to produce aggregate behaviour that enables delivery of service at agreed levels. The policy itself may be stored in a directory.

Mobility agent. This function provides support for handling personal mobility, terminal mobility and service mobility.

15

Interconnection

15.1 OBJECTIVES

To explain:

- what is meant by interconnect and interworking
- relationships between interworking and fundamental technical plans
- interconnect and roaming, roaming between technologies
- interconnect and VHE
- interconnect and e2e performance
- interconnect and signalling
- interconnect and charging and accounting
- interconnect and SLA
- the need for cooperation between technologies, such as UMTS and WLAN.

Figure 15.1 gives a hint on involved areas.

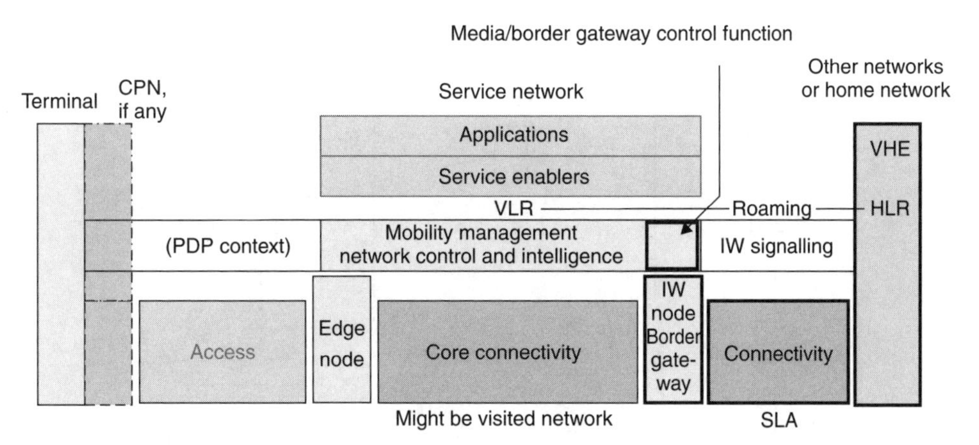

Figure 15.1 Dedicated interworking parts of the reference model – in a particular call other 'ordinary' parts will be involved as well

Understanding Changing Telecommunications – Building a Successful Telecom Business. Edited by A. Olsson
 ISBN: 0-470-86851-1

15.2 INTRODUCTION

It has been said before: it is the user perception that counts. The main user perceived areas in this book are services, QoS and security, all perceived end to end. The strong relationship between interconnect and user perception is therefore obvious.

In Europe the EC sees interconnection as a key element in the future competitive environment, allowing new market entrants access to existing end users, on a basis that will encourage increased investment and market growth in the telecommunications sector.

Among the main interconnection/interworking success factors are roaming or 'remote service access' including virtual home environment (VHE), representing a wider service access, e2e QoS, e2e security, and a concept that can be called 'always best connected'. Roaming will therefore include roaming between different technologies. The access network fragmentation regarding transport technique is counteracted by multi-technology terminals, moving through and between various technological islands. This is a condition for good utilization of the access network. The user services may be identical provided VHE is available.

In the good old monopoly days the interconnection issue stressed routing and hierarchical matters, such as position of international gateways, and mutual accounting between countries. See Figure 15.2. Among important quality matters the distribution of end-to-end loss was most important when the networks were analogue. The transmission plan catered for suitable values, also for national distribution.

In the deregulated telecom world the number of actors has increased tremendously. Further, the horizontalization creates possibilities to split the actor network into layers, for example to use the home service environment in a visited area (virtual home environment, see Section 15.6.1). Also the actors themselves can be split into layers, and/or interconnected in series. As mentioned, QoS and security have become crucial issues. See Figure 15.3.

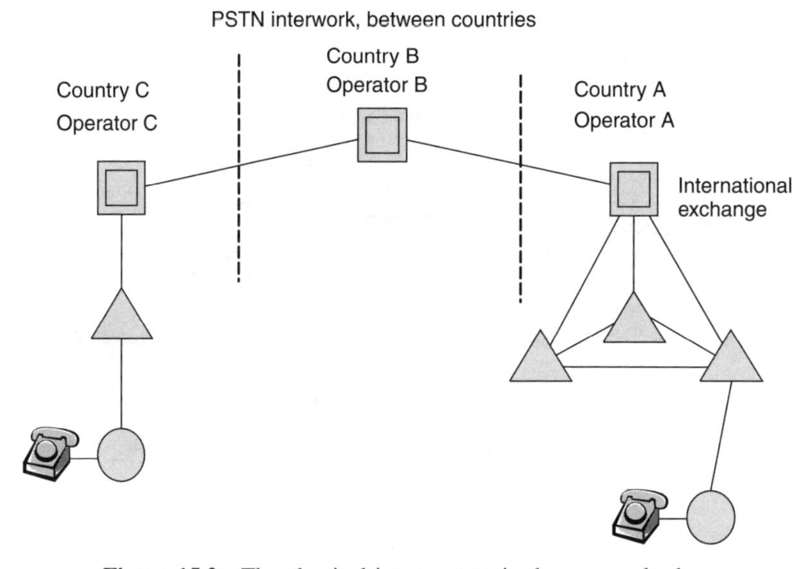

Figure 15.2 The classical interconnect in the monopoly days

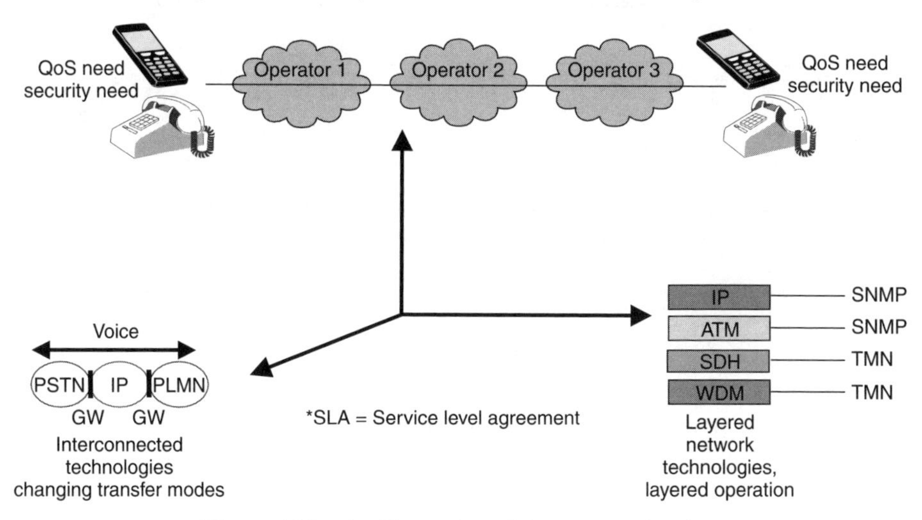

Figure 15.3 An illustration of the present situation

The introduction of 'multi-service core networks' may result in a core network role as mediating middleware between various accesses. Many access operators with different technologies may connect to few core/backbone network operators with fairly homogeneous backbone technologies. The core network develops towards a hierarchical IP transport with hierarchical servers according to Figure 15.4.

When many interworking actors cooperate to produce a service, achieve end-to-end QoS and end-to-end security, the real challenges will come. There is no dictator that sees to it that everything works, rather there is a kind of democracy that reigns. The situation must therefore be solved by means of agreements. Service level agreement (SLA) is thus a common expression in this chapter.

Whereas IP is the tool to interconnect users globally, web services using XML might become the glue at the service layer level. In future rather than software components being developed and bound together to form a single rigid solution, systems could be developed as a 'federation' of services at the point of execution. This will enable alternatively located software components to be substituted. A type of software interconnection appears. See also Appendix 1. XML and IP might then be accompanied by SIP signalling end to end. A target network?

This chapter can only cover a small amount of the important aspects. Since 'a picture says more than a thousand words' the chapter is rich in figures, and some contexts are probably left to the reader to finalize.

Basically, interconnection or interworking is a matter of relevant knowledge of the environment, including agreements. One could compare a large network, operated by actor A, and two networks, each half the size of the first one, operated by actors B and C. How can a B + C connection be as good as an A connection, provided the conditions are equal in all other aspects?

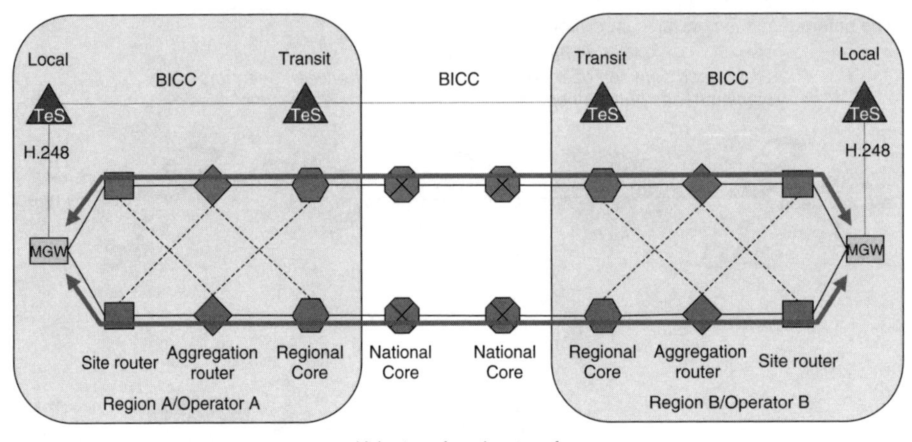

Figure 15.4 Example of two alternative connection paths between two MGWs in two interconnected core networks based on IP transport. Alternative paths create reliability

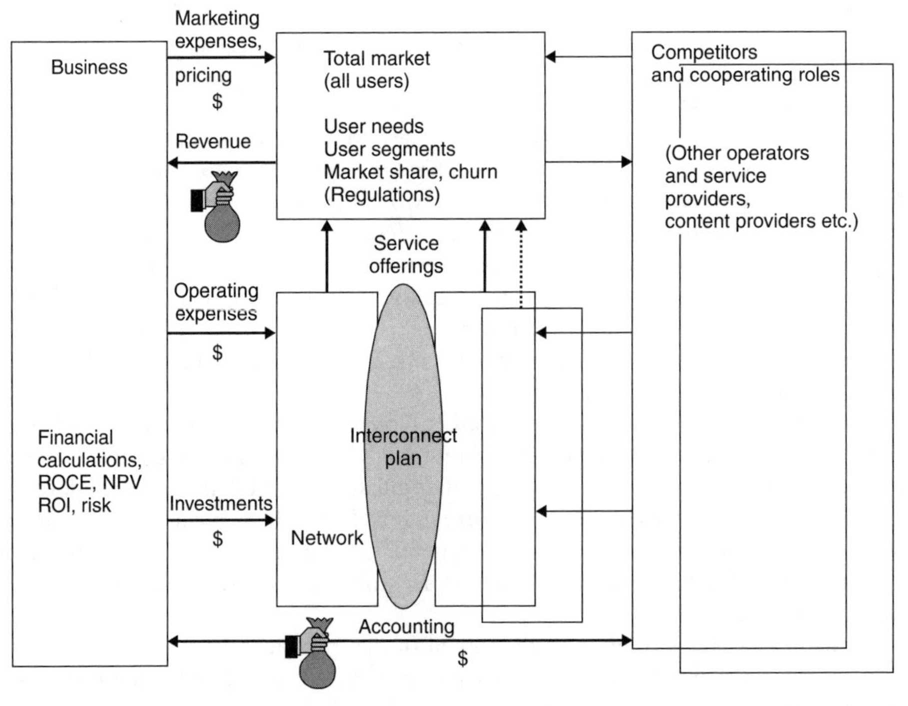

Figure 15.5 Position of an interconnect plan with vertical interconnect that specifies agreements and important mutual information. Because of all the possible roles according to Chapter 4, there is a need for many other interconnect plans as well

Obviously, if there is an efficient and relevant communication and written agreements between B and C, things should work. What type of communication is then relevant? Let us call it an interconnection plan or an interworking plan and integrate it among our fundamental plans. The position is seen in Figure 15.5. Not surprisingly, it turns out that the plan contains most of the other plans. Interfacing gateway locations and the values in the interfaces between the actors must be clearly defined. See also the paragraph on reference points below.

15.3 INTERCONNECTION IN TELE-CENTRIC FIXED VOICE NETWORKS

Critical issues are:

- Point of interconnect (POI)
- Services
- Subscriber access
- Number portability
- Charges and payments
- Billing
- Numbering
- Signalling and interface standards
- Network design and routing principles (e.g. emergency)
- Synchronization
- Network management, installation, operation and maintenance

Obviously the issues deal mainly with fundamental technical plans. It should be noted that mobility management has to be added for mobile networks, and security and QoS for the network dealt with in this book.

Among interworking FPs that will be briefly illustrated in this chapter are:

- Services, in the form of VHE.
- QoS, which is maybe the trickiest. Distribution of deteriorations.
- Security interworking examples. Indirectly security is concerned, when two networks utilize the same AAA in the HSS system.
- Routing. Gateways between networks, break out, routing via roaming operators, routing to own ISP.
- Charging. Charging when roaming.
- Mobility management (roaming). When describing success factors in the visions published by the Wireless World research forum, the forum stresses seamless service provisioning across a multitude of wireless systems.
- Signalling, in the form of PDP context between GPRS and a data network. Gateway protocols are mentioned.

See Figure 15.6.

Network management information in the actor interface is necessary for e2e supervision, handling of alarms etc. The connectivity oriented plans are more autonomous, particularly at the first layers of the OSI model, and therefore less critical.

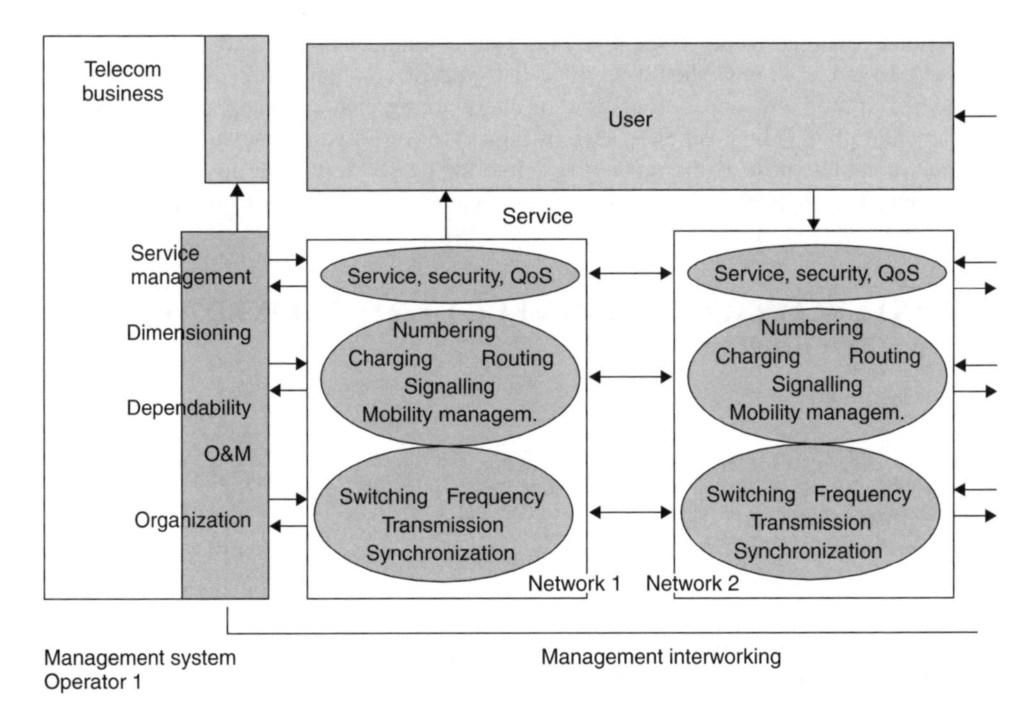

Figure 15.6 The interconnect plan contains interworking between other plans, in particular in control and service layers

E2e security is of particular interest to what are called e-workers, equipped with a laptop and a mobile. The e-worker workplace varies a lot: central office, local offices, the home, customer locations, airports and other hot spots.

However, let us start with reference points and SLAs.

15.4 DEFINITION OF AN ACTOR INTERFACE REFERENCE POINT

A reference point is a set of conformance specifications for end-to-end important properties for the service. It comprises a) the set of interfaces that describe the interactions that take place between the various entities responsible for performing a particular task, b) service level specifications (SLSs) between the entities and c) behavioural specifications between entities. Thus, reference points are a prescriptive part of the specifications which can be used to test an implementation's conformance.

A reference point is needed to enable the conforming entities to evolve independently of each other.

Figure 15.7 shows a reference point and an interface example. It also contains SLS and behavioural specifications. This example covers QoS, security, signalling, mobility management, and network management.

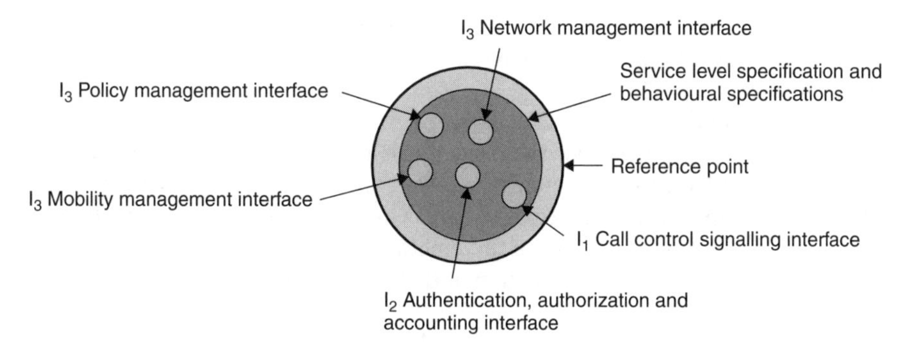

Figure 15.7 Example of properties in a reference point

15.5 SERVICE LEVEL AGREEMENTS

In this paragraph the expressions SLA and (Service) interconnection agreement are used. SLA has a wider interpretation. Let us start with the interconnection agreement, which is more of a peer-to-peer agreement. See Figure 15.8.

An interconnection agreement should contain at least the following:

- **Commercial criteria:** Make sure you know how interconnection impacts your business.
- **Service criteria:** Make sure that the interconnection agreement includes agreement on service provision and QoS in order to support your business objectives.
- **Technical criteria:** Select your points of interconnection to suit your business idea in the best possible way.

In a real interconnection situation, the two interconnecting operators may have to include both national and sometimes regional (e.g. EU) laws and regulations into the agreement.

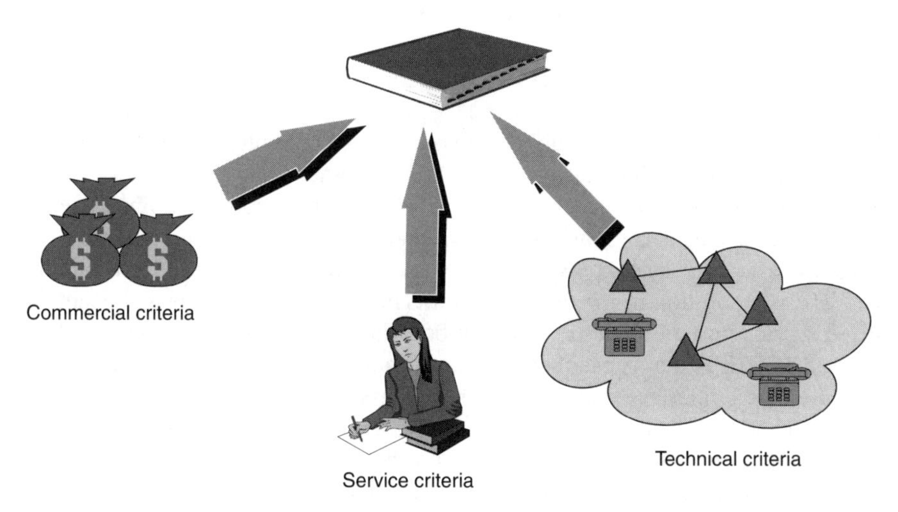

Figure 15.8 Main parts of a service interconnection agreement

A good help during the interconnection negotiations (at least in Europe) would be to consult the checklist 'Framework Interconnect Agreement' produced by the European Telecom Platform.

The original definition of an SLA is: The SLA is a legal contract between a service provider and a customer that specifies, usually in measurable terms, what services the service provider will furnish. It defines also the consequences (penalties) if the service provider fails to deliver the specified services according to the specified conditions. This means that all users/subscribers indeed are embraced by an SLA.

Today SLA also has a wider interpretation that covers contractual obligations in many types of interfaces, for example between two network operators. For such an interface much of the content in fundamental technical plans is relevant. Apart from technical parameters the monetary interconnection conditions are important.

SLA is a cornerstone in establishing end-to-end QoS and security.

Call termination/interconnect of data services is a big unresolved issue. For voice services call termination/interconnect is typically 25–30 % of total revenue. It will probably be a much lower percentage for Internet-based services. How do you charge for something that was free to send?

15.6 SERVICE INTERWORKING

15.6.1 VIRTUAL HOME ENVIRONMENT

In mobile systems like GSM, the home mobile network does not control the services when the subscriber roams to a visited PLMN. Instead, service information is transferred from the home to the visited PLMN for local service provisioning. The IN type service CAMEL, customized applications for mobile network enhanced logic, is an example of this. CAMEL is described in *Understanding Telecommunications*, Book 1. This approach poses some requirements on the home and visited PLMNs: both the visited PLMN and the protocol interface between the visited PLMN and the home PLMN must support the services the mobile station had subscribed to in the home PLMN. If these requirements were not fulfilled, service invocation would not be possible.

Roaming subscribers are no longer the exception; they are becoming the norm, which implies that the provision of seamless services for roaming subscribers becomes increasingly important.

The trend towards universal public network access continues with the globalization of mobile standards and Internet browsers. At the same time, services and content are becoming more personalized and local. Future applications require combined telecom and datacom services. This is combined with the market need to bundle services over different accesses, for example, 'one-stop-shop' and 'one-point-of-care', one bill and one brand.

The VHE is a solution in the new situation. The VHE is defined as a concept for personalized service portability across network boundaries and between terminals. The main idea with the VHE is that subscribers are consistently presented with the same personalized features, user interface and services, regardless of their access network or terminal (within the capabilities of the terminal). Bundling sets of services forms personalized service portfolios for subscribers belonging to a home environment.

The VHE concept is often referred to as 'personalized services for the masses'. The VHE can be implemented with service network solutions.

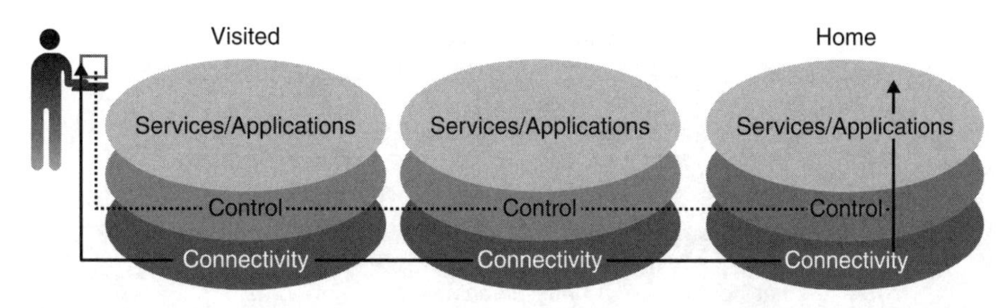

Figure 15.9 Portable services

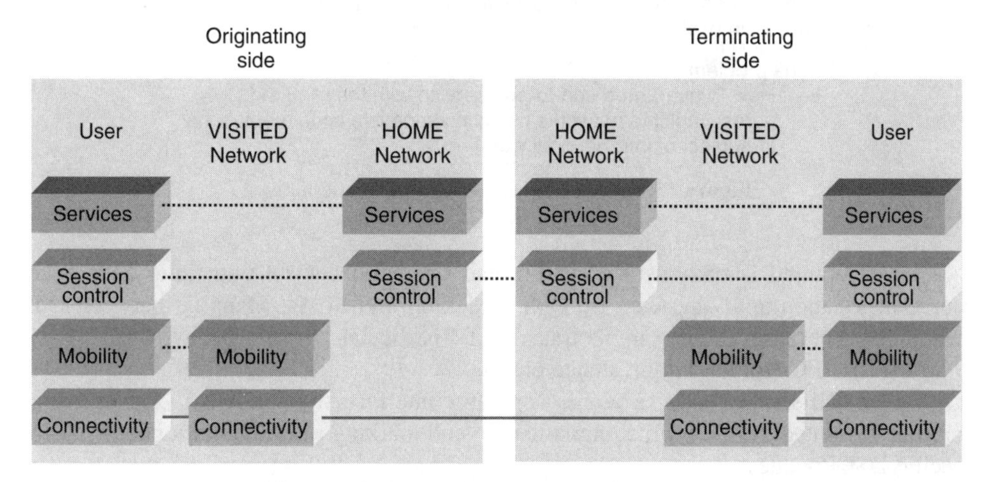

Figure 15.10 Fetching and returning services

The services are offered by the home environment service provider or an associated 'HE-VASP' (home environment-value-added service provider). Figure 15.9 shows how the home environment services are reached schematically.

To a significant extent VHE provides visited network independence. The relationship between visited and home network can be more basic (network oriented). This approach minimizes the restrictions on service deployment in the home network.

The end user depends only on the own service provider, not the 'foreign' (visited network) one, regarding service profiles, pin codes etc. Global credit card service is also solved by VHE. VHE will protect privacy issues.

The session also includes the return communication. Figure 15.10 provides both directions for a mobile subscriber.

15.7 QoS INTERWORKING

See Figure 15.11 that illustrates the situation. Many enablers for QoS are available (see Chapter 7). In the connectivity part. RSVP, Diffserv, MPLS and over-dimensioning are common. In the terminals and servers protocols for QoS such as CC/TP, RTP, RTSP and RTCP are available. What will the end-to-end QoS look like?

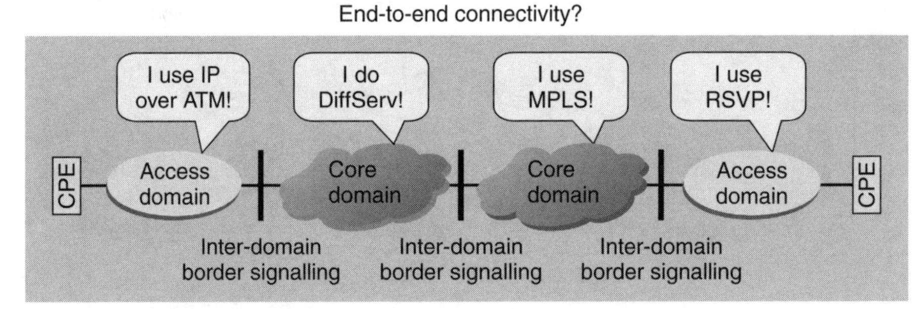

There is no shortage of possible approaches–and they are all in use!

The problem

- How to guarantee end-to-end service with the required QoS across multiple networks using incompatible implementations [a subject of international work]

Figure 15.11 QoS elements and some alternatives

What is required is obviously an end-to-end operation control with inter-domain resource negotiation and supported by QoS signalling and SLAs. Mapping between QoS techniques is necessary, for example transfer of a particular Diffserv behaviour aggregate to an MPLS FEC (forward equivalence class).

Provided Diffserv is used in a user–operator interface the following characteristics could be regulated: traffic profile, classification and marking rules, network behaviour for specific DSCP values.

PDP context shows a stage of QoS handling.

15.8 PDP CONTEXT ACTIVATION FOR CONNECTION TO A DATA NETWORK

In order for the user to be able to transfer data, a packet data protocol (PDP) context must be activated in the mobile station, SGSN and GGSN. The user initiates this procedure, which is similar to logging on to the required destination network. The process is illustrated and described below.

The user initiates the logging on process, using an application on the PC or mobile station.

The mobile station requests sufficient radio resources to support the context activation procedure.

Once the radio resources are allocated, the mobile station sends the activate PDP context request to the SGSN. This signalling message includes key information about the user's static IP address (if applicable), the QoS requested for this context, the access point name (APN) of the external network to which connectivity is requested, the user's identity and any necessary IP configuration parameters (e.g. for security reasons).

After receiving the activate PDP context message, the SGSN checks the user's subscription record to establish whether the request is valid.

If the request is valid, the SGSN sends a query containing the requested APN to the DNS server.

The DNS server uses the APN to determine the IP address of at least one GGSN that will provide the required connectivity to the external network. The GGSN IP address is returned to the SGSN.

The SGSN uses the GGSN IP address to request a connection tunnel to the GGSN.

Upon receiving this request the GGSN completes the establishment of the tunnel and returns an IP address to be conveyed to the mobile station. The GGSN associates the tunnel with the required external network connection.

15.9 SECURITY INTERWORKING

A mobile user has only access to, and is charged for, the services that his profile allows. Whenever a user moves into a foreign administrative domain, some kind of AAA signalling between the visited domain and the home domain is required to allow the visited domain to serve the user.

Security parameters are therefore transferred to the visited network for roaming subscribers. This enables authentication in the visited network. Typically, the AAA server is in charge of initially authenticating a user in an access system in an administrative domain, then authorizing him to access some specific services with certain QoS characteristics, in accordance with his profile. See also the AAA info in the reference point in Figure 15.7.

PDP context has both QoS and security tasks, such as setting up tunnels.

For interworking between UMTS and WLAN, the UMTS HSS capabilities can be used for authentication in both networks.

Figure 15.12 also illustrates the concept of *break-out point*. This is the point where the call leaves the originating network.

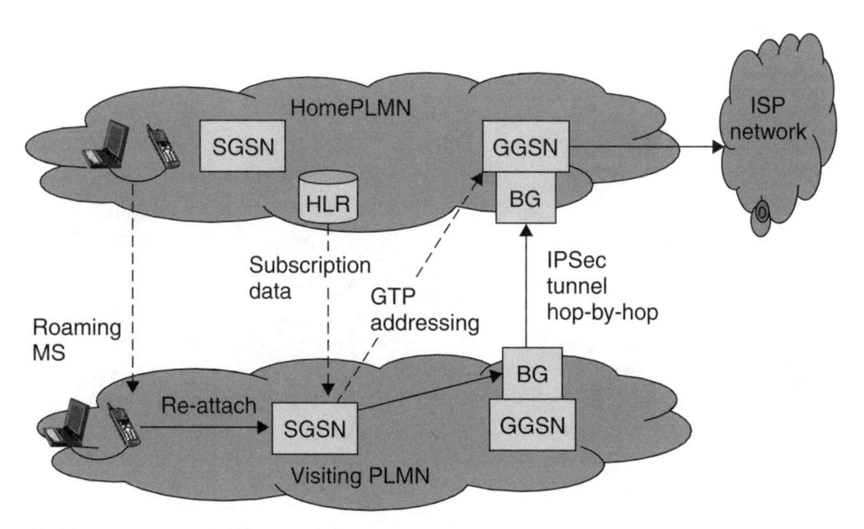

Figure 15.12 Two security features: fetching subscription data and setting up a tunnel between gateways

15.10 SIGNALLING INTERWORKING

15.10.1 A MOBILE STATION–MOBILE STATION CALL IN UMTS

Let us say that subscriber A wants to talk to subscriber B when both of them are in a visited network. Figure 15.13 depicts the concepts in an end-to-end scenario. The payload can still follow the optimal (shortest) path between visited networks to allow transmission efficiency and an optimal QoS. The signalling has to pass several routers.

Now we have to go back to Chapter 14 to check the tasks of nodes or functions P-(proxy), I-(interrogating) and S-(serving) CSCF.

By means of the explanations in Chapter 14 the call signalling according to Figure 15.13 can be understood. Observe the straight traffic path, which does not have to pass the home networks. The next example gives further clarification.

15.10.2 TRAFFIC CASE GPRS-ISDN WITH SIP SIGNALLING AND IMS CONTROL LAYER

Figure 15.14 shows a SIP call from a mobile terminal to a fixed ISDN phone.

The mobile terminal is already attached to GPRS in the SGSN server. We can also assume that the terminal has performed a SIP registration to the home network, that is, a S-CSCF server has been allocated to the terminal, and data relating to the user profile has been fetched from the HSS and stored in the S-CSCF server. When the mobile user wants to initiate a SIP call while roaming in another network, the GPRS bearer must first be established.

The network does this by activating a PDP context. When the GPRS tunnel has been established, a SIP message is sent to the P-CSCF server in the visited network. The P-CSCF server is able to resolve the address of the subscriber's home network and forward the SIP message.

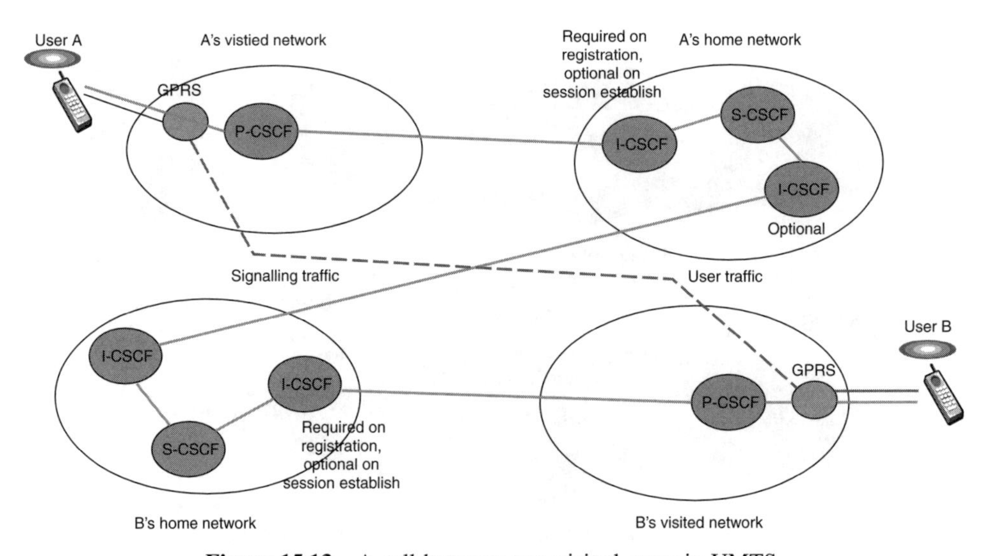

Figure 15.13 A call between two visited areas in UMTS

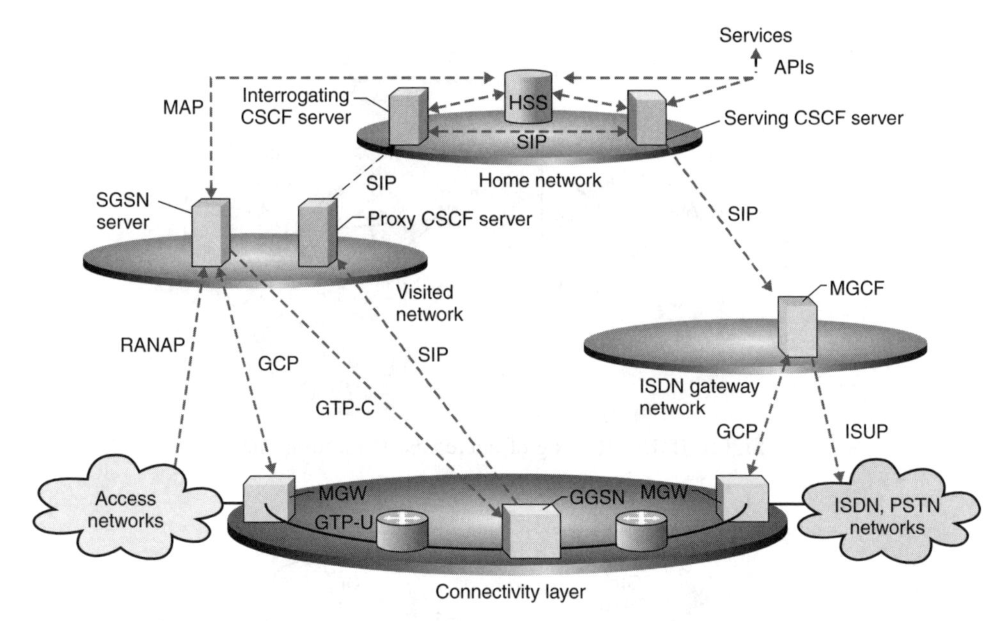

Figure 15.14 Example of a SIP call from a mobile terminal to the ISDN

The SIP message enters the home network in an I-CSCF server, whose task is to find the S-CSCF server in which the subscriber is currently registered. The I-CSCF server does this by sending a query to the location service in the HSS.

When it receives a reply from the HSS, the I-CSCF server forwards the SIP message to the S-CSCF server, which acts as host to the call-control logic. Because the destination address is located in the fixed ISDN network, the SIP message is forwarded to the ISDN gateway network, which may be physically located in either the home network or the visited network.

Within the ISDN gateway network, a media gateway control function (MGCF) converts the SIP message into appropriate ISDN messages and forwards them to the external ISDN network. While the control signalling messages travel from the visited network via the home network to the ISDN network, the user-plane stream is sent directly from the GGSN in the visited network to the media gateway in the ISDN gateway network, taking the shortest optimal path to the destination. The media gateway at the edge facing the external ISDN network is controlled by the MGCF.

15.11 ROUTING

15.11.1 ROUTING OF CONTROL TRAFFIC IN FIXED NETWORKS

When the control layer gets physically disconnected from the connectivity layer there is a need to organize pure server networks. The example below is based on a possible network for a very large fixed operator. The operator runs two distinct administrative areas and interconnects with a number of other operators. Figure 5.15 shows transit or interconnect telephony server locations between areas or operators. There are at least two

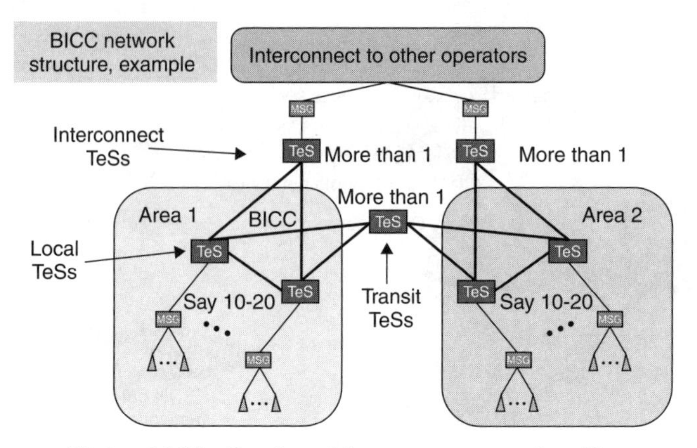

Figure 15.15 Routing of interconnect control traffic

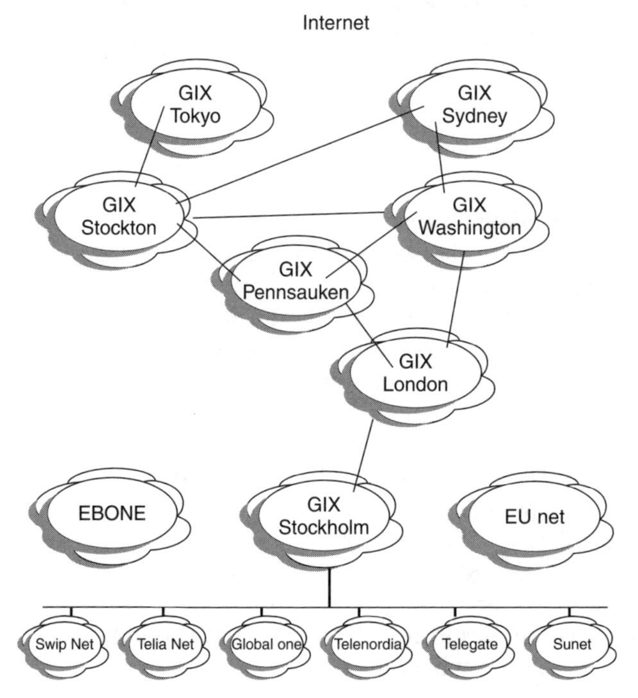

Figure 15.16 *Interworking GIX points in the Internet*

telephony server gateways for reliability reasons. A transit/interconnect telephony server has no underlying MGWs.

15.11.2 ROUTING OF EMERGENCY CALLS

As an example of a very important and critical issue, the routing of emergency calls can be mentioned. Calls to the police and fire brigade must be routed to the nearest geographical place and must contain the correct calling line identification (CLI) of the calling user.

15.11.3 INTERCONNECT IN INTERNET

See Figure 15.16. In the hierarchical Internet the most important interconnection points are called global interconnect points (GIX). A GIX can be compared with an international exchange or a national gateway.

15.12 MOBILITY MANAGEMENT

15.12.1 ROAMING AND HANDOVER

Handover from one technology/system in one cell to another technology/system in another cell is highly desirable, since it permits support from one network to the other one. The UMTS systems can for example be implemented more flexibly with good coverage from the start. This will of course also improve the UMTS business case.

Let us look at a handover from UMTS/WCDMA to GSM. We assume that the mobile station has a single radio receiver. The mobile station must in any case be able to communicate with both technologies to perform the handover. Further we look at the handover from WCDMA to GSM.

When there is a need for a handover to the GSM system the WCDMA device can for example use a technique called compressed mode. See Figure 15.17. Like the GSM system the WCDMA system has to measure the signal strength in the GSM cells while still communicating in WCDMA. By compressing some of the sent frames a small free time is created which can be used for GSM signal measurements. Provided the measurements are successful the GSM base station controller is informed via WCDMA radio network controller.

The handover from WCDMA to GSM is illustrated in Figure 15.18.

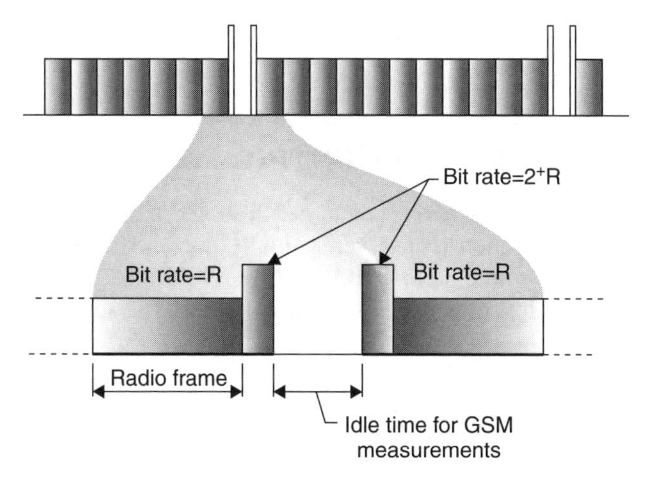

Figure 15.17 Compressed mode creates gaps or idle spaces in time that WCDMA mobile terminals use to perform measurements on GSM cells. Real-time services cannot use the sudden bit rate changes around the idle WCDMA time slot used for measurements of GSM

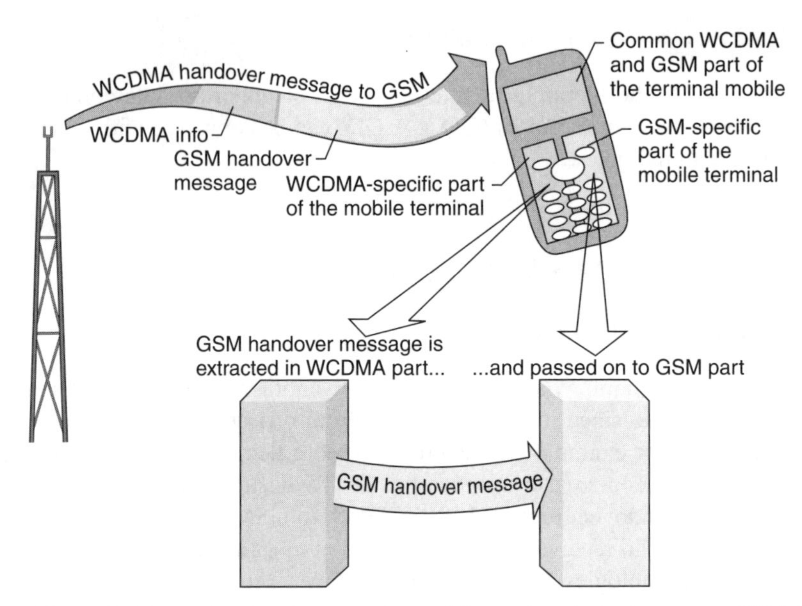

Figure 15.18 WCDMA orders handover to the mobile station, imitating a true GSM handover message. The GSM handover message is encapsulated in a 'container' that is part of the WCDMA handover message

Roaming using Roaming Exchange Operator

In order to interconnect GPRS and UMTS networks in different countries, one solution is to use intermediate trunking operators. A specification is defined in International Roaming Expert Group Rec. IR 34. It includes a GPRS roaming exchange (GRX).

GRX is a centralized and hierarchical solution for the roaming of data traffic in GPRS and UMTS networks. Packet-based data in mobile networks can be seen as an extension of the global Internet. The global Equant IP network is used to interconnect different backbone networks in a safe way.

15.13 CHARGING AND ACCOUNTING

See Figure 15.19. Charging interconnect can be illustrated based on messaging.

For SMS the interconnection issues were initially not satisfactorily considered. Sites on the Internet appeared where users could send free SMS, not generating any revenues for operators, but taking lots of capacity in their networks. The solution was interconnect settlements for SMS.

15.13.1 MMS INTERCONNECT SETTLEMENTS

Getting MMS Interconnect settlements in place enables the users to send MMS to other operators' users.

This will bring revenues from MMS that terminates in the operator's network.

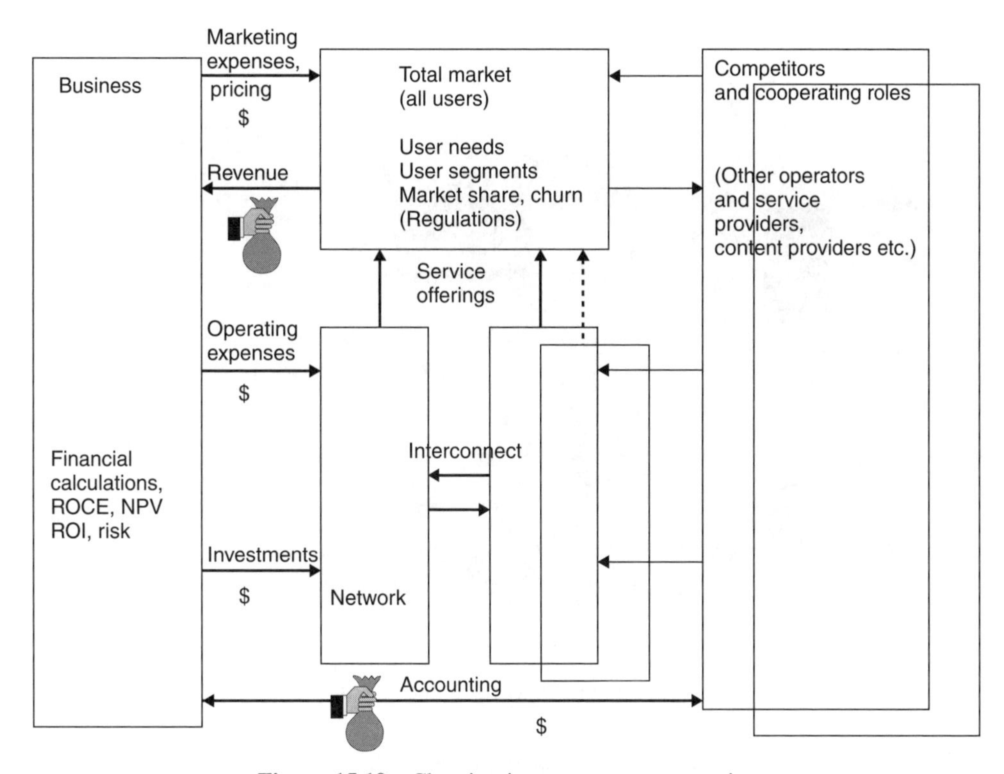

Figure 15.19 Charging interconnect – accounting

15.13.2 MMS ROAMING SETTLEMENTS

MMS roaming settlements enables the users to send MMS when travelling, and enables the operator to capture revenues from roaming users: its own and others. See Figure 15.20.

This will bring revenues from MMS sent by visiting subscribers. Prepaid roaming is an extra challenge.

Let us take a look at an MMS call from Singapore to Sweden. The home operator is Swedish. Peter's operator in Sweden will charge him for a person-to-person MMS.

The operator in Singapore wants to get paid for letting Peter send a MMS via their network. So there will be need for a roaming settlement between the Singaporean operator and Peter's operator in Sweden.

Somebody is transporting the MMS from Singapore, and wants to get paid for this. Meaning interconnect settlement for the MMS transiting.

And once again this might sound simple. But it is not. Operators must have this in place in order to make MMS take off and be able to capture the revenues.

For example: What is the right MMS price level? The relationships for buying and selling content, interconnection and roaming services keep increasing in complexity and values.

It is becoming increasingly important for operators to make sure that they can charge and get paid for everything they have delivered to their partners. Another requirement is

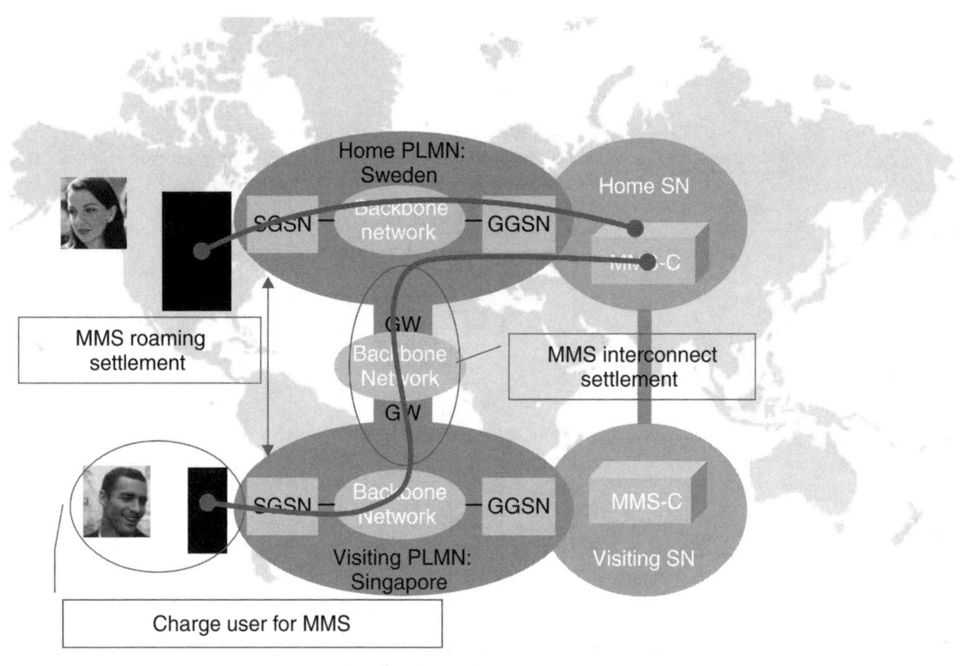

Figure 15.20 MMS roaming charging

to be able to verify and make sure that the claims from their partners are correct, in order to avoid being overcharged.

15.14 POSSIBLE INTERWORKING UMTS–WLAN

This type of interworking might become very common in the future, including roaming.

15.14.1 REQUIREMENTS

A cooperation of UMTS-WLAN combines the wide-area benefit of 3G with its unlimited roaming and mobility with the additional throughput and capacity by WLAN in hot spots. WLAN surf points are increasing quickly all over the world with a predicted number of 120 000 in 2007 (Source: Gartner-Dataquest). WLAN users can be part of the 3G-operator subscriber base and UMTS-WLAN operators are enabled to offer further 3G packet service expansion with small investment costs. However, UMTS-WLAN interworking standardization could be limited to a small set of standardization items, allowing reuse of 3G subscriber management mechanism, 3G authentication and security functions and billing functions. See Figure 15.21.

Not reusing the 3G-mobility management allows to use WLAN as an IP access network complementary to current UMTS packet-switched domains. Thereby the impact on the UMTS systems is minimized and the need for a common 3GPP and WLAN standardization is reduced. A common billing between the systems is already in use. See Figure 15.22.

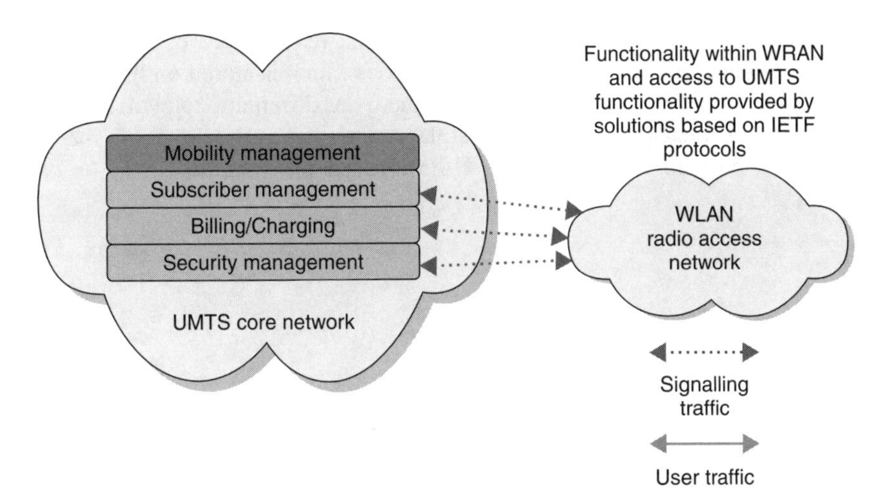

Figure 15.21 WLAN as an access to UMTS – related fundamental plans

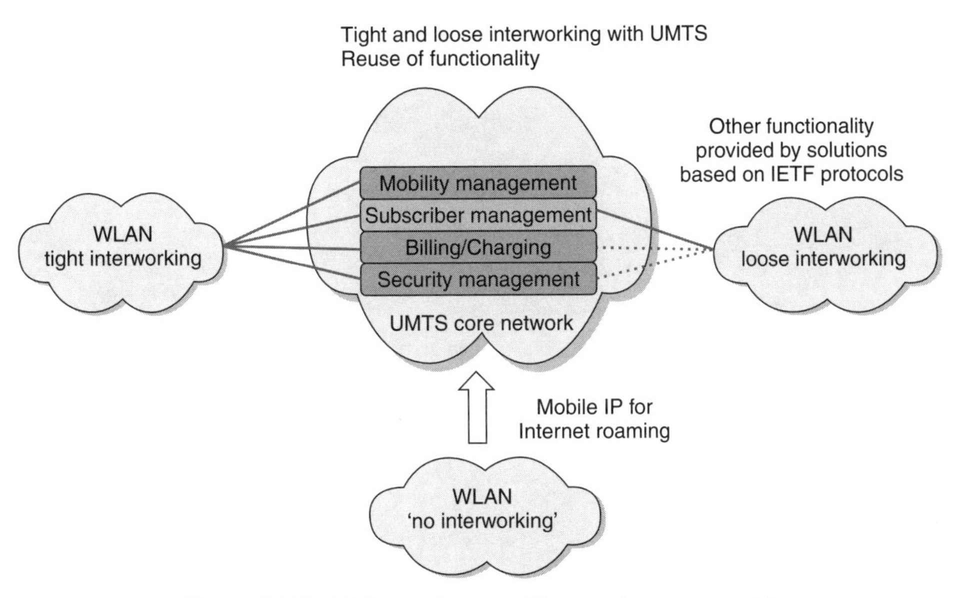

Figure 15.22 Different degrees of interworking are possible

Many mobile operators consider the SIM card as a way to own the customer, and would like to include the SIM function in the common functions.

Another interesting combination is xDSL and WLAN.

The end user needs to have two radio interfaces, one for the cellular radio access network and one for WLAN, to benefit from the advantages of complementing cellular access with WLAN. Different deployment alternatives for these two radio interfaces may appear on the marketplace: separated terminals for different radio interfaces and multi-mode terminals, whereby multi-mode terminals integrate both cellular radio interface and

WLAN into one physical device. In addition to the requirement of having two radio interfaces available, the user subscriber information, which is stored on the USIM, needs to be accessible from both radio interfaces. If there are different deployment alternatives for the two radio interfaces in the marketplace, then naturally there will be different alternatives for providing access to the USIM, based on a single USIM or based on two USIMs.

16

Telecom Management – Operations

> 'We have to learn to manage the disorder'. ('Debemos aprender a manejar el desorden'.)
> Juan Perón

16.1 INTRODUCTION

Besides covering telecom management operations at an overview level, this chapter summarizes the structure of the book by means of plenty of references to preceding chapters.

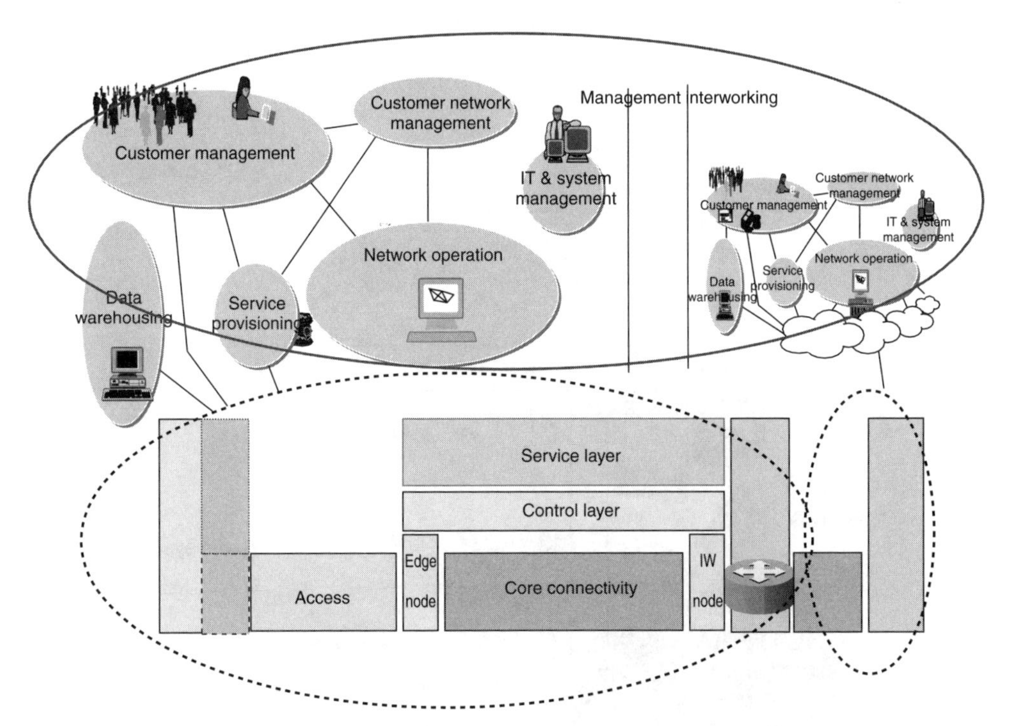

Figure 16.1 Position of telecom management – Operations

Understanding Changing Telecommunications – Building a Successful Telecom Business. Edited by A. Olsson
 ISBN: 0-470-86851-1

The aim is to take a final step in acquiring a holistic and consistent view of the subject. Unfortunately, the telecom management area often gets less attention than it deserves.

A reason for a holistic approach is that the extensively segmented network demands a subsequent integration to produce income. End users and telecom businesses need single, predictable and reliable network behaviour. An important requirement is to integrate segment behaviour into an end-to-end behaviour. Telecom management fulfils the wanted role as a network integrator, laid down in fundamental technical plans or similar rules. Telecom management allows configuration of the network and control of the network elements, as well as its overall behaviour, supporting the fulfilment of the business plan targets. Among the targets are conditions set and agreed between end users and the operator for service provisioning, activation of the required services and billing.

The traffic system in this chapter is of course identical to the system that is described in Chapters 9 to 15. However, there is greater focus on revenue, OPEX, CAPEX, processes, customer care, etc. Such aspects are already partly discussed in Chapter 9. The main network parts covered are access, terminals and service network management.

The extremely important user perception interface that is dealt with in Chapters 5 to 7 is also the same, but here the focus is on for example service configuration, management of QoS and security. See Figure 16.2.

In relation to the content of Chapters 2 to 4 the focus moves to a process focus, the technique of the traffic system moves to the technique of the management system with all its enablers, and the telecom business focus moves from 'SIP' (strategy, infrastructure and product, see Chapter 4) in Chapter 4 to operations in Chapter 16 with a retained ROCE focus.

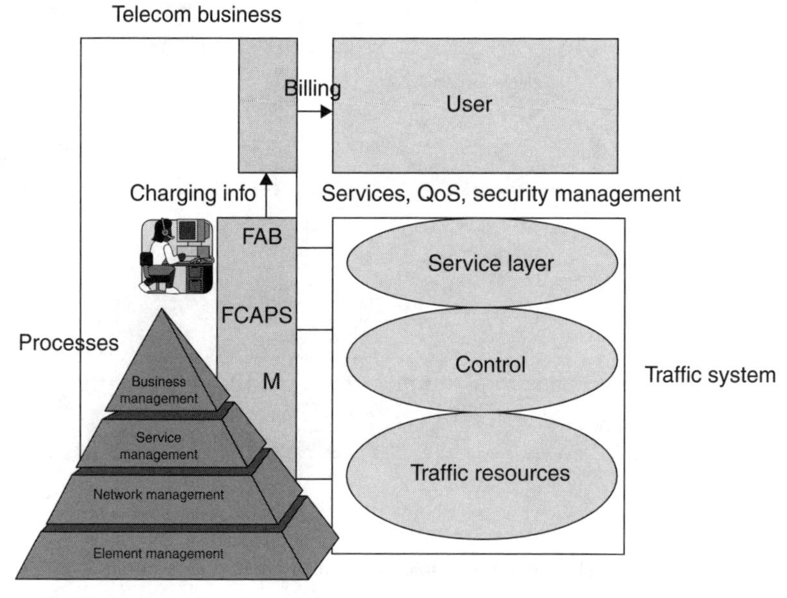

Figure 16.2 The focus moves to a management system and management processes

Chapter 8 covers fundamental plans, as do Chapter 5 (charging), Chapter 6 (security) and Chapter 7 (QoS). Since fundamental plans are in general more end-to-end oriented than network segments, they are even more crucial for the end-to-end behaviour. However, there is no explicit coverage in this chapter of the management of the plans in Chapter 8, such as numbering and signalling.

Like other subjects telecom management – operations is treated at an overview level in this book. The definition of operations, dominated by 'FAB', is presented in Figure 16.3.

The first reference concerns Chapter 3 where the split into a management system and a traffic system was presented. See Figures 3.3, 3.4, 3.5, 3.26 and 3.27 as an introduction to the following section.

16.2 THE MANAGEMENT SYSTEM

The modern management structure is shown in Figure 16.4.

16.2.1 THE TASK OF THE MANAGEMENT SYSTEM

There is a close relationship between management and the control performed in the traffic system. Management acts on a general basis (fixing policies or general directives), while control acts case by case and in real time. With increased personalization we will also see more of individual management.

A Brief on History

Two technically different management approaches have developed during the last decade. They address mainly the basic structure and data interchange within the emerging management network.

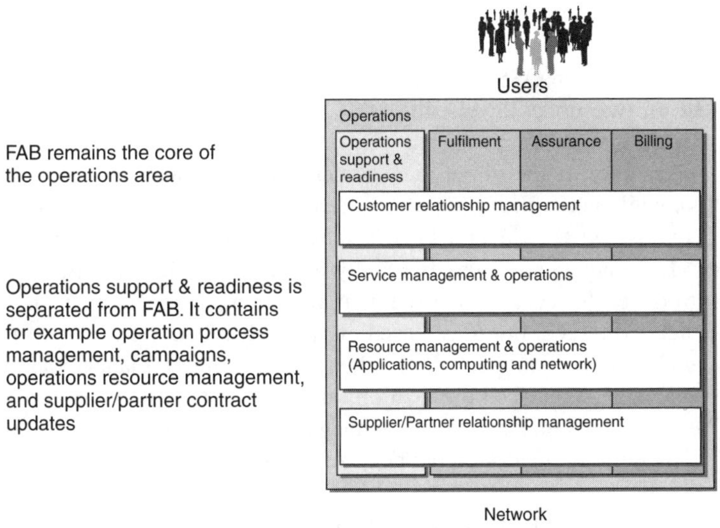

Figure 16.3 The TM operations area

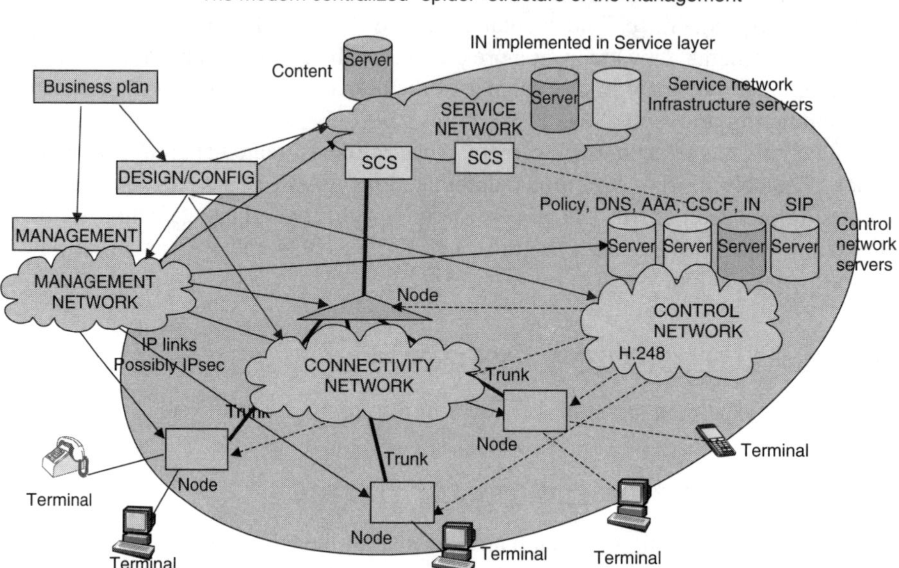

Figure 16.4 The management network applied to the network element based view presented in Chapter 3: the modern centralized 'spider' structure of the management

From the telecom side the need to concentrate and standardize many different types of management systems into an integral network management became apparent. The business side put forward a requirement to cope with the increasing operational costs derived from network expansion.

On the data side the explosion of Internet networking caused a tremendous expansion on the networks that brought many of them close to collapse. The network administrators needed solutions to smoothly adapt the networks to the pace of expansion.

Telecom: ITU-T developed the TMN concept (telecommunication network management). TMN is based on an OSS (operation support system) platform. It has a hierarchical structure based on two nodes levels, the OMC (operation and maintenance centre) and the NMC (network maintenance centre).

The OMC takes care of the direct management of the network nodes, called network elements (NEs), while network configuration, performance data, alarm and traffic statistics are managed by the NMC. Several OMCs may exist in large networks to regionally distribute the NE management. There is usually one single NMC. Additional centres have lately been introduced for service management and enterprise management.

The ITU-T defined the Q3 interface based on the SMI model (structure of management information) and a group of ISO standards like CMIP (common management information protocol) for manager–agent communication, and FTAM (file transfer access and management). The NE agents and the central manager communicate by means of a network called DCN (data communication network). For DCN the fairly secure X.25 technology was proposed.

TMN is a detailed model that has not become fully implemented. This is partly due to the strong advance of IP technology and the progressive obsolescence of X.25 networks, making IP a better solution (but less secure) for the DCN communication. The telecommunication management communication normally utilizes a specific telecommunication management network. This could be illustrated by means of the telecommunication management network model in Figure 16.5.

Datacom: Within the IP environment a flat management structure was used, based on the SNMP protocol for agent–manager communication. No intermediate hierarchical levels were devised. The communication was based on the TCP/IP protocol suite with FTP (file transfer protocol) for file transfer.

IP standards based on inherent network topology learning capabilities became a main tool that gave relief to the network administrators. Much of the traffic routing activities then became autonomously handled by the network elements by means of dynamic configuration.

A rather interesting paradox is that TM network technologies (data communication network, servers, etc.) strongly evolve towards IP, while the system structure becomes closer to the hierarchical TMN (OMCs and NMC) mainly due to the increasing size and diversity of managed networks.

A short summary of terms associated with both approaches on management is given below:

Enablers and terminology

TMN – Telecommunications Management Network

NMS – network management system
NMC – network management centre
OMC – operation and maintenance centre

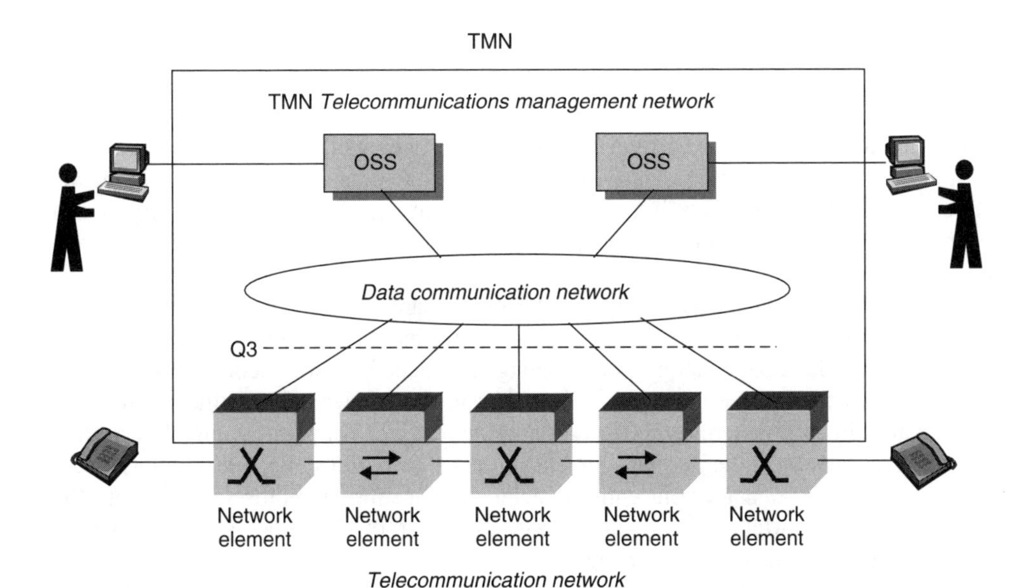

Figure 16.5 An illustration of the TMN model (OSS: operation support system)

OSS – operations support system
NE – network element
Q3 interface
MO – managed objects
SMI – structure of management information
MIT – management information tree
CMIP – common management information protocol
FTAM – file transfer, access and management

'Internet' standards

NEs, managed objects and MIT (as for TMN, but Q3 interface usually not mentioned)
SNMP – simple network management protocol
Web protocols – HTTP (hyper text transfer protocol) and HTML (hyper text mark-up language)
FTP – file transfer protocol
LDAP – lightweight directory access protocol
RADIUS – remote access dial in user service
L2TP and L3TP – label 2/3 tunnelling protocol
IP security

Standards from IT industry

CORBA – common object request broker architecture
EDI – electronic data interchange
XML – extensible mark-up language
SSL – secure socket layer
SQL – simple query language

Network Elements, Managers and Agents

When talking about TM we are referring to managed networks. Managed networks consist of a number of network elements or *managed devices* and a manager. Managed devices collect and store management information for the centralized management entity or manager (TM system). They can be routers, access servers, switches, cross-connects, computer hosts, etc.

Each managed device has an agent. The *agent*s are software modules residing in a managed device that handle local management information within the device.

Managed devices can contain several managed objects. Hardware, configuration parameters and statistics can be managed objects, if referring to the IETF definition. Managed objects are directly related to the operation of the device. They are arranged in a virtual information database called a management information base (MIB). See Figure 16.6.

The manager or management entity executes applications monitoring and controlling managed devices. Management entities provide most of the processing and memory resources required for network management.

C++ is used for 'agents' and 'managers' programming. CORBA is a standard for 'managed object' definition and all-to-all manager–agent communication. It is well suited

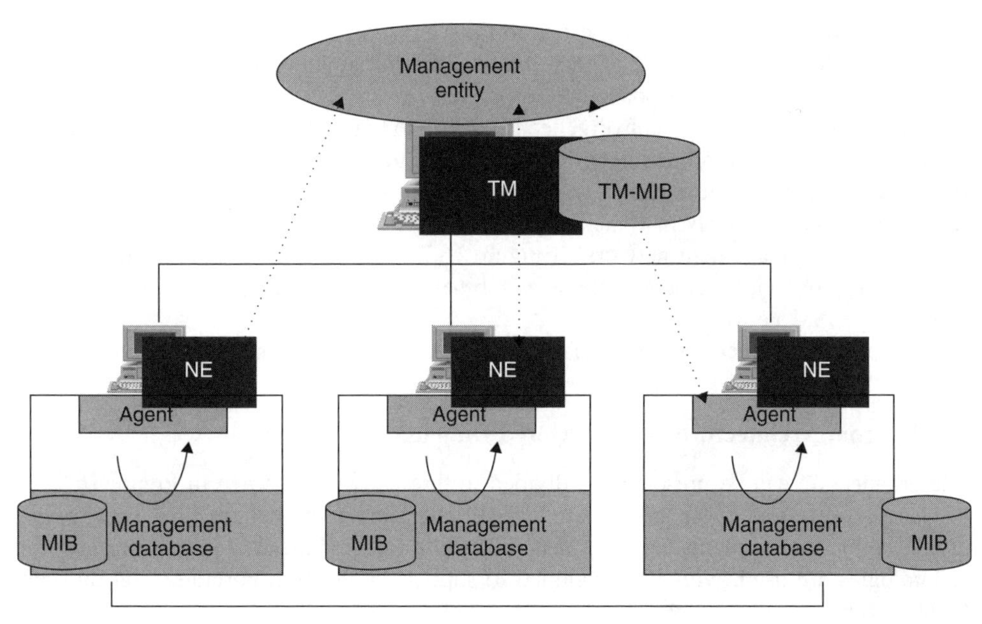

Figure 16.6 TM basic structure

for distributed systems running on different platforms (even written in different languages). This feature opens the possibility to integrate different vendor equipment and management systems in a managed network. CORBA has a drawback, though: the managed object definition is not fully compatible with TMN and IETF definitions.

CORBA can be seen as a 'software bus' used by clients and servers. All CORBA objects are described in an interface specification enabling other objects to communicate with them. It is used in OSS components handling traffic measurements, charging etc.

SNMP is a very simple protocol. It has a short list of commands, which basically are only read or write. It is used just for communication or data interchange between the managed objects in the management information base and the management entity (or manager). The job to translate the information set on the MIB into commands for the NEs or reports for TM, is performed by agents and manager, respectively.

TM Network Structure in the Distributed Dimension

Technically, TM extends all over the network like a spider, collecting charging information and information about the traffic conditions through agents and information databases. Configuration and corrective commands are sent in the reverse direction. The fundamental technical plans are a guiding tool. Updated subscriber data is sent to servers/databases. See Figure 16.4.

The basic management structure based on managers, agents, managed objects and MIBs, was simultaneously adopted both by IETF and ITU-T, during the early 1990s.

This network model might induce one to think of a flat management network, without intermediate levels. As a matter of fact, this was the initial approach in the IP world, and still endures to some extent today.

As mentioned, technology convergence is leading to bigger IP-based telecommunication networks but with hierarchical network structures (such as in TMN). The move towards IP in management communication is both the result of the Internet and packet switching success and improvements in IP security.

The UMTS management network in Figure 16.7 illustrates this evolution. It has three different network management systems at sub-network level (radio access, core network and legacy GSM network).

A simple management structure of a GPRS network is shown in Figure 16.8.

TM Internal Architecture in the Layered Dimension

The architecture can be presented as divided in three layers as shown in Figure 16.9.

The **presentation layer** takes care of the interface with the operator providing a user-friendly environment and easy network-handling capabilities. Internet presentation technologies are used, with Java included to support application portability on different IT platforms.

The **service layer** is responsible for the real interaction between manager and agents through the scheme command issuing or managed object parameters setting, and events or statistics collection. SNMP, CORBA and C++ have been explained before.

The **information layer** performs the administration of the manager and NE's management information base. SQL (simple query language) is a typical database administrator. LDAP handles the secure access to the databases and to the system and JDBC (Java data base connectivity) handles databases via Java coding.

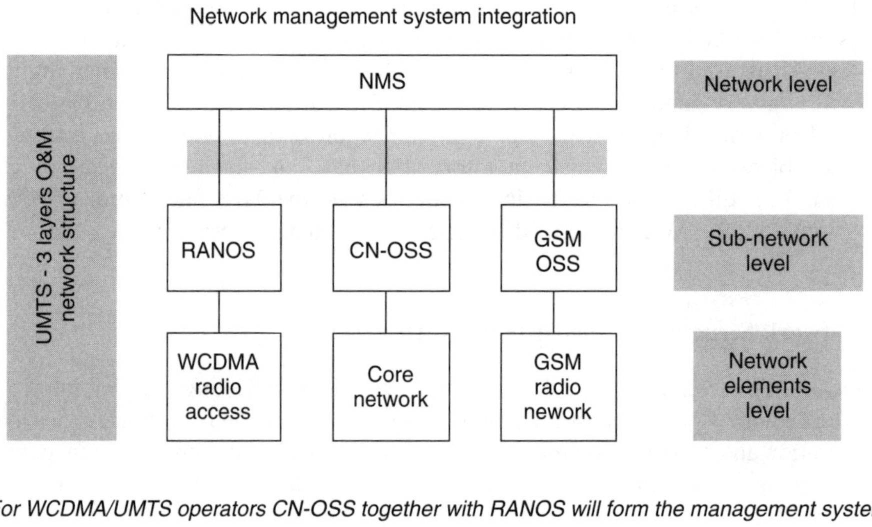

Figure 16.7 Example of management structure of a 3G system (RANOS: radio network operation system)

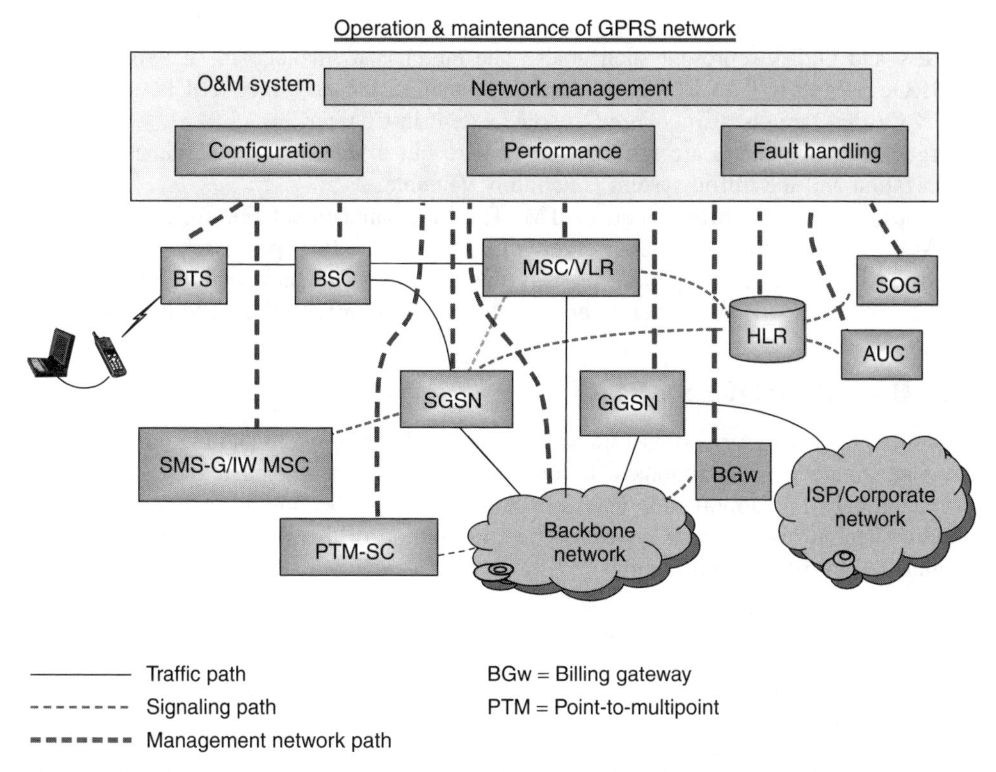

Figure 16.8 Management structure of a GPRS system, also showing some different types of management processes as a bridge to next part of this chapter

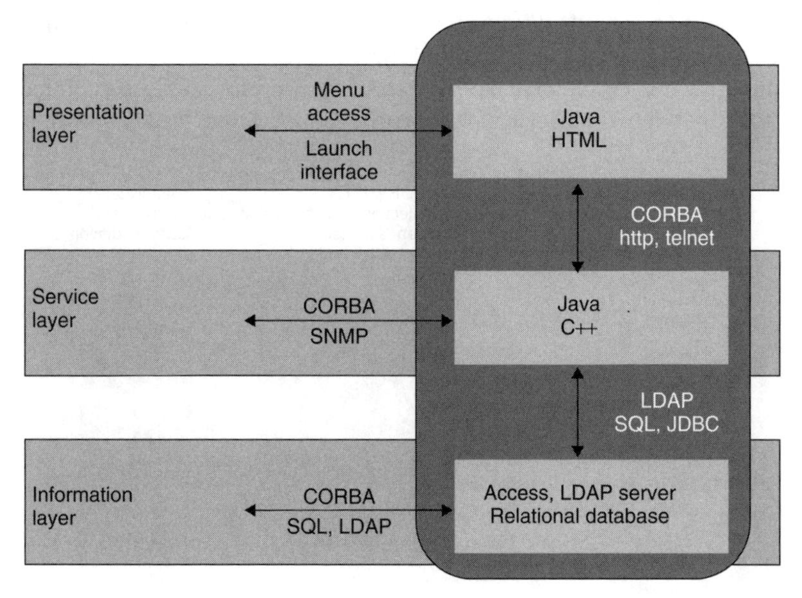

Figure 16.9 TM architecture including some enablers

The *overall goal* of the TM operations area is business oriented. ROCE, ROI, revenue, CAPEX and OPEX represent such goals. The target is to manage the network for the best ROCE or ROI. The ROCE impact is substantial, indeed, since TM is costly. The OPEX-CAPEX area is also covered to some extent in Chapter 9.

Organizational matters are not considered here but a significant congruence between organization and the traffic system is certainly valuable.

The processes are a crucial part of TM. The Telemanagement Forum, the TOM and eTOM processes and the TMN four-layer structure with the FAB processes were all introduced in Chapter 4, as the operator process plan and cost-efficiency section. The business and process areas will be further elaborated below, starting with the basic process part.

16.3 BASIC PROCESS PART

Processes play a very important role in defining, establishing and handling telecom management. Analysis, descriptions and management of processes pave the way for the introduction of IT support aids and applications that are becoming an integral part of modern telecom management systems. The contribution of the process approach may be summarized in the following points:

- Both customer satisfaction and telecom business greatly benefit from a clearly defined process environment, which also allows continuous improvement.
- The processes guide the operator's personnel, preventing damages derived from wrong operation or lack of knowledge. They provide an environment more independent of single personnel experience and expertise.
- The processes deliver continuous and clearly defined measurements on effectiveness and QoS, and provide proven methods for continuous improvement.

As explained in Chapter 4, this chapter will deal with subjects within what is called 'operations' by the Telemanagement Forum. See Figure 16.10.

The traditional heading of a chapter like this one used to be 'Operation and Maintenance' or possibly 'Network Management'. However, the scope has increased to embrace

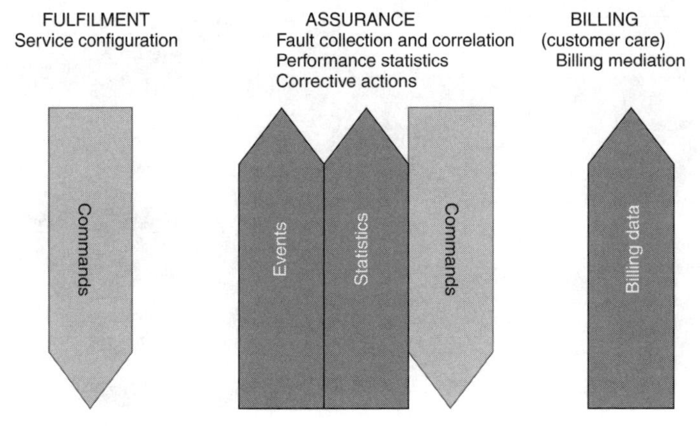

Figure 16.10 The main tasks within the TM area – the FAB operations

areas such as service management, customer/end-user management, business impact, management interconnection and much more.

The chosen heading 'Telecom Management – Operations' is a way to retain the word 'management' in the title. Telecom management as such without the elucidation 'operations' might also be interpreted in a wider sense.

The eTOM (enhanced telecom operation map) mentioned in Chapter 4 corresponds to a wider approach. It includes the:

- operations area
- strategy, infrastructure and product (SIP) area
- enterprise management area

The operations + SIP interpretation of telecom management corresponds roughly to the complete process cycle shown in Figure 16.11. Observe the considerations on investments regarding new or existing type solutions. The SIP area has no dedicated chapter in this book. Instead planning and investment aspects are distributed to many chapters.

When referring telecom management to reference models in the book, operations proves to be interface oriented. See the slightly darker area of Figure 16.12.

Overall telecom management interfaces like FAB, service management, network management, investments, QoS, and SLA are treated in this chapter. As related to telecom management, investments may need further clarification. In the view adopted in this book there are two kinds of investments: discretionary cost concerning new techniques or technologies introduction, and baseline cost (single network expansions). Only baseline cost belongs to telecom management. See also Chapter 4. The ultimate management–network

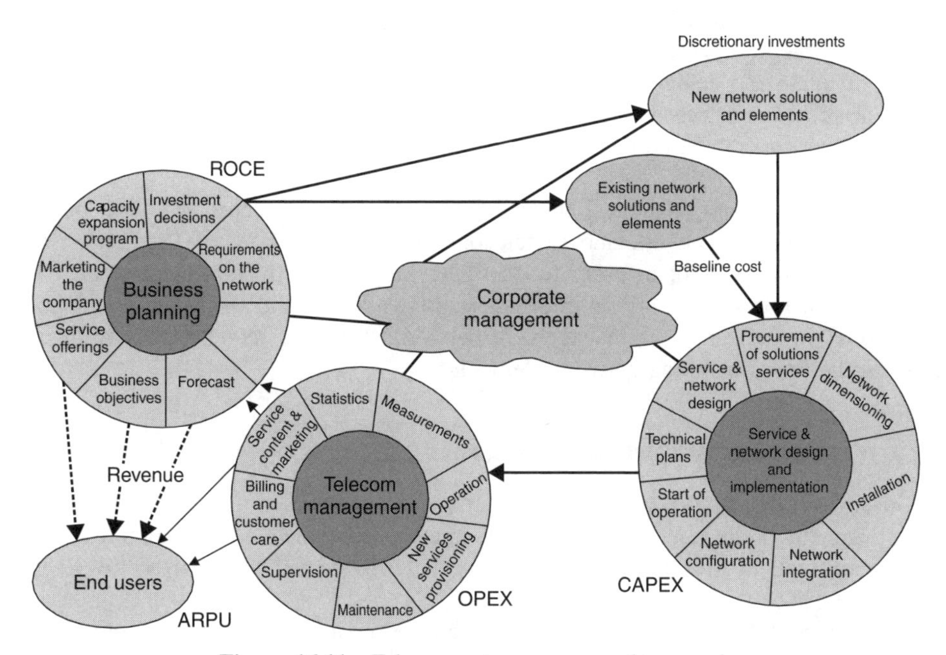

Figure 16.11 Telecom actor process cycle example

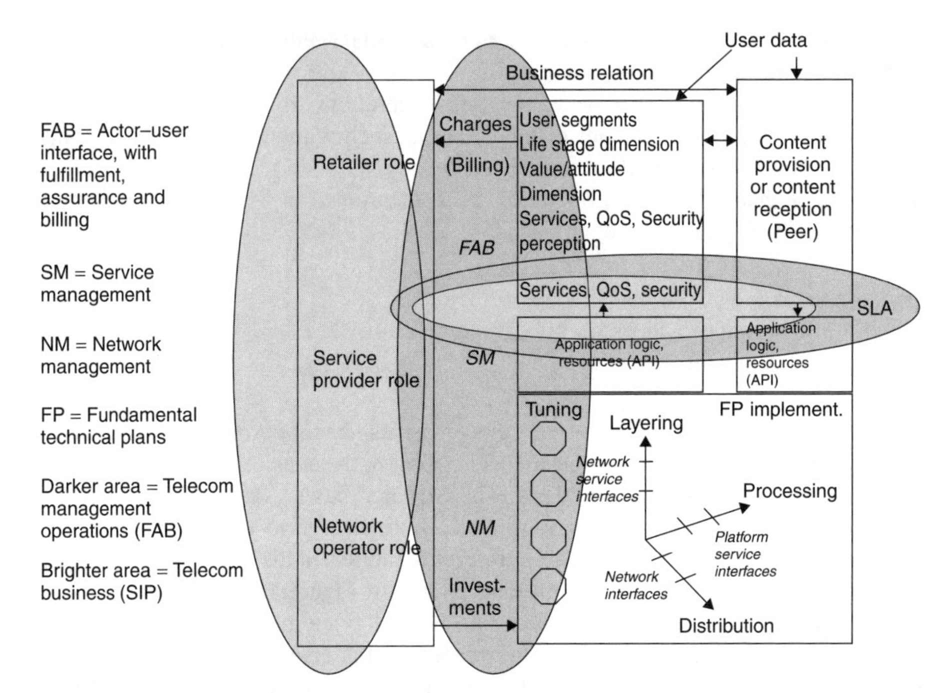

Figure 16.12 The main TM operations area in relation to one of the reference models (somewhat darker area)

interface is sometimes called the Q3 interface. The brighter elliptical area of Figure 16.12 is more internal. It falls roughly within the SIP area.

Regarding the interfaces, there must be a dialogue with the end users in order to sell, deliver, maintain and charge services with sufficient end-to-end QoS, security and availability, at attractive prices. The potential of new services in the new systems (including the service network) must be taken care of to increase the income. Marketing will play a key role here.

Towards the network telecom management becomes clearly dependent both on technologies used in the network and in telecom management systems. There must be a dialogue with the network in order to configure, upgrade, activate services, and to get information such as alarms, billing information, network load and statistics. This dialogue costs money in staff and in a management network. It requires control possibilities and agents in the traffic-handling network elements.

16.3.1 HANDLING OF VIRTUAL PRIVATE RESOURCES

Within an overall network run by an actor there are normally a number of resources which belong to, or are reserved for corporate intra-company networks. Common solutions are Centrex, call centres leased lines and virtual private networks based on IP (like the intranets).

The enterprises have natural reasons to take care of the business critical areas themselves, such as competition between the end users (in this case the enterprises). What areas are considered business-critical depends on the enterprise, but they might include streamlining the connections, managing a call centre and monitoring things like the QoS or the call-completion rate.

Therefore, large enterprises may need to control these resources in the same way as the operator. See also customer network management in Figure 16.1.

Summing up, the ability to create effective, highly automated processes for operation and service delivery, together with a pragmatic approach to technology choices is crucial.

16.4 THE TMN FUNCTIONAL AREAS

The TMN functional area segments follow the well-known network management model from ISO. The management functions are frequently known by the acronym FCAPS, which refers to:

- Fault management
- Configuration management
- Accounting
- Performance management
- Security management

Fault management covers all the necessary functions to show the network fault status in real time. It is known that a fault originating in one network element may lead to an avalanche of alarm indications. This is due to the incidence the original fault has on other network elements. In spite of working properly, they are not able to carry traffic due to the original fault, and therefore they are detected as faulty. For instance a transmission repeater fault may lead to fault indications at different levels on the transmission system, a rather high number of declared faults on associated routes, and also fault indications on switching devices utilizing those routes.

The process through which the original fault is obtained from a rather high number of alarm indications is called alarm (or fault) correlation. Alarm correlation becomes more complex with network heterogeneity and increased number of network elements. Some systems have the possibility to learn from previous network faults experiences and from network expert indications, to improve alarm correlation. Suppression of secondary alarms is a recognized method.

Fault management has advanced alarm correlation tools, which allow the operator a better understanding of the real alarm situation, and leads to shorter times for service recovery and fault repair.

Performance management portraits the network performance through diagrams and reports, either standardized or specially selected by the network administrator. The network status in each moment is usually compared with the performance corresponding to some level of network activity defined as a reference.

This reference consists of the network status when working without network element's faults and with normal traffic conditions. The reference is often mentioned as the 'network baseline' or the 'network normal condition'. Also comparison of the network status with the baseline raises alarms when the measured traffic or workload goes over certain threshold values, defined by the network administrator.

In this case, alarms are emitted to alert the network operator about abnormal traffic conditions. This information permits the operator to act in advance to any possible problem. The operator can then utilize the network management tools to solve or mitigate the congestion and its associated effects, or eventually plan new network expansions before the situation becomes critical.

Performance management is a key function to verify end-to-end QoS and network availability to the end user. It is extensively used to verify SLA compliance and to detect and measure possible SLA deviations. It also provides information about network performance and congestion levels, allowing fault detection. Finally it facilitates network re-engineering, monitoring on network element life cycles and network capacity supervision. All this gives the necessary input to future building and implementation activities, both for decision making and design.

Billing mediation or accounting management take care of the collection, buffering and delivering of information from the network elements or call control servers, either for billing or accounting purposes.

Accounting management could be illustrated by means of an international call. If a Swedish subscriber calls a friend in the UK, the Swedish operator sends the bill and gets the full income. However, the call has also passed the network of the British operator, and it may even have been transited through networks belonging to other operators. The call income must then be shared among the operators involved.

This is called accounting. How the money is shared is strictly defined by interconnection agreements. The accounting management takes cares of data collection in order to produce the call accounting and assist the income distribution among the different operators, either through accounting determination or checks.

Configuration management and corrective measures are downstream functions, based on command execution. The previous functions were based on event collection. Configuration management plays a major role when trying to set the network behaviour. Examples are configuration of installed equipment, capacity extension on a path and network protection configuration.

Configuration commands are issued to agents acting on the network elements, which are able to control the network behaviour. The control points in the network elements are called management control points.

For ordinary voice-type networks configuration is performed by commands on the central processor to introduce routing tables, charging tables, end of selection cases, etc., often in order to support the call control function.

This may be represented by management functions performed by management agents handling resources in the traffic plane and interconnected to it via management control points. The management actions are in this case ordered by centralized management functions.

IP networks are different. They handle packets rather than calls. IP networks are based on connectivity control, which is distributed in almost all network elements in the network. In IP, many control functions are autonomous. The typical case is routing where the IP routers learn by themselves the network topology to build up their routing tables. Similar principles are used by PNNI in ATM networks.

This case corresponds to decentralized management functions handling a given network resource through a management control point. This autonomous management performed in the network elements does not require central manager intervention.

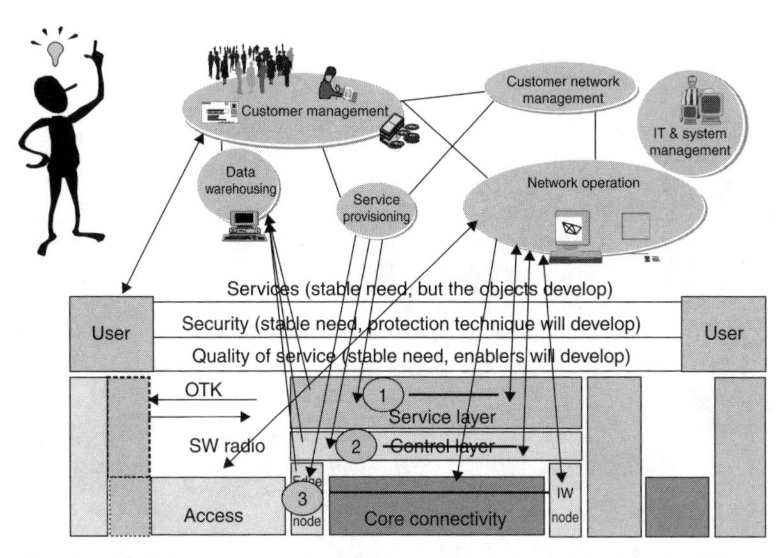

Figure 16.13 Configuration management handles configuration of services in different layers for IP, voice and services beyond voice (3, 2, 1 respectively)

These facts might mislead us to the thinking that neither centralized management functions nor management control points exist in IP networks, apart from the ever-existing static programming possibilities. This is far from the truth. Control and central management certainly do exist in IP networks, although they do not address the same functions. The key control points in IP become the edge routers or media gateways. See the functions covered in Chapter 11, such as QoS and security management.

Figure 16.13 summarizes the section on configuration.

16.5 SERVICE MANAGEMENT

This section is a bridge between the process-oriented part and a more business-oriented part. The user perception has a direct impact the service layer development. Centralized management for service provisioning becomes a key factor for increased operator agility and profitability. See Figure 16.14.

Server platforms are built to deliver services and media contents irrespective of the IP host's access type. Most future services beyond voice will be developed and driven into the marketplace by content-based servers.

The expression *service management* is used for service provisioning and service data management.

The managers that handle personal end-user service provisioning and service management are sometimes called personal service environment managers. It is an end-user contact point for managing his/her personal service environment.

The personal service environment contains secure personalized information, stored in user profiles, defining how subscribed services are provided and presented to the end user.

QoS and security become important prerequisites in the overall service management. As described in Chapter 6, QoS (and security) is a layered feature. Therefore the overall

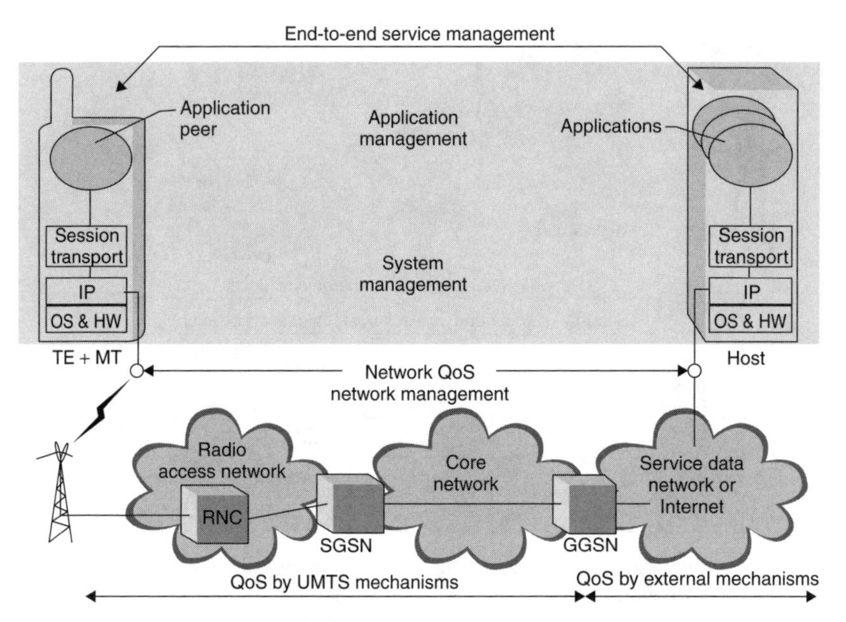

Figure 16.14 An end-to-end, all-layer view is necessary for service management

QoS, as perceived by the end user, is a result of the network QoS and the upper layer QoS protocols in the server and user equipment.

Since the service portfolio is a competitive tool, service creation, configuration, implementation and management are critical competencies. Service creation and management are possible areas for partnership. Telecom management performs the whole FAB (fulfilment, assurance and billing) processes necessary to provide the service to the end user.

The service layer includes vast possibilities but also particular problems and challenges. The overall goal is to capture revenue from new services 'beyond voice', with such cost of deployment, operation and management that the business case becomes attractive. Among targeted services are MMS, news, video calls, interactive messaging, games, music, entertainment and positioning services. It is a new area and the environment is in many aspects more complex and heterogeneous than the control and connectivity layer environment.

As mentioned before:

- There are many players and a growing number of applications.
- Messaging services and transactional services require much support from the service layer (Chapter 8).

To cope with the situation there is a need for partnerships between operators, vendors and content and application providers. See Figure 16.15. There is also a need for common enablers to control OPEX and CAPEX.

The middleware or framework functions are vital. They include support servers for service administration like registration in system component registry, discovery, common directory, common charging such as prepaid, and single sign on with authentication. Service development kits and API including service gateways make deployment easier.

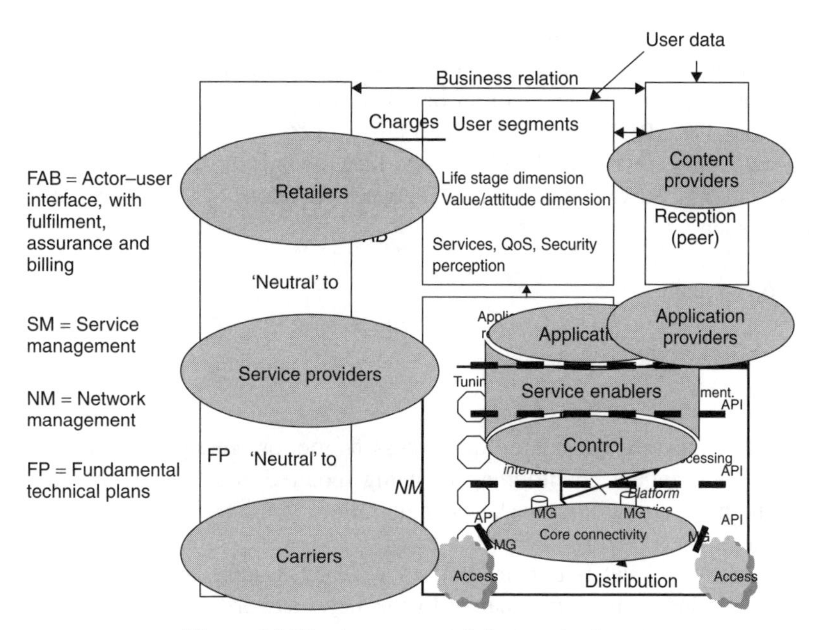

Figure 16.15 Actors around the service layer

Defined interfaces are important, and also to reduce complexity of the service layer architecture and evolve the service network in a structured manner.

Integration of all domains makes system integration crucial:

- Configuration and installation towards home environment applications, HE-VASP applications.
- Configuration and installation towards users including subscription management.
- Common processes shared by cooperating actors are desirable especially when using common resources.
- Provision of service enablers such as streaming, MMS and instant messaging to be included.
- Provision of security with AAA.
- Web service handling will come.

Other service layer management areas are:

- Performance management, e.g. QoS, alarms, monitoring, statistics, log list, performance criteria.
- Fault management with corrective actions.
- IP service network management.

16.6 TM OPERATIONS FROM A ROCE PERSPECTIVE

Chapters 4 and 5 contain a few sections on TM related business:

- Impact of management (Chapter 4)
- Cost-efficiency (Chapter 4)
- Telecom management and the service plan (Chapter 5)

This section treats telecom management from a ROCE view.

Telecom management is nowadays very much business oriented, focusing on increasing overall profitability. This target is often based on ROCE. Telecom management acts on all three terms of the ROCE equation. See Figure 16.16.

The *revenue side* is related to the billing interface towards the user. Apart from competitive and fair charging, user demands on the network operator/other actors are characterized by attributes such as:

- Service offerings
- Agility
- Quality of Service
- Security

All these demands are considered as success factors or prerequisites if you like. The factors have the stringent conditioning of being required end to end. In multi-operator environments this is not an easy task. A single failure in a link in the chain can spoil the whole service as perceived by the end user.

End user needs and behaviour are not static. They usually change over time. Many new service offerings will come thanks to the new network capabilities. The network itself changes, either due to scheduled extensions/new services and replacements or to unexpected events. This stresses the importance of telecom management as an integrator able to adapt to different end-user and telecom business environments.

A fast service provisioning is particularly important today. Two of the main competition issues frequently addressed in telecom business literature are TTM (time to market) and TTC (time to customer), which clearly demand actor agility. Old service provisioning involved heavy costs because of manual, non-standardized or non-integrated procedures. See Figure 16.17.

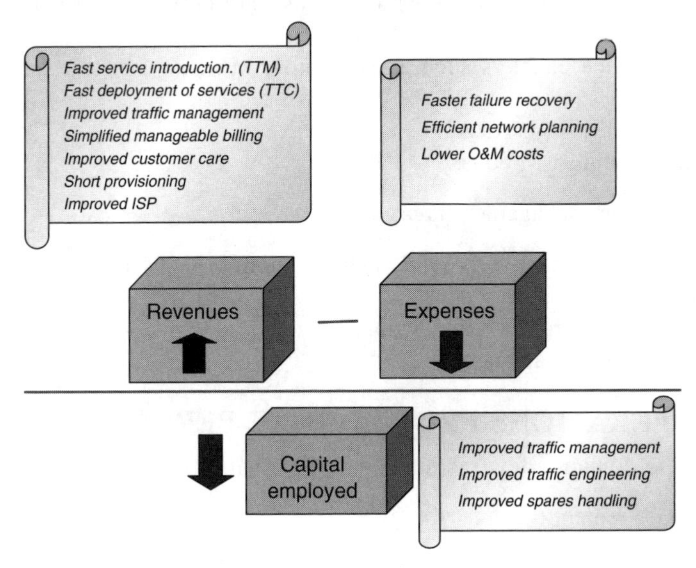

Figure 16.16 TM and ROCE

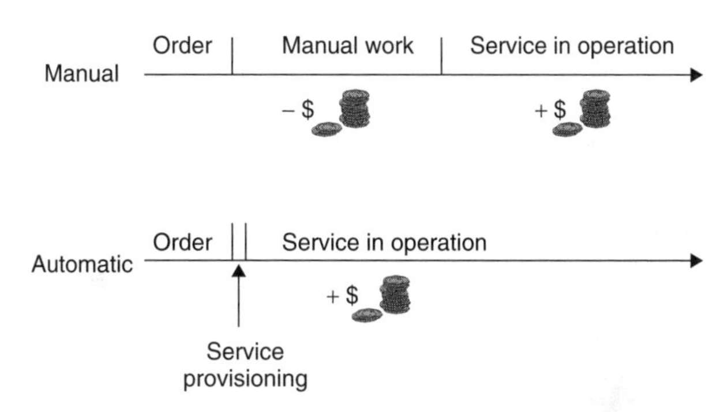

Figure 16.17 Automation is a key in reducing time to customer

It was quite complicated to connect a subscriber to a network or a new service, and data had to be set in several independent system databases (access network, local switch, billing, IN node, etc.) Also manual work or proceedings were often necessary.

Telecom management allows fully automated and standardized service provisioning. This is achieved in a user-friendly environment and from a single point in the network. This leads to satisfied customers, short TTC and faster time to revenue, favouring increased profitability.

For the operators themselves a fast and efficient handling of the billing makes a very strong impact on their result. The handling cost of one bill might well be in the range of $10–20, so being able to send a single bill to each customer instead of maybe three or four (e.g. fixed telephone, mobile phone, cable TV, Internet connection) saves money. See Figure 4.50 in Chapter 4.

Telecom management also contributes to revenue generation via what is called billing mediation. Billing mediation takes care of the collection, buffering and delivering of information from the network elements or call control servers, for billing or accounting purposes. Telecom management could possibly also assist in the detection and prevention of fraud, by capturing information on selected network element events, and through statistics and network management capabilities.

Heavy cost elements are marketing, network management and billing. See Figure 16.18.

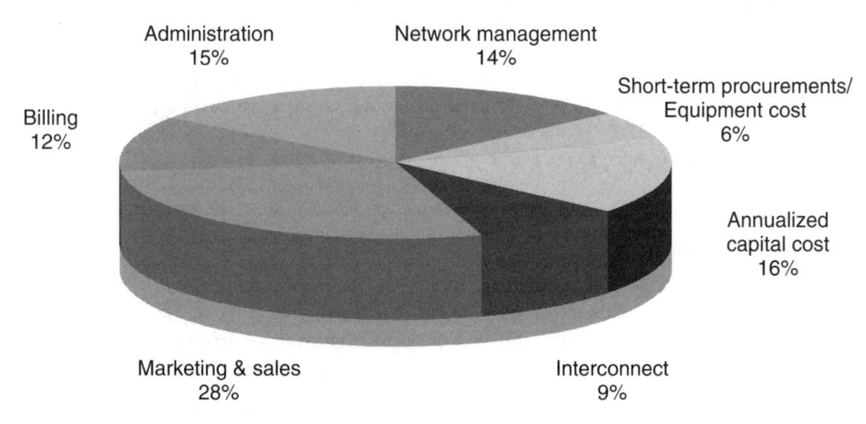

Figure 16.18 Example of cost elements for a network operator

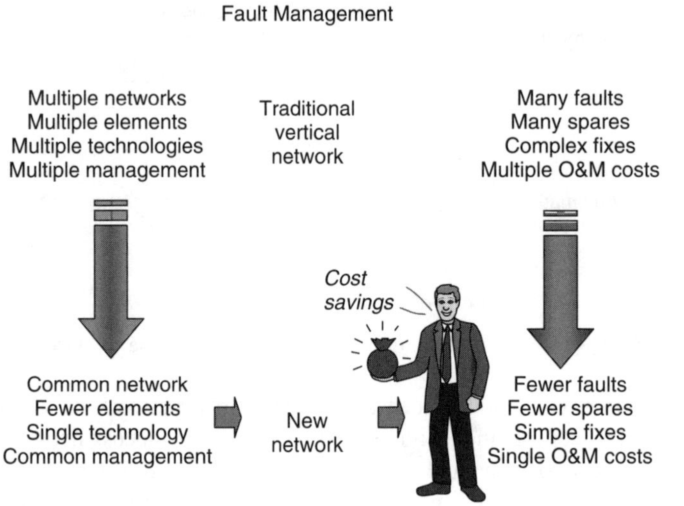

Figure 16.19 Fault management comparisons

The introduction of the new network architecture has an impact on cost as also discussed in Chapter 4. Cost reductions are achieved due to the sharing of common activities (surveillance to mention just a simple one). Let us here exemplify by means of fault management.

Fault management cost element savings are depicted in Figure 16.19, as a contribution to a better ROCE. There are reasons, however, to include an increased fault management cost at the point in time when the new network is taken into operation and until other technologies have been phased out.

The *capital employed* term is a little more complex to explain. Let us use the telecom management activity traffic management as a tool. In a given network situation an improved traffic management allows higher traffic handling, enabling a postponing of network expansions. Telecom management may be considered then as reducing the tied capital or network assets. Telecom management spare handling is another possible means to reduce capital employed through better administration, for example sharing store holding with suppliers.

16.7 CUSTOMER CARE AND DATA WAREHOUSING

16.7.1 BILLING AND CUSTOMER CARE

Customer care is important in winning and keeping customers, as it can be the differentiator in the marketplace. As more services are added to the operator portfolio the demands on customer care become more complex, involving more systems, more interfaces and more procedures. This complexity can severely inhibit the TTM goals.

Often customer care systems are the last to be developed or upgraded to cater for new products and services. The infrastructure and software for the new services are frequently available long before the customer care systems and procedures can be updated to handle them. This causes delays in launching the new services in the marketplace.

Billing is a vital area for operator business, since collection of billing revenues and customer billing satisfaction can build up competitive advantages. The position of billing at the end of the process chain is shown in Figure 16.20.

The IT development costs for a rapidly evolving operator is a dominant expenditure. It can even be greater than expenditures on the network infrastructure. Huge costs are frequently involved in adapting billing systems to new technologies and services. Furthermore, the billing adaptation can lead to fatal errors resulting in end-user annoyance. Therefore it is usual that most operators shudder when there is even a hint of a change on billing systems. Network evolution process must consider this, creating an environment as simple as possible, and preventing adaptations and redevelopment as far as possible.

Telecom management systems are the main input for billing and customer care, as they concentrate all the information available regarding status and events on network elements and call control servers. This theoretically opens the possibility to use any type of charging method or philosophy that might be imagined. Even if often limited in reality, there is a high degree of flexibility in defining the ways of charging that may be found suitable for customer billing satisfaction and business profitability.

Billing/accounting is based on event collection. All the events in the network, either call control or network element information, associated with billing are defined and informed via the telecom management system and sent to the billing mediation platform and to data warehousing. A common billing term is CDR (call detail record). The CDRs are sent to the billing mediation platform for reformatting (required if different vendors have different formats of the CDRs). See Figure 16.21. CDRs are also sent to data warehousing for further analysis of customer telecom habits etc. See Section 16.7.2 below.

The information received is processed to generate the actual bills. It can also be sent out directly to the customer for billing purposes. This feature is known as hot billing. Hot billing enables itemized call-charging information to be sent to, for example, a teleshop as soon as the call is finished.

For calls passing through different operator networks, billing information is also used by the accounting function regulating the flow of money between the operators.

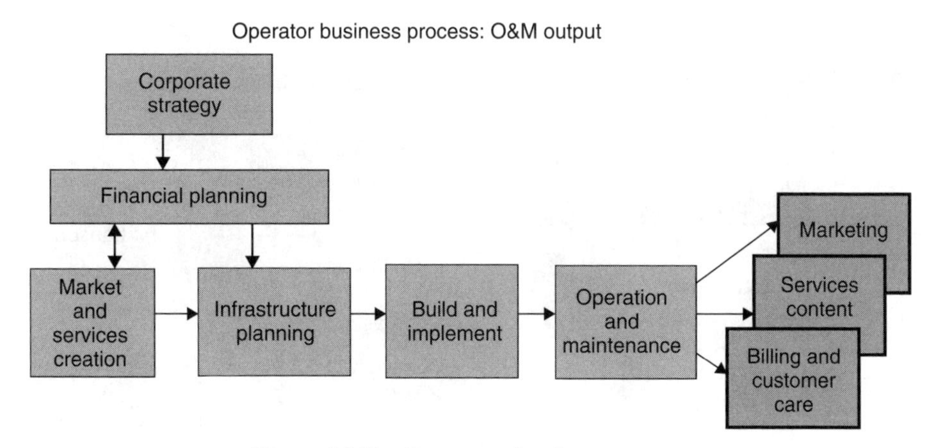

Figure 16.20 Revenue-related management

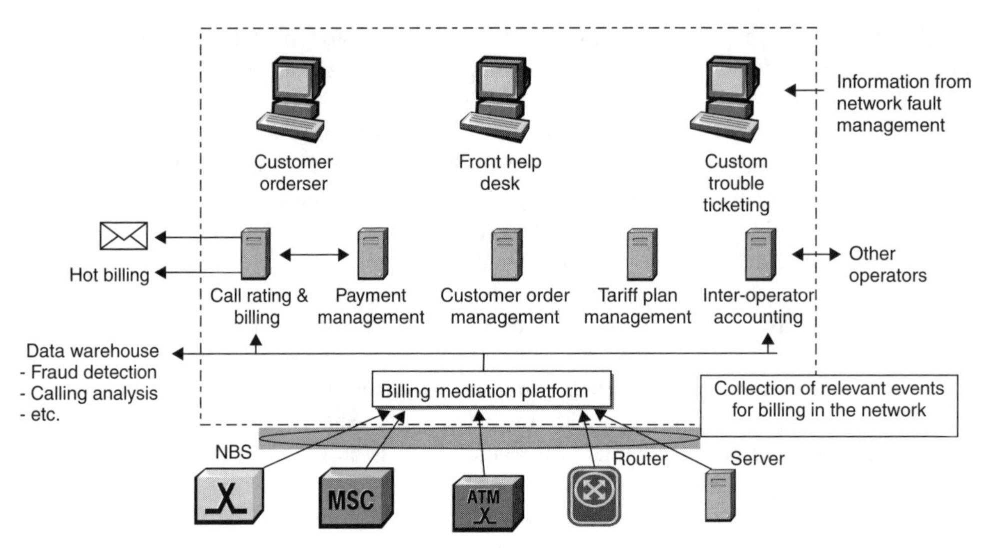

Figure 16.21 Billing mediation

It is worth mentioning the existence and discussion of two different philosophies regarding billing, as illustrated in Figure 16.22. The type of control, already treated in configuration management, also has important effects in terms of telecom business.

The use of pure call control or pure connectivity control in the network has important effects on billing methods and philosophy, as can be seen in Figure 16.22.

Connectivity-based technologies like IP have had limited capabilities regarding call control, which leads to a billing model not very attractive to many network operators. On the other hand smart billing requires a sophisticated call control, to clearly identify the different calls or sessions and to store the data associated with them for detailed billing.

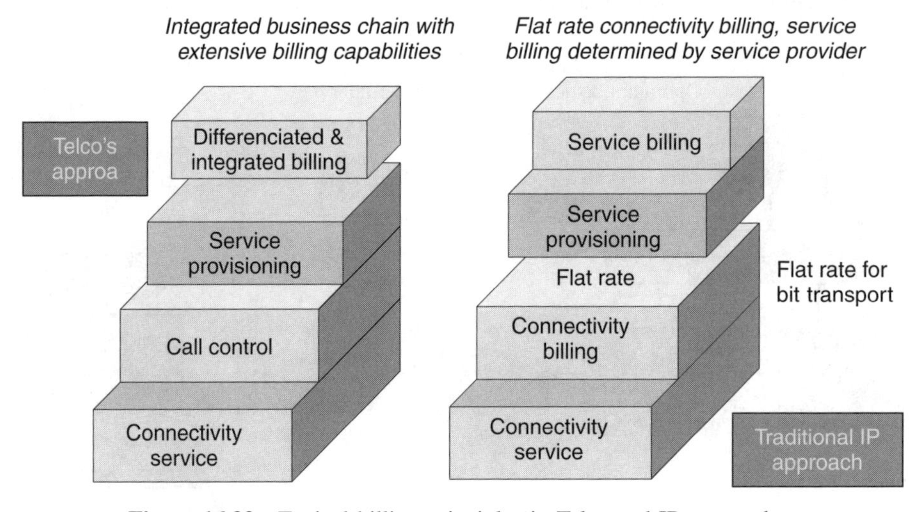

Figure 16.22 Typical billing principles in Telco and IP approaches

As already said IP is now evolving towards an increasing use of call control, where the deployment of SIP may play a very important role.

What will be the prevailing charging philosophy today remains an open question.

The convergent environment may be particularly beneficial regarding billing and customer care processes. When all services are provided on one common platform it is easier to migrate customers from one service to another, both in terms of procedures and of physical changes. It is no longer necessary to disconnect the customer from one network and reconnect her/him to another network.

This is especially relevant to keep operational costs (OPEX) down, particularly when hands-on intervention is required on remote or no longer manned network elements.

16.7.2 DATA WAREHOUSING

The term data warehousing means that lots of data are stored and post-processed to find many particulars about the subscribers and the network.

As a strategic marketing advantage, modern marketing trends known as 'personalized marketing' or 'marketing one-to-one', collect information about customer activity on the network and use it to promote particular offers suiting end-user needs and personality. Examples are use of individual applications and their geographical distribution.

Usually, most of the data originates from the CDR. CDR analysis can show many different things, ranging from the end-user call destinations and times (suitable to promote personalized telecommunication plans), to the existence of illegal practices like fraud.

A lot is won if the operator can discover customer needs or act upon problems before the customers or competing operators discover them. This will prevent other operators from seducing the customers with an attractive personal proposal.

This type of information, just as the customer database (customer management), is business critical. Therefore it is normally not outsourced, but handled by the operators themselves.

16.8 SECURITY MANAGEMENT

Security in telecom management usually addresses two different issues: telecom management systems intrinsic security and security management of end-user traffic.

Telecom management security is of particular concern for telecom operators regarding intentional damage and fraud detection and prevention. Telecom management has access and configuration capabilities to any network element in the network and is therefore an access point for security attacks. SNMP v3 provides an interesting set of security services, which include encryption to assure privacy, and manager–agent authentication to guarantee sender identity and message integrity.

Both telecom management internal security and end-user security make use of two different types of protection.

Protection against Unauthorized Access

This protection is performed through AAA (authorization, authentication and accounting) servers in the control layer, and it is addressed via password handling, access databases, authentication and encryption.

The main enabling technologies are LDAP allowing secure database access to the telecom management or the telecommunication system, and RADIUS, mostly used for cellular user access authorization in mobile systems.

Data Transfer Protection

Data transfer protection is mainly addressed through zonification, tunnelling and encryption.

Zonification is the determination of different boundaries between network segments called security zones. They usually correspond with natural boundaries observed in the network data flow or the network topology.

The traffic flowing between different security zones is tunnelled using protocols like L2TP and L3TP. The tunnelling process encapsulates the TCP/UDP or the IP datagram providing special means of protection against spoofing (injection of hacker's IP packets in the data flow). The data can also be encrypted in the tunnel using IPsec.

Before entering a security zone data contents of the traffic flow may be controlled at security inspection points. In these points there may be a firewall, a filtering router or even an IP security end-point aimed at control data integrity and privacy.

The existence of security policies, covering not only IT security, but general corporate security is essential to assure the system integrity. Security violations frequently come from sources not usually taken into account, such as employees' information, trivial or unchanged passwords, and even physical security attacks.

Security violation attempts have to be alarmed, reported and consistently analysed in order to prevent future damages. Integration between the telecom management alarm and performance management and the global security systems should be desirable although it is not always feasible.

16.9 QoS MANAGEMENT

The rather crowded Figure 16.23 exemplifies some of the ongoing initiatives and controversy around QoS. Delay and jitter are annoying to the end user for multimedia synchronous (isochronous) services, like voice or video. The associated and necessary QoS management consists mainly of:

- Admission policy (possibly shaping).
- Assignation of specific paths for isochronous real time traffic (preventing the competition from long IP data packets).
- Prioritization (certain packets can be dropped in case of congestion).

The isochronous traffic is identified at the entrance node (edge router or MGWs), eventually marked and allowed to flow through a special path, where the delay may be kept within acceptable values. Differentiation of traffic at certain entrance points is therefore a common procedure on modern networks. For all differentiation of traffic (quality, security etc.) there is a need for a policy. A policy server is a common solution.

The server provides mapping facilities for services, business and network configuration concepts. QoS bearer service provision, secure tunnel configuration or IPsec configuration

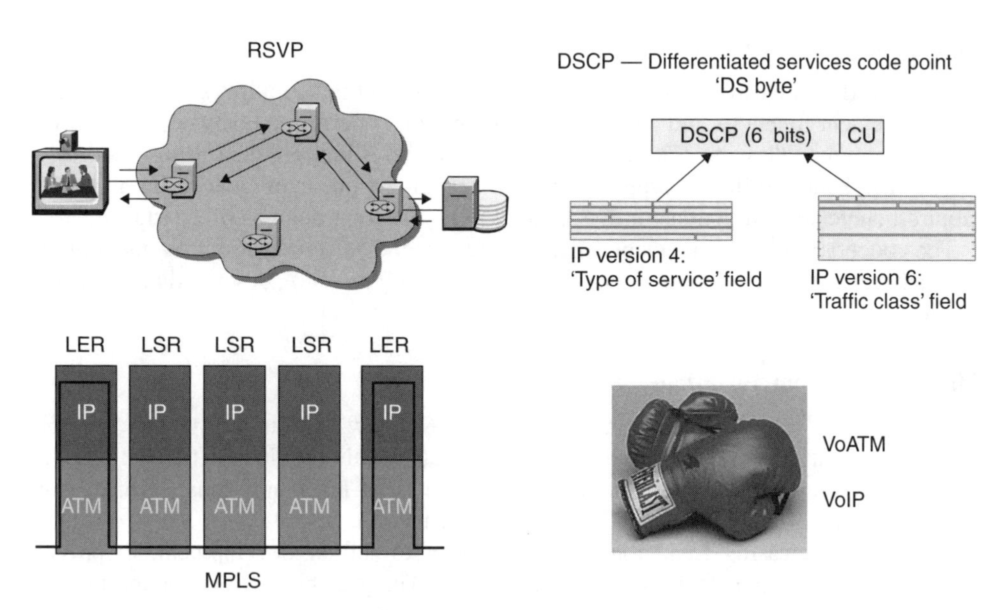

Figure 16.23 Some QoS standards and issues

can easily be specified in high-level terms. Example: 'Provide GOLD service for all SAP traffic from site A to site B'. The policy server creates logical rules on how this can be realized in the network, permanently or temporarily.

If a particular link fails, or QoS demands are not met under normal conditions, underlying network resources can be reconfigured, or low-priority connections can be dropped.

As described in Chapters 7 and 12, RSVP, MPLS and Diffserv are important QoS standards. RSVP, through resource reservation, MPLS through label switching or DiffServ with a defined per hop behaviour in each router, have the same goal, which is to open a differentiated lane for voice/video communications through a possibly congested area (the heavy traffic highway assigned to best-effort traffic).

The combination of IP and ATM contributes to QoS as IP can rely on the powerful ATM QoS handling capabilities. However, new standards offering interesting QoS alternatives are evolving for IP networks, such as MPLS and GMPLS.

Thus, telecom management plays a very important role in the satisfactory development of the network QoS. Performance management may lead to different traffic engineering initiatives and configuration management sets the admission policies, handles the establishment of special paths and fixes the prioritization rules within the network.

16.10 TERMINAL MANAGEMENT

Terminals are becoming one of the most relevant elements when considering end-user capabilities. The increasing distributed processing power has created terminals (typically PCs) with high processing power and management capabilities. However, dumb terminals are still frequently used.

Management of Terminals

Most networks take care of the resources on terminals, but this requires some distribution of management functions and entities. The distribution is performed between the network and the terminal, allowing centralized and/or autonomous management. It also addresses the need to handle different type of terminals, ranging from intelligent to dumb, which require maintenance functions on the network with different degrees of assistance.

The cooperation in management functions requires the existence of management protocols between network and terminals. Compare the CC/PP protocol, described in Chapter 7, Section 7.5.1.

Management from Terminals

A modern trend in telecommunication business is to allow and favour the end-user configuration of different services and network attributes. This provides extreme agility on service provisioning. The end user can also choose tailor-made solutions, selected from a defined menu of alternatives, on service implementation.

This is an increasing trend that will be subject to further development, to provide additional configuration facilities to the end users. This can be done either through the corresponding network and its traditional terminals, or via dedicated terminals capable of more complex configuration capabilities and information interchange.

Typical existing examples regarding this last case are operator's Internet pages, which allow service selection and configuration to the end users.

16.11 ACCESS NETWORK MANAGEMENT

The access networks are often decisive for the overall service quality.

Access is one of the most dynamic areas in telecommunications today, showing a strong growth, involving high levels of investments (often more than 50 % of the overall investment cost) and a myriad of promising technologies. Solutions for the 'last mile' to the subscriber have been a constant subject of debate in telecommunication forums for a long time.

Figure 16.24 describes a very likely state in the network development, where the core backbone will have to interconnect with many different types of access networks. This is what 'Agnostic Access' means. Most likely no one type of access technology or media will clearly prevail, and many different types of access will be selected and coexist, at least in the short run. Hybrid systems will also appear, such as UMTS-WAN.

Telecom management will have to manage a large number of different networks assuring seamless integration among all of them. They might be different access networks, existing legacy networks and the stratified core network. The proper and easy integration of the O&M activities on access, within the operator's telecom management, will then become a key element for assuring both business and operational success.

16.12 MANAGEMENT OF LAYERED AND SERIAL INTERWORKING

The complexity of modern networks, handling many different technologies simultaneously with a considerable number of layers and a fast pace of technology development, creates

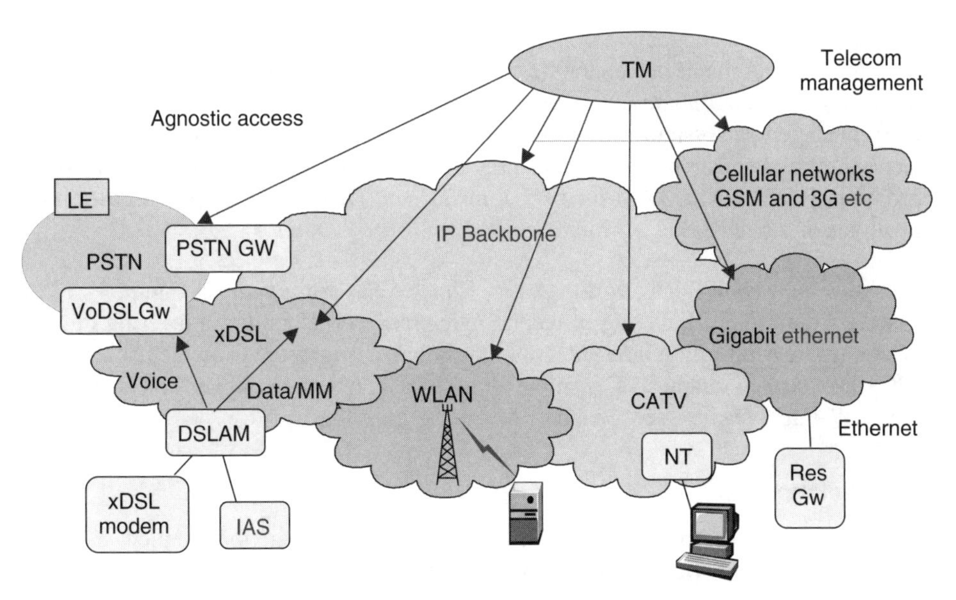

Figure 16.24 Access network management overview

difficulties in acquiring a proper telecom management understanding and management competence. The two considered cases are: single operator with many layers or many interconnected technologies, and interworking operators in layers or interconnected.

The interfaces and protocols are far more complex in interworking than in integration. The interworking protocols have to support the distribution along all planes (traffic plane and management plane) and layers (OSI-ISO layers) of the interconnected networks. In other words, they have to assure that the interconnection is held along all the layers composing the different networks, and that both traffic and management plane interconnection are included.

The complexity is even greater when the two interconnected networks do not provide exactly the same services (mobile and fixed telephony, LAN and WAN, X25 and IP, PSTN and VoIP). In this case the gateways have to be able to handle the asymmetries in service as well.

Why is my computer getting so slow when downloading files from the Internet?

The origin of the problem could be located either in the PC, the corresponding interworking modems, the copper, cable modem or wireless access network, the ISP and finally in the different nodes on the Internet network. Its determination will not be a simple management task, particularly considering that many operators are involved in the Internet communication and that they may have a null or at least very limited cooperation among their management systems.

1. The ITU-T (telecommunication network management – TMN) and the IETF (simple network management protocol – SNMP) approaches on network management, although based on similar principles, finally became pretty different (different protocols, different managed objects definitions, different terminology, etc.).

2. Usually isolated management systems already exist, either for particular layers, group of vendor's equipment or vendor's networks, or any combination of them.
3. Some vendors' equipment and networks, particularly old legacy networks, use proprietary management systems not based on the manager–agent principle.
4. Configuration management is frequently addressed by other means such as Telnet, proprietary CLI, FTP/FTAM for bulk transfer, and DHCP and LDAP in cases related to server or server-based configuration.

Examples of layered interworking: see Chapter 12, for example Figure 12.12 and Chapter 3, Figure 3.32. Layering of technologies may correspond to a layering of actors, but also a single actor might have integration problems, originated from the list above. The GMPLS standard in chapter 12 is interesting since it integrates many layers. Examples of 'vertical' interworking: see Chapter 15.

Possible Solution

A solution to possible problems is the use of mediation functions, called 'mediation devices' in TMN and 'proxy agents' in the Internet. The mediation function is located between the managed network and the central manager. It mediates on (or translates) the interchanged management information. It works under the selected TM management/agent standard towards the central manager, and under the corresponding proprietary management system towards the network elements.

To overcome TMN (ITU-T) and SNMP (Internet) integration drawbacks the introduction of management gateways, able to handle SMNP towards the network element and TMN towards the central general manager or vice versa, seems to be a solution.

The desired solution will very likely require the interconnection of management planes among different operators. This is a new trend emerging from the new technological and business environment but still at preliminary phases of R&D. Conceptually it poses problems, similar to those treated for the traffic plane interconnection. The addressed issues are:

- Determination of the management point of interconnection (POI)
- Admission policies
- Charges distribution
- Interfaces and protocols
- O&M responsibilities

See also Figure 16.1.

Determination of the management POI. Interconnection interfaces will probably be management gateways. Their location in the network will be a decision to make.

Admission policies. Determination of which requests or activities from other management systems will be accepted, and what kind of screening, if any, shall be made on output data. Determination of who will be authorized to program and decide this.

Charges distribution. The distribution of cost incurred in the development and adaptation of each management system or a general integrated ssystem, if available.

Interfaces and protocols. Interworking interfaces (how will the management gateways be?) and protocols for management planes cooperation.

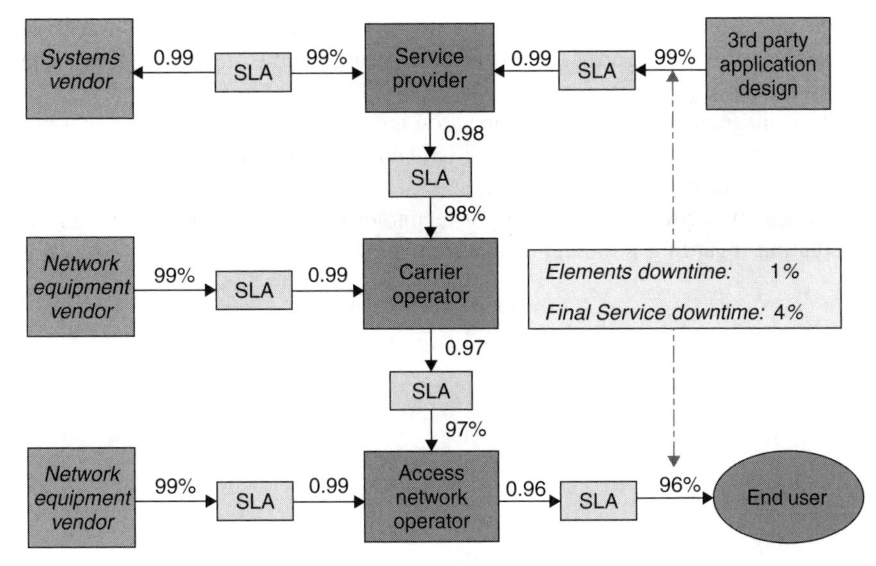

Figure 16.25 SLA management example. The SLA management includes maintaining the agreed values for each actor to keep the agreed value in the user interface. Among the users are banks, factories and selling units, which heavily depend on high quality communication

O&M responsibilities. Who will be responsible for the support and development of the global manager (if any) and the individual operator managers and the O&M of the 'management interfaces'.

Fundamental Plans

As mentioned in Chapter 15 the interconnected networks require a detailed planning to define the fundamental technical plans which should include the service plan. It is necessary to achieve a full alignment of the fundamental technical plan on the different interconnected parts, assuring a single behaviour along the whole network.

Since the end-to-end path often passes several actors/operators, SLA management must be introduced. A single failure in a link in the chain can spoil the whole service as perceived by the user. See Figure 16.25.

Different telecom management functions, particularly performance, fault and configuration management are associated to inter-operators SLA's monitoring and compliance.

16.13 CONCLUSIONS

Telecom management is a key element for network handling and business development. Telecom management is first of all a network integrator (using this word in a wide meaning covering both layer integration and network segments interworking). The segmentation, that the network has been subject to, requires a number of perspectives from where the network can be seen and considered as a whole.

Telecom management addresses the network in a holistic perspective, integrating its performance (performance and alarm management) and establishing the desired behaviour of the network (configuration management).

Telecom management has a strong impact on the operator business. Important business issues like in-service performance, TTC, TTM, billing, marketing and customer care are deeply related to telecom management.

To introduce any equipment, software or technology into the network demands a satisfactory solution regarding its integral management.

Appendix 1

Web Services and a Service-Oriented Architecture

A service-oriented architecture and web services enable re-use of software resources and information. A brief introduction is found in Chapter 3, Section 3.13.

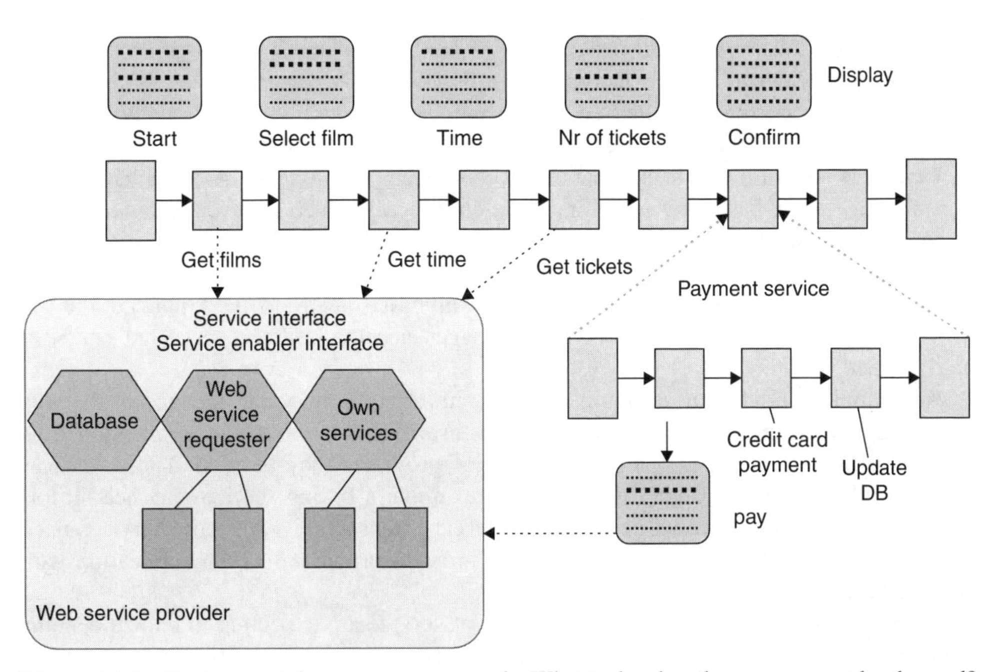

Figure A1.1 Business or telecom process example. What to do when the process must be changed? Or new processes are required? Can sub-processes or smaller or larger SW units be used from other sources?

Understanding Changing Telecommunications – Building a Successful Telecom Business. Edited by A. Olsson
 ISBN: 0-470-86851-1

Web services are part of a new architecture called a service-oriented architecture. This architecture eliminates the need for developers to repetitively design and build the same software components that solve the same set of recurring business problems, over and over again. In response to this, a service-oriented architecture enables greater use of existing IT applications by creating services that can be re-used across the enterprise or the telecom network. Such architecture thus minimizes the need for extensive rewriting of code each time a new business or telecom service is required or each time a business process is altered. Figure A1.1 illustrates a process flow which partly uses sub-processes from an external source, called web service provider.

The principle can for example be called:

- Software as a service
- Services on demand
- Service enablers on demand
- Business processes as a service (more advanced development)

A *web service* is a software application identified by a uniform resource identifier (URI), whose interfaces and binding are capable of being defined, described and discovered by XML artefacts and which supports direct interactions with other software applications using XML-based messages via Internet-based protocols.

Web services make software/services/service enablers/business processes accessible to any service provider to support activities in a process.

Web services are the tools that enable businesses to quickly and effectively integrate internal systems and share information and logic across the enterprise.

A *client* is software that makes use of a web service, acting as its 'user' or 'customer'.

A *message* is the basic unit of communication between a web service and a client; data to be communicated to or from a web service as a single logical transmission.

In order to make use of a web service, it is necessary to know what information is expected and in what form. Therefore, the interface needs to be explained for use by remote parties. The structure of the messages accepted and/or generated needs to be described.

An example can be seen in Figure A1.2. A number of services are defined in the web and by the different service providers and enterprises. For a specific application own services as well as services belonging to service providers may be used. Those services may use other services in other domains. For example a billing service may need information about the user account or his/her credit card transactions. All these services need to communicate and understand each other in order to provide a certain application with specific tasks.

Businesses/operators that purchase this type of service are expecting to realize benefits along the lines of shorter payback periods, higher return on investment (ROI) and a freeing of staff to manage projects they deem more strategic to the competitiveness of their business. The basic architecture model includes web services technologies capable of:

- Exchanging messages/requests
- Describing the web services
- Publishing and finding web service descriptions

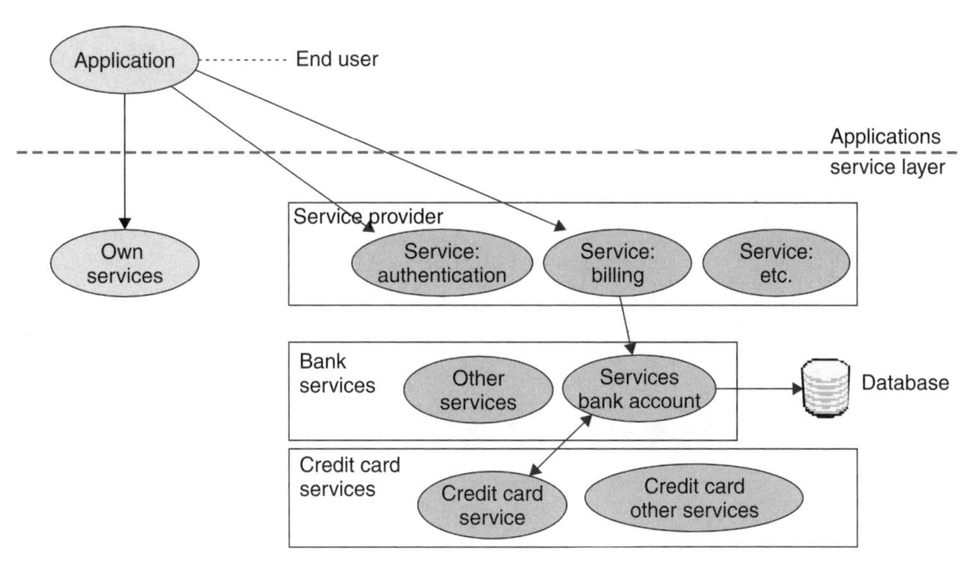

Figure A1.2 An application that is composed from different sources

Most web service applications will, most likely, be built on new or existing web technologies, including HTTP, XML, SOAP, SSL, WSDL, and UDDI. (See glossary).

SOAP (Simple object access protocol) allows a program running in one kind of an operating system to communicate with a program in the same or another kind of an operating system (e.g. Linux Vs Windows) by using HTTP and XML as the mechanisms for information exchange. In effect SOAP is a platform-independent access protocol.

The request for a web service is sent to an HTTP server. There are different methods to translate the information carried by HTTP (over TCP/IP) to XML protocol.

A service-oriented architecture may have impact on network design and QoS. When the end user requests a service the main issue is to analyse the required capabilities and choose proper process handling based on a viable architecture. Choice of appropriate service enablers is part of such process handling.

The 'software as a service' concept offers advantages and disadvantages. Sharing resources would normally imply cost benefits. Regarding QoS the remote access is a drawback, also for the security, but the possibility to provide alternative remote accesses as back-up will instead have a positive impact. Obviously the service-oriented architecture can solve many network architecture problems, when using service resources on demand across the network. This is an interesting area for network architects and designers. Among the main challenges are no doubt the QoS and security issues.

Appendix 2

Financial Calculations

FINANCIAL CALCULATIONS AND INVESTMENT ANALYSIS

Important financial terms and a tool for operators to decide if a business concept is profitable in the future.

SCOPE

The main purpose of this appendix is to explain some common financial terms and tools, used to assist an operator/service provider in making good *investment decisions*. Such tools and terms are case studies, cost/benefit studies, discounted cash flow studies, net present value and internal rate of return calculations. Further, the corporate policy should not be overlooked in this context.

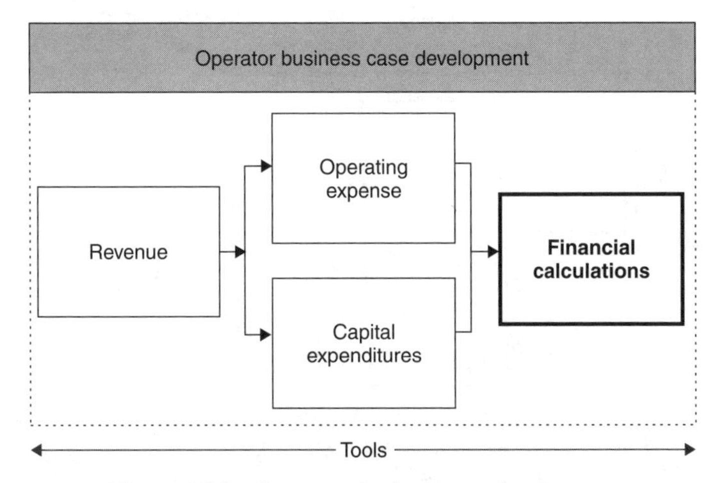

Figure A2.1 Operators business case development

Understanding Changing Telecommunications – Building a Successful Telecom Business. Edited by A. Olsson
 ISBN: 0-470-86851-1

Other related subjects are business accounting, business profitability and performance, and finally business risks.

The appendix constitutes the final step in the operator business case development otherwise covered in Chapter 4. See Figure A2.1.

BASIC TERMINOLOGY

REVENUE, COST AND PROFIT

Other basic terms are income statement and profit and loss account.

Let us consider a traditional operator. See Figure A2.2. In the middle of the figure is the service producing machine, the network. According to Section 4.7 on cost efficiency in Chapter 4 the investments in the network can be split into operational investments to meet customer growth (baseline cost), and discretionary investments for purposes such as the introduction of a new technology. The discretionary investments are often large and irregular. The cost of financing them, often through loans, is converted into annualized capital cost corresponding to interest and depreciation.

In the context of this chapter we simply add the two investment types. Then we add operating expenses and bought services including interconnection expenses. See Figure 9.3. This gives us the overall cost of operation.

The operating revenue side consists of revenue from the end-user market and/or income from sold services to other actors. The revenue from the end user is based on fixed and traffic-based revenues.

The difference between revenue and cost gives the operating income.

Adding financial posts (revenue and cost) and taxes to the operating income gives the net income or the (net) profit.

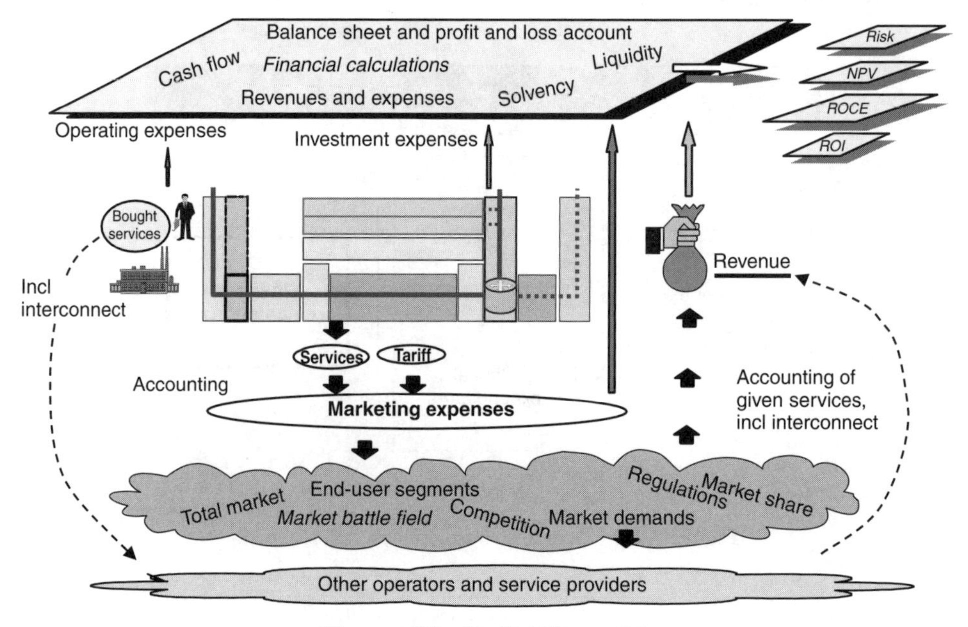

Figure A2.2 Profitability model

In Figure A2.2 the cost and revenue elements are brought to the 'financial calculation layer'.

The outcome from these calculations can be oriented towards comparisons between alternative discretional investments (NPV, net present value), performance measures (ROCE/ROI, return on capital employed/return on investment) or risk assessment.

BUSINESS ACCOUNTING

The Balance

Important terms are assets (working capital, and inventories such as buildings and machinery), liabilities (short- and long-term debts), and owner's equity (Assets – Liabilities).

The assets mentioned above are called tangible assets. In addition there are intangible assets such as patents, strong brand and goodwill.

The (net) profit from a certain period gives a corresponding change in equity for the same period.

The balance shows the financial strength of a company at a particular point in time. A balance sheet may look like Figure A2.3.

Working Capital

Working capital = Current assets – Short-term liabilities.

The Liquidity

The liquidity refers to current assets, that is to say cash or assets that can easily be converted to cash in the normal business situation, and short-term liabilities.

$$\textbf{Liquidity} = \frac{\textbf{Current assets}}{\textbf{Short term liabilities}}$$

The term cash flow describes the flow of money going in and out of a company. Payment coming in could be from sales, banks giving loans, owners etc. Payments going out could be salaries to employees, interests and repayments on loans, payments to suppliers or subcontractors.

ASSETS
How the capital is used

Fixed assets	60
Current assets	40
Total assets	100

LIABILITIES
How the capital is supplied

Owners equity	60
Long-term liabilities	20
Short-term liabilities	20
	100

Figure A2.3 Balance sheet

Negative Cash Flow

More money is going out than coming in to a company.

Positive Cash Flow

More money is coming in than going out of a company.

Net Cash Flow

Consists of the net effect of cash inflow and cash outflow.

BUSINESS INVESTMENTS

Corporate Policy for Investment Analysis

A corporate policy should normally be developed centrally and issued throughout the company.

It should prescribe the types of financial analyses required to justify investment proposals, for example that all capital expenditure proposals be supported by discounted cash flow (DCF) studies.

The corporate policy also sets the discount rate to be used, prescribes methods to consider inflation, and the need for sensitivity and risk analysis.

Evaluation Methods

Another possible heading would be indicators of merit.

In this appendix only the pure, short-term economical factors are used. Intangible effects such as goodwill, references etc. are not considered. It should in other words be recognized that there are other criteria than maximizing profits that a company may adopt, in determining whether to invest:

- The return on investment (ROI/or accounting rate of return (ARR))
- The payback method
- Discounted cash flow (DCF) methods
- The net present value (NPV)
- The internal rate of return (IRR)
- Earnings before interest, taxes, depreciation and amortization (EBITDA)

ACCOUNTING RATE OF RETURN (ARR)

Compare the estimated ARR of a proposed investment/project with the target ARR. If the estimate exceeds the target, accept the project. If it is lower, reject the project.

Depreciation should be accounted for.

$$\mathbf{ARR} = \frac{\textbf{Estimated average profit after depreciation} * \mathbf{100}}{\textbf{Estimated average investments}}$$

PAYBACK METHOD AND DISCOUNTED CASH FLOW (DCF)

Payback is normally defined as the period, usually expressed in years, which it takes the cash inflows from an investment project to equal the cash outflow.

Depreciation is not accounted for in this model.

The ARR method of investment appraisal ignores the timing of cash flows and the actual cost of capital. Payback considers the time it takes to recover the original investment cost, but ignores total cash flow over a project lifetime. The payback time for a particular investment is illustrated in Figure A2.4.

Discounted cash flow (DCF) takes into account both the time value of money and total cash flows over a project lifetime. It is therefore often argued that DCF is a superior method to both ARR and payback.

NET PRESENT VALUE

Present value (PV) can be defined as the present cash equivalent of a sum of money receivable or payable at the stated future date, discounted at a specified rate of return.

Compound interest:
P_0 = current investment
i = annual interest rate, compounded each year
P_n = value of the investment after n years
After one year the value of the investment increases to $P_1 = P_0 + iP_0 = P_0(1 + i)$
After two years, the investment value has increased to $P_2 = P_0(1 + i) + iP_0(1 + i) = P_0 (1 + i)^2$
In general, $\boldsymbol{P_n = P_0(1 + i)^n}$

Discounting a future value:
The present value P_0 of a future amount P_n is found from the above equation:
In general, $\boldsymbol{P_0 = P_n(1 + i)^n}$

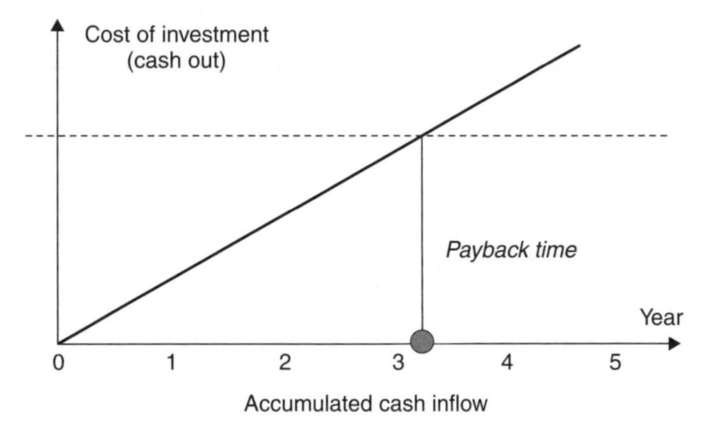

Figure A2.4 Pay back time

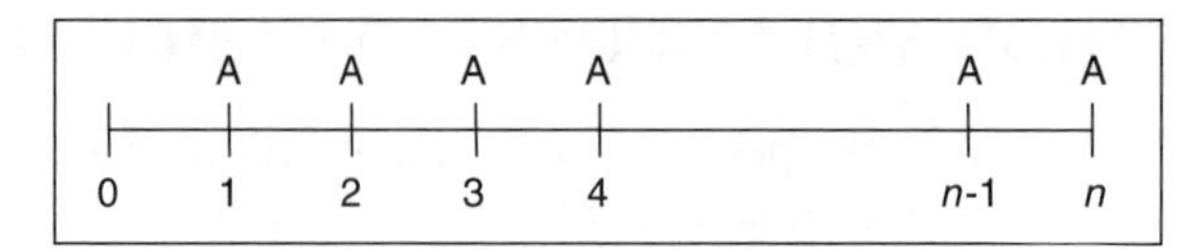

Figure A2.5 Assumed annual Telecom project (net) cash flows

PV of a series of regular equal cash flows:
n cash flows (payments or receipts), each of amount A, at the end of years 1, 2, 3 ... n. according to Figure A2.5.
$Po = A(1+i)^{-1} + A(1+i)^{-2} + A(1+i)^{-3} + \cdots + A(1+i)^{-n}$

Discounting can be applied to both money receivable and also to money payable at a future date. And so by discounting all payments and receipts from a capital investment to a present value, they can be compared on a common basis at a value that takes account of when the various cash flows will take place.

Example 1

The estimated cash flows on a project investment (100 000) over the next four years are (60 000, 80 000, 40 000 and 30 000). The company cost of capital (or interest) is 15 %. What is the NPV of the project, and should it be undertaken?

Year	**Cash flow**	**Discount factor (i = 15 %)**	**PV**
0	100 000(investment)	1	−100 000
1	60 000	0.870	52 200
2	80 000	0.756	60 480
3	40 000	0.658	26 340
4	30 000	0.572	17 160
NPV			56 160

Example 2

A telecommunication project is expected to have the following cash flows, where the figures in brackets represent costs and are negative (this is standard accounting practice).

The corporate discount rate for new investments is $i = 10$ %. The discount factor for each amount can be deduced from the formula $(1+i)^{-n}$.

The sum of the discounted cash flows will be the NPV for the project. The initial cost of $1 million is incurred at the end of year 0, that is; at the beginning of year 1, and hence the discount factor for it is 1.0. If paid at the end of year 1, it would have to be discounted and the PV of the expense reduced. To ensure that the timings of costs and income are correctly accounted for, it is useful to make a sketch to represent the various

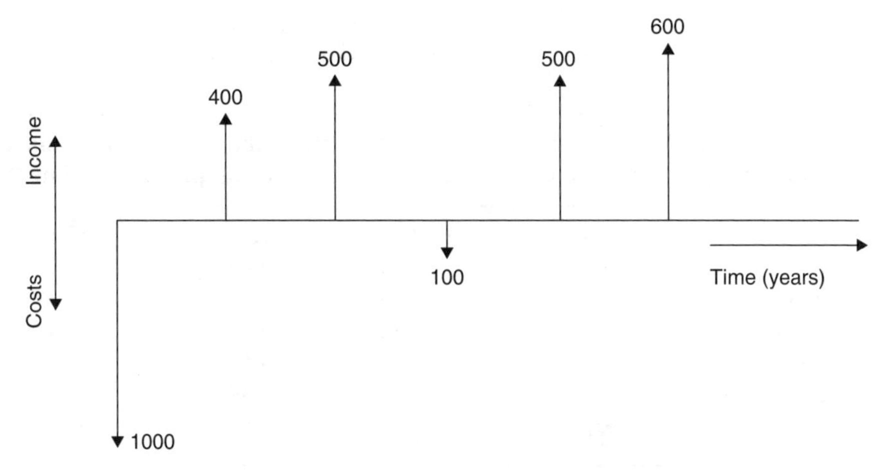

Figure A2.6 Annual cash flows are discounted in present value methods

amounts and when they occur. Such a sketch is shown in Figure A2.6. Income is shown above the line, costs below it.

End of year	Actual cash flow: $ 000	Discount factor @ 10 % per annum.	Discounted cash flow: $ 000
0	(1000)	1.000	(1000)
1	400	0.909	363.6
2	500	0.826	413.0
3	(100)	0.751	(75.1)
4	500	0.683	341.5
5	600	0.621	372.6
Total	**900**		**415.6**

(**Note**: The usual condition for the existence of an internal rate of return is that there must be at least one change of sign in the sequence of cash flows.)

INTERNAL RATE OF RETURN (IRR)

In the IRR study the aim is to find the discount rate which leads to an NPV of zero, that is; NPV of outgoing cash flows just balances NPV of incoming cash flows. The procedure to find IRR is iterative: guess values for IRR; calculate and plot values of NPV; find the value that gives a zero NPV by interpolation. These calculations can be done with computers (e.g. Excel).

IRR for a project can be compared with the corporate discount rate i: if IRR is less than i, the project is not profitable; the more IRR exceeds i, the more profitable the project.

Example

This example shows a trial-and-error method which also uses interpolation.

The table below shows a series of cash flows corresponding to an initial investment of $1 million, which results in yearly income of $263 800 per annum over a five-year period.

End of year	Actual cash flow ($M)	Discount factor @ 5 % pa	Discounted cash flow	Discount factor @ 15 % pa	Discounted cash flow ($M)
0	(1000)	1.0	(1000)	1.0	(1000)
1	263.80	0.952	251.10	0.867	228.70
2	263.80	0.907	239.30	0.756	199.40
3	263.80	0.864	227.90	0.658	173.60
4	263.80	0.823	217.10	0.572	150.90
5	263.80	0.784	206.80	0.497	131.10
		NPV	**142**.20	**NPV**	(**116**.30)

Deduce the IRR by calculating the NPV of the cash flows for two discount rates, one hopefully larger than the IRR, and the other hopefully smaller. Then interpolate to derive the discount rate for which the NPV is zero: this is the IRR.

(**Note**: The usual condition for the existence of an IRR is that there must be at least one change of sign in the sequence of cash flows.)

As desired, one NPV is positive and the other negative, so we then know that the IRR lies between the two discount rates 5 % and 15 %. A straight-line interpolation will give a satisfactory result (the actual curve is not quite straight). The point where the straight line crosses the discount rate axis gives the IRR, since this rate would give a zero NPV.

It is not easy to interpolate on Figure A2.7, but the line crosses the horizontal axis at the IRR = 10.2 %.

Using the IRR function of Microsoft Excel, the internal rate of return is calculated instantly as 10.27 %.

If this rate of return exceeds the weighted average cost of capital for the company, then the investment is assessed as favourable. Otherwise, the investment is not regarded as economic.

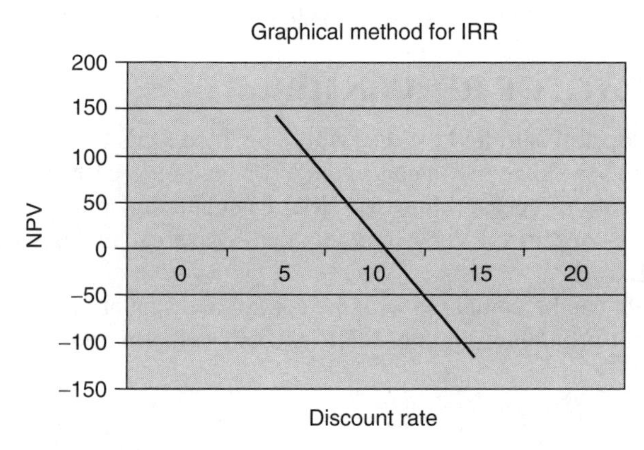

Figure A2.7 Interpolation of internal rate of return, IRR

BUSINESS PERFORMANCE MEASURES

In this chapter some financial measurements are accounted for. Note that there are other important performance measures that are non-financial, e.g. customer satisfaction measurements, people satisfaction measurements, efficiency, quality and time measurements, innovation measures etc.

Return on Investment (ROI)

The relationship between profit and money invested in a project or company, usually expressed as a percentage, ROI, is also called the accounting rate of return (ARR). ROI is normally used when evaluating a company or organization and ARR is used when evaluating an investment or a project.

Differences can occur: e.g. in some calculations the net profit is the numerator and the total assets are the denominators.

$$\mathbf{ROI} = \frac{\textbf{Result (profit)} * \mathbf{100}}{\textbf{Invested capital}}$$

Return on Capital Employed (ROCE)

Profit shown as a percentage of the capital in a business.

Capital employed = Current assets − Short-term liabilities

$$\mathbf{ROCE} = \frac{\textbf{Result (profit)} * \mathbf{100}}{\textbf{Capital employed (from the balance sheet)}}$$

See also Figure A2.8.

ROI/ROCE can often provide more insight into performance when it is divided into the following components.

Du Pont Model

$$\frac{\textbf{Revenues}}{\textbf{Total assets}} \times \frac{\textbf{Operating profit}}{\textbf{Revenues}} = \frac{\textbf{Operating profit}}{\textbf{Total assets}}$$

This approach is widely known as the *Du Pont method of profitability analysis*, which recognizes that there are two basic ingredients in profit making: using assets to generate more revenue, and increase income per unit revenue. An improvement in either ingredient without changing the other increases return on investment.

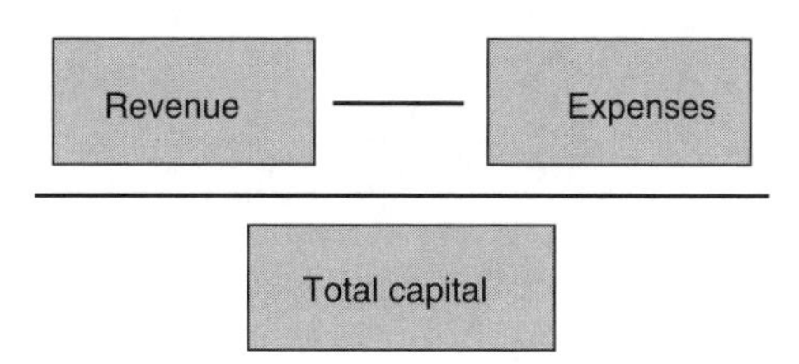

Figure A2.8 Return on capital employed

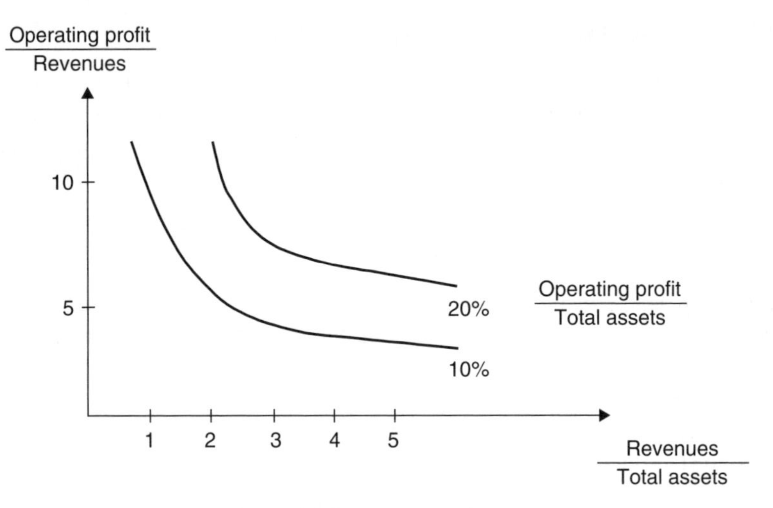

Figure A2.9 Du Pont diagram

Figure A2.9 shows this formula in a graphical manner (Du Pont diagram).

Return on Sales (ROS)

Profit shown as a percentage of revenue/sales.

$$\mathbf{ROS} = \frac{\mathbf{Operating\ profit} * \mathbf{100}}{\mathbf{Revenue}}$$

SENSITIVITY ANALYSIS

As a basis for decision making a sensitivity analysis is a supporting tool when relevant. You change the values for estimated expenses and revenues, for example by changing customer base development, tariffs and operation cost. Then the impact on cash flow or NPV is examined.

Appendix 3

Development Tracks

INTRODUCTION

Paradigm shift and migration are some typical subjects of this book. Many actors are already involved in such process of change, moving from different starting points towards a new network offering multimedia services as standard feature. The change can be fast or soft. A soft stepwise migration is typical for actors with considerable network assets in the network. The assets correspond roughly to the capital employed, which is part of the ROCE equation. When using ROCE as a tool, it is important to utilize the existing resources, such as copper, fibre, RBS and sites, switching and processing capabilities, in the best possible way, when converging the networks towards a target network shape. This gives the best return on capital employed. For incumbent telecom operators, the migration process for the telephony network is the most crucial. This is normally the biggest operated network where the operator has made the largest investments. Mobile operators with a large 2G network have a similar situation.

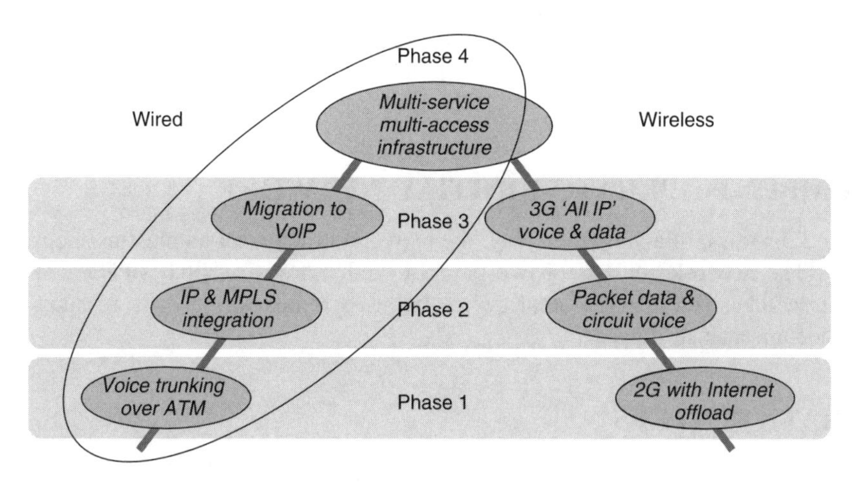

Figure A3.1 A common target network is assumed for different types of operators

Understanding Changing Telecommunications – Building a Successful Telecom Business. Edited by A. Olsson
 ISBN: 0-470-86851-1

The migration strategy is also a matter of actor competence. A gradual migration facilitates the actor task to manage the competence issue, keeping pace with the new technology, new processes and possibly new organization.

OBJECTIVES

The main objective of this appendix is to make the vertical to horizontal network migration process more concrete to the reader by means of illustrating a stepwise migration towards a target network for an incumbent operator.

It is assumed that the target network is fairly common for different actors. 'All roles go for multi-service and multimedia' (Chapter 4, Figure 4.25).

In this appendix we will therefore look at the left path of Figure A3.1, corresponding to an incumbent operator. The figure uses some expressions that might need clarification. Voice trunking means some kind of packetized transfer of voice in the trunk (= inter-exchange) network. Whether or when to use ATM or IP is under debate. For fixed voice centric operators migration experience is becoming increasingly available, which is the basis for this appendix.

Internet offload in the right path in the figure means that the data/Internet calls over modems in telephony networks are extracted from the telephony network.

In the following we will describe, in a simplified and pedagogical way, a possible migration of an existing circuit-switched telephony network into a packet-switched ATM based multi-service network.

TARGET NETWORK

Figure A3.2 shows a target connectivity network vision, exemplifying the future packet-switched multi-service network. Local switches, access nodes, enterprises, servers etc. are connected via multi-service gateways (MSG) to a common connectivity network. In this context 'service' means a bearer service such as frame relay, ATM, PSTN, PLMN or ISDN. The MSG is a specific type of MGW. Also connections to other networks are realized through MSGs.

INCUMBENT OPERATOR INITIAL NETWORK

Figure A3.3 shows, in a simplified way, the network structure of an existing metropolitan network. The network consists of two pairs of tandem/transit switches (TS) and around 25 local switches (LS). Every local switch has two routes for security reasons, one to each tandem/transit in a pair.

MIGRATION STEPS

Phase 1

As a first step, a telephony server and two MSGs could be introduced in the network. See Figure A3.4. This will be a first embryo to an ATM backbone network.

Two of the tandem switches are connected to the two MSGs.

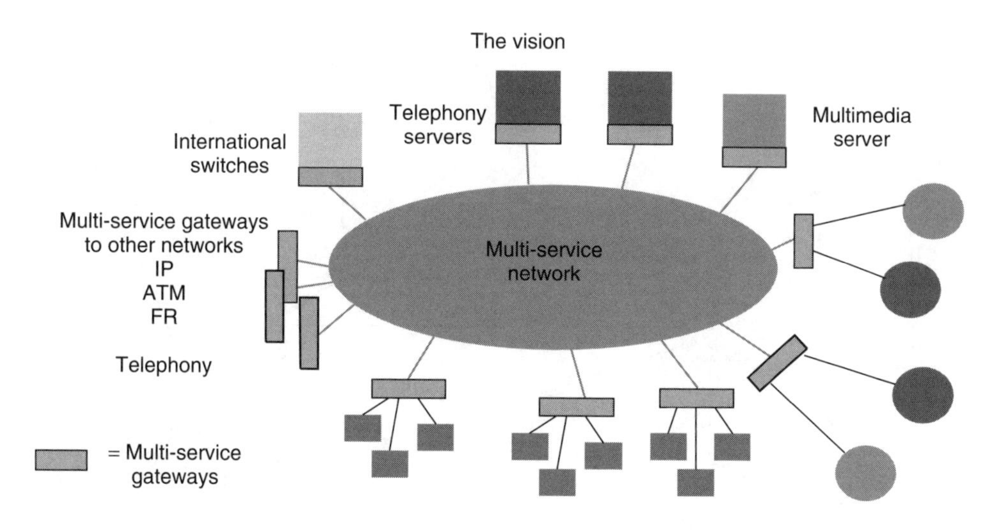

Figure A3.2 A target network variant for an incumbent operator

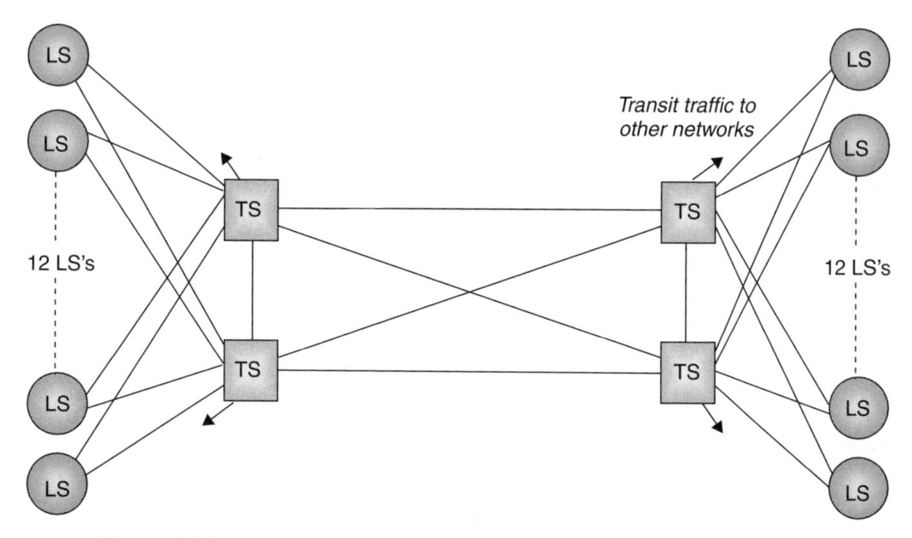

Figure A3.3 A simplified example of an existing metropolitan network

With this in place, we can then start to connect some of the local switches to the MSGs and begin routing some voice calls over the ATM network. When this arrangement works perfectly, we can remove the old routes to the tandems (marked with crosses in Figure A3.4). We still have the same reliability as before in the network. We can also connect IP routers or other data sources to the MSGs and thus transfer data traffic through the ATM network. In this way we can verify that it is possible to mix voice and data traffic in the same network.

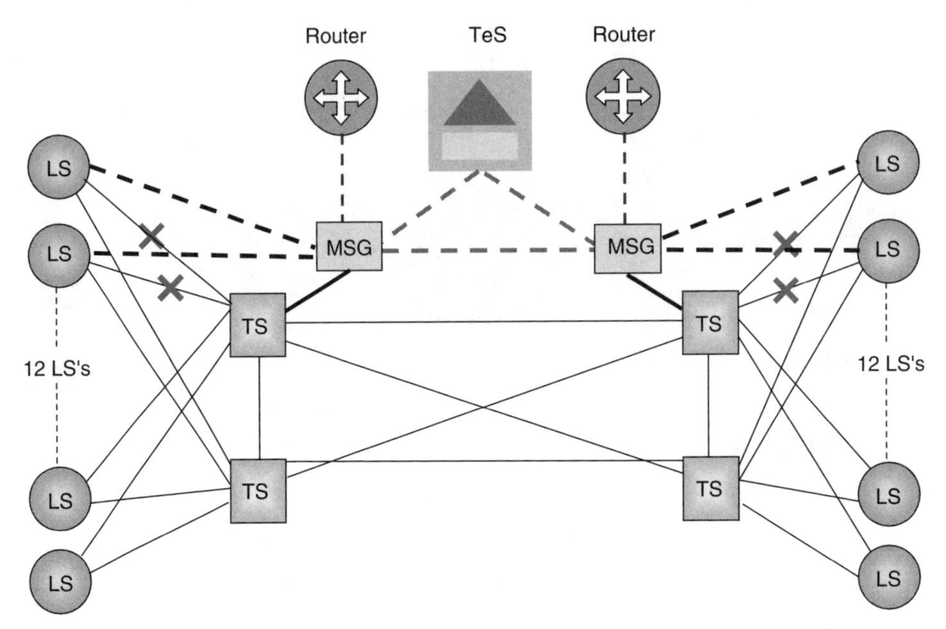

Figure A3.4 First phase in the migration towards a multi-service network

Phase 2

In the next step, we could add two more MSGs to expand the ATM network. We can connect the other two tandem switches to the two new MSGs and thus make it possible to continue to connect more local switches and data sources to the backbone network. When the new routes have been tested and are working perfectly, we can remove more of the old routes from the local switches. See Figure A3.5.

In this way we can, step by step, take more and more of the voice traffic through the ATM backbone network and reduce the traffic load on the tandem switches.

Phase 3

At a certain stage in this process, when a big part of the voice traffic is routed through the ATM backbone network, it could be wise to introduce a second telephony server – for reliability reasons. See Figure A3.6.

Phase 4

During this migration process there might be a need to expand the network with new subscribers. With this multi-service network solution it is possible to connect access nodes (AN) directly to the MSGs and thus there is no need to invest in any more local switches. See Figure A3.7.

In modern access nodes it is possible to combine narrowband and broadband access, so the operator can now also offer broadband services to subscribers who want this.

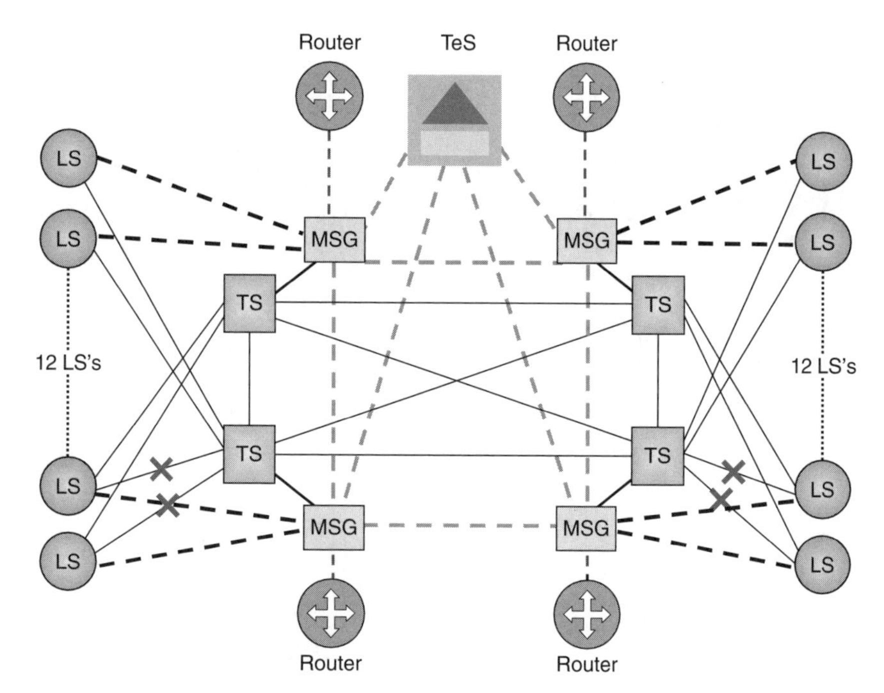

Figure A3.5 Second phase in the migration process

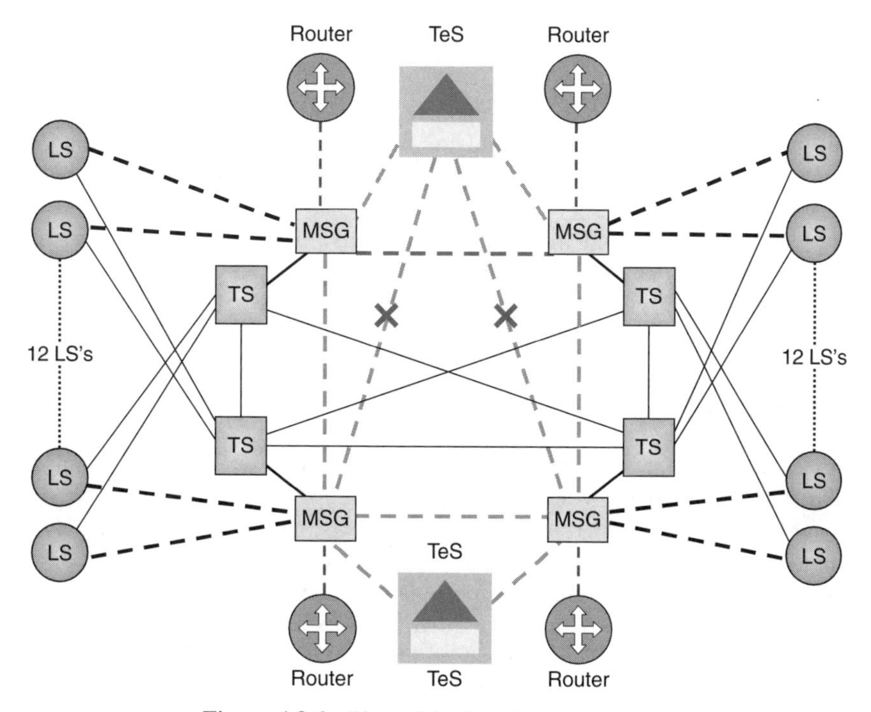

Figure A3.6 Phase 3 in the migration process

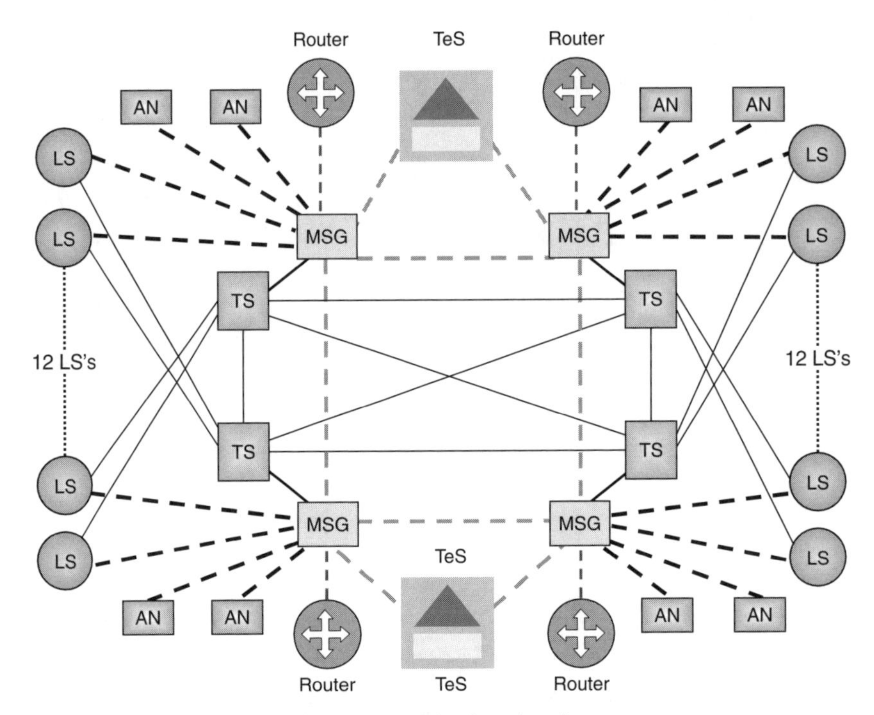

Figure A3.7 Phase 4 in the migration process

Phase 5

In the process of replacing routes to the tandem switches with routes to the MSGs, we finally get to a stage where the tandem switches are not needed any longer. They can actually be taken away, and maybe be used somewhere else in the network. See Figure A3.8.

Figure A3.9, where the tandem/transit switches have been removed, shows a network structure, which is very close to the vision, the packet-switched multi-service network. In the long run, the local switches will probably be replaced by access nodes.

MIGRATION FROM ATM TO IP

Today an ATM backbone network is a mature solution for voice traffic, keeping in mind that ATM was specified for voice from the beginning. It is a mature technology that gives roughly the same service quality as in the existing circuit-switched networks. It also allows managed provisioned bandwidth (e.g. for LAN interconnect).

However, an IP network could be a more natural choice from a data and multimedia service point of view. Thus it is important that an investment in an ATM network can be retained, if the telephony service is supposed to migrate from ATM to IP as a long-term goal, when the volumes and revenues for the data and multimedia services increase.

Let us take a look at a possible migration path from ATM to IP over ATM.

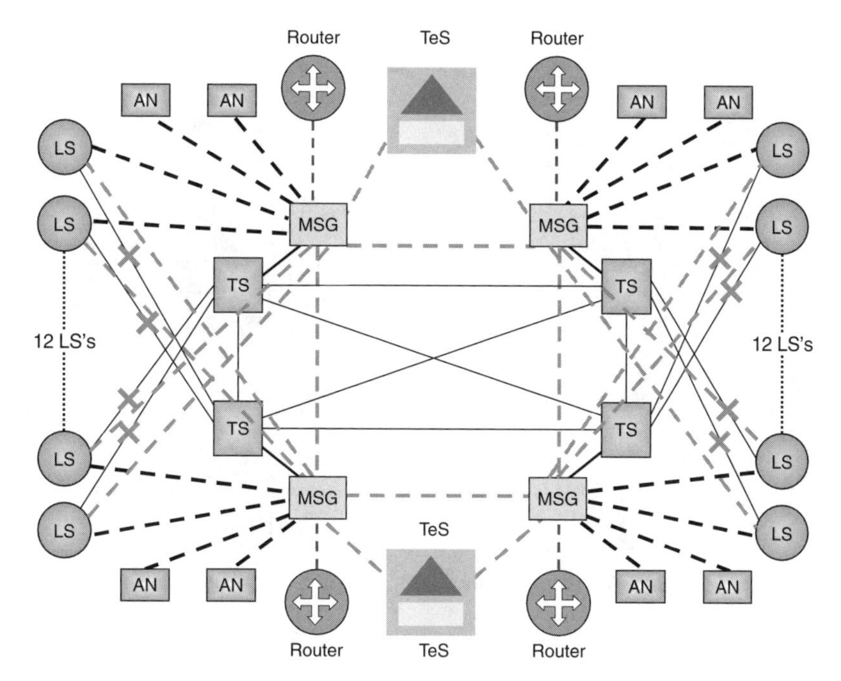

Figure A3.8 Phase 5 in the migration process

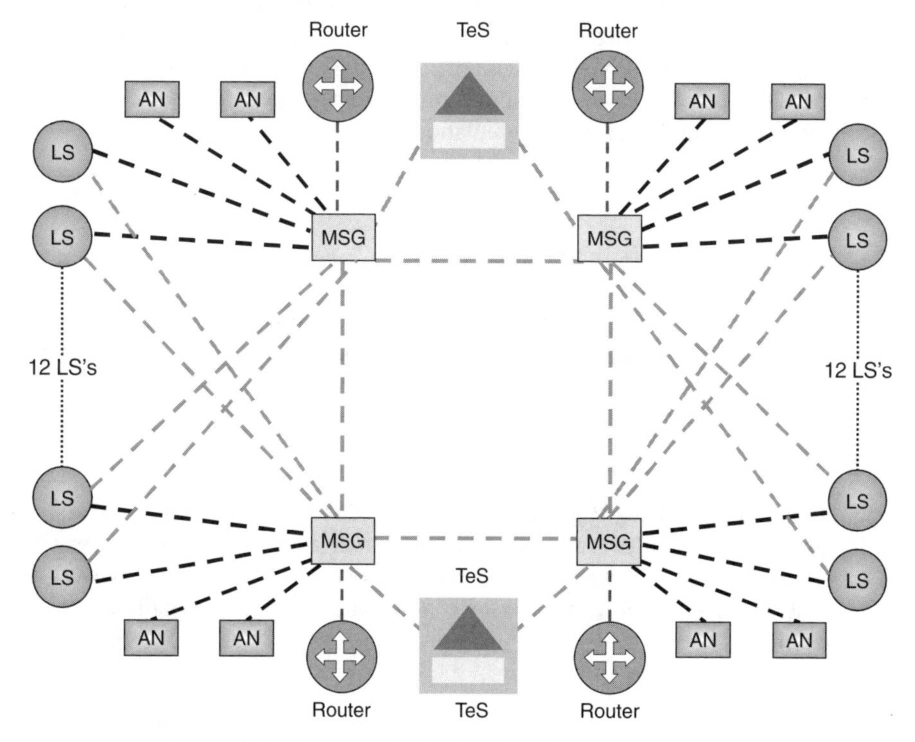

Figure A3.9 The result of phase 5 in the migration process

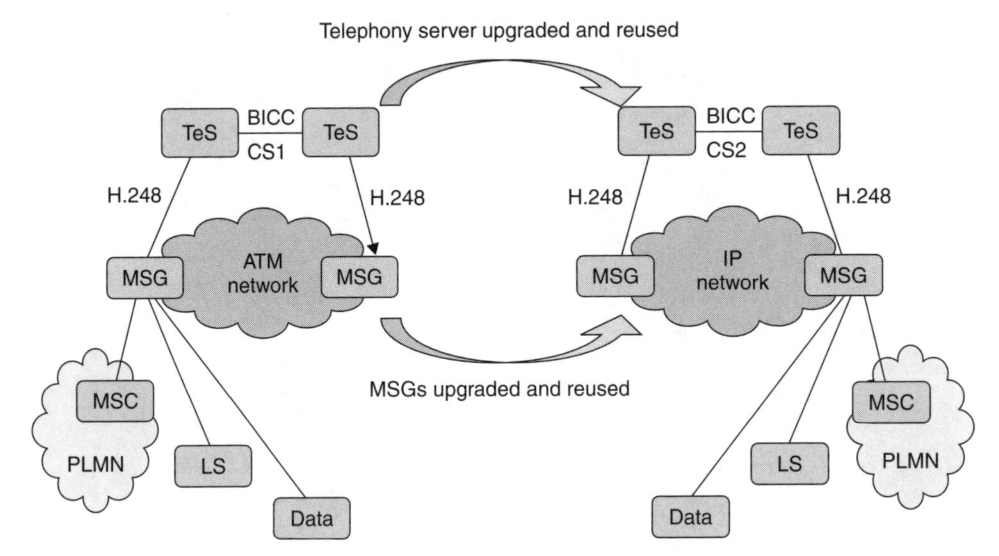

Figure A3.10 Migration from ATM to IP connectivity

Step 1

The signalling system BICC CS1 (bearer independent call control – capability set 1) is used as call control protocol between the telephony servers. The media gateway control is designed using the standardized H.248 protocol. The bearer control employs standardized ATM signalling (UNI, PNNI and A-INI). The signalling from the circuit-switched network (e.g. DSS1 or ISUP) is transported transparently through the ATM network to the telephony server via AAL1 circuit emulation.

Step 2

BICC CS2 will be used. The H.248 protocol will still be used for the control of media gateways from the telephony server. The basic applied principle is still that the call control and bearer control are separated in the network. Media gateways must convert the narrowband signalling information into IP packets and send them through the IP network. See Figure A3.10.

Upgrade Issues

The migration from ATM to IP connectivity can be done by means of some additional hardware, and change of software packages in the server and in the media gateways. The ATM backbone network can be reused to carry IP over ATM for voice as well as for data and multimedia services. Examples of changes are the BICC signalling and the IP address handling.

Appendix 4

Dimensioning Media Gateways and Associated Telephony Servers

MULTI-SERVICE NETWORK TRAFFIC ENGINEERING

A main difference between multi-service network (MSN) traffic engineering and engineering a traditional telephony network is that the MSN has to be dimensioned for a mixture of different types of traffic while the telephony network is dimensioned mainly for voice traffic. The fact that different types of traffic, e.g. voice and data, have totally different requirements on the network will of course have an impact on the dimensioning.

OBJECTIVES

The objectives of this chapter are to explain some general principles of dimensioning an MSN, and to give an example of a simplified dimensioning of an MSN, and in particular:

To dimension:

Interfaces: TDM circuit mode for voice and ATM for data and voice
Real-time processing capacity (CP): Telephony server and MSG

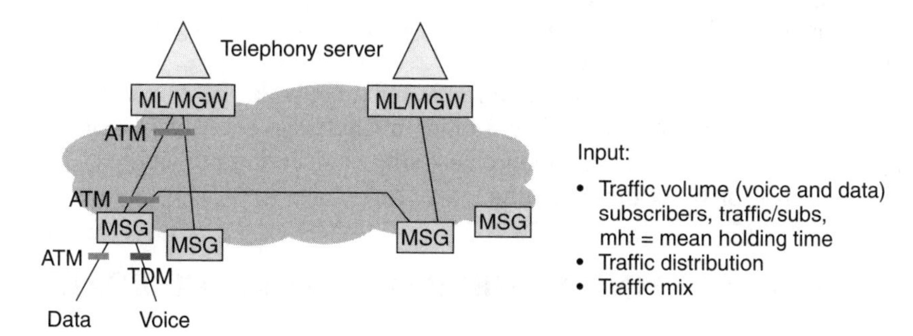

Figure A4.1 The studied area

Understanding Changing Telecommunications – Building a Successful Telecom Business. Edited by A. Olsson
 ISBN: 0-470-86851-1

- Calculate how many interface boards are needed in each multi-service gateway (MSG), based on the interface traffic from connected subscribers and data sources (Task 1).
- Calculate the MSG processor load (Task 2).
- Calculate the required processor capacity in the telephony servers (TeSs) (Task 3).

Figure A4.1 shows the studied area, which is served by a simplified ATM based MSN, with the subscriber network connections below the 'cloud'. Only ATM and TDM interfaces are indicated for simplicity, but a typical MSG has a number of other interface possibilities like frame relay, Ethernet and IP/MPLS (multi-protocol label switching).

The main components shown are MSGs and telephony servers with mediation logic (ML) and a MGW for switching signaling traffic, transit traffic (break in/break out to/from the area) and announcement machine traffic. The ML processor handles the gateway control communication, whereas the telephony server processor handles the normal 'horizontal' signalling. Within the cloud there might be a number of ATM core switches.

The TDM voice traffic is transformed into ATM by the boards at the TDM interface. This transformation from one transfer mode to another one is called emulation (CE) and the boards are called circuit emulation boards.

VOICE TRAFFIC VERSUS DATA TRAFFIC

The behaviour of the voice traffic is well known, and dimensioning voice networks using Erlang tables is a routine task that operators do regularly. Voice calls are normally randomly distributed in arrival time, enabling statistical calculations (such as Erlang formulas) to be used. The time arrival distribution is called a Poisson distribution.

The data traffic is more difficult to calculate and thus more rough estimates have to be done. The traffic is bursty with clustered arrivals. There might also be long range dependencies between the bursts. When carried in ATM we can use traffic expressions such as peak cell ratio (PCR) and average cell ratio (ACR) to measure the burstiness = PCR/ACR.

Voice and data have almost totally different QoS requirements for the end-to-end transfer. See Chapters 3 and 7. Voice is delay sensitive and loss tolerant, whereas data is error/loss sensitive and delay tolerant. Since ATM technology is designed for both traffic types it will be used in this appendix.

When we mix these two types of traffic in the same network, the dimensioning will of course become more complex than dimensioning a single-service network. However, we will show in the following that the traditional traffic engineering principles will apply to a large extent also for MSN dimensioning.

TASK 1: CALCULATING NUMBER OF INTERFACE BOARDS

Theory

Figure A4.1 also shows what type of input data we need in order to perform the dimensioning. For the voice traffic we need traditional input data such as number of subscribers, average traffic per subscriber during busy hour, and mean holding time.

We also need to know how the traffic is distributed, i.e. how much traffic is internal in an MSG, how much traffic stays within a domain (controlled by one TeS), how much traffic is going between two domains etc.

Local Switch Traffic Distribution

The traffic flow through a local switch is defined as *originating* and *terminating* traffic on the access side, and *incoming* and *outgoing* traffic on the network side. If it is a combined local and transit switch it can also handle *transit* traffic.

Traffic between subscribers connected to the same local switch is called internal traffic.

A typical traffic distribution (in percentage) is shown in Figure A4.2. The traffic is expressed in Erlang, which is the product of call intensity and mean holding time ($A = y \times s$).

Transit Switch Traffic Distribution

The required number of ports (input/output) in the group switch is related to the switched traffic and the utilization of the connected trunks. A typical utilization (u) of the trunks is 80 % (0.8 Erl/ch). As each call occupies two ports in the group switch, the number of ports in the group switch can be expressed as $A \times 2/0.8$.

The call intensity (y) can be calculated using the ordinary formula $A = y \times s$. See Figure A4.3.

Converting TDM Voice Traffic into Packets

To calculate the resources needed in the backbone packet-switched network, we have to be able to convert circuit-mode Erlang traffic in the access network into packet-mode Mbits/s in the backbone network. The preferred backbone packet modes are ATM or IP.

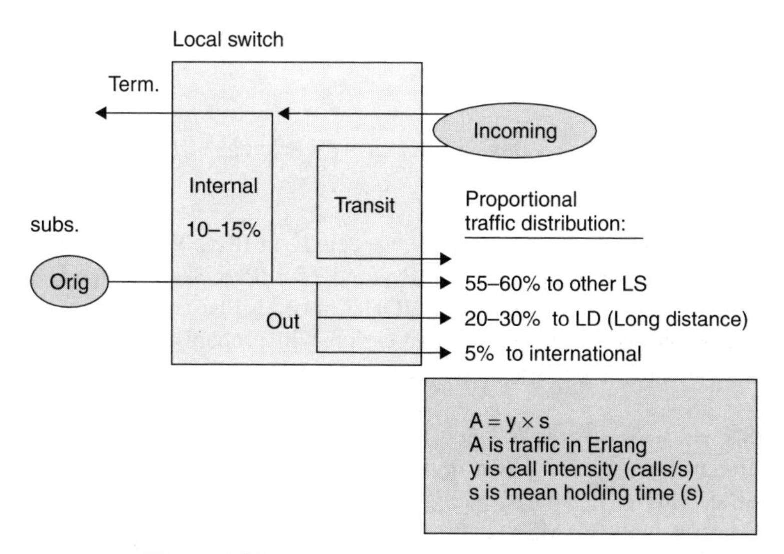

Figure A4.2 Traffic distribution in a local switch

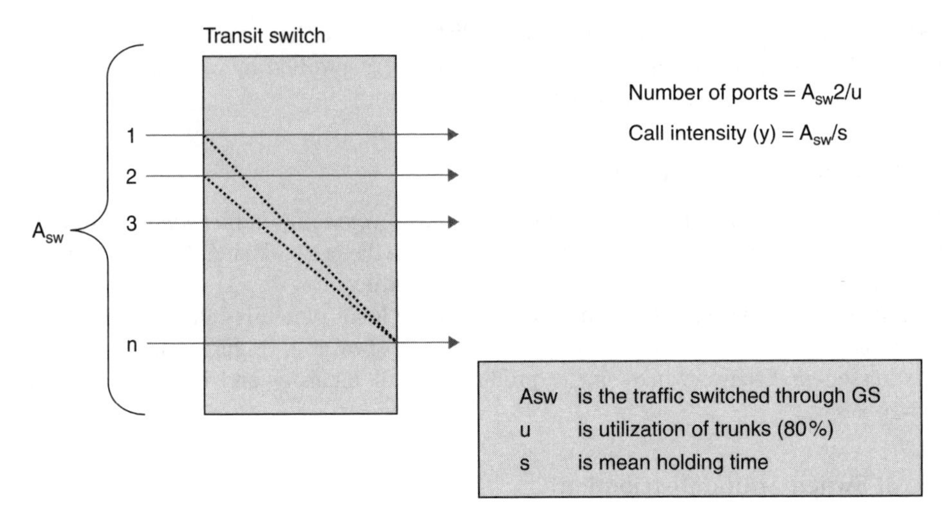

Figure A4.3 Traffic flows through a transit switch

ATM

ATM AAL1 is assumed, which means 64 kbit/s or 0.064 Mbits/s bandwidth per voice channel. As we have seen in the previous figures, the number of voice channels that we need in the circuit-switched network can be calculated by dividing the traffic (Erlang) with the utilization factor (u). Mapping (emulating) time slots into ATM cells means that we have to add a $5+1$ octets header information into each 53 octet ATM cell, which corresponds to roughly 13 % overhead.

Thus we get the conversion formula: Total traffic $=$ Traffic$/u \times 0.064$ Mbits/s $\times\ 1.13 =$ traffic$/u \times 0.0722$ Mbits/s.

When using ATM AAL2 the actual voice compression and overhead ratio must be considered.

IP

The same type of formula is valid also for an IP backbone network. The main differences are:

- The header in IP packets is much longer, especially in IPv6. When uncompressed, the overhead factor could be around 1.5 for IPv4 and 2 for IPv6. See Chapter 5, Section 5.7. If applied, header compression such as ROHC must also be considered.
- In an IP backbone network, voice compression will probably be used, and therefore the formula must include a voice compression factor.

In an MSN we will have both voice and data traffic. Typically voice and data traffic have different busy hours. When dimensioning, it is necessary to find the peak for the total traffic. This is illustrated in Figure A4.4, which also shows that we can save network resources with an MSN compared with having separate networks for voice and data.

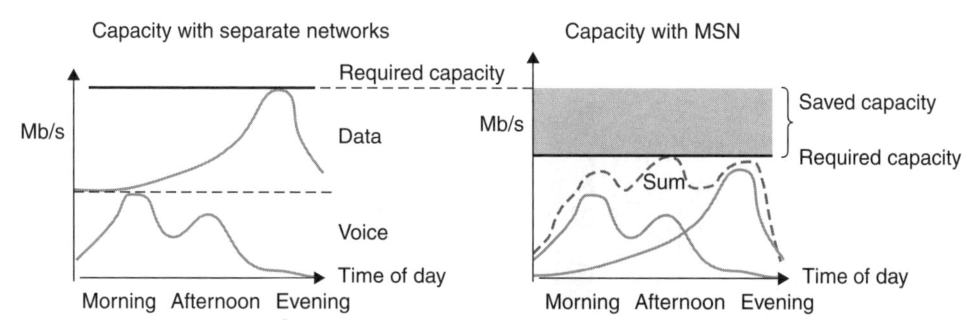

Figure A4.4 Reducing required capacity with a multi-service network

MSG Dimensioning

Let us now start looking at the dimensioning of the MSGs. This means to dimension the TDM/circuit emulation interfaces (voice traffic) on the access side, and the ATM interfaces on both the access side (data traffic) and the backbone side (voice and data).

An MSG is typically built in a modular way, where each module (or sub-rack) has a switching capacity of e.g. 10 Gigabits per second. One sub-rack is, for example, configured with 16 interface boards plus two processor boards (for reliability) and two switch boards (for reliability). See Figure A4.5.

The assumed *basic configuration* for an MSG sub-rack is defined as: two central processor boards, two switch core boards and two ATM boards (on the backbone network side). This leaves 14 slots for optional line interface boards, such as E1 CE, STM-1 CE or ATM boards.

Figure A4.6 shows some possible interface boards that can be used in an MSG.

The MSG will be equipped with a number of circuit emulation boards (E1 and/or STM-1), which are used as interfaces towards the TDM-based network.

The ATM part of the MSG should support the standardized ATM UNI 3.1 or 4.0 signalling. Also PNNI and AINI signalling should be supported, implying that an MSG can either be connected to the ATM backbone network or can be a part of it. The ATM

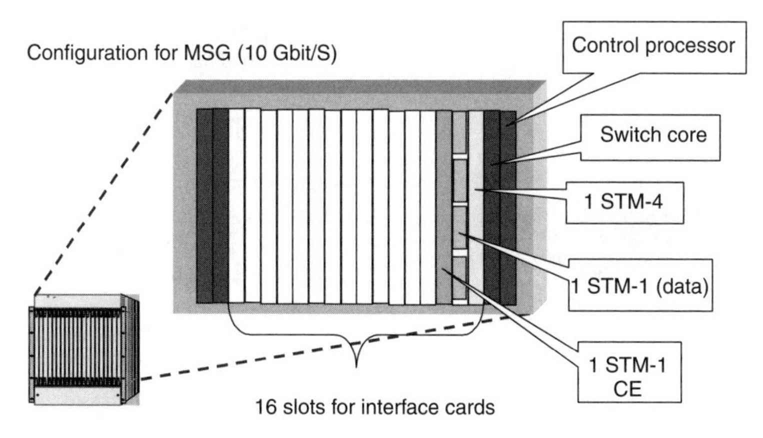

Figure A4.5 Typical configuration of a multi-service gateway

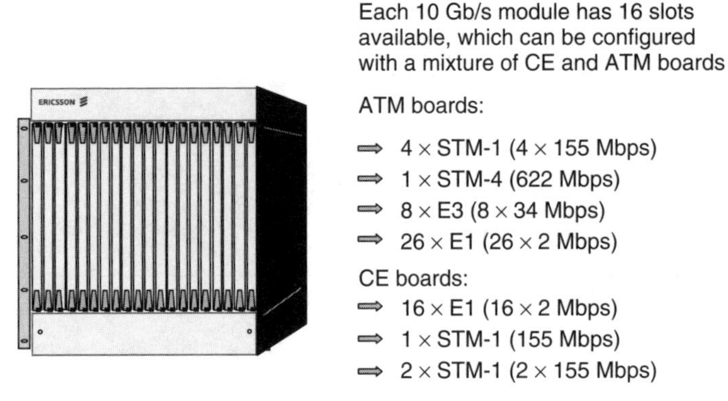

Figure A4.6 Possible interface boards in an MSG

backbone network consists of ATM core switches that support some basic requirements, such as setting up switched virtual circuits (SVCs) using the standardized signalling UNI 4.0 and PNNI 1.0. To calculate the required number of MSGs in a certain network solution the number of interface boards of different kinds must be examined.

TDM Interfaces and Circuit Emulation in an MSG

Let us start to calculate how many circuit emulation boards we need. We can use either interface boards with 16 E1 capacity, or we can use STM-1 interfaces with two assumed variants, either one or two STM-1 interfaces per board. In our example we will use the two alternatives 16 E1s or 1 STM-1 per board.

As input data we need to know the number of subscribers and the average traffic per subscriber, or as an alternative the number of E1s, if this information is available. For simplicity reasons we assume that the utilization of the E1s is 80 %. See Figure A4.7.

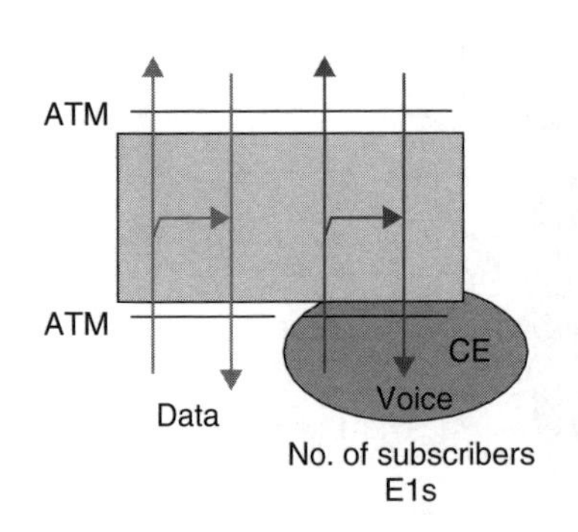

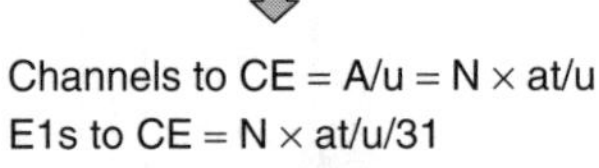

Figure A4.7 Calculation of the number of CE boards needed

If we do not know the number of E1s on the access side, we can easily calculate this by dividing the total access traffic with the utilization factor (80 %) and with the number of channels per E1 (31). Dividing this result by 16 gives us the number of circuit emulation boards. As an STM-1 trunk can carry 63 multiplexed E1s, we get the number of STM-1 interface boards by further dividing the number of E1s by 63.

If we do not have any data accesses, we can have maximum 14 circuit emulation boards connected to each sub-rack.

ATM Interfaces in an MSG, Access Side

Estimating the data traffic on the access side is quite a tricky task. We will here make some assumptions in order to show some basic principles. See Figure A4.8.

The interfaces for data traffic are usually STM-1, both on the access side and the network side. The data traffic is expressed in Mbits/s.

The payload in STM-1, for data traffic, is roughly 150 Mbits/s. 5 Mbits/s consists of overhead information (out of 155 Mbits/s). The utilization of STM-1 is approximately 90 %, so the dimensioning value of payload is $= 150 \times 0.9 = 135$ Mb/s.

Access side: we assume that the total data traffic is D Mbits/s. Dividing the total data traffic by the payload (135 Mbits/s) gives the number of STM-1s needed, and dividing the result by 4 ($4 \times$ STM-1 per board) gives the number of boards needed.

ATM Interfaces in an MSG, Network Side

On the network side the voice and data traffic are mixed together. Therefore the ATM interfaces on the network side are calculated by adding the traffic for voice (expressed in Mbits/s) and data. See Figure A4.9.

Network side:

Voice: if we assume that a certain portion of the voice traffic (pV) is switched internally in the MSG, the traffic that will continue out in the backbone network is $N \times at(1 - pV)$.

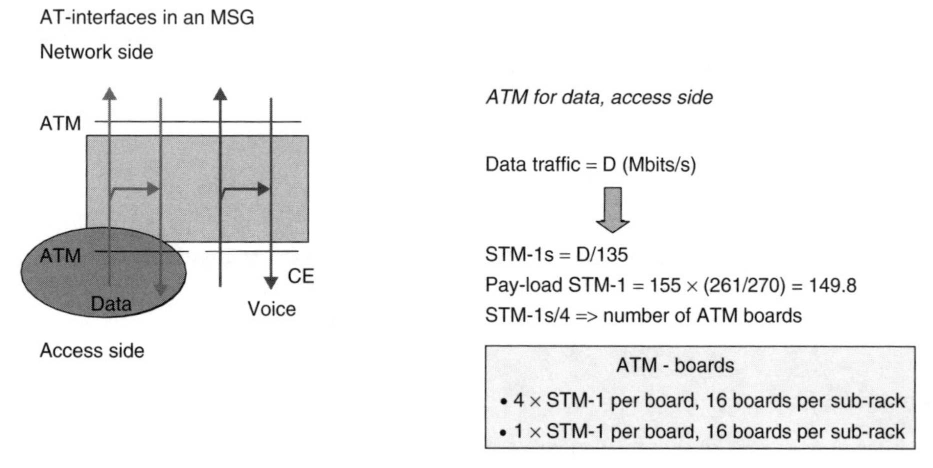

Figure A4.8 Calculation of the data traffic on the access side

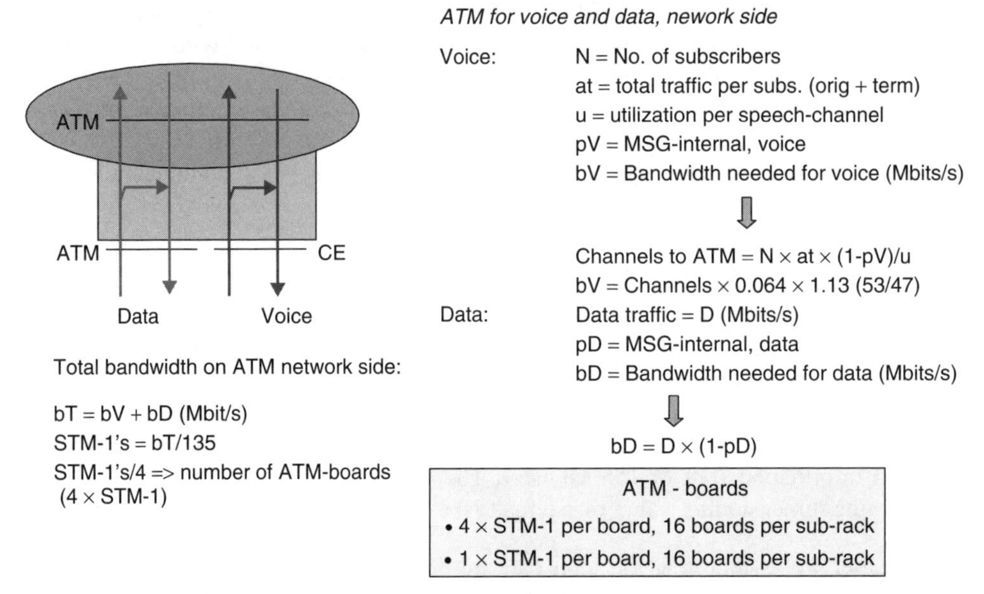

Figure A4.9 Calculation of the ATM interfaces on the network side

To calculate the corresponding number of channels interfacing the backbone network, we divide by the utilization factor $(u) : N \times at(1 - pV)/u$.

To convert the circuit-switched voice traffic, in Erlang, into packet-switched traffic (bV = bandwidth for voice), expressed in Mbits/s, we can use the formula: $bV =$ Channels × 0.064 × 1.13.

Data: if we assume that a certain portion of the data traffic (pD) will be switched internally in the MSG, the traffic that continues out in the backbone network can be expressed as $bD = D \times (1 - pD)$.

Note. As the voice and data traffic can be mixed in the backbone network, we add together the required bandwidth for voice and data $(bT = bV + bD)$. Dividing the total bandwidth (bT) by 135 (payload) gives the number of STM-1s, and dividing by 4 gives the number of boards (4 × STM-1) needed on the network side. Even if the result is one board, we should use two boards for reliability reasons.

CE/ATM Interfaces in an MSG – Summary

To calculate the number of sub-racks we need, we simply add together all the different interface boards we have calculated above and divide by 16. See Figure A4.10.

We have now calculated how many sub-racks we need, from an interface point of view. We also need to check the processor capacity in the MSG.

Numerical Example

In order to try to make things as clear as possible, we will illustrate the calculations with a numerical example. See Figure A4.11.

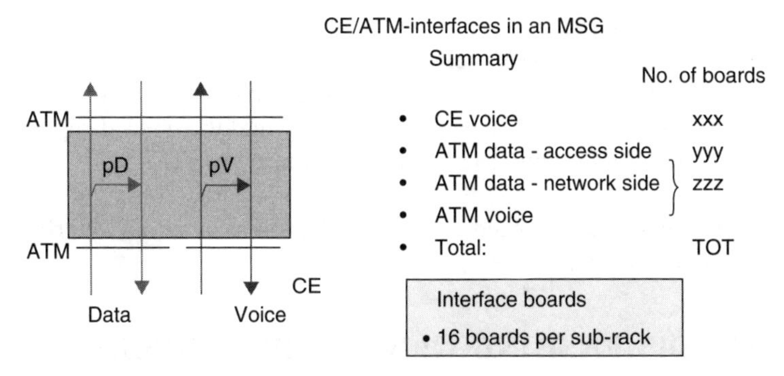

Figure A4.10 Calculation of total number of interface boards

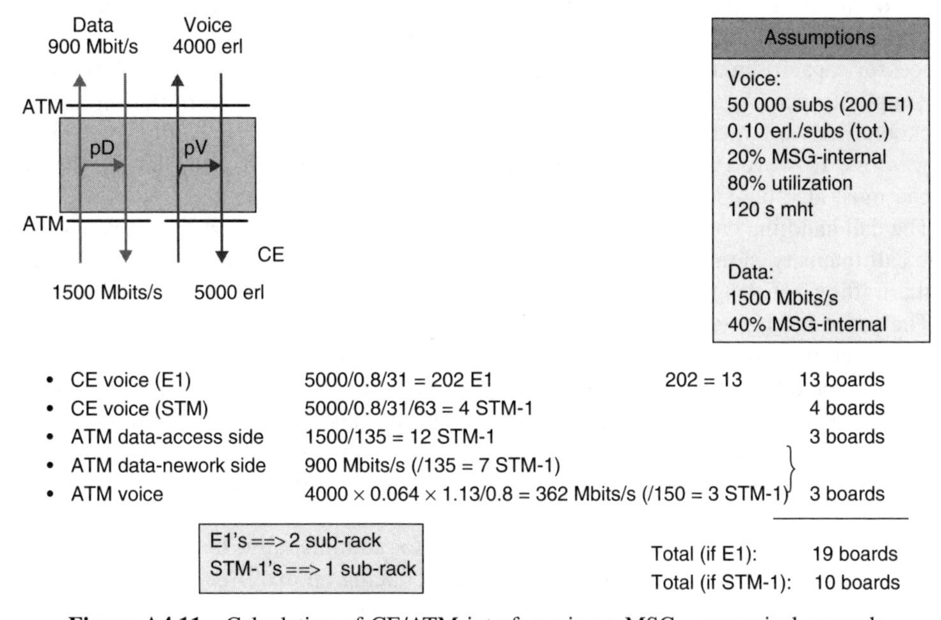

Figure A4.11 Calculation of CE/ATM interfaces in an MSG – numerical example

This figure shows a numerical example of calculating the number of interface boards in MSGs.

Voice traffic, access side:

The total voice traffic is $50\,000 \times 0.1 = 5000$ Erlang. Dividing the voice traffic by the utilization factor (0.8) and the number of speech channels (31) in a PCM-link (E1) gives the number of E1s needed.

If we are using STM-1 interfaces instead of E1 interfaces, we also divide by 63, which is the number of E1s that can be transferred in STM-1.

Data traffic, access side:

For the data traffic, we calculate the number of STM-1s by dividing the traffic with the payload (135 Mbits/s).

Voice + Data, network side:

For the voice (mapped into ATM cells) on the network side, we use the expression: traffic (in Erlang) × 0.064 × 1.13/0.8 in order to calculate the number of Mbit/s. As voice and data can be mixed in the ATM network, we simply add the needed capacity (Mbits/s) for voice and data and divide with 135 to get the number of STM-1s.

Result:

The result shows that we will need two sub-racks when using E1 interfaces, and one sub-rack when STM-1 interfaces are used.

PROCESSOR DIMENSIONING: BASIC PRINCIPLES

Goal and Methodology

To find out the number of telephony servers needed, we have to calculate the load on the processor (CP) and the ML/MGW in the telephony server. We also have to check the processor capacity in the MSG, which can be a limiting factor in some cases. For reasons of simplicity permanent virtual circuits (PVCs) are assumed for data traffic through the backbone network, which means that the data traffic will put very little load on the processors. However, when using data services that are switched using SVCs, the data traffic must also be considered when checking the processor capacity.

The call handling capacity of the processors is expressed in number of calls per second. The call intensity in the network can be calculated with the formula $A = y \times s$, where A is the traffic (in Erlang), y is the call intensity (calls/s) and s is the mean holding time.

The processor has to handle all the originating and incoming calls.

The overall possible load on a processor is normally expressed as a percentage of the processor total capacity. To estimate the loadability for traffic handling we have to take away the 'idle load' and the 'overload margin'. The 'idle load' is the part of the processor capacity that is used for running the operating system. The 'overload margin' is introduced to avoid instability problems that can occur if the processor is loaded close to 100 %.

A small part of the processor capacity is needed for administrative tasks like O&M. This is called usage load. Many operators are also introducing an extra security margin called 'dimensioning factor', in order to avoid problems in case of extreme traffic situations.

Typically, a traffic load of 90 % of the processor capacity is used for dimensioning purposes. Figure A4.12 illustrates the loads and margins.

Impact of Voice Traffic Mix

If we know how many milliseconds it takes for a processor to set up different types of voice calls, we can calculate the call handling capacity for a certain voice traffic mix.

As an example, we assume a traffic loadability of 90 % for the CP in the telephony server. Let's also assume that we have a traffic mix with 10 % IN calls and 90 % ordinary ISUP calls. To set up an IN call takes more processor capacity than to set up an ordinary ISUP call. Let's assume that an IN call takes 7 ms in the processor and an ordinary ISUP call takes 2 ms.

With the assumed traffic mix and corresponding processor load the average time it takes to handle a call with a mix of IN and ISUP is (0.1 × 7 ms + 0.9 × 2 ms) = 2.5 ms.

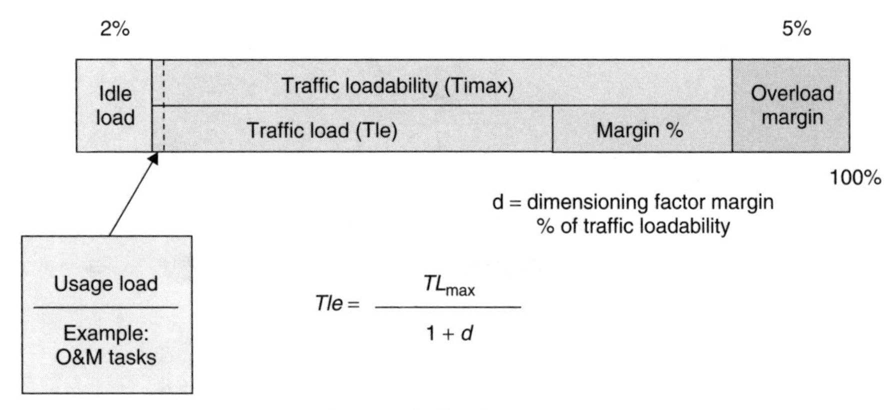

Figure A4.12 CP capacity

Since the processor has 900 ms per second available for traffic handling, the total number of calls/s the processor can handle is $900/2.5 = 360$ cps (calls per second). This corresponds to 1.296 million busy hour call attempts (BHCA).

Relationship between Group Switch Size and CP Capacity for a Telephone Exchange

Figure A4.13 shows an example of a transit switch with a 64 K group switch. With the assumptions shown in the figure, this group switch can switch roughly 25 000 Erlang traffic. In order to match this switching capacity, we need a processor that can handle approximately 200 calls/second, which corresponds to 720 000 BHCA.

With the assumptions that the MHT is 120 s, 200 cps corresponds to a total traffic load of 24 000 Erlang. ($A = y \times s$).

The total capacity of the switch is also given as the number of E1s that can be supported. In the example we can see that a 64 K group switch can connect approximately 2000 E1s, since an E1 has 31 speech channels. As a rule of thumb we can say that an E1 can carry roughly 24 Erlang of traffic (0.8 Erlang/channel).

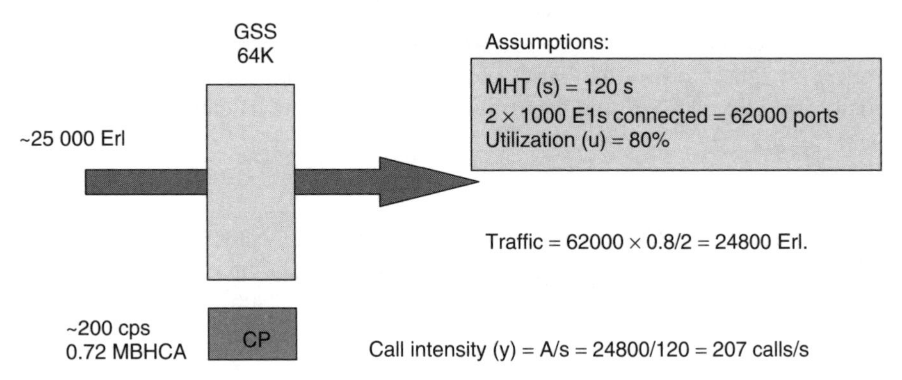

Figure A4.13 Example of switching and processing capacity in a transit switch

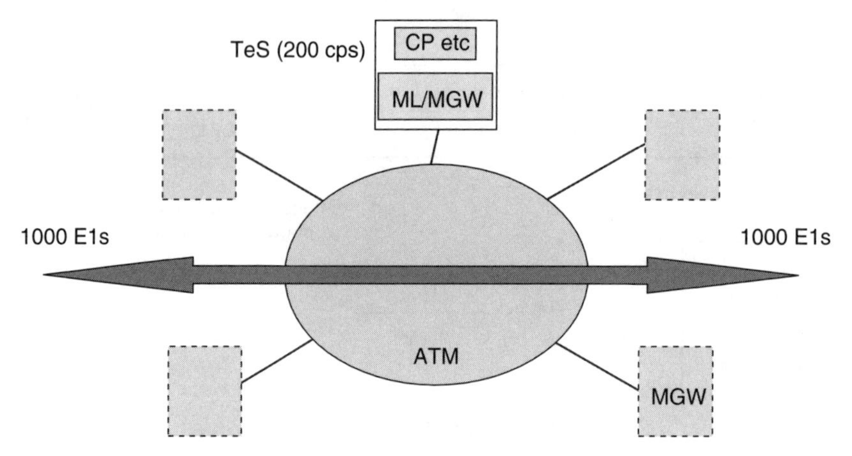

Figure A4.14 Capacity analysis of a telephony server domain

Relationship between a Virtual ATM Group Switch Size and Telephony Server Capacity

These basic principles for a telephony exchange also apply to a telephony server domain, where the group switch has been replaced by an ATM network.

Figure A4.14 shows a telephony server domain with distributed MGWs handling all the voice traffic. A MGW is here defined as the voice-handling part of an MSG. From the server view, the ports in the MGWs are seen as ports in a virtual group switch.

With 2×1000 E1s having 0.8 Erlang/channel connected to this 'virtual group switch', this corresponds to 24 000 Erlang. With a mean holding time of 120 s, 200 calls per second can be handled. This can be compared with a traditional transit switch with a 64 K group switch.

TASK 2: MSG PROCESSOR DIMENSIONING

Assumptions and Definitions

The processor in a typical MSG can handle approximately 55 calls/s in the MSG versions considered here. This means that we have to calculate the call intensity (calls/s) for the voice traffic. As mentioned, PVCs are assumed for data traffic through the backbone network with little load on the MSG processors. Therefore, as well as for reasons of simplicity, we will disregard the data traffic in the following calculations.

Voice

The processor in the MSG will handle all the originating calls plus all the incoming calls.

The assumed objective is that it shall be possible to handle 55 calls/s per call handler, where a call handler corresponds to one CP pair in one MSG sub-rack. A 10G MSG with two CP pairs can thus handle 110 calls/s. These values are valid for the reference call type, inter-MSG call and ISUP signalling.

In the calculation of the number of calls/s that the processor has to handle, we will use the originating traffic per subscriber. For simplicity reasons we assume that the originating and the terminating traffic are equal. Thus, if we have the total average traffic per subscriber available, we simply take half of this value as originating traffic.

We also assume that the outgoing and incoming traffic (on the network side) are equal. Then the incoming traffic can be expressed as: $N \times ao(1 - pV)$. See Figure A4.15. If we add the originating ($N \times ao$) and incoming traffic together, we get the formula: $N \times ao(2 - pV)$. If we then divide this traffic by the mean holding time we get the number of calls/s. The result of this calculation can then be compared with the value 55 calls/s for one processor pair, and we can conclude how many sub-racks we need, from the processor point of view.

Figure A4.15 includes an example where some assumed input data values have been inserted into the formula. The example shows that, with the assumed input values, the processor is not any limitation if the MSG is equipped with E1 interfaces. If STM-1 interfaces are used, the processor puts a limit at approximately four STM-1 interfaces per sub-rack for a MSG with one CP pair.

Furthermore, the maximum number of E1s will depend on the ratio between intra- and inter-MSG traffic.

Note: with higher mean holding time values, more STM-1 CE (or E1s) boards can be supported.

Access Signalling and Call Distribution Impact on MSG Workload

What has been shown so far is of course a simplified example. In reality the calculations are influenced to a large extent by factors like different access methods (ISUP, DSS1, V5.2 etc.) and the distribution of the calls (intra-domain, inter-domain, intra-MSG etc.).

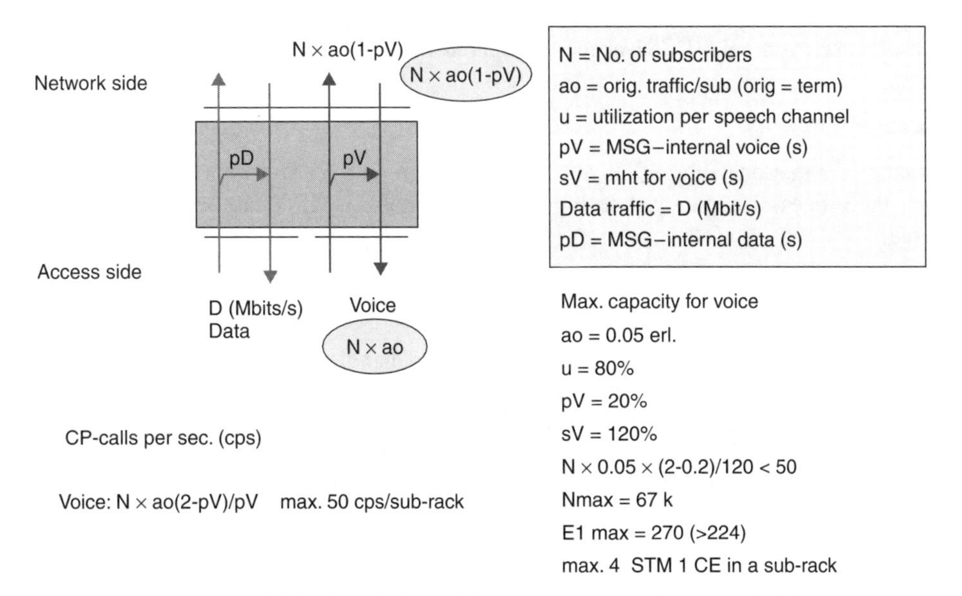

Figure A4.15 Calculation of the processor capacity in an MSG

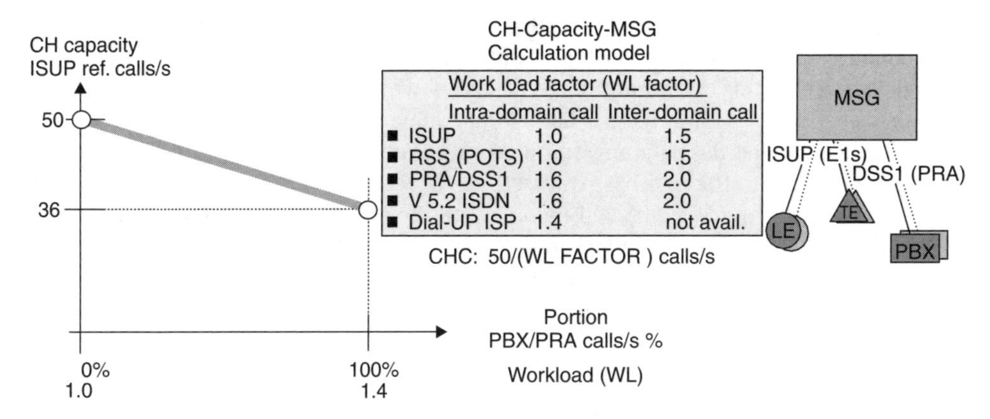

Figure A4.16 Example of work load factors

This is illustrated in Figure A4.16. A simple way to handle these differences is to define a reference type of call (ISUP inter-MGW intra-domain) and allocate different workload factors for the other types of call, in relation to the reference call type. In Figure A4.16 the work load factors for intra-MGW calls have not been included.

Different Access Types and Load Factors

The processor load, caused by the MGW application, on the call handler in the MSG, will depend on the amount and type of accesses (i.e. ISUP/PRA/V5.2 etc.) being connected. The assumed amount of traffic for each one of these access types will have impact on the CP load.

For example, the ratio between the number of messages needed for the set-up and call release phase for a PRA/DSS1 access compared to an ISUP access is approximately 1.6.

Processor Capacity

To make sure that one sub-rack is enough in the case with STM-1 interface, we must also check the processor capacity. See Figure A4.17. Note. A MSG can have more than one CP pair.

Voice

Regarding *voice* the processor in the MSG has to handle all the originating calls plus all the incoming calls. The number of calls per second (cps) can be calculated with the formula $A = y \times s$. So, dividing the traffic figures by the mean holding time (s) gives us the number of calls per second. The result of the calculation example in Figure A4.17 shows that one call handler has enough capacity to handle all the voice traffic.

Data

Regarding *data* it is assumed that the traffic is switched through the backbone network using PVCs, which means that the load on the processor in the MSG is very little.

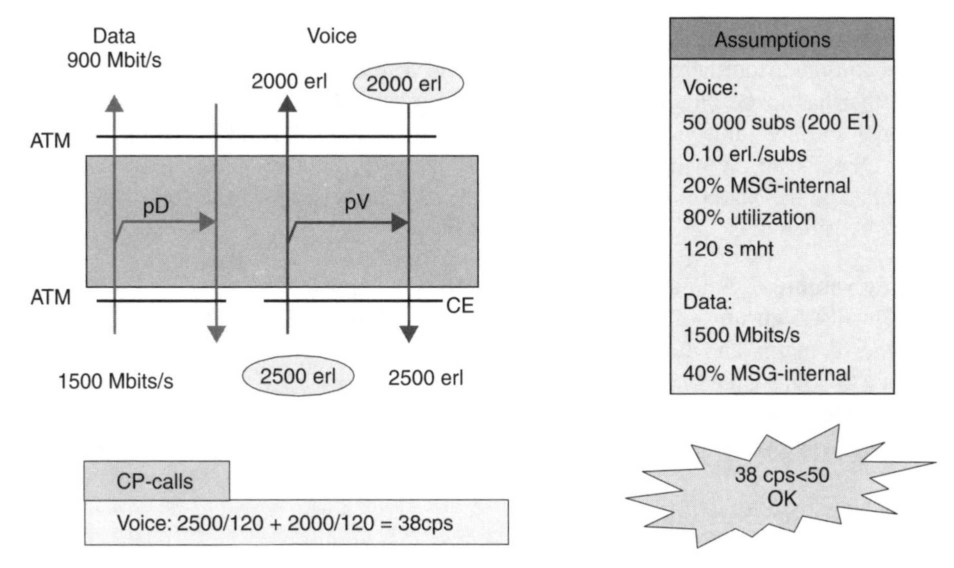

Figure A4.17 Processor capacity in an MSG – numerical example

TASK 3: TELEPHONY SERVER DIMENSIONING

Telephony Server Processors

In the telephony server there are two types of processors, both types having limitations when it comes to call handling. The central processor, exemplified in Figure A4.18, handles the normal signalling for setting up calls through the network, and the ML processor handles the communication with the involved distributed MSGs. Figure A4.18 shows an example of maximum number of ISUP calls per second that can be handled by the different processors.

If there is a mixture of IN calls and ordinary calls, as well as a mixture of accesses, the call handling figures will be lower. The maximum figures will also be reduced if there are calls including announcement devices and conference call devices.

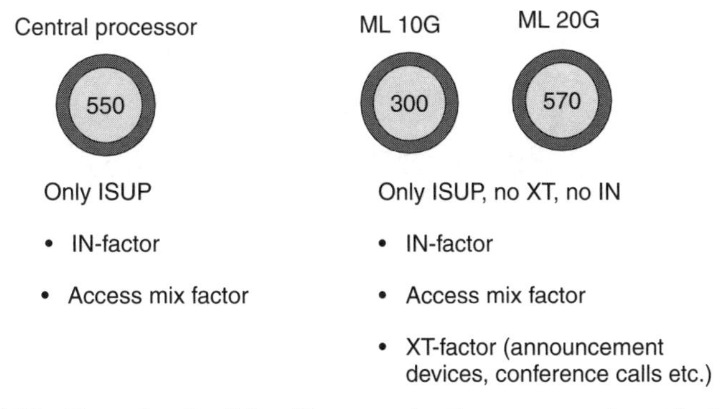

Figure A4.18 Example of call-handling capacity for processors in a telephony server

Input Data for Dimensioning

In order to dimension the telephony server we need to calculate the number of calls per second that has to be handled by the telephony server. The input data we need is the traditional traffic volume (in Erlang), the distribution of the calls and the complexity (e.g. portion IN calls) of the calls. See Figure A4.19.

To calculate the load on the processors in the telephony servers, the following factors must be considered:

- **Traffic volume –** same as for interface board calculations.
- **Traffic distribution –** how big portions of the traffic that goes to other networks, to other domains and how big part of the traffic that stays within the own domain (intra-MSG and inter-MSG).
- **Traffic mix –** how big part of the traffic is IN traffic (needs more complex signalling).
- **Type of processor –** different processors have different call handling capacity.

Impact of Traffic Distribution

As mentioned before, the processor has to handle all the originating calls plus all the incoming calls. See Figure A4.20.

With the assumptions that the traffic portion going to other networks is $p0$, the portion traffic going to other domains is $p2\ x(1 - p0)$ and the traffic staying in the own domain is $p1\ x(1 - p0)$, then the number of calls per second can be calculated with the formulas in Figure A4.20. For simplicity reasons, the intra-MSG traffic has not been considered in the example.

Numerical Example

Figure A4.21 shows by means of a numerical example how the calculation can be done.

In this example the total traffic in one domain is $200\,000 \times 0.1 = 20\,000$ Erlang. 10 000 Erlang is originating and 10 000 Erlang is terminating traffic.

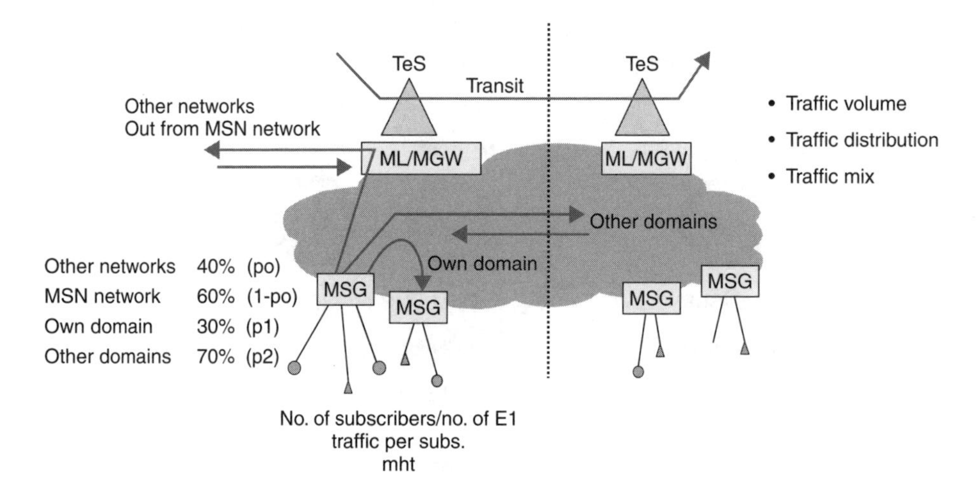

Figure A4.19 Call-handling capacity in a telephony server

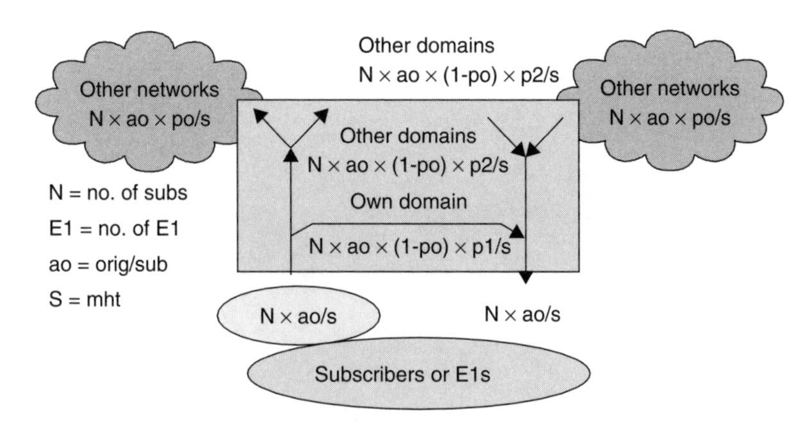

Figure A4.20 Call-handling in a telephony server

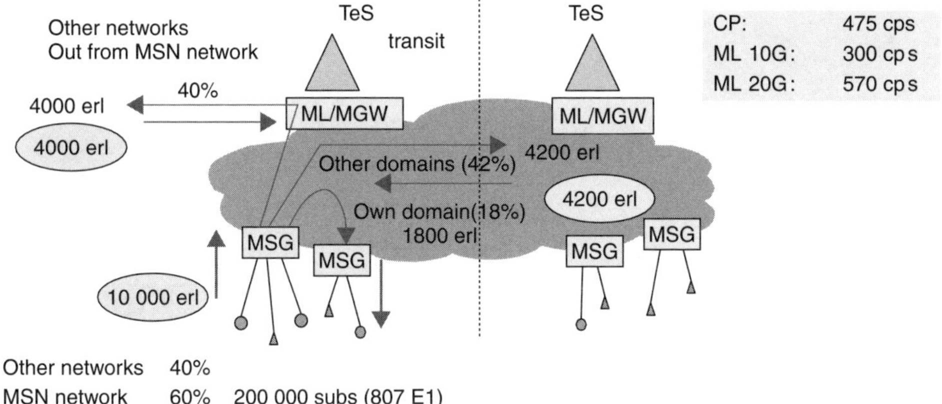

Figure A4.21 Calculation of the telephony server capacity – numerical example

The telephony server has to handle all the originating traffic in its own domain plus all the incoming traffic, i.e. traffic coming from other networks and from other domains. The result shows that with the assumptions made, one telephony server with ML (10G) can handle all the traffic from more than 200 000 subscribers.

Note: In this solution, we have assumed that there are two standard configurations of the telephony server. One configuration based on ML 10G, which can handle 300 cps, and one configuration with ML 20G, which can handle 570 cps.

The dimensioning examples described have been simplified in order to make it easier to follow. Impact of call mix, traffic between MSGs in the same domain, the different types of access and the mixture of voice and data traffic have not been included. In a real case, the work load factors for different types of traffic must of course be considered.

Index

0,1,2,3 solution 1, 295
1xEV-DO (data only) 373
1xEV-DV (data and voice) 373
3G 13, 14, 16, 42, 50, 83, 86, 130, 138, 144, 148, 156, 159, 169, 172, 183, 217, 265, 299, 354, 357, 371, 403, 426
3GPP/Parlay architecture 290
3GPP2 167, 367
4G 86, 172
802.11 198, 218, 219, 357, 379
  802.11a 380
  802.11b 86, 218, 379, 380
  802.11g 86, 379, 380
  802.1p 239, 240

AAA 88, 113, 186, 189, 197, 202, 203, 280, 295, 402, 407, 413, 419, 445, 451
AAL2 179, 239, 267, 273, 280–282, 325, 328, 335, 484
access node 323, 328, 360, 361, 364, 377, 396, 474, 476
access point name (APN) 256, 282, 388, 398, 418
accounting function 407, 450
accounting management 442
accounting rate of return (ARR) 466, 471
achievers 24, 25
ACR 482
ad hoc connections 382
ad hoc networking 97, 317
ad hoc networks 39, 306, 313, 315, 384
adaptation layer 237, 239, 328, 335, 340, 406
adding value (AV) 110, 120
address allocation 278, 280, 407
admission policy 240, 452, 454, 456
ADSL 87, 129, 138, 200, 223, 355, 357, 375, 377
advertising broker 116, 121
AF 346, 347
AH 214, 388
AINI protocol 341, 342, 485
AMR 79, 95, 168, 273, 280, 281, 320, 326
AMR wideband 168
API 50, 79, 80, 95, 199, 248, 287, 291, 292, 308, 315, 320, 386, 399, 444
APN 276, 419
APN address 276
application 12–14, 16, 30–32, 35, 39, 40, 42, 43, 45, 46, 50, 58, 60, 61, 63, 64, 68, 69, 71, 78–81, 83, 88, 89, 93, 95, 96, 110, 111, 114, 115, 121, 129, 139, 142, 151,

Understanding Changing Telecommunications – Building a Successful Telecom Business. Edited by A. Olsson

ISBN: 0-470-86851-1

application (*continued*)
152, 156, 157, 159, 161–165, 167–169, 172–175, 181, 184, 187–190, 194, 197–200, 204, 206, 208, 212, 217, 219, 222–224, 227–229, 231, 232, 234, 238, 244, 245, 247, 254, 259–263, 266, 271, 275, 276, 287, 289–297, 299, 301–304, 306, 307, 309, 311, 313–315, 317, 320, 326–329, 346, 348, 354, 382, 387, 390, 396, 405, 408, 416, 418, 435, 436, 438, 444, 445, 451, 460, 461
application programming interface (API) 79, 291
application service provider (ASP) 29, 100, 115
application support services 292, 293, 303, 304
AR 152, 173
ARPU 19, 105, 130, 154, 352, 354, 357
ARR 466, 467, 471
ASP 115, 129
assured forwarding (AF) 346
ASuS 260
ATM 1, 13, 56, 57, 60, 76, 77, 83, 91, 95, 117, 129, 141, 160, 177, 179, 183, 197, 215, 231, 232, 237, 239, 240, 245, 254, 255, 267, 268, 271, 273–275, 280, 303, 304, 317, 320, 322–327, 331, 332, 334, 335, 337, 338, 340–342, 347, 349, 354, 355, 360, 361, 363, 376, 386, 395, 396, 404, 453, 474–476, 478, 480, 482–488, 490, 492
ATM adaptation 239, 328
augmented reality (AR) 152, 173
augmented/virtual reality applications 173
authentication 186, 187, 190–192, 196, 199, 202–204, 207, 209, 212, 214, 217–219, 272, 277, 278, 280, 293, 296, 301, 307, 381, 390, 392, 399, 402, 419, 426, 444, 451
authentication header (AH) 214
authenticity 186, 199, 202, 209, 220
AV 111, 113, 116, 120
availability 12, 67, 130, 153, 161, 186–188, 193, 194, 200, 218, 295, 296, 351, 404, 440, 442
average cell ratio (ACR) 482
average revenue per user (ARPU) 19, 21, 100, 129

BA 344, 346
back end 303
background class 162, 164, 165, 229
balance 136, 166, 351, 465, 469
baseline cost 136, 137, 439, 464
bastion host 206
bearer signalling 336, 337, 341, 396
behaviour aggregates (BA) 344
BER 160, 161, 205, 235
best effort 256, 281, 348, 386
BGCF 388, 399, 400
BGP 53, 94, 343, 344
BICC (bearer independent call control) 94, 97, 268, 274, 281, 336, 387, 394, 396, 397, 406, 480
BICC CS1 (bearer independent call control – capability set 480
BICC CS2, 480
BICC IP Bearer Control Protocol (IPBCP) 385, 397
B-ICI 267, 268, 341, 342
billing 17, 53, 70, 71, 103, 118, 120, 121, 123, 132, 227, 228, 256, 272, 292, 293, 297, 327, 385, 388–390, 399, 426, 430, 440, 442, 444, 446–451, 458, 460
billing mediation 442, 447, 449
bit error rate (BER) 50, 159, 177, 223
bluetooth 39, 261, 306, 308, 311, 313, 315, 322, 353, 382–384
border gateway routing protocol (BGP) 343
break-out point 362, 419

broadband 14, 61, 65, 131, 143, 152, 166, 173, 183, 188, 322, 328, 335, 339, 340, 353, 354, 360, 375, 381, 396, 476
broadband access server 327, 360
BSP 32
Buddy locator 262

C++ 434, 436
CA 165, 214, 301
cable modem 339, 354, 374, 378, 455
cable modem termination system (CMTS) 378
caching 166, 244, 258
CAGR 131
call session control server (CSCF) 53
call state control function (CSCF) 399
CAMEL 250, 254, 267, 416
CAPEX, capital expenditure 106, 123, 136, 140, 144, 248, 297, 332, 339, 430, 438, 444
cash flow 124, 128, 463, 465–470, 472
CATV 63, 374, 378
CAU 359, 360
CC/PP 243, 244, 316, 454
CDMA 159, 364–367, 373, 374
  CDMA2000 167, 366, 373
  CDMA2000 1X 373, 374
  cdmaOne 367, 373
CDR (call detail record) 272, 327, 390, 450, 451
CE 308, 482, 485
certification authority (CA) 209
Churn 19, 47, 58, 134, 141, 202, 222
circuit emulation 320, 360, 482, 485–487
client 60, 61, 71, 79, 83, 89, 111, 115, 162, 163, 168, 170, 177, 199, 203, 211–213, 218, 238, 243, 244, 262, 272, 280, 287, 295, 305, 307, 312, 315, 349, 379, 391–393, 407, 435, 460
CMIP (common management information protocol) 432, 434
CMTS 378
code division multiple access (CDMA) 365
comfort noise 227
common business language, CBL 296
communicator 21, 28, 29, 157, 173, 307, 311
composite capabilities/preferences profiles 243
compound annual growth rate (CAGR) 131
confidentiality 47, 78, 186–188, 190, 198, 202, 207, 210, 212–214, 392
configuration management 442, 450, 453, 456, 457
connectionless 50, 52, 63, 210, 234, 242, 336, 354, 386, 396
content and application (C&A) provider 70, 109, 290, 444
content-based routing 258
control plane 51–53, 195, 248, 256, 267
convergence 12, 16, 49, 50, 56–61, 63, 65, 68, 70, 71, 75, 79, 80, 100, 105, 111, 119, 130, 144, 167, 183, 222, 309, 312, 321, 436
conversational class 160, 164, 271
CORBA – common object request broker architecture 80, 96, 198, 199, 434–436
corporate 38, 39, 42, 106, 126, 127, 136, 155, 164, 166, 199, 200, 203, 210, 211, 214–216, 218, 254, 275, 301, 327, 335, 339, 346, 356, 361, 367, 381, 440, 452, 463, 466, 468, 469
CPN 313, 314, 317
CR-LDP 348, 350
CRM 39, 103, 166
cryptography 207, 217
CSCF 53, 388, 390, 397, 399–402, 420, 421
customer access unit (CAU) 359, 360
customer care 19, 103, 105, 117, 120, 137, 141, 154, 430, 448, 449, 451, 458

customer premises network (CPN) 313
customer relations management (CRM) 28
customized applications for mobile network enhanced logic (CAMEL) 98, 250, 416
CWDM (coarse WDM) 335

D-AMPS (digital AMPS) 367
data offload 255
data over cable service interface specification (DOCSIS) 378
data warehousing 32, 449, 451
DCN (data communication network) 432, 433
dejitterization 179, 226, 234
delay and jitter 226, 232, 452
delay variation 159, 161, 177, 179, 223, 224, 226, 231
demilitarized zone (DMZ) 206
denial of service 187, 188, 193, 196, 218, 220, 267
dense wavelength division multiplex (DWDM) 331, 363
dependability 50, 67
DESP 32
DHCP 92, 94, 276–278, 456
diameter 203
differentiated services 237, 240, 242, 326, 344, 346
diffserv 53, 242, 245, 267, 326, 344–347, 417, 418, 453
diffserv code point (DSCP) 242, 344
diffserv, RSVP 94
dimensioning factor 490
directory server 407
discretionary cost 136, 138, 439
distance vector routing 343
DMZ 206
DNS 94, 164, 195, 199, 276, 277, 301, 393, 402, 407, 419
DNS protection 199
DOCSIS 378
domain name server (DNS) 277
domain name system (DNS) 195, 301
DRM 150, 151
DSCP 245, 344, 345, 418
DTMF 326
Du Pont model 471
DWDM 323
dynamic host configuration protocol (DHCP) 92, 278, 301

E.164 275, 276, 278, 361
echo 223, 225, 226, 282, 326, 396
e-commerce service 188, 263
EDGE 138, 182, 322, 356, 369, 370
edge node 83, 84, 89, 199, 240, 253, 254, 267, 319–322, 328, 334, 336, 352, 360, 361, 363, 368, 377, 386, 395, 399
edge routers 216, 443, 452
EDI (electronic data interchange) 296, 434
EF (expedited forwarding) 346, 347
EGP 342
egress LER 347
electronic signatures 209
ELL (emulated leased line) 346
empowerment 34, 35
EMS 158, 311
EMSP 32
enabler 11, 14, 44, 50, 59, 79, 89, 94, 110, 127, 130, 184, 190, 198, 201, 210, 215, 220, 222, 226, 237, 240, 260–264, 287, 289, 292, 293, 297, 299, 313, 315, 316, 323, 353, 390, 417, 430, 444
encapsulating security payload (ESP) 214
encryption 30, 91, 96, 172, 191, 192, 196–199, 201, 204, 207–209, 214, 215, 217–220, 227, 267, 308, 327, 451, 452
enhanced message service (EMS) 158, 261
enhanced telecom operation map 102
enterprise resource planning (ERP) 39
ERP 39, 286
ESP 32, 214

Ethernet 57, 95, 145, 191, 313, 337, 340, 347, 354, 355, 360–363, 376, 377
eTOM (enhanced telecom operation map) 102, 103, 438
extensible markup language (XML) 81
exterior gateway routing protocol, (EGP) 342

fast active queue management scalable TCP/IP 236
FAST/IP 236
fault management 44, 441, 445, 448
FDD 366
FDMA 364
feature phone 309, 311
FEC 347
federated applications 301
fibre switching capability (FSC) 336
fibre to the home (FTTH) 376
file transfer, access and management 432, 434
firewall 29, 30, 32, 189, 196, 199, 205–207, 213, 218, 293, 301, 315, 452
flexible numbering register (FNR) 276
FNR 276, 402
forwarding equivalence class (FEC) 347
FP 50, 66, 68, 71, 72, 79, 91, 94, 219, 220, 248, 254, 258, 259, 266, 267, 381, 385, 388, 389, 407, 413
framework 4, 59, 70, 75, 79, 123, 258, 289, 290, 293, 295–297, 416, 444
frequency division duplex (FDD) 366
frequency division multiple access (FDMA) 365
front end 303
FTAM – file transfer, access and management 433, 434, 456
FTTH 376
FTTx 376
fundamental technical plan 4, 50, 65, 66, 74, 133, 165, 188, 194, 258, 267, 413, 416, 430, 435, 457

G.723.1 Algebraic CELP 169
G.726 ADPCM (Adaptive Differential PCM) 169
G.728 Low delay CELP 169
G.729 Conjugate structure CELP 169
G.Lite 375
gatekeeper 255, 272, 387, 391
gateway control protocol, GCP/H.248, 53, 394
gateway GPRS support node (GGSN) 278, 368
gateway location 407, 413
gateway tunnelling protocol-control (GTP-C) 398
generalized MPLS, GMPLS 339, 348
generalized multi-protocol label switching (GMPLS) 83, 97, 231, 317, 339, 348–350
Generic Route Encapsulation (GRE) Protocol 211
GEO 382
geostationary orbit satellite 382
GGSN 217, 218, 245, 250, 280, 282, 283, 368, 386, 397–399, 402, 418, 419, 421
GIF 59, 168
GIX 423
global interconnect points (GIX) 423
GMPLS 83, 231, 339, 348–350, 453
GMSK modulation (Gaussian minimum shift keying) 369
GPRS 11, 14, 16, 50, 54, 111, 148, 159, 169, 170, 175, 177, 182, 199, 217, 256, 260, 261, 265, 276, 322, 325, 367–371, 381, 385, 387–389, 397, 413, 420, 424, 436
GPRS roaming exchange (GRX) 424
GPRS tunnelling protocol (GTP) 217, 399
GRX 424
GSM 54, 56, 63, 64, 137, 143, 159, 182, 183, 198, 223, 226, 227, 230, 250, 260, 301, 312, 322, 354, 365–372, 389, 398, 416, 423, 436
GSM codecs 169

GSTN, general switched telephone network 403
GTP (gateway tunnelling protocol) 217, 218, 282, 283, 325

H.248 50, 89, 90, 97, 268, 274, 320, 387, 480
H.263 168
H.323 79, 90, 97, 238, 267, 269, 271, 281, 328, 354, 358, 387, 391, 392, 403, 404
HDSL 134, 357, 376
header compression 14, 75, 79, 166, 236, 258, 484
HE-VASP (home environment-value-added service provider) 110, 286, 290, 417, 445
HFC 378
higher radio access network 361
HLR 56, 165, 245, 276, 278, 280, 394, 398, 402
home environment 110, 290, 291, 416, 417, 445
home location register (HLR) 165, 266, 303, 402
home subscriber server (HSS) 386, 399
HSP 32
HSS 399, 400, 402, 413, 419–421
HTTP 61, 88, 89, 94, 96, 169, 170, 198, 204, 206, 214, 238, 243, 269, 275, 294, 392, 434, 461
hybrid fibre-coaxial (HFC) 378

IBSP 32
ICMP 206, 220
idle load 490
IGMP 53, 94
IGP 342, 343
I-mode 36, 121, 156, 158, 172
IMS 78, 354, 362, 387, 397
IN 56, 79, 83, 90, 95, 98, 112, 113, 165, 250, 258, 386, 390, 416, 447, 490, 495, 496
IN (intelligent network) 12, 397
IN services 56, 112, 254, 390
incoming traffic 493, 497
incumbent operator 118, 133, 254, 255, 474
ingress label edge router (ingress LER) 347
integrity 45, 47, 78, 160, 186–188, 192, 199, 204, 207–209, 212, 214, 218–220, 289, 392, 451, 452
intelligent network services 54, 71
interactive class 162, 163, 165, 229, 274
interior gateway routing protocol, (IGP) 342, 343
internal rate of return (IRR) 145, 463, 466, 469, 470
international mobile telecommunications-2000 (IMT-2000) 366
Internet control message protocol (ICMP) 53
Internet offload 255, 327, 474
Internet protocol security (IPsec) 92, 212
Internet service provider (ISP) 29, 110, 111, 227, 277, 327
interrogating CSCF (I-CSCF) 400, 401, 421
intrusion 32, 187, 206, 218, 267
IP 11, 13, 18, 30, 44, 47, 53, 54, 56, 58, 60, 61, 63, 65, 68, 75, 77, 78, 81, 88, 91, 92, 94, 95, 100, 106, 111, 116, 118, 131, 145, 160, 161, 169, 171, 174, 175, 177, 179, 180, 187–189, 195, 197–200, 203, 206–208, 210–216, 220, 223, 224, 226, 227, 230–234, 237, 238, 240–242, 245, 249, 250, 254–256, 258, 268, 270–272, 275, 276, 278, 280–282, 286, 289, 290, 296, 299–301, 322, 325–327, 331, 332, 334–338, 340, 342–347, 349, 350, 352–355, 359, 360, 362, 367, 368, 376, 377, 379, 385–387, 390–393, 395, 396,

398, 399, 402–406, 411, 418, 419, 424, 426, 432–434, 436, 440, 442, 443, 445, 450–453, 455, 474, 475, 478, 480, 483, 484
IP multicast routers 329
IP multimedia subsystem 78, 385
IP over ATM 106, 243, 335, 340, 347, 480
IP routing 94, 342
IP-Centrex 254, 271
IPsec 196, 198, 199, 208, 210, 212–215, 218–220, 231, 235, 236, 263, 301, 327, 452
IPv6 50, 76, 91, 92, 94, 210, 212, 248, 267, 275, 278, 361, 484
IRR 145, 469, 470
ISP 111, 113, 121, 134, 199, 203, 211, 215, 217, 218, 227, 236, 277, 287, 288, 327, 343, 344, 346, 413, 456

JAIN – Java API for integrated networks 95, 387
Java 78–80, 87, 95, 97, 198, 206, 254, 261, 290, 295, 307, 313, 315, 317, 436
JDBC (Java data base connectivity) 436
JPEG 95, 168
JPEG pictures 12

L2TP (layer 2 tunnelling protocol) 210, 215, 325
label edge routers (LER) 243, 329
label switched path (LSP) 243, 347
label switching routers (LSR) 347
lambda light-wave (wavelength) switching capability LSC 349
LDAP (lightweight directory access protocol) 202, 204, 277, 434
LER 243
lightweight directory access protocol (LDAP) 202, 204, 277, 434
link state routing 343
liquidity 465
LMDS 340, 361, 374, 382
loadability 490
local multipoint distribution system (LMDS) 357, 382
low earth orbit bandwidth 382
lower radio access network 360, 361, 382
LSC 349
LSP 243, 347–350
LSR 329

M2M (machine to machine) 17
M2PA 405
M3UA 406
machine input function (MIF) 308
machine output function (MOF) 308
management 5, 30, 32, 49–51, 53, 63, 65, 66, 68, 72, 74, 75, 79, 80, 83, 88, 89, 94, 101, 102, 122, 123, 127, 132, 135, 137, 142, 145, 158, 164, 183, 184, 195, 197–200, 209, 218, 243, 266, 267, 276, 280, 287, 289, 295, 297, 299, 326, 331, 336, 339, 348–350, 360, 374, 381, 398, 413, 426, 429–436, 438–448, 451–458
management gateway 456
management information base (MIB) 434–436
management system 51, 183, 184, 187, 199, 248, 301, 303, 364, 430–432, 435, 436, 438, 440, 449, 451, 455, 456
MAP control signalling 398
materialists 24, 25, 40
MCS (multimedia core system) 399
media gateway (MGW) 12, 65, 68, 89, 90, 217, 218, 243, 268, 274, 319, 320, 322, 326, 352, 360, 363, 386, 396, 399, 402, 403, 421, 443, 480
media gateway control protocol (MGCP) 387
media stream 12, 89, 151, 238, 320, 325, 326, 393
Megaco 89, 90, 387, 399

message 35, 53, 71, 158, 159, 163, 165, 190–193, 199, 204, 207–209, 227, 261, 263–266, 276, 294, 296, 301, 303, 311, 385, 392, 393, 397, 399, 403–405, 418, 420, 421, 460, 494
message transport protocol 2 – peer-to-peer adaptation (M2PA) 406
messaging 21, 32, 34, 35, 41, 42, 130, 131, 151, 152, 156, 158–160, 165, 172, 249, 254, 255, 260, 265, 266, 309, 311, 326, 444, 445
messaging class 165, 264
messaging services 18, 32, 160, 165, 262, 444
MExE 290, 294, 295, 303
MGCF 388, 399, 400, 402, 421
MGCP 90
MGW 245, 246, 281–283, 320, 322–327, 329, 338, 348, 386, 387, 394, 396, 403, 406, 422, 452, 474, 482, 490, 492, 494
MIB 435
middleware 14, 68, 79, 81, 88, 148, 166, 248, 254, 258, 287, 289, 411, 444
MIF 308
MIPv6 92
MIT – management information tree 434
MM 61
MMS 41, 97, 131, 155, 156, 158, 159, 165, 255, 261, 262, 264, 265, 311, 424, 425, 444, 445
  MMS interconnect settlements 424
  MMS relay 265
  MMS roaming settlements 425
  MMS server 265, 266
  MMS user agent 266
  MMS user databases 265, 266
MO – managed objects 434
mobile enterprise 39, 286
mobile execution environment (MExE) 295
mobile IP 94, 210, 267, 278
mobile media 58, 70, 71, 121, 130, 312
mobile media (MM) 157
mobile virtual network enabler (MVNE) 121
mobile virtual network operator (MVNO) 110, 117
mobile web services 289
mobility agent 407
mobility management 67, 92, 389, 398, 413, 426
  handover 389
  roaming 389, 413
MOF 308
MP-3 – MPEG-1, layer 96
MPEG 83, 95, 270
MPEG video 12
MPEG-4 79, 168, 169, 254, 258
MPLS 94, 177, 180, 237, 240, 242, 243, 245, 255, 259, 262, 317, 322, 337, 344, 346–349, 418, 453, 454, 482
MPLS-LER (multi-protocol label switching-label edge router) 325
MRF 388, 399, 402
MSC server 281, 389, 394, 398
MSG 474–476, 478, 482, 483, 485–490, 492–497
MSN 481, 482, 484
MSP 32
MSSP 32
multimedia 2, 5, 12, 17, 21, 30, 33, 42, 46, 57, 59, 61, 65, 68, 78, 79, 91, 96, 111, 119, 138, 139, 150–153, 156, 160–162, 172, 173, 222–224, 228, 230, 244, 262, 265–267, 269, 280, 281, 289, 292, 311, 320, 322, 328, 329, 332, 333, 353, 354, 367, 368, 382, 385, 387, 388, 391, 392, 396, 399, 401, 402, 452, 473, 474, 478, 480
multimedia message service (MMS) 18, 159, 261, 265
multi-protocol label switching (MPLS) 53, 177, 242, 326, 329, 337, 347

multi-service gateways (MSG) 13, 322, 474
multi-service network (MSN) 58, 71, 139, 474, 476, 478, 481
MVNE 121
MVNO 117, 121, 286

name server 407
Napster 17, 44, 61, 313
narrowband access servers 327, 328
NAS 200, 203, 215, 216, 255
NAT 91, 94, 206, 275, 277, 362
NE agent 432
net present value 463
net present value (NPV) 126, 145, 465, 466
network access server (NAS) 200, 301
network address translation (NAT) 91, 206, 275, 301
new generation operation support systems (NGOSS) 103
next generation network (NGN) 13, 85, 103
NGN 14, 136, 371
NGOSS 103
NMC (network maintenance centre) 432, 433
NMT 357
noise 161, 202, 222, 223, 225, 227, 378
non-repudiation 186, 193, 209, 214, 218
NPV 100, 468–470, 472
numbering plan 67, 94, 210, 275, 388

OMA 12, 262, 288–290, 294, 296, 387
OMA service enablers 260
OMC (operation and maintenance centre) 432, 433
open market alliance (OMA) 12, 288
open service architecture (OSA) 14, 80, 290
open shortest path first (OSPF) 53, 243, 343
OPEX, operational expediture 106, 123, 136, 138, 140, 248, 297, 332, 339, 430, 438, 444, 451
originating traffic 493, 497
OSA 289–292
OSPF 94, 268, 343, 348
OSPF-TE 348, 350
OSS (operation support system) 198, 432, 434, 435
OTA 294, 295, 316
outgoing traffic 483, 493
overload margin 490

P2C 70, 71
P2P (peer to peer) 17, 71
packet classification 345, 346
packet switching capability (PSC) 336
parlay 80, 95, 96, 287, 290, 387, 400
  parlay SCS 260
  parlay/OSA 260
particular (fundamental) technical plans (PTP) 68
pay as you grow 134, 142
payback method 467
payload compression 258
payment transaction handler 116, 121, 177
PCM G.711 169
PCR 482
PDA 31, 166, 168, 234, 244, 311, 312, 379, 382
PDC (personal digital communication) 276, 367
PDP (packet data protocol) context 175, 256, 258, 280, 388, 398, 399, 418
PDP context activation 245, 280, 282, 418
peak cell ratio (PCR) 482
peer-to-peer (P2P) 61, 70, 71, 89, 90, 111, 177, 262, 312, 313, 375, 415
peering 134, 174
performance management 44, 442, 446, 452, 454
per-hop behaviour (PHB) 242, 344

permanent virtual circuit (PVC) 327, 342, 490
personal service environment 290, 296, 444
personal service environment management (PSEM) 296, 299
person-to-content (P2C) 70, 71
person-to-person (P2P) 31, 70, 71, 332, 425
PGP 208
PHB 344–347
piconet 315, 384
pioneers 23, 24, 40, 43
PKI (public key infrastructure) 205, 209, 214
plug and play 97, 317
PNNI 267, 341, 342, 442, 480, 485, 486
  PNNI routing protocol 342
  PNNI signalling 341, 342
  PNNI signalling protocol 342
POI 132, 456
point of interconnect (POI) 74, 132, 413, 456
point-to-multipoint 267, 281, 342, 361, 382
point-to-point 210, 211, 213, 214, 281, 335, 337, 342, 360, 361, 382
policing 240, 241, 267, 345, 347
policy function 407
portal 28–30, 100, 113, 114, 117, 121, 154, 156, 157, 287, 289, 295, 299, 301
positioning 21, 34, 45, 69, 70, 95, 119, 151, 160, 164, 165, 220, 254, 260–262, 303, 311, 444
positioning application 303
PPP/SLIP 363
pretty good privacy (PGP) 197, 207
processor capacity 488, 490, 494
proxies 244, 301
proxy CSCF (P-CSCF) 401, 420
proxy firewall 206, 275, 400
PSC 336
PSEM 296, 299
PTP 68
public key infrastructure 209
PVC 115, 117, 342, 492, 494

Q.931 387, 392, 396
Q3 interface 432, 434, 440
QAMS quantitative assured media playback service 346
quality of service (QoS) 5, 16, 17, 46, 47, 72, 92, 110, 150, 221, 326, 347

radiation 12, 354
radio access network operation support (RANOS) 371, 372
radio base station 323, 359, 365, 371, 372, 374, 376, 382
radio link control (RLC) 235
radio network controller (RNC) 371
RADIUS (remote authentication dial-in user service) 94, 186, 203, 218, 219, 277, 278, 280, 282, 301, 390, 434, 452
RADSL 357
RANAP control signalling 394, 398
RANOS 371, 373
real-time transfer protocol (RTP) 271, 402
remote authentication dial-in user service (RADIUS) 202, 277
retailer 18, 19, 103, 117, 118, 174
return on capital employed (ROCE) 3, 465, 471, 473
return on investment (ROI) 124, 460, 465, 466, 471
return on sales 472
RIP 94, 343
RNC 281–283, 371, 372, 394, 398
roaming and handover 423
robust header compression (ROHC) 167, 236, 248, 258, 484
ROCE (return on capital employed) 100, 105, 107, 124–126, 133, 136, 145, 147, 177, 222, 228, 248, 297, 332, 351, 352, 359, 361, 371, 430, 438, 446, 448, 465, 471, 473

ROHC 167, 236, 248, 258, 484
ROI 100, 438, 465, 471
round trip delay 234, 235
round trip time, (RTT) 234
routing information protocol (RIP) 342, 343
RSVP 53, 96, 97, 240–242, 245, 262, 316, 344, 392, 453
RSVP-TE 344, 348, 350
RTCP 79, 237, 238, 271, 316, 326, 417
RTP 76, 79, 83, 237, 238, 270, 316, 325, 326, 392, 403, 417
RTSP 79, 237–239, 271, 316, 326, 417

S/MIME 214, 219
SAT 97, 254, 294
SCCP 406
SCCP user adaptation layer (SUA) 406
SCM 39, 286
SCP (service control point) 56, 250, 275, 386
SCS 14, 80, 97, 292, 293
SCSs 80
SCTP 97, 404–406
SDP 79, 83, 96, 195, 267, 270, 392, 393
SDSL 357, 376
secure shell (SSH) 212
secure socket layer (SSL) 198, 263, 434
security function 30, 189, 205, 217, 218, 280, 301, 308, 327, 407, 426
security gateway 301
security plan 91, 187, 196, 389
serialization process 232
service capabilities 179, 290, 291
service capability feature 80, 290, 292, 293
service capability server (SCS) 14, 80, 83, 111, 184, 260, 290, 292, 293, 299, 303
service control point (SCP) 56, 250, 275, 386
service description protocol (SDP) 83
service enabler 14, 69, 79–81, 88, 160, 165, 184, 222, 249, 250, 259, 260, 262, 289, 292–294, 305–307, 316, 317, 329, 445, 460, 461
service layer 4, 14, 63, 65, 70–72, 78, 81, 83, 121, 183, 184, 222, 254, 258, 259, 262, 263, 267, 286–288, 296, 297, 305, 307, 385, 386, 411, 436, 443–445
service level agreement (SLA) 19, 105, 109, 115, 133, 230, 267, 346, 411, 415
service level specifications (SLSs) 414
service management 72, 79, 150, 184, 287, 296, 297, 303, 432, 439, 443
service provider 12, 18, 31, 32, 39, 47, 57, 61, 63, 101, 103, 109–111, 115, 120, 121, 126, 154, 160, 175, 177, 183, 184, 262, 289, 295, 303, 399, 416, 417, 460, 463
service-oriented architecture 459–461
service-oriented programming 13, 88
serving CSCF (S-CSCF) 400
serving GPRS support node (SGSN) 218, 245, 246, 250, 280, 282, 283, 325, 368, 386, 397–399, 402, 418–420
session description protocol (SDP) 267, 269, 393
session initiation protocol (SIP) 83, 269, 328, 391, 392
session management 139, 280, 398
SGSN 218, 245, 246, 250, 280, 282, 283, 325, 368, 386, 398, 400, 403, 418–420
SGW 403, 404
SHDSL 134, 357, 363, 375, 376
short message service (SMS) 18, 158, 367
S-HTTP 263
signal transfer point (STP) 404
signalling gateway (SGW) 68, 326, 386, 387, 403

signalling system No 7 (SS7) 53, 390, 396
signalling transport (Sigtran) 405
silence encoding 222
SIM 80, 88, 216, 217, 294, 295, 427
SIM application toolkit (SAT) 88, 216, 294
single sign on 295, 296, 309, 444
SIP (session initiation protocol) 79, 83
SLA 133, 267, 290, 409, 415, 416, 439, 442, 457
slow start mechanism 234, 235
SLAs 414, 418
small and medium enterprise 32, 33, 38, 39, 376
small dish satellite internet 381
small office/home office (SOHO) 33, 38, 39, 134, 375
smartphone 311
SMI – structure of management information 432, 434
SMI model 432
SMS 19, 21, 131, 156, 158, 159, 164, 217, 256, 260, 261, 265, 266, 294, 311, 398, 402, 424
SNMP – simple network management protocol 199, 433, 435, 436, 451, 456
SOAP, Simple object access protocol 262, 461
sociables 25, 26, 40
softswitch 320, 387, 403
software defined radio 313
SOHO 339
SQL – simple query language 434
SS7 63, 390, 396, 404–406
SSL 204, 205, 263, 461
SSP 32
STP 404
strategy, infrastructure and product (SIP) 102, 103, 430, 439
stratification 65, 76–78, 106, 256, 334, 335
streaming class 161, 162, 270
SUA 406
success factors 4, 11
supplementary services 112, 367, 396
supply chain management (SCM) 39
support servers (ASuS) 111, 156, 184, 260, 263, 264
SVC 490
switched virtual circuit (SVC) 342
SyncML 261

take five model 21, 23, 26
tandem/transit switches (TS) 474
TCAs 346
TCP/IP 56, 63, 65, 83, 189, 203, 234, 236, 271, 292, 306, 317, 433, 461
TCP/IP networks 56
TDM 77, 320, 325, 326, 396, 404, 482, 485
TDMA 56, 364, 365, 367, 369
TD-SCDMA 366
TE 344
telecom operations map (TOM) 99, 101
telecommunications network, (TMN) 66, 101, 455
telemanagement forum (TM Forum) 101, 438
telephony server 255, 268, 274, 395, 421, 422, 474, 476, 480, 482, 490, 492, 495–497
terminating traffic 493
third generation partnership project 2 (3GPP2) 373
time division multiple access (TDMA) 365
time division multiplexing TDM 349
TINA-C 117
TMN 101, 296, 432–436, 438, 441, 456
TMN concept 432
TOM 101–103, 438
TOM classic 102
tools for radio access management (TRAM) 371
traditionalists 26, 41
traffic 34, 39, 42–44, 51, 53, 54, 60, 61, 63, 66–68, 75, 78, 91, 92, 111, 113, 115, 116, 120, 129,

134, 136, 138, 140–142, 153, 157, 159, 162–166, 174–176, 183, 187, 188, 191, 193, 195–199, 201, 204, 206–208, 210, 213, 214, 217–219, 228, 229, 231, 232, 234, 237, 240, 241, 243–246, 254, 255, 259, 266, 268, 271, 272, 274, 275, 280, 286, 287, 303, 306, 314, 322, 323, 327, 328, 331, 332, 334–336, 339, 340, 343–348, 350, 354–356, 358–361, 363, 368, 375, 376, 382, 385–390, 394, 396, 403, 406, 418, 420, 424, 432, 433, 435, 440–442, 448, 451–453, 455, 456, 475, 476, 478, 481–494, 496, 497
traffic conditioner 347
traffic engineering 96, 141, 142, 237, 240, 243, 331, 332, 337, 339, 344, 347, 348, 453, 481, 482
traffic shaping 240
traffic system 44, 51, 187, 248, 430, 431, 438
traffic mix 490, 496
TRAM 371
transactional class 165
transactional services 40, 157, 160, 172, 188, 249, 262, 263, 312, 444
transfer mode 13, 55, 68, 75, 95, 230, 239, 255, 256, 320, 322, 325, 332, 334, 355, 361, 482
transport layer security (TLS) 203
transport mode 212–214
TS 167, 168
TSC/GMSC server 394
tunnel mode 213, 214
tunnelling 78, 106, 197, 199, 201, 210, 211, 215, 217, 220, 227, 256, 262, 267, 275, 280, 301, 327, 452

UDDI 262, 461
UMTS-WLAN 1, 5, 11, 426
UNI 267, 341, 349, 480, 485, 486
UNI signalling 341
untrusted network 190
user plane 51, 53, 248, 256, 280, 394
USIM (universal subscriber identity module) 217, 301, 312, 428
UTRAN (universal terrestrial radio access network) 250, 371, 372

V5 interface 396
value-added service provider 14, 110, 115, 289, 290
value-added services 112
VAS (value-added services) 165, 250
VC (virtual circuit) switching 340
VCI/VPI (virtual circuit identifier/virtual path identifier) 340
VDSL 86, 87, 134, 322, 357, 376
very-high-bit rate DSL 87
VHE 97, 289–291, 409, 410, 413, 416, 417
video playback 166, 261
virtual home environment 289, 410, 416
virtual home environment (VHE) 265, 290, 291, 410
virtual private network (VPN) 29, 31, 35, 201, 250, 275, 328, 440
VoATM 55, 61, 232, 256, 272
VoATM gateway 328
VoDSL gateway 328
voice terminal 311
voice trunking 255, 272, 387, 474
VoIP 30, 55, 61, 94, 113, 118, 126, 131, 167, 169, 207, 223–225, 229, 232, 256, 271, 272, 314, 320, 344, 346, 353–355, 390, 403, 405, 455
VoIP gateway 328
VoIPoWL 223
VPN 39, 96, 200, 201, 210–212, 215, 216, 243, 254, 275, 293, 301, 347, 349, 381, 399
VSP 32

WAP 16, 40, 95, 148, 157, 158, 165, 166, 169–172, 177, 188, 219, 244, 254, 260–262, 265, 294, 295, 301–305, 309, 311–313, 315
WASP 32
WCDMA 65, 235, 366, 371, 381, 398, 423
WCDMA (wideband CDMA) 366
WCDMA radio access network (UTRAN) 371, 423
WCDMA/UMTS 217, 278, 280, 282, 367, 371
wearables 312
web services 13, 87–89, 150, 459–461
WEP 198, 218, 219
WI 157
wired equivalent privacy (WEP) 198, 218, 219
wireless application protocol (WAP) 148, 157, 158, 165, 166, 169–172, 177, 188, 219, 244, 254, 260–262, 265, 293, 294, 295, 301, 303, 304, 309, 311–313, 315
wireless Internet (WI) 58, 71, 157
wireless Internet service provider (WISP) 111, 112, 389
wireless session protocol (WSP) 169
wireless transport layer security (WTLS) 204
WISP 111, 112, 389
WLAN 4, 7, 74, 83, 86, 87, 111, 172, 198, 218, 255, 278, 312, 313, 322, 354, 356, 357, 359, 379–381, 389, 426–428
WLAN roaming 389
working capital 465
WSDL 262, 461
WSFL 262
WSP 169

xDSL 4, 133, 182, 200, 255, 339, 354, 357, 363, 427
XHTML 96, 168, 243
XML (extensible markup language) 79, 88, 96–98, 243, 254, 262, 287, 290, 296, 317, 411, 434, 460, 461
XML data 12

zonification 452